HIGHWAY ENGINEERING

Saed Rahwanji

HIGHWAY ENGINEERING

FIFTH EDITION

Paul H. Wright

Radnor J. Paquette

Georgia Institute of Technology

JOHN WILEY & SONS

NEW YORK CHICHESTER BRISBANE TORONTO SINGAPORE

Library of Congress Cataloging in Publication Data:

Wright, Paul H.
   Highway engineering.

   Includes indexes.
   1. Highway engineering.   I. Paquette, Radnor Joseph.
II. Title.

TE145.W74   1987   625.7   86-15720
ISBN 0471-82624-3

10 9 8 7 6 5 4 3 2 1

ABOUT THE AUTHORS

PAUL H. WRIGHT is Professor of Civil Engineering at Georgia Institute of Technology, where he has taught since 1963. He is a registered Professional Engineer in Georgia and a member of many professional societies, including the American Society of Civil Engineers, the Institute of Transportation Engineers (ITE), and the American Society for Engineering Education. Dr. Wright received a Sigma Xi Faculty Award and ITE's Technical Paper Award (Southern Section) in 1976. He has been active in continuing education programs and has served as a consultant to transportation agencies in the United States and Central America. The author of numerous papers, he is the coauthor of two other books: *Airport Engineering* (first edition, 1979; second edition, 1984), and *Transportation Engineering: Planning and Design* (first edition, 1972; second edition, 1982). He earned his Ph.D. in civil engineering (transportation planning) from Georgia Tech.

RADNOR J. PAQUETTE holds the rank of Professor Emeritus at Georgia Institute of Technology, where he taught before his retirement in 1969. He received the Bachelor of Science and Master of Science degrees from the Michigan College of Mines and Technology. A former Contract Engineer for the Michigan State Highway Department, Professor Paquette taught at the University of Florida before his tenure at Georgia Institute of Technology. He is a former President of the Education Division, American Road Builders Association. In addition to the first three editions of this book, Professor Paquette coauthored the textbook *Transportation Engineering: Planning and Design* (first edition, 1972).

PREFACE

This book is the successor to *Highway Engineering* by Professor R. J. Paquette and the late Leo J. Ritter. The first three editions of this publication served as a valuable textbook for thousands of students over a period of 28 years. The fourth edition, prepared by Professor Wright, appeared in 1979. That edition incorporated materials in areas of emerging significance and included new chapters on driver and pedestrian characteristics, traffic characteristics, transportation planning, land acquisition, and traffic engineering and safety. New material was also included on the social, economic, and environmental impacts of highways, highway maintenance management, and the planning and design of terminals and facilities for bicyclists and pedestrians.

This, the fifth edition, incorporates major changes in design standards from *A Policy on Geometric Design of Highways and Streets*, published by the American Association of State Highway and Transportation Officials (AASHTO) in 1984. New approaches to highway capacity analyses from the 1985 edition of the *Highway Capacity Manual* are included. Two entirely new chapters have been added: Computer Applications in Highway Engineering, and Highway Mass Transit Facilities. This edition also has a new chapter entitled Highway Materials which brings together parts of earlier chapters on bituminous materials and soil engineering as well as new material on aggregates and portland cement. New sections have been added on engineering fabrics, reinforced earth embankments, and pavement recycling and rehabilitation. The chapter on property acquisition has been omitted.

In a number of instances, because of the breadth and extent of information available, certain areas of interest are only outlined or briefly described. For those who wish to broaden and extend their knowledge in specific areas, up-to-date bibliographies are included at the end of each chapter.

Although the book uses English units throughout, it contains several problems that use metric (SI) units. These problems are even-numbered and are identified by an asterisk. Where appropriate, table footnotes have been included that give conversion factors from English to SI units.

It is hoped that this book will serve as a basic textbook for introductory courses in highway engineering as well as a useful source of reference for practicing professionals.

Many people contributed to the publication of this edition. Dr. Peter S. Parsonson and Mr. John Hibbard wrote the chapter on Computer Applications in Highway Engineering. Virginia Tiernan, Loren Jack, and Carol Bowler assisted with research and revision. Mr. R. H. Kuhlman, P.E., reviewed and recommended changes to the chapter on Concrete Pavements, and Dr. Q. L. Robnett helped with the revision of the chapter on Flexible Pavement Design. Mr. Peter Malphurs and Ms. Ellen Whitlock made valuable suggestions for improving the sections on environmental studies. Mr. Henry L. Ashmore prepared the plan and profile sheet which appears as Figure 13-6. All of these contributions are gratefully acknowledged.

In addition, many organizations and individuals supplied information, photographs, and sketches for this book. Direct credit is given in the text for these sources. The photograph for the dust jacket was supplied by the Georgia Department of Transportation.

PAUL H. WRIGHT

CONTENTS

HIGHWAY ENGINEERING

chapter one

INTRODUCTION

In the United States and other developed nations, the influence of motor vehicles is both profound and pervasive. More than 11 million U.S. workers (one in every six) are employed in the manufacture, distribution, service, and commercial use of motor vehicles. Highway passenger and freight transportation expenditures in 1983 exceeded $546 billion or about 16.5 percent of the total gross national product (1). Twenty-five percent of all U.S. retail sales are automotive.

By 1985, there were over 129 million automobiles and 36 million trucks and buses registered in the United States, which comprise 36 percent and 35 percent of the world totals, respectively. Motor carriers operating on the nation's highways transport 36.4 percent of the intercity tonnage of freight and account for 23.6 percent of the total intercity ton-miles. Highway transportation overwhelmingly dominates the transportation of people, accounting for 91 percent of all person trips. In 1984, highway travel in the United States exceeded 1.7 trillion vehicle-miles (2).

Good highways are so interwoven with every phase of our daily activities that it is almost impossible to imagine what life would be like without them. We depend on highways for the movement of goods, for travel to and from work, for services, for social and recreational purposes, and for many other activities necessary to the functioning of our complex society. The planning, design, construction, operation, and maintenance of highways depend largely on the efforts of the highway engineer, who must translate the desires of the people for better highway transportation into physical being.

THE DEVELOPMENT OF HIGHWAYS

Before beginning the study of highway engineering, it is interesting and informative to briefly trace the development of highways and highway systems from early historical periods to the modern era.

1-1 Ancient roads

The great highway systems of our modern civilization have their origin in the period before the dawn of recorded history. Even before the invention of the wheel, which is popularly supposed to have occurred some 10,000 years ago, individual and mass movements of people undoubtedly took place. The earliest travel was on foot; later, pack animals were utilized, crude sleds were developed, and simple wheeled vehicles came into being. Many of the migrations of the early historical period involved large numbers of people and covered relatively great distances. More or less regularly traveled routes developed, extending to the limits of the then-known world.

As various civilizations reached a higher level, many of the ancient peoples came to a realization of the importance of improved roads. The streets of the city of Babylon were paved as early as 2000 B.C. History also records the construction of a magnificent road to aid in the building of the Great Pyramid in Egypt nearly 3,000 years before the birth of Christ. Traces of early roads have been found on the island of Crete, and it is known that the early civilizations of the Chinese, Carthaginians, and Incas also led to extensive road building.

By far the most advanced highway system of the ancient world was that of the Romans. When Roman civilization was at its peak, a great system of military roads reached to the limits of the empire. Many of these roads were built of stone and were 3 ft or more in thickness. Traces of this magnificent system are still in existence on the European continent; in fact, some of these roads still serve as bases for sections of modern highways.

After the decline and fall of the Roman Empire, road building, along with virtually all other forms of scientific activity, practically ceased for a period of 1,000 years. Even as late as the early part of the eighteenth century, the only convenient means of travel between cities was on foot or on horseback. Stagecoaches were introduced in 1659, but travel in them proved exceedingly difficult in most instances because of the extremely poor condition of rural roads.

1-2 Later European development

Interest in the art of road building was revived in Europe in the late eighteenth century. During this period, Trésaguet, a noted French engineer, advocated a method of road construction utilizing a broken-stone base covered with smaller stones. The regime of Napoleon in France (1800–1814) gave a great impetus to road construction, chiefly for military purposes, and led to the establishment of a national system of highways in that country.

At about the same time in England, two Scottish engineers, Thomas Telford and John L. McAdam, developed similar types of construction. Telford urged the use of large pieces of ledge stone to form a base with smaller stones for the wearing surface. McAdam advocated the use of smaller broken stone throughout. This latter type of construction is still in extensive use, being the forerunner of various types of modern macadam bases and pavements.

DEVELOPMENT OF HIGHWAYS IN THE UNITED STATES

During the early history of the American colonies, travel was primarily local in character, and rural roads were generally little more than trails or cleared paths through the forests. (See Figure 1-1.) Toward the end of the eighteenth

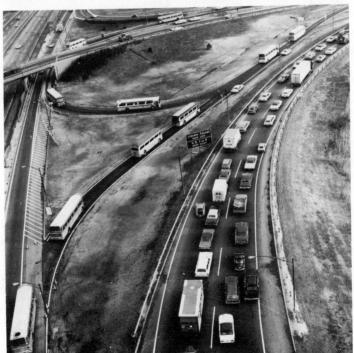

FIGURE 1-1 During the pre-Revolutionary War period, the journey from Philadelphia to New York required 2 days; today, it requires only 2 hours. (Courtesy Federal Highway Administration and Port Authority of New York and New Jersey.)

century, public demand led to the improvement of various roads by private enterprise. These improvements generally took the form of toll roads or "turnpikes" and were located principally in areas adjacent to the larger cities. The first American turnpike made use of existing roads in Virginia to connect Alexandria with settlements in the Blue Ridge Mountains. Another important toll road of this period was the Philadelphia and Lancaster Turnpike Road, from Philadelphia to Lancaster, Pennsylvania.

Another famous early road in American history was the Wilderness Road, blazed by Daniel Boone in 1775, which led from the Shenandoah Valley in Virginia through the Cumberland Gap into Kentucky.

In 1806 the federal government entered the field of highway construction for the first time with the authorization of construction of the National Pike, or Cumberland Road. The first contract on this route was let in 1811, and by 1816 the road extended from Cumberland, Maryland, the terminus of the Chesapeake and Ohio Canal, to Wheeling, West Virginia. When construction was completed (1841), it had reached Vandalia, Illinois. This route, which was surfaced largely with macadam, was nearly 800 miles in length and was built at a cost of approximately $7 million.

During the middle of the nineteenth century the great western expansion of the country began, and many roads or "trails" figured prominently in this development. Probably the most famous of the great trails was the Oregon Trail. This route began at Independence, Missouri, went overland to the Platte River, near Grand Island, Nebraska, and thence along the Platte River and its North Fork to Fort Laramie, Wyoming. Continuing westward, the trail followed the Sweetwater and crossed the Continental Divide through South Pass, in Wyoming. West of this point travelers followed the path of the Snake River into Oregon. Other famous western roads, each of which has a fascinating history, included the Santa Fe Trail, the Mormon Trail, the California Trail, and the Overland Trail.

In 1830, however, the steam locomotive demonstrated its superiority over horse-drawn vehicles, the interest in road building began to wane. By the time the first transcontinental railroad (the Union Pacific) was completed in 1869, road-building activities outside the cities virtually ceased until near the beginning of the twentieth century.

By 1900 a strong popular demand for highways again existed. The principal demand came from farmers, who clamored for farm-to-market roads so that they might more readily move their agricultural products to the nearest railhead. Some demand was also being felt for improved routes connecting the larger centers of population. During this period certain states began to recognize the need for state financial aid for road construction. The first state-aid law was enacted by New Jersey in 1891, and by 1900 six other states had enacted similar legislation.

1-3 Advent of the motor vehicle

The year 1904 marked the beginning of a new era in highway transportation in America with the advent of motor vehicles in considerable numbers. Almost overnight an enormous demand was created for improved highways, not only for farm-to-market roads but also for through routes connecting the metropolitan areas. Additional state-aid laws were enacted, and by 1917 every state participated in highway construction in some fashion. By this time, also, most states had established some sort of highway agency and had delegated to these bodies the responsibility for the construction and maintenance of the principal state routes. Figure 1-1 shows the evolution of one U.S. road from those early days to the present time.

This period was also marked by radical changes in road construction methods, particularly with regard to wearing surfaces. The early roads of American history were largely natural earth. Since timber was readily available, plank and corduroy roads were numerous, while later, wood blocks were used. Some gravel was used on early surfaces, while many city streets were paved with cobblestones. The invention of the power stone crusher and the steam roller led to the construction of a considerable mileage of broken-stone surfaces. Development of this type of surface generally paralleled that in Europe.

In the cities, relatively large concentrations of wheeled vehicles and the need for abatement of noise and dust brought various improved surfaces into being. The first brick pavement in this country is supposed to have been built in Charleston, West Virginia, in 1871, and asphalt was used for paving Pennsylvania Avenue, in Washington, D.C., in 1867. Concrete pavements were introduced about 1893, and the first rural road of concrete was built in 1909 in Wayne County, Michigan.

1-4 Development of the federal-aid program

World War I intensified highway problems in the United States and led the federal goverment to again actively enter the field of highway construction. Actually, federal participation in highway affairs on a continuing basis began in 1893, when Congress established the Office of Road Inquiry, an agency whose work was primarily educational in nature. This agency later became the Bureau of Public Roads of the Department of Agriculture, and in 1939 it became the Public Roads Administration under the Federal Works Agency. In 1949, this agency was transferred to the Department of Commerce, and its name again changed to the Bureau of Public Roads. In 1967, a National Department of Transportation (DOT) was established, and the Bureau of Public Roads was made a part of the Federal Highway Administration within that Department. The name Bureau of Public Roads was dropped in 1970, and the organization was merged with the Federal Highway Administration (FHWA).

The modern era of federal aid for highways began with the passage of the Federal Road Act of 1916, which authorized the expenditure of $75 million over a period of five years for improvements of rural highways. The funds appropriated under this act were apportioned to the individual states on the basis of area, population, and mileage of rural roads, the apportioning ratios being based on the ratio of the amount of each of these items within the individual state to similar totals for the country as a whole. States were required to match the federal funds on a 50-50 basis, with sliding scale in the public land states of the West.

The Federal-Aid Highway Act of 1921 extended the principle of federal aid in highway construction and strengthened it in two important respects. First, it required that each state designate a connected system of interstate and intrastate routes, not to exceed 7 percent of the total rural mileage then existing within the state, and it further directed that federal funds be expended on this designated system. Second, the act placed the responsibility for maintenance of these routes on the individual states.

During this period many states, experiencing difficulty in securing the necessary funds to match the federal appropriations, looked for new sources of revenue. This fact led to the enactment of gasoline taxes by several states in 1919, and other states rapidly followed suit.

During the ensuing years, appropriations for federal aid were steadily increased, while the basis of participation remained practically the same. Funds were provided (as now) for the improvement of roads in national parks, national forests, Indian reservations, and other public lands. The principle of federal aid was broadened to allow the use of federal funds for the improvement of extensions of the federal-aid system into and through urban areas and for the construction of secondary roads.

Another important step taken during this period was the authorization by Congress, in 1934, of the expenditure of not more than 1.5 percent of the annual federal funds by the states in making highway planning surveys and other important investigations. The first planning survey was inaugurated in 1935, and work of this nature was being conducted by all the states by 1940.

World War II focused attention on the role of highways in national defense, and funds were provided for the construction of access roads to military establishments and for the performance of various other activities geared to the war effort. Normal highway development ceased during the war years.

The Federal-Aid Highway Act of 1944 provided funds for highway improvements in postwar years. Basically, the act provided $500 million for each of the three years. Funds were earmarked as follows: $225 million for projects on the federal-aid system; $150 million for projects on the principal secondary and farm-to-market roads; and $125 million for projects on the federal-aid system in urban areas. The law also required the designation of two new highway systems. One is the National System of Interstate Highways, which will be discussed later in this chapter, and the other is one composed of the principal secondary routes (Federal-Aid Secondary System).

The 1952 act, for the first time, contained specific authorization of funds for the National System of Interstate Highways. In 1954, funds for the Interstate System were greatly increased, with apportionment based one-half on the traditional formula for the primary system and one-half on population. The Secretary of Commerce was directed to make studies of highway financing, with special attention to toll roads and to problems of relocating public utilities services. In 1955, the President's Advisory Committee on a National Highway

Program (Clay Committee) reported the need for a greatly increased program for modernization of the Interstate System. Intensive efforts by all interested groups in 1955 and 1956 led to the passage of the Federal-Aid Highway Act of 1956.

1-5 Federal-aid highway act of 1956

The Federal-Aid Highway Act of 1956 was a milestone in the development of highway transportation in the United States. It marked the beginning of the largest peacetime public works program in the history of the world. The act authorized completion, within a period of 13 to 15 years, of the 41,000-mile National System of Interstate and Defense Highways. Federal funds to be expended on this system were set at $24.825 billion, beginning July 1, 1956, and ending June 30, 1969.

Federal funds were made available on a 90-10 matching basis; state expenditures were expected to be approximately $2.6 billion, making a total of more than $27 billion available for the program. Subsequent increases in cost have led to upward revision of estimates of the cost of completing the system. By 1984, the Interstate System, which has been extended to 42,500 miles, was 96.3 percent complete. It is expected that the cost of the completed system will exceed $100 billion.

Several new provisions were included in the 1956 act that are applicable only to the Interstate System. One of these provisions denies federal funds to any state that permits vehicles with excessive weight and length to use the Interstate System. Approved design standards, including control of access, must be used, and the design was to be adequate to accommodate traffic expected in 1975. (In 1963, Congress amended the highway act to require designs to be based on 20-year traffic forecasts, dating from the time when the plans for each project are approved.) Toll roads could be included in the Interstate System, if suitably located and adequately designed. Congress indicated its intent to consider at a future time whether or not, and how, the states should be reimbursed for previously constructed toll and free portions of the Interstate Sytem.

Title II of the 1956 act, cited as the Highway Revenue Act of 1956, required revenues derived from taxes levied by the federal government to be earmarked and put into a special trust fund (the Highway Trust Fund).

Increases in the estimated cost of completing the Interstate System and a desire to complete the system within the time period originally established led Congress to increase highway user taxes in subsequent years.

1-6 Highway legislation of the 1960s

The highway legislation of the 1960s recognized the growing transportation needs of urban areas. The 1962 act required that, after July 1, 1965, all federal-aid highway projects in urban areas with a population of more than 50,000 be based on a continuing, comprehensive, and cooperative transportation planning process. The act permitted more extensive use of federal-aid secondary funds in urban areas and required that urban highway improvements be "an integral part of a soundly based, balanced transportation system" for the involved area.

In 1965, Congress authorized a $1.2 billion, five-year program to build 2350 miles of "developmental highways" in the 11-state Appalachian area, plus another $50 million to build access roads. The purpose of the program, in which the federal government paid 70 percent of the highway costs, was to spur economic development of the region. Congress also inaugurated a highway beautification program in 1965 under which funds are provided for the control of billboards and junkyards adjacent to highway rights of way and for roadside landscaping.

The Federal-Aid Highway Act of 1968 reflected the continuing concern of Congress for urban transportation problems. It authorized the use of federal-aid urban funds for parking facilities in urban fringe areas and established Traffic Operation Programs to Increase Capacity and Safety of urban facilities (TOPICS). The act also increased the compensation to families and businesses displaced by highway projects.

1-7 Highway legislation of the 1970s

The highway legislation of the 1970s reflected the desire of Congress to cope with the trends of increasing urbanization, traffic congestion, environmental degradation, and shortages of fuel.

The Federal-Aid Highway Act of 1970 recognized the growing importance of urban highways by providing for the establishment of a federal-aid urban highway system.

The 1970 legislation provided that under certain circumstances federal highway funds could be used for exclusive or preferential bus lanes, traffic control devices, and fringe and corridor parking facilities that serve public mass transportation.

The 1970 act also required the Secretary of Transportation to promulgate guidelines designed to ensure that possible adverse economic, social, and environmental effects are properly considered during the planning and development of federal-aid highway projects. The intent of this requirement is to ensure that decisions on federally funded projects be made in the best overall public interest, taking into consideration any adverse effects due to air, noise, and water pollution, the displacement of citizens, farms, and businesses, and the disruption or destruction of natural resources, public facilities, community cohesion, and aesthetic values.

The Federal-Aid Highway Act of 1973 greatly expanded the federal-aid uban system and provided for a major realignment of the federal road systems based on anticipated functional usage in 1980. (See Section 1-15.) Meanwhile, efforts were being made to revitalize the economy of rural areas and smaller communities and to reverse the pattern of migration from rural to urban areas by the judicious use of highway construction funds. To further these goals, the 1970 act authorized demonstration grants to states for the construction and improvement of *economic growth center development highways* on the federal-aid primary system. The development of economic growth center development highways was upgraded to a standing program by the 1973 act and expanded to include all federal-aid highway systems except the Interstate.

The 1973 act made it possible for local officials in urban areas to decide not to use revenues from the Highway Trust Fund for road improvements, but instead to receive an equal amount of public mass transit aid from the general fund. The law gave conditions under which trust funds could be used for mass transit and placed a limit on the amount of such transfers.

Public Law 93-643, entitled Federal-Aid Highway Amendments of 1974, included the following landmark provisions:

1. It required the highway projects receiving federal aid be planned, designed, constructed, and operated to allow effective utilization by elderly and handicapped persons.
2. It established a rural highway mass transportation demonstration program to encourage the development, improvement, and use of public mass transportation systems utilizing highways that serve rural and small urban areas.
3. It authorized the Secretary of Transportation to make grants for demonstration projects designed to encourage the use of carpools in urban areas. Under the act, funds may be provided for establishing carpooling systems and for marking, signing, and signalizing preferential carpool or carpool/bus highway lanes.
4. It authorized similar grants to states for demonstration projects for the construction of bikeways. The federal share for such projects is 80 percent of the total costs.
5. It increased the maximum permissible weight restrictions that states may impose on vehicles which use the Interstate System. Single axle maximum weights were increased from 18,000 to 20,000 lb. The overall gross weight limit was changed from 73,280 to 80,000 lb. The act also gave a formula for the calculation of the allowable overall gross weight of two or more consecutive axles.
6. It established a uniform national maximum speed limit of 55 mph on public highways.

1-8 Surface transportation assistance act of 1982

The Surface Transportation Assistance Act of 1982 (STAA) was passed to address the problem of a deteriorating highway infrastructure. The act extended authorizations for the highway and transit programs from 1983 to 1986 and raised highway user fuel taxes from 4 to 9 cents per gallon. The act also instituted substantial increases in truck user fees. It specified that the additional funds were to be used for (1) accelerating the completion of the Interstate Highway System, (2) increasing the Interstate resurfacing, restoration, rehabilitation, and reconstruction (4R) program, (3) expanding the bridge replacement and rehabilitation program, and (4) providing greater funding for other primary and secondary projects.

The Surface Transportation Assistance Act allowed larger trucks to operate on the Interstate Highway System and certain other federal-aid highways. The law specified that states must allow twin-trailer combination trucks and may not establish overall length limits on tractor—semitrailer or tractor—twin-trailer combinations. The law established a width limitation of 102 in. and required that states allow

1. Semitrailers in a tractor—semitrailer configuration 48 ft in length.
2. Trailers in a tractor—semitrailer—trailer configuration 28 ft in length.

The law required that states allow the longer vehicles on interstate highways and on "those classes of qualifying Federal-aid Primary System highways as designated by the Secretary" (of Transportation). It provided that the Secretary designate as qualifying highways those Primary System highways that are capable of safely accommodating the longer vehicle lengths. The implementation of this part of the law has been accompanied by controversy, and by the mid-1980s research was being performed to determine the classes of highways that could safely accommodate the larger trucks.

HIGHWAY SYSTEMS IN THE UNITED STATES

The highways of the United States now total about 3.9 million miles. These highways have been divided into various administrative systems for the purpose of allocation of highway funds and to define the authority for planning, design, construction, maintenance, and operation of specific highways. Administratively, highways may be grouped into three broad systems: federal-aid systems, state systems, and local systems.

1-9 Federal-aid highway systems

Since the passage of the Federal-Aid Highway Act of 1921, Congress has required that federal road funds be spent on specified systems of roads. The federal-aid highway systems are segments of state and local mileage eligible for federal aid. The federal-aid systems consist of the interstate,* primary, secondary, and urban highway systems.

National system of interstate and defense highways

There is no federal system of highways as such in the United States, that is, no system of highways built and maintained by the federal government, except for relatively small mileages in national parks and forests, Indian reservations, and so forth. The closest thing to a federal road system is the Interstate System; however, selection, construction, and maintenance of these roads are prerogatives of the individual states.

Designation of the National System of Interstate Highways was authorized by the Federal-Aid Highway Act of 1944, with a limitation of 40,000 miles. The 1956 act changed the name to the National System of Interstate and Defense Highways and added an additional 1000 miles to the system. The mileage has since been increased to 42,500 miles (see Figure 1-2).

The Interstate System connects, as directly as practical, the nation's principal cities and industrial centers and connects at suitable border points with

---

*The Interstate System is a part of the Federal-Aid Primary System. Because of its importance, it is discussed separately.

FIGURE 1-2 National System of Interstate and Defense Highways. (Courtesy Federal Highway Administration.)

routes of continental importance. It comprises only about 1.0 percent of the total road and street mileage and carries 20 percent of all traffic. Access is controlled on all sections of the system. On most of the system, grade separations carry cross traffic over or under the interstate route. There are no railroad grade crossings, no traffic signals, and no stop signs. Many miles of the system comprise divided highways with four, six, or eight lanes of traffic. High design standards prevail uniformly throughout the system.

Federal-aid primary system

As the name implies, the federal-aid primary system is comprised of the most important highways in the nation's highway network. Typically, highways in the primary system are heavily traveled and are characterized by wide pavements and modern design features. The primary system totals about 300,000 miles, including the Interstate System, and about 32,000 miles in urban areas. The system was authorized by the Federal-Aid Highway Act of 1921, which limited it to 7 percent of a state's rural mileage existing at that time. Although states have been permitted to exceed this proportion under certain conditions, the primary system is still occasionally referred to as the "7 percent system."

Federal-aid secondary system

The federal-aid secondary system, which was authorized by Congress in 1944, is comprised of the main secondary and feeder roads that connect farms and smaller communities with the federal-aid primary system. The secondary system is made up of 398,000 miles of highways.

Federal-aid urban system

Originally, federal aid for highways was provided exclusively for rural highways. Beginning in 1944, limited federal funds were provided for the construction and improvement of "urban extensions" of federal-aid highways. Urban portions of the federal-aid primary and secondary highways are now considered to be integral and important parts of the system (2). In 1970, Congress established a federal-aid urban system.

> The urban system was established in each urbanized area as designated by the Bureau of the Census, and in such urban areas as the State highway departments may designate. The system serves major centers of activity and includes high traffic volume arterial and collector routes, including access roads to airports and other transportation terminals, so as to serve the goals and objectives of the community (2).

As of 1984 there were approximately 140,000 miles of federal-aid urban highways.

1-10 U.S.-Numbered highways

Every tourist is familiar with the comprehensive network of interstate highways which are designated as "U.S.-Numbered Highways." These routes are marked by characteristic shield-shaped emblems bearing the initials "U.S.," the name of the state, and a route number. They have been designated by joint action of the various state highway departments, acting through the American Association of State Highway and Transportation Officials. Highways included in this system are normally a part of the federal-aid and state systems. Some have been incorporated in or replaced by elements of the Interstate System. Routes traversing the country from east to west bear even numbers, while those with characteristic north and south directions have been given odd numbers. Among the more famous of the routes are U.S. 1, "from Maine to Miami," and U.S. 30, the "Lincoln Highway."

1-11 State systems

Each state has its own "state-designated" system, made up principally of primary and secondary routes of statewide importance. In some states, however, county roads and municipal streets have been placed under state control. In 1984 state highway systems included approximately 920,000 miles of rural routes and urban extensions (routes within each of the nationwide systems previously discussed are included in this total).

1-12 Local road systems

By far the greatest percentage of mileage of roads in the United States is under the control of local govermental units. Included in this classification are more than 2,700,000 miles of county, township, town, and village roads. County systems are predominant, with approximately two-thirds of the above mileage being under county jurisdiction. Principal areas in which the county is not the primary unit of local road administration include the New England states and Pennsylvania.

Four states, North Carolina, West Virginia, Virginia (except for two counties), and Delaware, have assumed responsibility for all local roads. In a few other states some counties have transferred responsibility for local roads to the state highway department. Certain routes under local control are included in the Federal-Aid Secondary System, as previously described.

1-13 City streets

The picture of highway systems in the United States is completed by recognition of the fact that there exist about 517,000 miles of city streets and alleys that are constructed and maintained by the individual municipalities.

1-14 Toll roads

During the period 1950 to 1958, many toll roads or turnpikes were built and placed in service in various sections of the country. The second "toll road era" of American history was brought about by the necessity for providing high-type facilities capable of relieving pressing traffic problems and the corresponding inability to construct such facilities from current revenue. Toll roads were, in general, built by special authorities created by the various state governments. Revenue bonds were sold and the income from tolls collected used to retire the bonds and interest charges.

The initial toll road of this era was the original Pennsylvania Turnpike, completed in 1940. Toll roads were built and placed in operation in Colorado, Connecticut, Florida, Illinois, Indiana, Kansas, Kentucky, Maine, Massachusetts, New Hampshire, New Jersey, New York, Ohio, Oklahoma, Pennsylvania, Texas, Virginia, and West Virginia. Total mileage of toll roads in operation was 3210 miles, built at an estimated cost of more than $5 billion.

With passage of the Federal-Aid Highway Act of 1956, interest in the construction of new toll roads waned in most states. However, toll road mileage has been built in the intervening years in several states, including Delaware and Maryland in addition to the original list. The purpose generally has been to speed construction of intrastate routes not on the Interstate System. By 1986, there were aproximately 3,000 miles of toll roads in operation in the United States. About three-fourths of these routes had been incorporated in the Interstate System.

1-15 Functional classification of highways

To facilitate orderly highway development and efficient fiscal planning and to ensure logical assignment of jurisdictional responsibility, highways should be classified according to function. Functional classification involves grouping streets and highways into classes or systems according to the character of service they are intended to provide. Highways generally serve a dual role in a highway system, providing both travel mobility and access to property. Thus, highways may be viewed as a hierarchical system consisting of highway classes with differing purposes and levels of importance. The dominant function of arterial highways is mobility, while the principal function of local roads is to provide access to abutting property. Collector roads provide more-or-less balanced service for both of these functions. The approximate percentages of highway mileage and daily vehicle-miles of travel for various functional classes are given in Table 1-1.

It may be useful to employ a more refined functional classification system that subclassifies arterial and collector highways:

Principal arterials	Minor collectors
Minor arterials	Local roads and streets
Major collectors	

FUTURE HIGHWAY DEVELOPMENTS

The highway system of the United States has reached a level of development without parallel in the modern world. In terms of highway mileage, the system is approaching a state of completion. Yet, the nation's highway needs are great, and the challenges of future highway engineers surpass those of their predecessors.

As the Interstate System nears completion, there is a special need to improve other road systems. A 1983 Report of the Secretary of Transportation to the U.S. Congress (3) indicated that a total of $456 billion (1980 dollars) would be required during the period 1981 to 2000 to bring all U.S. roads and

TABLE 1-1
PERCENTAGES OF HIGHWAY MILEAGE AND VEHICLE-MILES
OF TRAVEL BY FUNCTIONAL CLASS

Functional class	Percentage by Mileage	Percentage by Vehicle-Miles of Travel
Rural		
Principal arterial		
Interstate	0.9	8.9
Other	2.1	8.7
Minor arterial	3.9	8.5
Collector		
Major	11.4	10.0
Minor	7.8	2.7
Local roads	57.6	5.5
Rural total	83.6	44.3
Urban		
Principal arterial		
Interstate	0.2	10.8
Other freeways and expressways	0.2	5.3
Other	1.2	15.1
Minor arterial	1.7	11.2
Collectors	1.8	5.3
Local roads	11.3	8.1
Urban total	16.4	55.7
Total rural and urban	100.0	100.0

SOURCE: *The Status of the Nation's Highways: Conditions and Performance.* Report of the Secretary of Transportation to the United States Congress, U.S. Government Printing Office, Washington, D.C. (July 1983).

streets to an acceptable level of design, based on an estimated 2.8 percent average annual growth in traffic. The report indicated that $162 billion would be needed for improvement of the urban road system and $294 billion would be required for the rural system.

As we look to the last years of the century, shortages of energy will likely be accompanied by greater demands for separate access-controlled "busways" and express lanes for public mass transit vehicles. Selected improvements in highway gradient and curvature will be required. In order to lessen the impact of urban congestion, more extensive use will be made of sophisticated traffic control measures, and some physical improvements will be needed along urban streets. Heavier permissible wheel loads for commercial vehicles will likely bring a need for stronger pavements, improved pavement design techniques, and better maintenance. Deficient bridges will need to be repaired or replaced. Special efforts will need to be undertaken to decrease the number of highway accidents and to decrease the number of injuries and fatalities by controlling the energy dissipated during crashes.

As these improvements are made, extraordinary efforts will be needed to provide highways that are aesthetically pleasing, socially acceptable, and environmentally compatible. Highway engineers will continue to make a vital contribution to providing the nation with safe, efficient, and convenient transportation.

REFERENCES

1. *MVMA Motor Vehicle Facts and Figures '85*. Motor Vehicle Manufacturers Association, Detroit, Mich. (1985).

2. *Highway Statistics 1984*. U.S. Department of Transportation, Federal Highway Administration, Washington, D.C. (1985).

3. *The Status of the Nation's Highways: Conditions and Performance*. Report of the Secretary of Transportation to the United States Congress, 98th Congress, 1st Session (July 1983).

chapter two

HIGHWAY ADMINISTRATION

Administration of public highways in the United States is a governmental function, responsibility for which is delegated, in whole or in part, to appropriate agencies of the federal government, to the various state governments, and to numerous local governmental units. Although there is and must be a separation of function and responsibility among the governmental units mentioned, close contact and cooperation among all governmental units concerned with road and street construction are necessary if the development of the overall highway system is to proceed in an intelligent, adequate, and economical fashion.

Highway transportation is influenced to some extent by all of the three major divisions of government: legislative, judicial, and executive. Three standing congressional committees have jurisdiction over highway-related legislation: the Commerce, Science and Transportation Committee and the Environment and Public Works Committee in the Senate and the Public Works and Transportation Committee of the House of Representatives.

The economic regulation of interstate carriers (bus companies, trucking firms, brokers, and freight forwarders) is the responsibility of the Interstate Commerce Commission, a quasi-judicial arm of the Congress. Similar state commissions regulate intrastate carriers. Decisions of these regulatory bodies may, of course, be appealed in the courts.

THE FEDERAL HIGHWAY ADMINISTRATION

In the executive branch, the Federal Highway Administration (FHWA) acts as the representative of the federal government in all matters relating to public highways. FHWA is one of nine* principal organizational units that comprise the U.S. Department of Transportation (DOT). Prime responsibility within the Department of Transportation rests with the Secretary of Transportation, a member of the President's cabinet.

2-1 Organization
As Figure 2-1 indicates, the Federal Highway Administration is headed by an administrator under whom serve the executive director and six associate ad-

---

*The other eight units are the Coast Guard, Federal Aviation Administration, Federal Railroad Administration, National Highway Traffic Safety Administration, Maritime Administration, Urban Mass Transportation Administration, St. Lawrence Seaway Development Corporation, and the Research and Special Programs Administration.

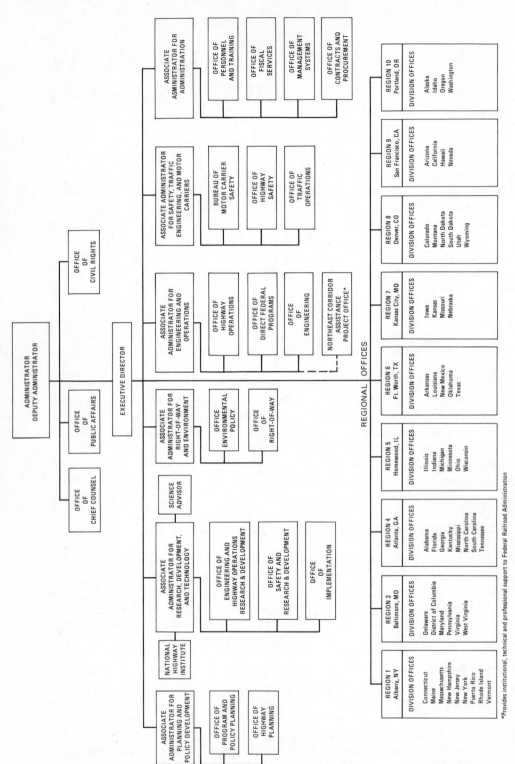

FIGURE 2-1 Organization chart for the Federal Highway Administration. (Courtesy Federal Highway Administration.)

*Provides institutional, technical and professional support to Federal Railroad Administration ministrators. The Washington headquarters staff is responsible for policy formulation and general direction of public roads operations in engineering, finance, management, and legal fields.

The Federal Highway Administrtion has nine regional offices located across the country, each supervising the federal-aid program in from four to eight states. The regions have division offices in every state, Puerto Rico, and the District of Columbia. It is through this field organization that most relations with the state highway departments are carried on.

2-2 Administration of the federal-aid program

Some of the principal functions of the Federal Highway Administration in its administration of federal aid to the states are described in the following paragraphs.

Authorization and apportionment

The history of a federal-aid project begins with the authorization by Congress of federal funds for highway construction. The federal funds for each year are apportioned among the states, usually from 6 to 12 months before the beginning of the indicated fiscal year, and are available for that fiscal year and for two years thereafter. Within this time limit the funds apportioned to a state must be committed to specific construction work by formal agreement between the state and the FHWA; once such commitments are made, the actual expenditures may continue beyond the three-year limit. Federal-aid authorizations are proportionally divided among the states by formulas that take into account the area, population, and postal route mileage in each state, as provided by law. Amounts available to the states also are limited by mandatory operation of the program on a pay-as-you-go basis within the limits of revenue available to the Highway Trust Fund.

Programming and subsequent steps

Knowing the amount of federal aid available to it in each fiscal year, each state draws up programs of projects to be built with those funds, based on preliminary surveys and cost estimates. Since there is close contact between the Federal Highway Administration and the states, questions as to the selection of projects and their location, design, and cost are settled in the early stages of the program planning.

After acceptance of the program by the Federal Highway Administration, the state makes detailed surveys for each project and prepares plans, specifications and final cost estimates. These must be approved by the FHWA division engineer. The state then advertises for bids and awards a contract for constructions subject to concurrence by the Federal Highway Administration. The day-to-day supervision of the work is performed by state engineers resident on the job, but an FHWA engineer or investigating team makes periodic inspections on the site.

FHWA does not prescribe detailed design standards and construction specifications for federal-aid work. The general policies of design standards adopted by the American Association of State Highway and Transportation Officials (AASHTO) are used by most of the states and are endorsed by the Federal Highway Administration. Details of design and construction specifications are prepared by each state to fit its individual needs and are subject to review and approval by the federal agency.

Payments

As the work progresses on a project, the state periodically pays the contractor and claims reimbursement from FHWA for the allowable federal share of the cost. When the project has been completed, inspected, and accepted, a final detailed accounting of costs is made by the state and reviewed by federal auditors, after which final payment of the federal share is made. Thus the federal-aid funds are needed over a period of time rather than all in advance, and the federal aid for each fiscal year is initially made available only as an authorization. After completion of a federal-aid project, FHWA engineers continue to inspect it at intervals to see that it is being properly maintained.

Certification acceptance

FHWA provides an alternative procedure for administering certain federal-aid highway projects in lieu of the detailed procedures that have traditionally been employed. The process is know as "certification acceptance." A "state certification" is a written statement prepared by a state highway agency that sets forth the state's organization and capability and the laws, regulations, and standards it will use in the administration of certain federal-aid projects. Once the process is approved, a state highway agency may use state inspectors and auditors to ensure and certify that established procedures are followed on applicable federal-aid projects.

2-3 Road building by the Federal Highway Administration

In addition to its work in the supervision of the federal-aid program to the states, the FHWA cooperates with other federal agencies in the construction of roads in national parks and forests, Indian reservations, and other federally controlled lands. These projects are generally constructed entirely with federal funds.

Many routes of outstanding engineering accomplishment and scenic beauty have resulted from this work of the federal agency. Included among these are the Blue Ridge Parkway in Virginia and North Carolina; the Trail Ridge Road in Rocky Mountain National Park, Colorado; and the Natchez Trace Parkway, which extends from Nashville, Tennessee, to Natchez, Mississippi.

2-4 Other activities of the Federal Highway Administration

Research was one of the purposes for which the Office of Road Inquiry was created in 1893 and is one of the important functions of the Federal Highway

Administration today. In its own offices and laboratories, and in cooperation with the state highway agencies, universities, and other organizations, both public and private, FHWA conducts research and development programs in all phases of highway engineering.

Highway research and development programs were stimulated by the Federal-Aid Highway Act of 1962, which provided for mandatory, rather than permissive, use of 1-1/2 percent of the funds apportioned to each state for federal-aid highway projects, for planning and for research and development projects. An additonal 1/2 percent of federal-aid funds may be used by the states for such projects.

The Federal Highway Administration, AASHTO, and the Transportation Research Board perform large-scale research projects under the National Cooperative Highway Research Program (NCHRP). Inaugurated in 1962, the NCHRP involves the expenditure of about $3.5 million per year in a planned research program.

Research activities of the Federal Highway Administration are very broad in scope, including economic studies, design investigations, environmental research, and physical research of many kinds. Work of the administration has led to many advances in design, construction, and management.

Planning activities of the FHWA include the development of multi-modal long-range transportation plans as well as short-range improvement programs. FHWA planners develop estimates of overall highway needs and coordinate the preparation of annual lists of capital improvement projects by the various states.

The Federal Highway Administration also gathers and disseminates highway statistics, research reports, and planning and design manuals. Its quarterly publication *Public Roads* contains articles describing the results of recent research, describes research in progress, and lists highway documents available from the FHWA and the National Technical Information Service.

Through its National Highway Institute, the FHWA conducts a program that provides educational opportunities for engineers and other employees of state and local highway agencies.

FHWA's Bureau of Motor Carrier Safety has jurisdiction over the safety performance of approximately 212,000 motor carriers engaged in interstate commerce. Its field personnel make vehicle inspections, control the transportation of hazardous cargoes, analyze truck and bus accidents, and enforce federal regulations regarding driver qualifications and their hours of service on the road.

STATE HIGHWAY AND TRANSPORTATION DEPARTMENTS

State governments, through the various state highway and transportation departments, occupy the key position in the development of highway systems in the United States at the present time. The relation of the Federal Highway Administration to highway development has already been discussed, but it must be emphasized that federal-aid programs are undertaken at the option of the individual states and that the states are responsible for the planning, design, construction, operation, and maintenance of routes constructed with federal participation, subject only to review and approval by the FHWA.

In the majority of states, the state highway agency has the responsibility for the development of roads included within the state-designated system, which is generally made up of primary and secondary routes of statewide importance, including urban facilities. As previously mentioned, certain states have assumed responsibility for all local roads.

2-5 Functions of state highway and transportation departments

Functions of the various state highway organizations vary widely because of differences in the laws, directives, and precedents under which they are organized and operated. However, the principal function of any state highway agency is the construction, operation, and maintenance of highways contained within the state-designated system, including those built with federal participation. The departments are also usually responsible for the planning and programming of improvements to be made on the various components of the state highway system and for the allocation of state funds to the various local units of government. They are also usually charged with the duty of distributing information regarding highways to the general public, and they may be responsible for motor vehicle registration. In some states the highway agency exercises police functions relative to safety on the highways, usually through the state highway patrol.

2-6 Organization of state highway agencies

As might be expected, state highway and transportation departments vary a great deal in their administrative organization. In terms of administrative control, the departments generally fall into one of three classes, those with:

1. A single executive.
2. A single executive with a board or commission acting in an advisory or coordinate capacity.
3. A board or commission.

The single executive type of organization usually has a "Commissioner," "Director," or "Secretary" who reports directly to the Governor. The executive may be elected but is more commonly appointed by the Governor.

A large number of state highway organizations are under the administrative control of elected or appointed boards or commisions. The board may have total administrative control and involve itself with the day-to-day management and operations of the department, or it may serve in a purely advisory capacity. In a few states, the routine management functions have been shared by a single executive and a board or commission.

Since the establishment of the U.S. Department of Transportation, most states have established multimodal departments of transportation. Certain of these departments have been organized by transportation mode, as illustrated by Figure 2-2*a*. Others have been organized by function, with major divisions provided for planning and research, engineering operations, design, and so forth (Figure 2-2 *b*). Still others have a mixed organizational structure (Figure 2-2*c*).

An organization chart for a traditional state highway department is shown as Figure 2-3.

LOCAL ROAD ADMINISTRATION

As previously pointed out, some 2.7 million miles of highways are currently under the administration of local authorities. Administration of this vast mileage of local roads presents an extremely varied pattern.

2-7 Types and general features

In most sections of the country, the county is the prevailing unit of local road administration. Exceptions to this exist principally in the New England states and Pennsylvania, where the township continues to be the most important unit having responsibility for local roads. In some of these states there are township systems of "feeder" roads, tributary to the main county routes.

In considering the administration of local roads by the important county unit, considerable differences in practice are also to be noted. In some counties road affairs are administered for the county as a unit; this is the so-called

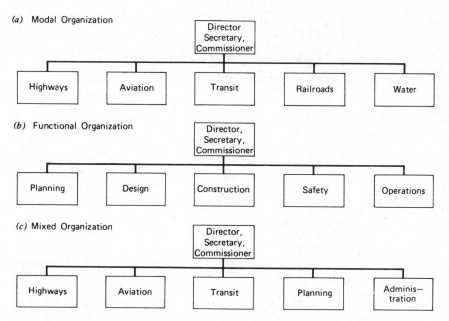

FIGURE 2-2 Basic organizational structures of state transportation departments. (Courtesy Transportation Research Board.)

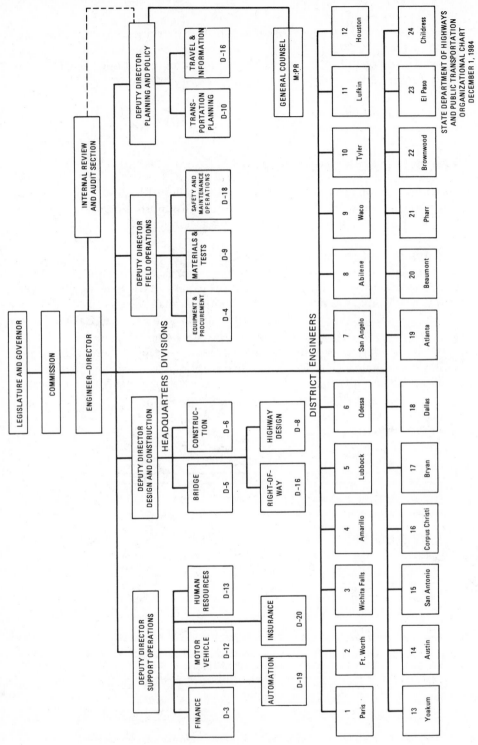

FIGURE 2-3 Organization chart for Texas State Department of Highways and Public Transportation. (Courtesy Texas State Department of Highways and Public Transportation.)

county unit plan. In others, responsibility for roads is divided among various precincts or commissioners' districts, each with its own funds, personnel, and equipment. Extreme variations in area (from a few square miles up to more than 20,000 square miles in one county in California) and population (from a few hundred or less up to more than 5 million in Cook County, Illinois, and over 7 million in the city and county of Los Angeles) further account for the various types of county road administrations.

It is generally believed that the county-unit plan is the most efficient unit of local road administration. Advantages claimed for the county-unit plan over smaller units of local road administration include the creation of a more efficient and capable organization, a more equitable distribution of the tax burden, higher standards of engineering and construction, and increased ability to cooperate with state highway departments. There has also been some indication that consolidation of the road-building activities of adjacent counties may be advisable in certain instances where the individual counties do not possess sufficient resources to purchase and maintain adequate equipment or to secure necessary engineering services.

County engineering organizations are as varied in character and quality as the county administrative organizations. Generally speaking, the larger and more populous counties have entirely adequate engineering organizations, many of which are on a par with those of the state highway departments. Many other counties have inadequate engineering forces, while hundreds of counties operate with little or no engineering aid.

The Federal-Aid Secondary Program, with its large allocations of funds for the secondary road program, has stimulated local road affairs and led many counties to establish engineering organizations capable of cooperating with the state and federal agencies in the planning, design, and construction of local roads.

The following criteria have been enumerated for good highway administration at the county level (1):

A road commission whose functions are limited to policy formulation, with an engineer responsible for all management functions.

Maintenance of adequate and uniform records for guidance in management, for developing and carrying out a long-range road improvement program, and for use in making such reports as the state legislature may direct.

Cooperation with other county and city agencies in the development of integrated regional highway programs.

Operation under an annual improvement program, based on a long-range plan and related to a definite annual budget.

Preparation of construction plans and specifications for all contractual work.

Maintenance of roads and bridges to uniform standards.

Extension of contractual highway services to small cities and villages upon request.

Installation of traffic-control and safety measures in accordance with uniform standards.

Adoption of uniform accounting practices.

Adoption of standard specifications for purchasing, in accordance with sound business practice.

MUNICIPAL STREET ADMINISTRATION

Administration of the 517,000 miles of city streets in this country is also quite variable in nature. This is not surprising when it is realized that this mileage exists in thousands of incorporated communities, the large majority of which have populations of less than 5000.

Generally speaking, responsibility for city streets is vested in the city engineer, director of public works, city manager, or some similar official, depending on the form of government used. Functions of the city street department, or similar body, include design and construction of street improvements, formulation of long-range plans for street improvement, the making of traffic surveys, installation of traffic-control and safety devices, and the supervision of maintenance.

As in the case of the counties, street departments in the larger cities operate at a high level of efficiency, performing all necessary engineering functions. In many smaller cities, engineering organizations are inadequate or nonexistent. However, it may be noted that there is an increasing appreciation of the desirability of providing engineering control in the planning, design, and construction of city streets, with the result that many municipal engineering organizations have been created or expanded in recent years.

NECESSITY FOR INTERAGENCY COOPERATION

It is recognized that the various components of the highway system are not separate entities but are, or must become, integrated parts of a coordinated system if highway development is to keep pace with the demands of the modern motorist. In recognition of this fact there must be close cooperation among all the units of govverment concerned with highways and streets, including the federal government, the states, counties, and cities.

In the planning, design, construction, operation, and maintenance of highways, the state highway and transportation departments are the key agencies. They are usually responsible for the distribution of state funds to the various units of local government; they are in a position to supply the technical knowledge and engineering skill so badly needed in many local agencies; they possess facilities for performing special surveys and analyses, such as urban traffic studies; and they possess a wealth of information that should be made available to local units of government for planning and programming purposes. In view of these and other facts, several state highway departments have established special county (secondary road) and urban divisions to cooperate more closely with the appropriate local governments and to further the development of an integrated highway system. In most urban areas, cooperative transportation planning agencies have been established.

Planners of important federally assisted highway projects must now coordinate the planning activity through a regional or state clearinghouse (2). In metropolitan areas, the clearinghouse is an areawide agency (e.g., a regional planning commission) that has been recognized by the Office of Management and Budget as an appropriate agency to evaluate, review, and coordinate federally assisted projects. In nonmetropolitan areas, the clearinghouse is a comprehensive planning agency designed by the Governor or by state law to review and coordinate such activities.

RELATED ORGANIZATIONS

In addition to the governmental agencies described, many other organizations contribute to the development of the highway system. One of the most important of these groups is the American Association of State Highway and Transportation Officials (AASHTO). The membership of this group consists of the principal executive and engineering officers of the various state highway agencies and the Federal Highway Administration. The principal objective of the association is the fostering of "the development, operation, and maintenance of a nation-wide system of highways to adequately serve the transportation needs of our country." Work of the various committees of the association has resulted in a large number of publications which are widely used by other groups.

The American Road and Transportation Builders Association (ARTBA) is another large and powerful group in the highway field in this country. The membership of ARTBA comprises six divisions, including persons and organizations from every branch of the highway industry and representatives from many foreign countries. As a result of the work of the divisions and the committees included therein, the organization issues technical bulletins and reports at intervals; it also publishes a monthly magazine, *The American Road Builder* and a weekly newsletter. ARTBA is active in furthering highway legislation and in disseminating information about highways to the general public. ARTBA holds annual and regional meetings, with a "Road Show" that is held at periodic intervals.

The Institute of Transportation Engineers (ITE) is a society of transportation professionals with a membership of approximately 6800. Formerly called the Institute of Traffic Engineers, ITE has a membership that consists predominantly of traffic engineers and others interested in vehicular traffic control. The organization meets annually and publishes a number of handbooks and procedural guides, as well as a monthly magazine called the *ITE Journal*.

The Transportation Research Board (TRB), a division of the National Research Council, is a vital factor in the advancement of technical knowledge in transportation. The board does not, in general, perform research itself, but has its primary function in the correlation of research efforts of educational institutions, governmental agencies, and industry. The TRB holds an annual meeting and issues many technical publications.

Many other groups are engaged in activities related to highway development, including organizations of highway users, materials producers, equipment manufacturers and dealers, automotive manufacturers, engineering groups, colleges and universities, and many others. Many such groups could be named, and publications of some of them have been cited at various points in this book.

HIGHWAY FINANCE

This section is concerned with the sources and distribution of the money that is obtained for highway purposes; in other words, with the financing of highway systems and their improvement. Attention will be focused principally on current methods of providing money for highway expenditures and the distribution of these sums. Certain general principles of highway finance and taxation will also be presented.

There are two general methods of financing highways. The first of these is the "pay-as-you-go" method, which involves paying for all highway improvements and the costs of maintaining and operating the highway system from current revenue. This is the method that is currently in use by many governmental agencies, and rightly so in many cases, because of constitutional prohibitions or similar legislative provisions against borrowing for highway purposes. The second method is, of course, that of borrowing money to pay the costs of highway improvement, with the borrowed sums plus interest being repaid over a period of time from future income. This latter method may be generally termed "credit financing." In most cases, it is also "bond financing." Both methods have merit under certain circumstances, and strong advocates of either method may be found among the various authorities in the highway field.

Attention will first be given to the various sources of revenue for highway purposes. These sources may, in many cases, be regarded as sources of current revenue, although they may also be used for debt service, either on obligations incurred in the past or for obligations incurred at the present time and against which some of these revenues may be pledged for future payment. Bond financing is considered separately in Section 2-12.

2-8 Sources of revenue for highway expenditures

The funds necessary for the improvement of the nation's highway system are obtained from a variety of sources. The principal sources are

1. Highway user taxes.
2. Property taxes.
3. Tolls.

Highway user taxes

Highway user taxes comprise the dominant class of funds for highways at both the state and federal levels. As Table 2-1 indicates, income from federal highway user imposts totaled more than $11 billion in 1983, while state user taxes amounted to more than $21 billion. A breakdown of state highway user tax revenues for 1984 is given in Table 2-2. User taxes, typically in the form of vehicle fees or "wheel taxes," are also imposed by local governments in a relatively small number of instances.

Highway user taxes are of three general types: (a) fuel taxes, (b) registration taxes and related fees, and (c) special taxes on commercial vehicles.

Fuel taxes   Taxes were first levied on motor fuels in this country in 1919 by Oregon and three other states. By 1929, all states levied a gasoline tax, that is, a stated tax on each gallon of gasoline sold as motor vehicle fuel. In 1984, state gasoline taxes ranged from 5 to 17 cents per gallon in the United States and averaged about 10 cents per gallon. Other motor fuels are similarly taxed. This tax is by far the greatest revenue producer of all highway user taxes, and, in most cases, it is far and away the biggest single source of income for the operation and improvement of the state highway system. Fuel taxes constitute nearly two-thirds of the total obtained from state highway user taxes.

TABLE 2-1
TOTAL RECEIPTS FOR HIGHWAYS IN 1984 FOR ALL UNITS OF GOVERNMENT

	Millions of Dollars			
Item	*Federal*	*State*	*Local*	*Total*
Motor fuel and vehicle taxes	11,364	19,426	463	31,253
Tolls	—	1,825	201	2,026
Property taxes and assessments	—	—	3,031	3,031
General fund appropriations	1,806	1,031	6,013	8,850
Other taxes and fees	81	1,291	225	1,597
Investment income and other receipts	1,251	1,215	1,393	3,859
Bonds	—	1,715	1,589	3,304
Total receipts	14,502	26,503	12,915	53,920

SOURCE: *Highway Statistics 1984*, Federal Highway Administration *(3)*.

TABLE 2-2
STATE HIGHWAY REVENUES AND RECEIPTS, 1984

Item	*Millions of Dollars*
Motor fuel taxes	11,962
Motor vehicle registration and taxes	7,464
Tolls	1,825
Appropriations from general funds	761
Other state imposts	996
Miscellaneous receipts	1,287
Total	24,295

SOURCE: *Highway Statistics 1984*, Federal Highway Administration *(3)*.

Registration taxes and allied fees   Laws pertaining to the registration of motor vehicles have a somewhat longer history than those relating to the imposition of a fuel tax, since the first registration law became effective in New York in 1901. By 1917 all states had enacted laws of this type. Although the procedure of registration was first adopted as a regulatory device, it soon became recognized as a ready source of highway revenue. The practice of graduating registration fees with the weight or capacity of the vehicle also began early in the modern era of highway development in this country, being first occasioned by the great increase in the number of trucks using the highway system after World War I. This practice is now followed in nearly all states. Registration fees are collected principally by the state governments. Several other taxes and fees are closely associated with registration fees and are levied by the various state governments. These special fees include such things as operators' and chauffeurs' license fees, fees for certificates of title, title-transfer fees, and fees charged for duplicate license plates. In many states income derived from these latter sources is used primarily for the operation of the highway system rather than for actual construction and maintenance. For example, in some states money derived from operators' and chauffeurs' license fees goes wholly or in part to pay the expense of maintaining a highway police organization.

Special taxes on commercial vehicles   The practice of assessing special fees against motor vehicles used for commercial purposes, including both passenger and freight carriers, has developed somewhat more gradually and on a less uniform basis than those taxes that have been previously discussed in this section. Special fees are levied against this class of motor vehicles in proportion to their use of the highway system and, roughly at least, in proportion to their effect on the physical design of the highway. These imposts have generally taken the form of graduated fees computed on a total mileage basis, on the basis of ton-miles or passenger-miles of travel, or taxes on gross receipts. The special nature of these vehicles is also taken into account by the graduated registration fees previously described.

The general basis on which these taxes are levied seems to be sound as there is little question that certain classes of commercial vehicles, for example heavy trucks and tractor−trailer combinations, have a disproportionate effect on the design of the highway. The thickness of a pavement that is required to carry heavy trucks may be considerably greater than that required for the adequate support of passenger automobiles. Similarly, a direct effect may be seen in the necessity of providing extra lanes for heavy trucks on long grades. They are often found necessary in highways designed to carry both passenger cars and heavy commercial units in order to provide some relief from congestion. A long and continuing argument has been going on for many years between those who represent the interests of commercial haulers, on the one hand, and highway administrators and engineers, on the other, relative to the imposition and collection of these special taxes.

Property taxes

Two types of property taxes are of importance when sources of funds for highways are being considered. These are (1) ad valorem property taxes and (2) special assessments against real estate lying contiguous or adjacent to a highway improvement. The first of these types of property taxes is extensively used in the financing of highway improvements by local units of government, including counties, townships, incorporated places, and similar administrative units. The second type is now principally used in towns and cities. Neither of these types is an important source of highway revenue to state governments at the present time.

The common concept of the use of general property taxes for highway improvements, which may be levied on real property or on both real and personal property, is that such a procedure is justifiable if the improvements which are to be effected are of general benefit to the community. The same line of reasoning is applicable if the improvements are expected to enhance the value of property located throughout the area. In practice, of course, in many local governmental units, property taxes are not levied specifically for highway improvements and are not earmarked for this purpose. Money derived from this source goes into the general revenue fund and is then budgeted, generally on an annual basis, for highways, education, public health, welfare, and so on.

General property taxes are, of course, levied on the assessed or "fair" value of the land and improvements located thereon, or on a similar value of personal property. Although general property taxes have numerous faults, they will almost certainly remain a source of income for highways and other public services for many years to come.

Special assessments are simply what the name implies. They are special assessments that are made directly on the owners of property which lies on or adjacent to a highway or street improvement in accordance with the benefits which are expected to accrue directly to the parcels of land thus affected. As has been indicated, this form of property tax is now used principally in cities and is not in widespread use for the financing of rural roads. It has been most generally used in recent years in the provision of street improvements in new residential areas and is less generally used for reconstruction. Special assessments are usually made on a frontage basis rather than on the property value. The assessments may be designed to recover the entire cost of the improvement over a period of time, or the initial cost may be shared among the property owners and the city government. Assessments may be made on abutting property only or on property located on adjacent side streets and even parallel streets, depending largely on the nature and amount of the benefits that are expected to accrue from the proposed improvement. Assessments against property located on adjacent streets would generally be made on a reduced or sliding scale. As is the case with general property taxes, special assessments have some serious inherent disadvantages. It seems quite certain, however, that this method of financing street improvements will be continued.

Tolls

Funds derived from the assessment of tolls for the use of a particular facility have become an important source of revenue as a result of the construction of a large number of toll roads following World War II. Prior to that time, tolls were generally limited to assessments for particular structures, such as a bridge or tunnel. Tolls are levied by public agencies, including special authorities of various types, to repay costs incurred in the construction of the facility (see Section 2-12 for a discussion of bond financing). As soon as the debt incurred by the construction of the facility has been paid, the facility is then made "free." In 1984, toll receipts totaled approximately $2.0 billion.

Miscellaneous funds

Brief attention may be given at this point to general revenue funds as a source of money for the financing of highway expenditures. Generally speaking, funds from general revenue have not been of major importance in the improvement of highways by the state governments in recent years. In a few states, however, income derived from taxes levied for highway improvement, consisting chiefly of highway user taxes, goes into the general fund from which the funds required for operation and improvement of the state highway system are budgeted periodically. In states where this system is used, the amount of money appropriated from the general fund would usually be substantially the same as that which would normally accrue to the state highway department from highway user taxes, so that the end result is about the same. Students of taxation are frequently strong advocates of this method of putting income derived from all tax sources into the general fund, with appropriations being made at intervals for all public services on the basis of need. From the viewpoint of the highway administrator, this concept has a serious disadvantage in that highway improvements are best made on a planned, orderly basis with the assurance of a steady and predictable annual income. In 1984, combined receipts from property taxes and general revenues totaled nearly $12 billion.

Another source of revenue that is worthy of mention here is that derived from parking meters, as used in many towns and cities. Although parking meters were originally envisaged as regulatory devices, the income derived from them has proved to be very substantial and has become quite important in many urban areas. Many thousands of parking meters are now in use in cities and towns over the country; in 1980, revenues from parking fees totaled $85 million.

2-9 Federal funds

For many years, funds collected from certain federal excise taxes (principally on gasoline) went into the general revenue fund of the federal government. Appropriations were then made from the general fund for federal aid to highways and the construction, maintenance, and operation of roads in national parks, national forests, and so on. However, the Highway Act of 1956 changed this system and established the Highway Trust Fund, from which expenditures are made directly for federal aid and for roads on federally controlled lands. Into the Trust Fund go monies received from federal taxes on gasoline and diesel fuel, tires and inner tubes, tread rubber, new trucks, truck–trailers, and buses; there is also a highway use tax on heavy vehicles. Estimated federal taxes paid by private and commercial highway users during 1984 are given in Table 2-3.

TABLE 2-3
FEDERAL HIGHWAY TRUST FUND RECEIPTS
ATTRIBUTABLE TO HIGHWAY USERS IN 1984

Type	Millions of Dollars
Gasoline	7,699
Special fuels	1,470
Lubricating oil	-10^a
Federal use tax	179
Trucks and trailers	865
Parts and accessories	-28^a
Tires	320
Tubes and tread rubber	12
Total	10,507

[a]Certain highway user taxes were eliminated by the Surface Transportation Assistance Act of 1982. The negative values represent refunds of taxes that had been paid at point of manufacture.

SOURCE: *Highway Statistics 1984*, Federal Highway Administration *(3)*.

2-10 Highway cost allocation

One underlying concept should govern all levies for highway purposes. This concept is that the cost of highway improvement should be borne on an equitable basis by the three principal groups who benefit from highway improvement, namely (1) highway users, (2) property owners, and (3) the general public. Every effort should be made to assess the cost of improved highways in an equitable maner against these groups. The effort should be made at every level of government. The accomplishment of the equitable distribution of the tax burden in proportion to benefits received is, of course, much more difficult than the statement of this underlying objective.

Highway cost allocation is a thorny fiscal (and philosophical) problem that has occupied the attention of highway economists for many years. It has been the subject of a number of studies including at least two required by federal laws *(4, 5)*.

Usually, the first consideration is the division of cost responsibility among highway users and nonusers. One study recommended that about 8 percent of the cost of the federal-aid highway program be charged to revenue sources other than motor vehicle taxes (i.e., to non-highway users) *(4)*.

Several methods of cost allocation have been proposed, the more prominent of which are the incremental cost method and the differential benefit method.

The incremental cost method is the traditional way of allocating highway user charges. In this approach, each element of highway design affected by the size or weight of vehicles is broken down into a series of additions (increments);

the cost of providing each increment is assigned only to those vehicles whose size and weight require them. Thus, all vehicles share in the cost of the basic design, but the cost of each succeeding increment is borne only by the larger vehicles.

The differential benefit method assigns cost responsibility to the various users in direct proportion to the vehicular benefits they receive by their use of the highway system. Four types of vehicular benefits are generally considered: reductions in operating costs, in time costs, in accident costs, and in the strains and discomforts of driving. The last (impedance costs) are normally calculated only for passenger cars.

2-11 Disbursement of highway revenues

Table 2-4 presents information relative to the disbursement of highway revenues in 1983 for federal, state, and local governments. Two of the items given in this table deserve additional explanation. These are the items listed as "Interest on debt" and "Debt retirement." These items reflect the money expended in the payment of obligations previously incurred for highway purposes. The total highway debt of the state governments, counties, and municipalities was estimated by the Federal Highway Administration to be nearly $29 billion at the end of 1984; much of this was for toll facilities. Although the sums required for debt service of these obligations represent a sizable portion of the total income for highway purposes, it is generally felt that the various governmental units, taken as a whole, are still in a favorable position with regard to the relation between the existing highway debt and the total capital investment that has been made in the various units of the highway system. Several of the toll roads built since World War II have been highly successful from a financial point of view, as are many toll bridges and tunnels.

Another factor in the distribution of highway revenues is diversion. By

TABLE 2-4
DISBURSEMENT FOR HIGHWAYS IN 1983 FOR ALL UNITS OF GOVERNMENT

Item	Millions of Dollars			
	Federal	State	Local	Total
Capital outlay	480	14,772	5,270	20,522
Maintenance and traffic services	134	5,562	8,323	14,019
Administration, research, miscellaneous	269	1,902	1,105	3,276
Debt retirement	—	1,160	993	2,153
Interest on debt	—	1,208	695	1,903
Highway law enforcement and safety	—	2,312	1,945	4,257
Total	883	26,916	18,331	46,130

SOURCE: *Highway Statistics 1983*, Federal Highway Administration.

"diversion" is meant the expenditure of funds collected from highway user taxes for nonhighway purposes. Although these diverted funds are in general expended for very worthwhile purposes, with expenditures of this sort being principally made for education at the present time, the diversion of highway funds is regarded in a very unfavorable light by the large majority of highway administrators and engineers. The principal basis for condemnation of diversion arises from the concept that since highway user tax schedules are predicated on benefits received by the taxpayer in the form of highway improvements, these taxes become unjust and inequitable if funds derived from this source are expended for purposes other than highway improvement. The amount of diversion, expressed in terms of percentage of highway user tax receipts, has decreased substantially in recent years. There is every reason to believe that this trend will continue. A large number of states now have constitutional provisions that prevent the diversion of highway funds.

2-12 Bond financing for highways

It sometimes becomes necessary or advisable for a governmental unit to obtain funds for highway improvements by credit financing. That is, the government must make use of its credit in order to borrow money, which is then repaid, with interest, usually over a period of years. Necessary improvements of a highway system or a specific facility may be financed in this manner. By far the greater amount of credit financing for highways is done through the issue of bonds.

One form of bond that may be used for the financing of highway improvements is the general-obligation bond. This type is backed by the entire faith and credit of the issuing agency. In recent years, general-obligation bonds have been regarded with less favor than in the early period of highway improvement, and the trend has been toward the use of the so-called limited-obligation bond. Limited-obligation bonds are usually secured by the anticipated revenues from a particular source of income. For example, many limited-obligation bonds are now secured by the pledging of future income from gasoline tax receipts to their repayment. Still another form of bond is what may be termed here a revenue bond. A bond issue of this type may be employed, for example, in the construction of a toll road or bridge, with the bond issue secured by the pledging of future tolls to repayment of the bonds. Limited-obligation bonds are also sometimes termed, somewhat loosely perhaps, "revenue bonds."

The mechanics of a bond issue and its subsequent repayment may be generally described as follows. The total amount of the bond issue is decided on by the governmental agency involved and then approved by referendum or legislative action. Bonds are then issued and are sold, frequently on a competitive bid basis, to banks, investment companies, or similar organizations, or in some instances to the general public. The amount of interest carried on the bond issue will generally be determined by the type of bond issue, the term of issue, and the credit standing of the issuing agency. The total amount of the bond issue, plus interest, is then repaid by the issuing agency in any one of several ways, as briefly described later. The bonds are said to have been "retired" when they are repaid.

Highway bonds have been issued for varying periods, with typical periods ranging from 10 to 50 years. Ten years is probably somewhat too short a period of issue for most highway bonds, while a period of 50 years might be regarded as proper under some circumstances in the financing of some relatively long-lived structure, such as a road, bridge, or tunnel.

Any one of several general methods may be employed to retire bonds which have been issued for highway improvements. Two of these general methods will be described here. One of these may be termed the "sinking-fund" method and the other the "serial" method. Bonds that are to be retired by the use of a sinking fund are generally term-issue obligations. That is, the entire amount borrowed falls due at the end of a period of time that is fixed when the bonds are issued. Money that will be required to pay the obligation when it falls due or when the bonds "mature" is provided by setting aside an annual sum toward the payment of the borrowed amount. Payments that are made to the sinking fund must be large enough so that their sum, plus whatever income may be derived from their investment, will be sufficient to retire the bonds when they mature. The sinking-fund method has several inherent disadvantages, chief among which is the temptation that the sinking fund continually presents to public officials over a period of years as it gradually accumulates. This and other factors have generally led to the abandonment of the sinking-fund method in favor of serial bonds. Serial bonds are retired by periodic (annual) payments on the principal plus periodic payments of interest. In their most common form, constant payments are made on the principal each year, with annual interest payments which decrease as time passes. No sinking fund is required when this method of payment is used, and the bonds are completely retired at the end of their term of issue.

There is little doubt that bond financing is entirely proper and desirable as a financing method for highways, at least under certain circumstances. Bond issues of this type may generally be justified by application of the benefit theory which has been previously emphasized throughout this chapter. Credit financing may also be resorted to in an effort to bolster the general financial condition of a state through increased employment resulting from construction activity.

The chief disadvantage accruing from bond financing is the added cost occasioned by the payment of interest on the obligation. This consideration may frequently be eliminated by the benefits resulting from the improvement and the resulting reduced costs of highway transportation. Bond issues should not, however, be used for the payment of current obligations. This is an unsound policy in public finance as well as in private finance, where an individual borrows money to pay current bills. Other disadvantages that may be associated with bond issues include the generation of a volume of work beyond the capacity of the contractors in the area, with a resultant lack of competitive bidding with accompanying price increases, and the fact that a volume of work may be involved which cannot properly be handled by the engineering force of the agency issuing the bonds.

PROBLEMS

2-1 Determine how state highway affairs are administered in your state. Is the highway agency administered by a single executive or by a commission? Draw an organization chart for the headquarters organization, showing lines of responsibility for the various primary functions of the department. (Consult the annual or biennial reports of the highway department in your college library.)

2-2 How are local road affairs administered in your state? Is the county, township, or some other form of local government responsible for roads? Is there an engineer responsible for such functions as design, construction, and maintenance in the place in which you live or attend school? If so, draw a typical organization chart for this engineering organization.

2-3 Has your state highway department set up special divisions to cooperate with local governmental units in the improvement of secondary roads and urban extensions of state routes? If so, briefly describe the functions of these divisions.

2-4 Prepare a brief report describing an important project now under construction (or recently completed) in your state that required close cooperation among representatives of federal, state, and local highway agencies.

2-5 Prepare a report outlining the organization and functions of the Transportation Research Board.

2-6 Prepare a report describing the organization, membership, and activities of the American Association of State Highway and Transportation Officials. List the current books of standards and policies adopted by this association. What do you think is this group's most valuable contribution to the highway program?

2-7 Determine who is responsible for the planning, design, construction, and maintenance of streets in your community. How much money is expended in a typical year? What are the sources of street funds? Is there a long-range program of street improvement? If so, how did the program come into being? Briefly outline the main points of the program.

2-8 Determine the annual income of the government of your state from direct highway user taxes. Consult the annual report of your state highway agency, or some other source, and determine how this money is distributed to the various governmental units in your state.

2-9 Estimate the annual contribution which the owner of an average light passenger car makes in the form of direct highway user taxes levied by your state government. Assume 10,000 miles of travel and 25 miles to a gallon of gasoline. How much does this motorist contribute to the Federal Highway Trust Fund each year?

2-10 Prepare a brief report on a major highway (or street) improvement project, which is now under way or planned in your state, for which bond financing is to be used. Describe the nature of the project, the total amounts of money involved, and the financing arrangements.

REFERENCES

1. *Highway Needs in Michigan—An Engineering Analysis.* Michigan Good Roads Federation, Highway Study Committee, Lansing (1948).

2. U.S. Office of Management and Budget. *Intergovernmental Review of Federal Programs; Implementation of Executive Order 12372, Federal Register,* Vol. 48, No. 70 (April 11, 1983).

3. *Highway Statistics 1984.* U.S. Department of Transportation, Federal Highway Administration, Washington, D.C. (1985).

4. *Final Report of the Highway Cost Allocation Study,* House Document No. 72. 87th Congress, 1st Session, U.S. Government Printing Office, Washington, D.C. (1961).

5. *Final Report on Federal Highway Cost Allocation Study.* Report to Congress from the Secretary of Transportation, Washington, D.C. (May 1982).

chapter three

DRIVER, PEDESTRIAN, AND VEHICLE CHARACTERISTICS

The highway engineer must design for a wide range of vehicle operating characteristics and allow for great differences in driver and pedestrian characteristics. Most highway facilities must be designed to accommodate the smallest subcompact automobile as well as the largest tractor–trailer truck. In many instances, the design must also accommodate motorcycle and bicycle users and pedestrians. It should be remembered that within each class of users there is great variability: differences in vehicle sizes, weights, and operating characteristics; differences in pedestrian walking speeds; and differences in drivers' and pedestrians' ability to comprehend and react to highway features and traffic events. In the first part of this chapter, we will examine some of the driver and pedestrian characteristics that are of consequence and interest to the designer. In the latter paragraphs, we will discuss certain physical characteristics of passenger cars and trucks that have a bearing on the design process.

DRIVER CHARACTERISTICS

Drivers in the United States are licensed to drive under laws and regulations of the various states. There is a great deal of variability among the states in licensing standards and presumably in driving skills, especially for beginning drivers. The vast majority of U.S. drivers have received no formal driver training but were taught by family or friends or were self-taught.

Despite efforts to improve driver licensing procedures spanning some 75 years, many observers doubt that driver licensing makes much of a contribution to safe motor-vehicle operation (1). State reports, spot checks, and special surveys have suggested that "on a given day there are at least 10 million unlicensed drivers on the road" (2). A report from one state indicated that 9.6 percent of drivers involved in fatal accidents were unlicensed at the time of the crash (2).

In 1985 there were over 152 million licensed drivers in the United States: about 52.4 percent male and 47.6 percent female. About 7.3 percent are under the age of 20 and about 6.5 percent are aged 70 and older. The number of elderly drivers has increased dramatically in recent years. In 1940, drivers aged 65 years and older comprised only 3 percent of all licensed drivers (3); by 1974, drivers of that age bracket comprised 11 percent of the total. The number of elderly drivers increased from 1.36 million in 1940 to 5.71 million in 1959 (*3*) to 17.57 million in 1984 (*4*).

It is important that highway engineers keep in mind that street and highway facilities must be designed to accommodate drivers with a wide range of ages and skills—the young and the elderly, and the novice as well as the experienced professional.

3-1 The human sensory process

A driver's decisions and actions depend principally on information received through the senses. This information comes to the driver through the eyes, ears, and the sensory nerve endings in the muscles, tendons, joints, skin, and organs. In general order of importance, the senses most used by drivers are

1. Visual (sight).
2. Kinesthetic (movement).
3. Vestibular (equilibrium).
4. Auditory (hearing).

Visual information may come to a driver by means of foveal or peripheral vision. In foveal vision, images are concentrated in a small area of the eye near the center of the retina at which visual perception is most acute. At a given moment, a person's sharpest vision is concentrated within a cone with a central angle of about 3°. For most persons, visual acuity is reasonably sharp within a conical angle up to about 10°. Beyond this, a person's vision is less well defined. A person with normal vision can perceive peripheral objects within a cone having a central angle ranging up to about 160° (*5*).

A driver increases the amount of visual information received by movement of the head and eyes. The driver's eyes search and scan the field of view, moving the area with sharpest acuity. The direction of foveal vision repeatedly shifts and the eyes fixate, each such event requiring respectively 0.15 to 0.33 sec and 0.1 to 0.3 sec (*5*).

Mourant and Rockwell (*6*) found that experienced drivers scanned a wider range of horizontal fixation locations than novice drivers. The average horizontal fixation range for the experienced group varied from 30° to 48°, depending on the type of driving task. The novice drivers concentrated their eye fixations in a smaller area and looked closer in front of the vehicle and more to the right. By pursuit eye movements, the novice drivers tended to observe the guardrail and lane edges continuously for periods of about 1.0 sec to gain information on lateral position. Similar findings have been reported for drivers who have been deprived of sleep (*6*).

Visual acuity declines and the field of vision narrows with advancing age, especially when lighting conditions are poor (*3*). Older drivers also tend to have more difficulty in judging distances and distinguishing colors than young drivers.

Drivers receive information by kinesthesia, the sensation of bodily position, presence, or movement that results primarily from stimulation of sensory receptors in muscles, tendons, and joints. Lateral or longitudinal accelerations exert forces on the driver that are transmitted by the seats, steering wheel, brake pedal, arm rest, and so forth. Proprioceptors in the muscles and joints are stimulated by the tendency of the body to move and shift as a result of the forces. This provides a feedback that may cause a driver to brake, slow down, or take some other action. This phenomenon has been termed "driving by the seat of the pants."

Drivers also receive messages from the vestibular nerve within the inner ear. Three fluid-filled semicircular canals in the inner ear enable a person to recognize the direction of movement and maintain balance or equilibrium. Movement of the head or body causes the fluid in the canals to move, affecting delicate fibrous cells located at one end of each of the canals. Impulses from the vestibular nerve are then transmitted to the brain, which, in turn, sends messages to the muscles, which take action to maintain balance.

The kinesthetic and vestibular senses provide important information to a driver about forces associated with change of direction, steering, braking, vibrations, and stability of the vehicle (7).

Sounds of horns or skidding tires may alert a driver to an impending collision, and engine noise aids the driver as well as pedestrians in judging vehicle speeds. When traveling along a curve or turning at an intersection, tire noises may indicate to the driver a need to slow down. Highway engineers employ sound in the form of special pavement textures, raised pavement markers, or "rumble strips" to signal to drivers to slow down or to avoid certain paved areas. Sounds within a vehicle such as loud conversation or radio music may mask out the sounds of sirens, bells, or other important warning devices.

Drivers may detect a fire or engine malfunction by their sense of smell. Hunger, thirst, or discomfort may cause drivers to change their trip pattern or stop. Drivers may also occasionally receive information through the sensory nerve endings in the head, skin, and internal organs and experience sensations of touch, coldness, warmness, fatigue, or pain.

3-2 Driver perception and reaction

Driver perception–reaction time is defined as the interval between seeing, feeling, or hearing a traffic or highway situation and making initial response to what has been perceived. Take, for example, a simple braking situation. The "clock" for perception–reaction time begins when the object or condition first becomes visible and it stops when the driver's foot touches the brake.

Traditionally, perception–reaction time has been couched in terms of perception, intellection, emotion, and volition, termed PIEV time (5). Perception is the process of forming a mental image of sensations received through the eyes, ears, or other parts of the body. It is recognizing and becoming aware of the information received by the senses. Intellection is another word for reasoning or using the intellect. Emotion is the affective and subjective aspect of a person's consciousness. It has to do with how a person feels about a situation. More often than not, emotion is detrimental to safe motor vehicle operation. Volition is the act of making a choice or decision. Perception–

reaction times vary from less than 0.5 sec to 3 or more sec. Greenshields (*8*) has indicated that, in general:

1. The speed of all forms of reaction varies from one person to another and from time to time in the same person.
2. Reaction time changes gradually with age, very young and very old people being slower in their reactions.
3. People generally react more quickly to very strong stimuli than to weak ones.
4. Complicated situations take longer to react to than simple ones.
5. A person's physical condition affects his reactions. For example, fatigue tends to lengthen a person's reaction time.
6. Distractions increase the time of all reactions except the reflex.

Johansson and Rumar (*9*) measured brake-reaction times for a group of 321 drivers who had some degree of braking expectation. The median brake-reaction time was 0.66 sec and the range was from 0.3 to 2.0 sec. The frequency distribution of the brake-reaction times is shown in Figure 3-1. The researchers performed a second experiment to see how brake-reaction time in response to a completely unexpected signal compares with such time in response to a somewhat anticipated signal. They reported that unexpected brake-reaction times were larger by a factor of 1.35.

Elderly drivers have more difficulty than the young in perceiving traffic situations and tend to pause longer between successive acts. Data presented by Marsh (*3*) suggest that on the average, simple reaction times for persons aged

FIGURE 3-1   Frequency distribution of brake-reaction times. (Source: Reference 9.)

65 are about 16 percent higher than for those aged 20; complex reaction times are about 33 percent higher for the older subjects. However, because of the wide variability among drivers, "chronological age is not a sound criterion for appraisal of driving competence" (*3*).

As Figure 3-2 illustrates, a person's response time depends on the sense stimulated. Generally, a person responds quickest to stimuli of touch and hearing and requires only slightly more time for response to visual stimuli. Response times for the kinesthetic and vestibular stimuli are considerably longer.

Figure 3-3 shows that response time increases with the complexity of a situation, as does the error rate.

Perception–reaction time tends to be detrimentally affected by fatigue, the use of alcohol or other drugs, and certain physiological and psychological conditions.

For the purpose of computing stopping sight distances, AASHTO recommends the use of a combined perception–reaction and brake–reaction time of 2.5 sec (*10*).

FIGURE 3-2 Mean response time due to various types of sensory stimulus. (Source: T. W. Forbes and N. S. Katz, Summary of Human Engineering Data and Principles Related to Highway Design and Traffic Engineering Problems; sketch courtesy of the Institute of Transportation Engineers.)

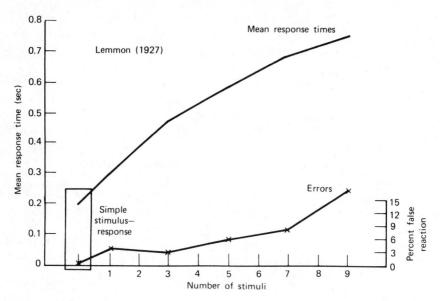

FIGURE 3-3   Increase in mean response time and errors with increasing number of stimuli. (Source: T. W. Forbes and N. S. Katz, Summary of Human Engineering Research Data and Principles Related to Highway Design and Traffic Engineering Problems; sketch courtesy of the Institute of Transportation Engineers.)

3-3 Driver information needs

Given the ever increasing demands of the driving task, it is important that highway and traffic engineers understand the driver's need for information and the means of its transmission. Research was undertaken in the late 1960s by AIL, a division of Cutler–Hammer, to identify types of information needed by drivers and the principal factors and interactions that affect the reception and use of this information. The results of one aspect of that work have been described in a paper by Allen, Lunenfeld, and Alexander (*11*), which forms the primary basis for the material presented in the following paragraphs.

In order to make sound decisions, a driver must receive reliable and understandable information to reduce uncertainty. A driver gathers information from a variety of sources, uses it as a basis for decision making, and then translates those decisions into actions to control the vehicle (*12*).

As Table 3-1 indicates, the various activities involved in driving a motor vehicle may be grouped into three categories or levels of performance: (1) the control level, (2) the guidance level, and (3) the navigation level.

The control level of driver performance is related to the physical operation of the vehicle. It includes lateral control that is maintained by steering and longitudinal control that is exercised by accelerating and decelerating. To maintain vehicle control, the driver requires information on the vehicle's position and orientation with respect to the road as well as feedback on the vehicle's response to braking, accelerating, and steering. Such information is received primarily through the visual, kinesthetic, and vestibular senses.

TABLE 3-1
A FRAMEWORK FOR DRIVING TASK CONCEPTUALIZATION

Subtask Category	Related to	Examples of Sources of Information	Importance of Information	Likely Consequence of Failure
Control (Microperformance)	Physical operation of vehicle Steering control Speed control	Road edges Lane divisions Warning signs Kinesthesia	Highest	Emergency situation or crash
Guidance (Situational performance)	Selecting and maintaining a safe speed and path	Road geometry Obstacles Traffic conditions Weather conditions	Intermediate	Emergency situation or crash
Navigation (Macroperformance)	Route following Direction finding Trip planning	Experience Directional signs Maps Touring service	Lowest	Delay, confusion, or inefficiencies

Guidance level of driver performance refers to the driver's task of selecting a safe speed and path on the highway. While performance at the control level is overt action, performance at the guidance level is a decision process. The driver must evaluate the immediate situation, make appropriate speed and path decisions, and translate these decisions into control actions needed to survive in the traffic stream. Activities include decisions relating to lane positioning, overtaking, and passing . . . .

The driver is constantly time-sharing activities at the guidance level with those at the control and navigation levels. Furthermore, two or more guidance level processing requirements may occur simultaneously or in close enough time proximity to require simultaneous action, which may not necessarily be compatible.
This points to the importance of experience and prior knowledge throughout the driving task, since the driver may be required to make several decisions under heavy time pressure. (*12*)

The navigation level of driver performance refers to the driver's ability to plan and execute a trip. It includes trip preparation and planning and direction finding. The former activities may be performed before the trip begins, while the latter occurs en route. Information at the navigation level comes from maps, directional signs, landmarks, and from the driver's memory based on previous experience.

Allen et al. (*11*) have listed five basic principles for the systematic presentation of information required by drivers:

1. First things first—primacy.
2. Do not overload—processing channel limitations.
3. Do it before they get on the road—*a priori* knowledge.
4. Keep them busy—spreading.
5. Do not surprise them—expectancy.

Primacy
The concept of primacy is based on the realization that at any given instant certain driver information is more important than other information. In evaluating the primacy of information, the effect of failure to receive that information is considered. Lack of certain types of information may lead to an emergency situation or a crash; lack of other types of information may only cause momentary confusion or delay. Avoiding a car emerging from a driveway (a control subtask) is more important than deciding whether to pass a slow-moving vehicle (a guidance subtask). Either of these subtasks would take precedence over observance of directional signs (a navigation subtask).

In the guidance category, where failures tend to be catastrophic, the question is one of degree. Generally, colliding with a guardrail is less serious than colliding with a bridge pier. Colliding with

a vehicle moving in the same direction is less serious than a
head-on collision . . . .

A second aspect of primacy judgment is based on distance criterion. A hazard close by would have a higher primacy than a
more distant hazard of the same or similar type. . . . Assessing
the primacy (both distance and severity potential) of a single
hazard is a task that drivers probably do well. Assessing the relative primacy of two competing hazards in a dynamic situation is
a difficult task that is more prone to driver error. (*12*)

Processing limitations

As the complexity of a driving situation increases, so does the driver's
response time. Obviously, if a driver is confronted with several complex situations within a short time span, a point may be reached when the driver is unable
to receive and process the amount of information required for error-free
response. Information needed by the driver should be presented systematically, requiring a series of simple choices rather than a smaller number of
complex choices.

A priori knowledge

Drivers should be given as much information as possible before they get
on the road. To the extent feasible, *a priori* information should be communi-
cated to drivers on distances and directions to destination points, average travel
times, highway types and design features, and other such information that will
facilitate the planning of trips.

Spreading

During the course of trips, drivers may be confronted with highly variable
demands for their attention, too much information being provided in certain
areas and too little in others. Efforts should be made to spread the information
load in order to avoid boredom of drivers in certain areas and overloading of
their processing capability in others. This is accomplished by performing a
primacy analysis and transferring the less important information from the
high-information "peaks" to the low-information "valleys."

Expectancy

A driver's readiness to respond to roadway conditions, traffic situations,
or information systems is referred to as driver expectancy. Based on previous
experience, drivers are conditioned to expect certain characteristics of road
design and operation. For example, drivers generally expect exit and entrance
ramps along freeways to be on the right-hand side. Drivers do not generally
expect lane drops or other changes in the roadway cross section. At locations
where driver expectancies are violated, drivers may require a longer response
time and will have a greater tendency to make an inappropriate or hazardous

response. Although expectancies occur at all levels of the driving task, their importance is probably greatest at the guidance level.

PEDESTRIAN CHARACTERISTICS

The pedestrian is the major user of the roadway system; when the system fails, he is a major victim. Approximately 7000 pedestrians are killed in the United States each year and about 120,000 are injured. The problem is most serious in urban areas, where pedestrian fatalities comprise about 48 percent of all traffic deaths.

Certain segments of the pedestrian population—notably the very young and the very old—are either unaware of rules of safe pedestrian behavior or unresponsive to efforts to enforce pedestrian traffic regulations. For example, Wiener (13) studied the responses of elderly pedestrians to an education and enforcement program directed toward eliminating jaywalking behavior. He concluded:

> A large percentage of elderly pedestrians are totally unrespon-
> sive to campaigns, legal sanction, enforcement symbols, or traffic
> regulation. No amount of campaigning will reach this group,
> for through lifelong habits, ignorance of traffic hazards and reg-
> ulations, senility or willful neglect, their behavior is unmodifiable
> by the threat of sanctions. To make cities safer for pedestrians,
> officials must look for means which do not require their
> conscious cooperation.

Engineers are challenged to design safe and convenient pedestrian facilities that will function well even for those persons who willfully or ignorantly disobey rules of safe walking behavior. Such facilities serve small children, as well as elderly and physically handicapped persons, including those who are blind or ambulatory.

The designers of pedestrian facilities require knowledge of the space requirements and walking speeds of individual walkers, as well as an understanding of traffic flow characteristics of groups of pedestrians. The former subjects are treated in the following paragraphs, while the latter is discussed in Chapter 5.

3-4 Space needs for pedestrians

The design of pedestrian facilities depends to a large extent on the shoulder breadth and body depth of the human body. One study of fully clothed male laborers indicated, for the 95th percentile, a shoulder breadth of 22.8 in. and a body depth of 13.0 in. (14). An elliptical shape with a 24-in. major axis and an 18-in. minor axis has been used for determining the capacity of subway cars (14). The ellipse, encompassing an area of 2.3 ft², is slightly larger than the area occupied by the 95th percentile male. It allows for body sway, the tendency to avoid bodily contact with others, and for packages or personal articles that many pedestrians carry. It should be emphasized that the 18- by 24-in. ellipse is useful primarily for the determination of space needs or capacities for elevators or other conveyances or locations where the pedestrians are standing rather than walking. For design of sidewalks or other pedestrian corridors, one needs to be concerned with the dynamic spatial requirements for avoiding collisions with other pedestrians.

> The space required for locomotion may be divided into a pacing zone, the area required for foot placement, and the sensory zone, the area required by the pedestrian for perception, evaluation, and reaction. The length of the pacing zone is dependent on the age, sex, and physical condition of the pedestrian, and has been shown to have a direct linear relationship with speed. Both the pacing and sensory zones can be affected by external influences such as terrain and traffic conditions. Pedestrian pacing lengths may be physically measured, but sensory zone requirements are comprised of many human perceptual and psychological factors. (*14*)

Research has shown that one pedestrian following another prefers to leave an average distance-spacing between himself and the lead pedestrian of about 8 ft. This corresponds to an average time-spacing of about 2 sec between pedestrians walking in line.

In a remarkably efficient manner, pedestrians monitor the varying speeds and angles of other approaching pedestrians and make adjustments to their pace and speed in order to avoid collisions.

3-5 Walking and running speeds

Empirical research has shown that under free-flow conditions pedestrian walking speeds tend to be approximately normally distributed. Under such conditions, pedestrian walking speeds range from about 2.0 to 6.0 ft/sec. Typical mean walking speeds are in the 4.0 to 4.5 ft/sec range. A pedestrian walking rate of 4.0 ft/sec is generally assumed for the timing of pedestrian traffic signals. In areas where there are large numbers of elderly pedestrians, AASHTO (*10*) recommends the use of a 3.0 ft/sec walking rate. The presence of significant numbers of handicapped persons would also dictate the use of a lower rate of movement.

Free-flow walking speeds vary with the pedestrian's age and sex as well as trip purpose. Grades less than about 5 percent seem to have little effect on walking speed, and pedestrians carrying baggage tend to walk about as fast as those without baggage (*14*).

Walking speeds decrease with increases in pedestrian density. Empirical studies indicate that for an average of 25 or more ft^2 per pedestrian, walking speeds are only slightly affected by pedestrian conflicts (*14*). As the available space per pedestrian drops below 25 ft^3, the average walking speed decreases sharply. Traffic flow characteristics for pedestrian flow are further examined in Chapter 5.

Speeds beyond about 470 ft/min (7.8 ft/sec) may be construed as running

(*14*). The fastest movement afoot is the 100 yard dash in 9.1 sec, a speed of 33 ft/sec.

VEHICLE CHARACTERISTICS

During the decade 1970 to 1980, automobile registrations in the United States increased 36 percent, while the number of registered trucks increased 80 percent. By 1984, there were over 168 million registered motor vehicles in the United States including 130 million passenger cars, 37 million trucks, and 611,000 buses (*4*). In 1984, the United States had about 36 percent of the world's passenger cars and about 35 percent of the world's trucks and buses.

Nineteen percent of the automobiles in use in the United States are less than 3 years old, while nearly 30 percent are 8 or more years old. There are more than 2 million cars on the road that are 16 years or older (*4*). The average age of automobiles in use in the United States is about 7.4 years, while the average age of trucks is 8.1 years.

Retail sales of passenger cars in the United States totaled 10.39 million in 1984, including 2.44 million imports. Sales of the intermediate size comprised 25.5 percent of retail sales. (See Table 3-2).

Total retail sales of new trucks in the United States in 1984 exceeded 4.0 million. Table 3-3 gives a breakdown of retail truck sales by franchised dealers by weight and type.

3-6 Vehicle size

The size of motor vehicles influences clearances for bridges, tunnels, and grade separation structures and the geometric characteristics of streets, roads, and parking facilities. Street and highway facilities are used by a wide variety of vehicles ranging from bicycles to (in certain states) articulated truck units up to

TABLE 3-2
PERCENTAGE OF
U.S. NEW CAR SALES
IN 1984, BY CLASS

Class	Percent
Subcompact	14.0
Compact	18.1
Intermediate	25.5
Standard	10.6
Luxury	8.3
Import	23.5
Total	100.0

SOURCE: *MVMA Motor Vehicle Facts and Figures '85.* Motor Vehicle Manufacturers Association, Detroit, Mich. (1985).

TABLE 3-3
RETAIL SALES OF NEW TRUCKS BY FRANCHISED
DEALERS OF U.S. MANUFACTURERS, 1984

Gross Vehicle Weight/Type	Sales
Trucks with GVW 10,000 lb or less	
Utility	473,141
Car-type pickup	25,699
Compact pickup	
Domestic	560,378
Import	52,894
Van	386,371
Minivan	23,727
Van, cutaway chassis	54,954
Conventional pickup	1,204,996
Station wagon (truck chassis)	147,679
Minipassenger carrier	168,774
Passenger carrier	99,088
Multistop	57,540
Subtotal	3,255,241
10,001–14,000 lb	6,019
14,001–26,000 lb	60,903
26,001–33,000 lb	78,479
33,001 lb and over	137,693
Grand total	3,538,335

SOURCE: *MVMA Motor Vehicle Facts and Figures '85.* Motor
Vehicle Manufacturers Association, Detroit, Mich. (1985).

75 ft in length and weighing over 100,000 lb. Since the highway system must be
designed to accommodate all classes of vehicles, the designer requires information on maximum and minimum dimensions, for example, the class of vehicle
with the greatest overall width, the longest turning radius, the smallest dimension from the bottom of the bumper to the ground, and so forth. Table 3-4
gives maximum and minimum dimensions of standard American passenger
cars manufactured in 1985. These dimensions are considered to be reasonably
representative of the total population of such cars since the dimensions have
changed little during the past decade.

The maximum size of trucks and buses is established by state laws, and
there is considerable variation among the states regarding the maximum allowable vehicle weights and dimensions. The range of state limitations is given by
Table 3.5. The maximum weight and width of vehicles that use the Interstate
System are set by federal law. Title 23 of the United States Code specifies a
maximum width of 102 in. for such vehicles. The code specifies a maximum–
overall gross vehicle weight for the Interstate System of 80,000 lb, including all
enforcement tolerances. The law also specifies a limit of 20,000 lb for any one

TABLE 3-4
**MAXIMUM AND MINIMUM DIMENSIONS OF STANDARD
AMERICAN PASSENGER CARS MANUFACTURED IN 1985 (IN.)[a]**

Dimension	Minimum	Maximum
Overall length	160.6	220.5
Overall width	61.8	79.8
Overall width, doors open	126.0	169.5
Overall height	46.4	60.0
Front overhang distance	30.9	46.5
Rear overhang distance	30.1	60.3
Wheelbase length	93.4	134.4
Front tread width	51.2	62.2
Rear tread width	51.2	64.0
Bottom of front bumper to ground	4.9	20.7
Bottom of rear bumper to ground	9.1	18.12
Ramp breakover angle	10.0	20.5
Turning diameter, outside front		
Wall to wall	32.5	51.1
Curb to curb	30.3	47.8
Turning diameter, inside rear		
Wall to wall	7.4	28.1
Curb to curb	7.6	28.4

[a] 1 in. = 2.54 cm.

SOURCE: *Parking Dimensions, 1985 Model Year Passenger Cars.* Motor Vehicle
Manufacturers Association, Detroit, Mich. (1985).

TABLE 3-5
**SUMMARY OF STATE LIMITATIONS ON
VEHICLE WEIGHTS AND DIMENSIONS**

Item	Minimum in Any State	Maximum in Any State
Width (in.[a])	96	108[c]
Height (ft[b])	12.5	14.0
Length (ft)		
Single unit truck	35	60
Bus	35	55
Truck tractor–semitrailer	55	85
Other combination	55	108
Maximum practical gross weight on any vehicle, lb	73,000	139,000

[a] 1 in. = 2.54 cm.
[b] 1 ft = 0.3048 m.
[c] 120 in. on Indiana and Pennsylvania turnpikes.

SOURCE: *NCHRP Report 198.* Transportation Research Board, Washington, D.C. (1979).

axle, including tolerances, and gives a formula for calculating the maximum overall gross weight on any group of two or more consecutive axles.

Vehicle performance characteristics

Vehicle performance limitations and capabilities affect required lengths of speed-change lanes, weaving lanes, and no-passing zones, and the design of traffic control systems. If highway systems are to operate safely and efficiently, highway and traffic engineers need to understand performance characteristics of the various vehicles that use such systems. In the following paragraphs, we will describe some of the more important vehicle operating characteristics.

3-7 Resistances

A vehicle's motion tends to be retarded by at least five types of resistance:

1. Inertia resistance.   4. Curve resistance.
2. Grade resistance.   5. Air resistance.
3. Rolling resistance.

Inertia Resistance

It will be recalled from the study of physics that inertia is the tendency of a body to resist acceleration: the tendency to remain at rest or to remain in motion in a straight line unless acted upon by some force. The force, F_i, required to overcome a vehicle's inertia is described by the familiar relationship

$$F_i = ma = \frac{W}{g} a, \text{ lb}$$

(3-1)

where
m = vehicle mass
a = acceleration (ft/sec^2)
W = vehicle weight (lb)
g = accelerative force due to gravity (32.2 ft/sec^2)

When acting to move a vehicle forward, this force is positive. When the force is slowing or stopping a vehicle, it and the corresponding acceleration (deceleration) are negative.

Grade resistance

Grade resistance is the component of the vehicle weight acting down a frictionless inclined surface. As Figure 3-4 illustrates, the magnitude of the force is directly proportional to the gradient. By similar triangles,

$$\frac{G}{100} = \frac{F_g}{F_n}$$

$$F_g = F_n \frac{G}{100} = W \cos \theta \frac{G}{100}$$

where
G = gradient (percent)
W = vehicle weight (lb)

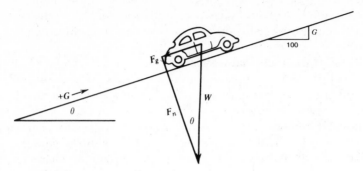

FIGURE 3-4 Forces on a vehicle on a frictionless inclined surface.

F_g = grade resistance (lb)
F_n = normal force (lb)

For even the steepest highway gradients encountered in practice, $\cos \theta \approx 1.0$. For practical purposes, the gradient resistance is

$$F_g = \frac{WG}{100}$$

$$(3\text{-}2)$$

Rolling resistance

A vehicle does not operate on a smooth frictionless surface. There is resistance to motion as the tires roll over irregularities in the surface and push through mud, sand, or snow. This resistance, termed rolling resistance, includes that caused by the flexing of the tires and the internal friction of the moving parts of the vehicle. Empirical tests have shown that, for speeds up to 60 mph, the rolling resistance of modern automobiles is practically constant at about 27 lb/ton of weight (15). Table 3-6 gives rolling resistances of passenger cars on low-quality road surfaces.

TABLE 3-6
ROLLING RESISTANCES OF PASSENGER CARS ON LOW-GRADE ROAD SURFACES (LB/TON)[a]

	Uniform Speed (mph)[b]			
Type of Surface	20	30	40	50
Badly broken and patched asphalt	29	34	40	51
Dry, well-packed gravel	31	35	50	62
Loose sand	35	40	57	76

[a] 1 lbf = 4.448N.
[b] 1 mile = 1.6093 km.

SOURCE: Claffey, Paul J. Vehicle Operating Characteristics. *Transportation and Traffic Engineering Handbook*, Institute of Transportation Engineers, Arlington, Va. (1975).

TABLE 3-7
TYPICAL CURVE RESISTANCES
OF PASSENGER CARS ON
HIGH-TYPE ROAD SURFACES (LB)[a]

Degree of Curve	Speed (mph)[b]	Resistance (b)
5	50	18
5	60	36
10	30	18
10	40	54
10	50	108

[a] 1 lb = 0.536 kg.
[b] 1 mile = 1.6093 km.

SOURCE: Claffey, Paul J. Vehicle Operating Characteristics. *Transportation and Traffic Engineering Handbook*, Institute of Transportation Engineers, Arlington, Va. (1975).

Curve resistance

As was stated earlier, once a vehicle is set in motion, it tends to remain in motion in a straight line unless acted on by some force. The forces changing the direction of a vehicle are imparted through the front wheels. Components of these forces tend to impede a vehicle's forward motion. Curve resistance then is the force required to cause a vehicle to move along a curved path. It is a function of the radius or degree of curvature and the vehicle speed. Typical curve resistances for passenger cars are shown in Table 3-7.

Air resistance

Air resistance includes the force required to move air from a vehicle's pathway as well as the frictional effects of air along its top, sides, and undercarriage. It is a function of the frontal cross-sectional area of the vehicle and the square of the vehicle speed. According to Claffey (*16*), the air resistance on passenger cars can be estimated by the following equation:

$$F_a = 0.0006AV^2 \tag{3-3}$$

where
A = frontal cross-sectional area (ft^2)
V = vehicle speed (mph)

Power requirements

Power is defined as the rate at which work is done. The power P required to overcome the various resistances and to propel a vehicle may be computed by the following equation:

$$P = \frac{Rv}{550} \text{ hp} \qquad (3\text{-}4)$$

where

R = sum of the various resistances = $F_i + F_g + F_r + F_c + F_a$ (lb)
v = vehicle speed (ft/sec)

It should be emphasized that the power computed by Eq. 3-4 is only that required for propulsion. It does not include the power required for a vehicle's automatic transmission, power steering, air conditioner, and other such accessories.

3-8 Acceleration and deceleration performances

A motor vehicle moves according to fundamental laws of motion; relationships between distance, time, velocity, and uniform acceleration are given by the following equations:

$$v_f = v_0 + at \qquad (3\text{-}5)$$

$$d = v_0 t + \tfrac{1}{2}at^2 \qquad (3\text{-}6)$$

$$v_f^2 = v_0^2 + 2ad \qquad (3\text{-}7)$$

where

v_f = final velocity (ft/sec)
v_o = initial velocity (ft/sec)
a = acceleration or deceleration rate (ft/sec)
t = time (sec)
d = distance (ft)

Maximum acceleration rates vary with the size of the vehicle and its operating speed. Vehicles are capable of greatest acceleration at lowest speeds. From a standing start to a speed of 15 mph, maximum acceleration values range from about 2 mph/sec (2.9 ft/sec^2) for tractor–semitrailer trucks up to about 10 mph/sec (14.7ft/sec^2) for large cars (*16*). For a speed change of 0 to 30 mph, typical maximum accelerations are

For tractor semitrailer truck	1.5 ft/sec^2
For large car	10.3 ft/sec^2
For small high-performance sports car	14.2 ft/sec^2

Figure 3-5 shows typical maximum acceleration rates for 10 mph increases at various running speeds along level roads. Wide variations are shown for various vehicle types and running speeds.

Without braking, a vehicle will decelerate when the driver releases the accelerator due to the "drag" of the engine, air resistance, grade resistance, and so forth.

A passenger car operating in the range of 50 to 60 mph will decelerate about 3.0 ft/sec^2 without braking; in the range of 20 to 30 mph, an automobile will decelerate about 1.5 ft/sec^2.

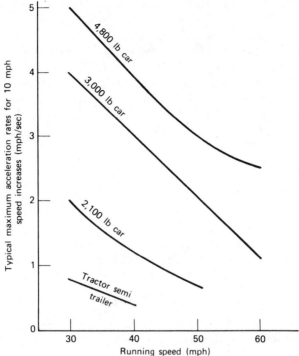

FIGURE 3-5 Typical maximum motor-vehicle acceleration rates for 10-mph speed increases at various running speeds on level roads. (Source: Paul J. Claffey, Vehicle Operating Characteristics, *Traffic and Transportation Engineering Handbook,* Institute of Transportation Engineers, Washington, D.C., 1975.)

Under normal braking conditions, the levels of deceleration developed usually do not reach the limit of a vehicle's braking capability nor that of the pavement–tire interface. The magnitude of decelerative force developed under normal or nonskid braking depends on the force the driver applies to the brake pedal. That force, in turn, is related to driver comfort. Claffey (*16*) has reported that observed normal deceleration rate for passenger cars (with braking) are about 3.3 mph/sec (4.8 ft/sec^2) in the range of 30 to 70 mph and about 4.6 mph/sec (6.7 ft/sec^2) in the range of 0 to 30 mph.

The deceleration may be limited by either the condition of the brakes or that of the tires and roadway surface. A vehicle with extremely worn brakes will produce little deceleration no matter what the condition of the tire tread or the pavement surface. On the other hand, a vehicle with new tires and excellent braking capability is extremely apt to skid on an ice-glazed pavement. Assuming the vehicle has adequate brakes, the maximum deceleration depends primarily on the coefficient of friction between the tires and the roadway surface. A further discussion of the effect of the coefficient of friction on stopping distance is given in Chapter 8.

In braking tests from speeds of 70 mph, Wilson (*17*) found that the average deceleration rate for comfort did not exceed 8.50 to 9.00 ft/sec^2. Drivers generally consider decelerations of about 11 ft/sec^2 to be undesirable and of 14 ft/sec^2 to be uncomfortable and possibly alarming (*17*). The maximum deceleration developed under loaded-wheel conditions on a dry pavement with good skid characteristics is about 25 ft/sec^2. In panic situations, most drivers tend to apply the brakes to the extent that a locked-wheel skid develops, even though greater frictional force is developed before skidding occurs.

PROBLEMS

3-1   A driver in a vehicle traveling 50 mph shifts her eyes from left to right and focuses on a vehicle emerging from a side street. Estimate the distance in feet the vehicle travels as the driver's eyes shift and fixate.

3-2*   A driver in a vehicle traveing 80 km/hr shifts his eyes from right to left and focuses on a child about to dart across the street. Estimate the distance in meters the vehicle travels as the driver's eyes shift and fixate.

3-3   In an intersection collision, one of the vehicles leaves 160 ft of skid marks. A skid mark analysis indicates the the vehicle was traveling 50 mph at the onset of braking. Assuming an average driver (i.e., median brake-reaction time), estimate the distance from the point of impact to the vehicle position when the driver initially reacted.

3-4*   A braking distance of 152 ft is assumed for a design speed of 45 mph. Show that a stopping distance of 97m is appropriate for that design speed.

3-5   Estimate the horsepower required to accelerate a 2500-lb vehicle traveling 30 mph up a 5.0 percent grade at a rate of 6 ft/sec^2. The vehicle has a frontal cross-sectional area of 20 ft^2. The roadway has straight alignment and a badly broken and patched asphalt surface.

3-6*   Develop an equation for determining vehicular air resistance in kilograms given a frontal cross-sectional area in square meters and a speed in kilometers per hour.

3-7   A 2100-lb car enters an acceleration lane at 25 mph and merges into a traffic lane at 55 mph. Estimate the desired length of the acceleration lane.

3-8*   On the basis of Figure 3-5, prepare a table showing the typical maximum acceleration rates in meters per square second for a 3000-lb car at 50, 60, 70, 80, and 90 km/hr.

REFERENCES

1. Goldstein, Leon G., On the Future of Driving Licensing and Driver Licensing Research. Transportation Research Board Special Report No. 151 (1974).

2. Kerrick, John C. What Have We Learned to Date? Transportation Research Board Special Report No. 151 (1974).

3. Marsh, Burton W. Aging and Driving. *Traffic Engineering*, Arlington, Va. (November 1960).

4. *Motor Vehicle Facts and Figures '85*. Motor Vehicle Manufacturers Association, Detroit, Mich. (1985).

5. Matson, Theodore M., Smith, Wilbur S., and Hurd, Frederick W. *Traffic Engineering.* McGraw-Hill, New York (1955).

6. Mourant, Ronald R., and Rockwell, Thomas H. Strategies of Visual Search by Novice and Experienced Drivers. *Human Factors*, Vol. 19, No. 4, pp. 325–335 (1972).

7. Greenshields, Bruce D. The Driver. *Traffic Engineering Handbook*, John E. Baerwald (ed.). Institute of Traffic Engineers, Washington, D.C. (1965).

8. Greenshields, Bruce D. Reaction Time in Automobile Driving. *J. Appl. Psych.*, Vol. 20, No. 3, p. 355 (1936).

9. Johansson, Gunnar, and Rumar, Kare. Driver's Brake Reaction Times. *Human Factors*, Vol. 13, No. 1, pp. 23–27 (1971).

10. *A Policy on Design of Urban Highway and Arterial Streets.* American Association of State Highway and Transportation Officials, Washington, D.C. (1973).

11. Allen, T. M., Lunenfeld, H., and Alexander, G. J. Driver Information Needs. *Highway Research Record 366* (1971).

12. Alexander, Gerson J., and Lunenfeld, Harold. *Positive Guidance in Traffic Control*, U.S. Department of Transportation, Washington, D.C. (1975).

13. Wiener, Earl L. The Elderly Pedestrian: Response to an Enforcement Campaign. *Research Review* (December 1968).

14. Fruin, John J. *Pedestrian Planning and Design.* Metropolitan Association of Urban Designers and Environmental Planners. New York (1971).

15. Kessler, Jay C., and Wallis, Stanley B. Aerodynamic Test Techniques. *SAE Trans.*, Vol. 75, Sec. 3, p. 12 (1967).

16. Claffey, Paul J. Vehicle Operating Characteristcs. *Transportation and Traffic Engineering Handbook.* Institute of Transportation Engineers, Arlington, Va. (1975).

17. Wilson, Ernest E. Deceleration Distances for High Speed Vehicles. *Proceedings, 20th Annual Meeting of the Highway Research Board*, pp. 393–397, Washington, D.C. (1940).

SUGGESTED READINGS

Driver Expectancy Checklist, A Design Review Tool. American Association of State Highway and Transportation Officials, Washington, D.C. (1972).

Forbes, Theodore W. Human Factors in Highway Design Operation and Safety Problems. *Human Factors*, Vol. **2**, No. 1 (1960).

Hulbert, Slade. Driver and Pedestrian Characteristics. *Transportation and Traffic Engineering Handbook.* Institute of Transportation Engineers, Arlington, Va. (1975).

Proceedings, National Conference for the Aging Driver. American Medical Association and the American Association of Motor Vehicle Administrators, Washington, D.C. (1974).

Physician's Guide for Determining Driver Limitation. American Medical Association, Chicago, Ill. (1973).

Tignor, S.C. Braking Performance of Motor Vehicles. *Public Roads*, Vol. **34**, No. 4 (1966)

chapter four _____

HIGHWAY EVALUATION

In the United States up until about 1950, highway engineers often were able to select desirable highway projects and plan, design, and construct them primarily on the basis of engineering and travel demand factors. Engineers, and the public as well, understood the need for extensive improvements to the highway system and recognized the considerable benefits from such projects.

During the 1950s and 1960s, thousands of miles of new highways were constructed, many in heavily populated areas, causing social disruption, public hostility, and, in certain instances, militant opposition. Some segments of the public expressed concerns about possible environmental and ecological harm from highway construction. The partial fulfillment of the need for new highways and changes in public attitudes have brought about higher standards for economic, social, and environmental review.

As the Interstate Highway System nears completion, highway agencies are increasingly focusing attention on reconstruction and improvement of existing highway systems rather than building new facilities. Now, new highway systems and improvements to existing facilities are being subjected to greater scrutiny; and formal evaluation of the economic impact as well as social and environmental impacts of highway projects is becoming commonplace.

In this chapter, we will examine some of the techniques and procedures for making such evaluations. As is the case with many of the subjects covered in this text, space will not permit a detailed discussion of all aspects of highway evaluation. It will be necessary to simplify some of the material presented, with emphasis being placed on certain principles of fundamental importance.

ECONOMIC ANALYSIS FOR HIGHWAY IMPROVEMENTS

Although economic analysis for highway projects was proposed as a decision-making tool in the mid-nineteenth century, the technique was adopted by highway officials much later. About 1920, the first adequate motor vehicle running cost data appeared in the literature, but economic analyses for highways did not gain wide acceptance until after the publication of the AASHTO "Red Book" (1) in 1952. The Red Book was updated in 1960, and others (2–5) have since updated highway user cost factors and refined economic analysis methodology. A Stanford Research Institute report (6) prepared for the National Cooperative Highway Research Program, which contains unit highway user cost factors based on 1975 levels of vehicle performance characteristics and prices, has replaced the original AASHTO Red Book as the basic reference source. The material in the following paragraphs is based to a large degree on the Stanford Research Institute report.

Economic studies for highway purposes are done principally for one or more of the following reasons:

1. To determine feasibility of a project.
2. To compare alternative locations.
3. To evaluate various features of highway design, for example, the type of surface to be used.
4. To determine priority of improvement.
5. To allocate responsibility for the costs of highway improvement among the various classes of highway users (and nonusers, in some cases).
6. Occasionally, to compare proposals for highway improvement with proposals for other public projects such as education.

Before describing analytical concepts and approaches for economic studies, let us briefly consider the costs and benefits of highway transportation.

4-1 Costs of highway transportation

Highway transportation cost is defined as the sum of the highway investment cost, the maintenance and operating costs, and the highway user costs. Where bus transit is present, the transportation cost includes the transit capital costs, the costs of operating the transit system, and the costs that accrue to transit users (fares and travel time costs). Examples of highway transportation costs are given in Table 4-1.

The highway investment cost is the cost of preparing a highway improvement for service including the cost of rights-of-way, engineering design, construction, traffic control devices, and landscaping. Maintenance cost is the cost of preserving a highway and its appurtenances and keeping the facility in serviceable condition. Operating costs include the costs of traffic control, lighting, and the like.

Transit capital costs are the investments in fixed facilities for bus transit such as terminals, shops, bus parking facilities, and passenger benches and shelters. Special highway facilities for buses such as busways may also be included. Reference 6 treats the costs of bus vehicles as an operating rather than a capital cost. Other transit operating costs include drivers' wages, bus fuel and tires, bus maintenance, and the costs of management, administration, and insurance.

Transit user cost consists of the sum of bus fares paid to make a transit trip, the passenger car running costs experienced by users in getting to a stop or terminal, and the costs of waiting and walking (6).

Highway user costs

A major component of the costs of highway transportation, highway user costs include motor vehicle operating costs, the value of travel time, and traffic accident costs. Usually, only those operating costs that depend on the mileage

TABLE 4-1
EXAMPLES OF HIGHWAY TRANSPORTATION COSTS

Type of Cost	Examples
Highway investment cost	Engineering design, right-of-way, grading and drainage, pavement
Highway maintenance cost	Mowing, care of roadside parks, lighting
Highway user cost	
(a) Motor vehicle operating cost	Fuel, lubrication, tires
(b) Travel time	Total vehicle-hours of travel times unit value of time
(c) Accident costs	Estimated accident rate times unit accident cost
Transit capital costs	Terminals, shops, administrative offices
Transit operating costs	Drivers' wages, operating and maintenance cost of buses
Transit user costs	
(a) Transit money costs	Sum of fares paid, auto running costs to get to terminal
(b) Transit travel time costs	Time waiting for bus times unit value of time
	Time riding bus times unit value of time

of travel are included in highway economic analyses, including the cost of fuel, tires, engine oil, maintenance, and a portion of depreciation. Registration and parking fees, insurance premiums, and the time-dependent portion of depreciation may be excluded when estimating the reduction in user costs due to a highway improvement. Anderson et al. (6) have published a set of nomographs by which estimates of operating costs can be made. Examples of these nomographs and a discussion of their use are given in Section 4-4.

The value of travel time is the product of the total vehicle-hours of travel (by vehicle type) and the average unit value of time. The magnitude of travel time depends on average running speed and the number and duration of stops. Studies have indicated that the perceived value of travel time is sensitive to trip purpose and time savings per trip. Generally, travelers place little value on time savings less than 5 min. For passenger cars, Ref. 6 recommends a travel time value of $3.00 per vehicle per hour based on a value of $2.40 per person-hour and 1.25 persons per vehicle. The report recommends time values of $7.00 per hour for single-unit trucks and $8.00 per hour for combination trucks based on truck driver wages and fringe benefits.

The accidents that occur every year on the streets and highways of America represent a great economic loss besides the pain and suffering that they cause. Total economic loss resulting from motor vehicle accidents in 1984 was estimated to be $47.6 billion (7). This total included losses incurred in fatal accidents, accidents involving nonfatal personal injury, and those involving only property damage. It follows that from an economic standpoint highway improvements that result in decreased accident rates result in monetary saving to the road user and the public in general.

The National Safety Council issues periodic reports (8) on the costs of vehicular accidents. Such costs include wage losses, medical expenses, property damages, and insurance administration costs. In 1983, the costs of these items per case were

Death	$210,000
Nonfatal disabling injury	$8,600
Property damage accident (including minor injuries)	$1,150

For each death in motor vehicle accidents, there are approximately 36 nonfatal injuries and 390 property damage accidents (8).

To account for savings in accident costs in highway economic studies, accident rates before and after the proposed improvement should be estimated. In many instances, the analyst is able to rely on empirical data that will likely account for regional differences in vehicle mix, driver behavior, and climate. General guidance on typical accident rates for various classes of roadways is given by Table 4-2. A number of researchers (9–11) have also developed multiple regression equations by which estimates of accident rates can be made on the basis of roadway and traffic characteristics.

4-2 Benefits of highway improvements

It is obvious that many benefits result from highway improvement or, to put it more broadly, from improved highway transportation. Some of these benefits are direct and readily apparent; others are indirect and more difficult of discernment. Likewise, some benefits may be readily evaluated in terms of dollars and cents; others defy evaluation in this fashion, although they are nonetheless as real and lasting as monetary returns. Sometimes, of course, the consequences of a given highway improvement are not entirely beneficial; for example, construction of an urban expressway may produce a decline in ridership on a commuter railroad; and public funds spent from the general treasury for highways cannot be used, say, for education. Such consequences should be considered in comprehensive economic analyses.

For purposes of discussion, it is convenient to group highway benefits into the following categories: (1) direct benefits that result from a reduction in highway user costs and (2) indirect benefits, including benefits to adjacent property and to the general public. The most quantifiable and, to the analyst, the most significant highway benefits are those that result from a reduction in user costs. As the previous section implies, such benefits result from decreased operating costs, higher operating speeds, fewer delays, and decreased accident losses. The procedures for estimating such benefits are described later in this chapter.

Those benefits that accrue to the owner of property located adjacent to or in the vicinity of improved highways are fairly simple to observe and delineate but are somewhat more difficult to evaluate. In the vast majority of cases, high-

TABLE 4-2
ACCIDENT RATES BY ROAD TYPE

	Rate (Number per MVM[a])			
Location and Road Type	Fatal Accidents	Injury Accidents	Property Damage Only Accidents	Total Accidents
Rural				
No access control				
2 lane	0.070	0.94	1.39	2.39
4 or more lanes, undivided	0.047	0.89	1.95	2.89
4 or more lanes, divided	0.063	0.77	1.25	2.09
Partial access control				
2-lane expressway	0.051	0.52	0.76	1.33
Divided expressway	0.038	0.44	0.76	1.24
Freeway	0.025	0.27	0.49	0.79
Urban				
No access control				
2 lanes	0.045	1.51	3.38	4.94
4 or more lanes, undivided	0.040	2.12	4.49	6.65
4 or more lanes, divided	0.027	1.65	3.19	4.86
Partial access control				
2-lane expressway	0.033	6.65	1.05	1.73
Divided expressway	0.022	1.08	2.04	3.14
Freeway	0.012	0.40	1.01	1.43
Suburban				
No access control				
2 lanes	0.048	1.26	2.56	3.88
4 or more lanes, undivided	0.037	1.58	3.31	4.93
4 or more lanes, divided	0.030	1.10	2.24	3.37
Partial access control				
2-lane expressway	0.096	0.82	1.42	2.34
Divided expressway	0.060	0.82	1.29	2.16
Freeway	0.015	0.32	0.74	1.07
Statewide				
No access control				
2 lanes	0.066	1.01	1.66	2.74
4 or more lanes, undivided	0.041	1.88	3.97	5.88
4 or more lanes, divided	0.031	1.50	2.89	4.43
Partial access control				
2-lane expressway	0.053	0.54	0.81	1.40
Divided expressway	0.037	0.61	1.08	1.73
Freeway	0.015	0.36	0.86	1.23

[a]1 mile = 1.6093 km.

Source: *A Manual on User Benefit Analysis of Highway and Bus-Transit Improvements.* American Association of State Highway and Transportation Officials (1977).

way improvements result in an increase in the value of land located adjacent to or served by the highway involved. A simple case of direct property benefit is that of a farm or a number of farms located on an unimproved earth road, which is rough, muddy in wet weather and dusty in dry. Any degree of improvement of a road of this sort would be of direct benefit to the property owners located along it since they would be provided with easier access to market, schools, and recreational facilities. Similar direct benefits may accrue to city dwellers who live on an improved residential street. In many other cases, property benefits may be more difficult to discern and to evaluate.

Benefits of improved highway transportation that are enjoyed by the general public may well be regarded as the most important of the three classes of benefits which have been listed. These benefits include many of the advantages commonly associated with modern civilization, including such things as efficient fire and police protection, rapid postal deliveries, improved access to educational, recreational, and social facilities, decreased costs of commodities, improved national defense, and so on ad infinitum. The general advantages of improved highways are so interwoven with every phase of modern life in this country that it is virtually imposible to imagine what existence would be like without them.

Although it is apparent that indirect benefits are of great importance, these benefits cannot usually be evaluated except in a general sense and are therefore seldom used in economic calculations.

4-3 Present value concepts

In economic studies, it is important to recognize the time value of money. Because of the existence of interest, a quantity of money is worth more now than the prospect of receiving the same quantity at some future date. Therefore, in order to compare costs and benefits of highway improvements on a sound basis, they must be converted to equivalent values at some common date. This procedure, which is known as discounting, is accomplished by using a suitable interest rate in accordance with established principles of compound interest and present value concepts. Such concepts are more fully described in standard textbooks on economic analysis. For example, see Winfrey (2) and Grant and Ireson (12).

The present value (also termed present worth) P of some future single payment F can be calculated by the folowing equation:

$$P = \frac{F}{(1 + i)^n} \tag{4-1}$$

where
i = interest or discount rate per period
n = number of interest periods, usually years

Alternatively, present worth can be determined by use of present worth factors (PW) such as those given in Table 4-3.

TABLE 4-3
**PRESENT WORTH (PW) AND SERIES PRESENT WORTH (SPW) COMPOUND
INTEREST FACTORS FOR $1 AND VARIOUS INTEREST RATES**

Years	4%		6%		8%		10%	
n	PW	SPW	PW	SPW	PW	SPW	PW	SPW
5	0.8219	4.4518	0.7473	4.2124	0.6806	3.9927	0.6209	3.7908
10	0.6756	8.1109	0.5584	7.3601	0.4632	6.7101	0.3855	6.1446
15	0.5553	11.1184	0.4173	9.7122	0.3152	8.5595	0.2394	7.6061
20	0.4564	13.5903	0.3118	11.4669	0.2145	9.8181	0.1486	8.5136
25	0.3751	15.6221	0.2330	12.7834	0.1460	10.6748	0.0923	9.0770
30	0.3083	17.2920	0.1741	13.7648	0.0994	11.2578	0.0573	9.4269
40	0.2083	19.7928	0.0972	15.0463	0.0460	11.9246	0.0221	9.7791

SOURCE: *A Manual on User Benefit Analysis of Highway and Bus-Transit Improvements.* American Association of State Highway and Transportation Officials (1977).

EXAMPLE 4-1.   In 20 years, the residual or salvage value of a certain highway will be $250,000. Determine the present value of that sum, using an interest rate of 6 percent.
By Eq. 4-1,

$$\text{present value} = \frac{\$250,000}{(1 + 0.06)^{20}} = \$77,951$$

The same result is obtained by multiplying $250,000 by the appropriate present worth factor *PW* from Table 4-3.

The present worth P_s of a series of uniform annual end-of-period payments *A* can be determined by the following equation:

$$P_s = A \left[\frac{(1 + i)^n - 1}{i(1 + i)^n} \right] \tag{4-2}$$

or

$$P_s = A(SPW) \tag{4-3}$$

where
SPW = series present worth factor (Table 4-3)

EXAMPLE 4-2.   The user benefits for a certain highway are estimated to be a uniform $85,000 per year. Determine the present worth of those benefits assuming an interest rate of 8 percent and an analysis period of 25 years.
By Eq. 4-2,

$$P_s = \$85,000 \left[\frac{(1 + 0.08)^{25} - 1}{0.08(1.08)^{25}} \right] = \$907,358$$

In many instances, the annual values of user benefits (or highway costs) change from year to year. In those cases, the present worth cannot be calculated by Eq. 4-2 or 4-3 but can be computed for each year of the analysis period by Eq. 4-1 and then summed. An equation given in Ref. 6 makes it possible to closely approximate the present value of a series of annual values based on the average annual rate of growth of the values. The equation, which follows, is based on the assumption that the annual values increase (or decrease) at approximately an equal annual percentage. The present worth factor

$$PW_g = \frac{e^{(r-i)n} - 1}{r - i} \tag{4-4}$$

where
i = assumed interest rate (decimal fraction)
r = rate of growth of annual value (percent per annum/100)
n = analysis period (years)

The average rate of growth r can be approximated on the basis of estimates of 2 years' values:

$$r = \frac{\ln(\alpha)}{Y} \tag{4-5}$$

where
α = future annual value estimate/first year value estimate
Y = number of years between estimates

EXAMPLE 4-3.   A certain highway has user benefits of $17,500 during the first year that uniformly increase at a rate of 4 percent each succeeding year. Determine the present value of the benefits, given a 12-year analysis period. Use an interest rate of 6 percent.
By Eq. 4-4, the present worth factor,

$$PW_g = \frac{e^{(0.04-0.06)12} - 1}{0.04 - 0.06} = 10.67$$

present worth = 10.67 × $17,500 = $186,725

4-4 Economic analysis techniques

Economic calculations intended to demonstrate economic justification for a particular project, to permit comparison of alternative schemes or locations, to determine priority of improvement, and so on, may be carried out by one of several methods. Methods used in engineering economy studies include (1) benefit–cost ratio, (2) net present value, (3) comparison of annual costs, and (4) determination of the interest rate at which alternatives are equally attractive. The benefit–cost ratio (B/C) and net present value (NPV) methods, described here, are commonly used by highway engineers and administrators. For additional information on the latter two techniques, the reader is referred to standard textbooks on economic analysis.

Benefit–cost ratio

Traditionally, in the benefit–cost ratio method, analysis was made by comparing the relations of the *annual* road user costs to the *annual* highway costs for logical alternatives in location and design. The following formulation of the benefit–cost ratio, which utilizes present values, is now recommended for all highway and transit applications as an approximate indicator of project desirability (6). The benefit–cost ratio

$$\frac{B}{C} = \frac{PV(\Delta U)}{PV(\Delta I) + PV(\Delta M) - PV(\Delta R)} \qquad (4\text{-}6)$$

where

PV = present value of the indicated amount or series

ΔU = user benefits, the reduction in highway or transit user costs due to the investment (costs without the improvement less costs with the improvement)

ΔM = increase in annual maintenance, operating, and administrative cost due to the investment (costs with the improvement less costs without the improvement)

ΔR = increase in residual value due to the project at end of project life

ΔI = increased investment costs due to the project

Net present value method

In the net present value method, the procedure is to reduce all benefits and costs to their present value in accordance with the principles given in Section 4-3. The net present value NPV is defined as the difference between the present values of the benefits from the project and the costs of developing the project. The net present value may be computed as follows:

$$NPV = PV(\Delta U) + PV(\Delta R) - PV(\Delta I) - PV(\Delta M) \qquad (4\text{-}7)$$

Stated another way:

$$NPV = \frac{B_1 - C_1}{(1 + i)} + \frac{B_2 - C_2}{(1 + i)^2} + \ldots + \frac{B_n - C_n + R}{(1 + i)^n} \qquad (4\text{-}8)$$

where B_1 and C_1 are the benefit and cost for the first year, B_2 and C_2 for the second year, and so forth, R is the residual value, i is the discount rate, and n is the analysis period.

Decision rules in economic analysis

Rules of application of the B/C ratio and the net present value techniques depend on whether there are budgetary constraints and whether the projects under consideration are independent or mutually exclusive. Consider first the case where there are no budgetary constraints and independent projects are being compared; that is, the selection of one project would not preclude the selection of other projects. In this instance, all projects with positive net present values or benefit–cost ratios greater than one would be economically feasible.

Where the investment budget is constrained,* the analyst should select the combination of projects that produces a maximum net present value but does not exceed the available budget. Alternatively and approximately, projects may be selected in order of decreasing B/C ratios, adding projects until the budget is exhausted.

In cases in which several mutually exclusive locations or designs are being compared, the alternative with the highest net present value would be chosen. If benefit–cost ratios are used to select the preferred project, the selection must be made in increments. Beginning with the lowest-cost alternative having a B/C ratio greater than one, each increment of additional cost is justified only if the *incremental B/C* ratio is greater than one. This is illustrated by the following example.

EXAMPLE 4-4.   Three alternative designs (X1, X2, and X3) are being considered for a major interchange. For each of the alternatives, the present values are shown below for user benefits, investment costs, maintenance costs, and the residual value. By means of Eqs. 4-6 and 4-7, the B/C ratios and NPVs are computed as indicated. Which is the preferred alternative?

On the basis of net present values, alternative X2 is chosen.

To make the selection by the B/C ratio method, it is necessary to use incremental ratios, that is, based on the differences in costs of pairs of alternatives. Alternative X1, which has the lowest investment cost, is tentatively selected. The next most costly alternative, X2, is tested by calculation of the B/C ratio for the increment X2−X1. Since that ratio is greater than one, alternative X2 is selected to replace X1. Alternative X3 is then tested by the incremental B/C ratio for X3−X2. Since that ratio is less than one, alternative X3 is not justified, and X2 is selected.

	X1	X2	X3	(X2 − X1)	(X3 − X2)
$PV(\Delta U)$	95	127	128	32	1
$PV(\Delta I)$	12	13	20	1	7
$PV(\Delta M)$	7	20	17	13	−3
$PV(\Delta R)$	2	1	3	−1	2
B/C ratio	5.6	4.0	3.8	2.1	0.5
NPV	78	95	94	17	−1

4-5 Measurement of user benefits

One of the most difficult aspects of economic analyses for highways is the measurement of highway user benefits. First, consider highways that do not serve transit vehicles. The unit highway user cost for a given section of highway HU expressed in dollars per thousand vehicles is the sum of the basic section costs, accident costs, and delay costs:

*A linear or dynamic programming procedure is recommended for those situations where both the investment budget and the future maintenance and operating budget are constrained (6).

$$HU = (B + A)L + T + D \qquad (4\text{-}9)$$

where

B = basic section costs, consisting of the unit cost (time value and vehicle running costs) associated with vehicle flow and the basic geometrics (grades and curves) of the analysis section

A = unit accident costs in the analysis section

L = analysis section length in miles

T = transition costs—additional unit user time and running costs incurred due to changes in speeds between accident sections

D = additional unit time and running costs caused by delays at intersections. at traffic signals, stop signs, or other traffic control devices.

The highway user benefit ΔU in Eq. 4-6 is the difference in highway user costs for any two alternatives:

$$\Delta U = HU_0 - HU_1 \qquad (4\text{-}10)$$

Highway user costs are determined largely on the basis of empirical data, and Ref. 6 provides data for the estimation of the basic section costs, transition costs, and delay costs for various highway and traffic conditions.

The revised manual includes nomographs that give one-way basic section costs for three classes of vehicles (passenger cars, single unit trucks, and 3-S2 combination trucks) using four types of highways (freeways, multilane highways, two-lane highways, and arterials). One of the 12 nomographs that shows the basic section costs for passenger cars on multilane highways is reproduced as Figure 4-1.

To use Figure 4-1, the nomograph is entered at the lower left with average running speed or with the volume–capacity ratio for the representative hour of operation for the analysis section. Travel time, the inverse of running speed, can be read on the left ordinate of the lower left graph. A horizontal line at the level of average running speed is then projected to the right to determine the tangent running cost (along a specified grade) and added running costs due to curves. Finally, added running costs due to speed change cycles are determined by entering the upper left graph with the volume–capacity ratio and reading the added cost on the ordinate. The added costs due to speed change cycles are relatively small except under level of service F (or queuing) conditions. An example of the use of the nomograph is shown on the figure.

The basic section costs given by Figure 4-1 apply only to paved roadways. The costs of operting vehicles on gravel and stone roadway surfaces can be estimated by multiplying the basic section costs from the nomographs (e.g., Figure 4-1) by an adjustment factor. For passenger cars, such conversion factors range from about 1.1 for a speed of 5 mph to about 1.6 for a speed of 60 mph (2). Larger values are recommended for single-unit trucks and truck combinations.

FIGURE 4-1 Basic section costs (*B*) for passenger cars on multilane highways.

As Section 4-1 details, unit accident costs can be estimated on the basis of empirical accident rates that reflect regional effects of vehicle mix, driver behavior, roadway type, and climate, or can be estimated from the average rates (Table 4-2) and costs (by accident type) previously given.

Empirical studies have made it possible to estimate the costs of speed changes when vehicles pass between sections with different physical or traffic characteristics. These section transition costs, symbolized by the letter T in Eq. 4-9, depend on the magnitude of average speed change. Figure 4-2 provides a means of estimating the transition costs for one-way traffic based on the speeds on adjacent sections. Reference 6 gives adjustment factors that can be applied to the graph values to estimate transition costs for single unit and combination trucks.

Perhaps the most complex aspect of determination of unit highway user costs for a highway section is the estimation of the costs of intersection delay (referred to as D in Eq. 4-9). Two nomographs have been published by Anderson et al. (6) to determine intersection delay costs: one to estimate the unit costs of stopping at a traffic signal and the other to estimate the unit costs of idling. These nomographs are reproduced as Figures 4-3 and 4-4. The figures require data on the following traffic and signalization parameters for the intersection being studied.

1. **Green-to-cycle time ratio** λ. The ratio of effective green time of the signal (generally taken to be the actual green time) to the cycle length of the signal, both expressed in the same unit of time (usually seconds).
2. **Saturation flow s.** The approach volume in vehicles per hour of green that is found for the intersection when the load factor is 1.0 and the appropriate adjustment factors are applied. In the absence of Highway Capacity Manual (13) solutions, recommended values for saturation flow are 1700 to 1800 vehicles per hour times the number of approach lanes.
3. **Capacity c.** Where the Highway Capacity Manual is used, capacity is the service volume of the approach at a load factor of 1.0. It is also equal to the saturation flow times the green-to-cycle time ratio.
4. **Degree of saturation** χ. The ratio of the volume of traffic approaching the intersection (usually in vehicles per hour) to the capacity of the intersection, in the same units.
5. **Approach speed.** Also termed "midblock speed," this is the average running speed at which the signalized intersection is approached by the vehicle stream.

The examples that accompany Figures 4-3 and 4-4 illustrate their use. Adjustment factors are provided in the manual to account for the effects of single-unit trucks and combination trucks.

One-way versus two-way traffic

The costs given by the highway user nomographs relate to one-way traffic on a highway facility. In the usual case where the improvements affect both sides of the roadway and the traffic is balanced in direction over the course of a

ADJUSTMENT FACTORS FOR TRUCK TRAFFIC

SINGLE UNIT TRUCKS (percent)	3-S2 COMBINATION TRUCKS (percent)											
	0	1	2	3	4	5	6	8	10	20	50	100
0	1.00	1.10	1.19	1.29	1.38	1.48	1.57	1.77	1.96	2.92	5.79	10.58
1	1.02	1.12	1.21	1.31	1.40	1.50	1.59	1.79	1.98	2.94	5.81	-
2	1.04	1.14	1.23	1.33	1.42	1.52	1.61	1.81	2.00	2.96	5.83	-
3	1.06	1.16	1.25	1.35	1.44	1.54	1.63	1.83	2.02	2.98	5.85	-
4	1.08	1.18	1.27	1.37	1.46	1.56	1.65	1.85	2.04	3.00	5.87	-
5	1.10	1.20	1.29	1.39	1.48	1.58	1.67	1.87	2.06	3.02	5.89	-
6	1.12	1.22	1.31	1.41	1.50	1.60	1.69	1.89	2.08	3.04	5.91	-
8	1.16	1.26	1.35	1.45	1.54	1.64	1.73	1.93	2.12	3.08	5.93	-
10	1.20	1.30	1.40	1.49	1.58	1.68	1.77	1.97	2.14	3.12	5.97	-
20	1.41	1.50	1.60	1.70	1.78	1.88	1.94	2.17	2.34	3.32	6.17	-
50	2.02	2.12	2.21	2.31	2.39	2.49	2.58	2.78	2.95	3.93	6.78	-
100	3.04	-	-	-	-	-	-	-	-	-	-	-

EXAMPLE

GIVEN:

A stream of traffic consists of 5% single unit trucks and 10% combination trucks. The traffic is travelling at 60 mph and enters a slower section on which the speed is 30 mph.

SOLUTION:

(a) Passenger Car Cost - $6.75

T - $6.75 × 2.06

= $13.91 per 1000 Vehicles

FIGURE 4-2 Section transition costs.

ADJUSTMENT FACTORS FOR PERCENT TRUCKS IN TRAFFIC STREAM

TIME COST

APPROACH SPEED (mph)	SINGLE UNIT TRUCKS (percent)	3 S-2 COMBINATION DIESEL TRUCKS (percent in traffic stream)				
		0	5	10	20	100
5-20	0	1.00	1.15	1.30	1.61	4.03
	5	1.07	1.22	1.37	1.67	
	10	1.13	1.28	1.43	1.74	
	20	1.26	1.41	1.57	1.87	
	100	2.31				
21-40	0	1.00	1.25	1.51	2.01	6.05
	5	1.10	1.35	1.60	2.11	
	10	1.20	1.45	1.70	2.21	
	20	1.40	1.65	1.90	2.41	
	100	2.99				
41-60	0	1.00	1.41	1.82	2.63	9.17
	5	1.11	1.56	1.93	2.74	
	10	1.22	1.61	2.04	2.85	
	20	1.44	1.85	2.26	3.07	
	100	3.20				

RUNNING COST

APPROACH SPEED (mph)	SINGLE UNIT TRUCKS (percent)	3 S-2 COMBINATION DIESEL TRUCKS (percent in traffic stream)				
		0	5	10	20	100
5-20	0	1.00	1.35	1.70	2.40	8.02
	5	1.08	1.43	1.78	2.49	
	10	1.16	1.51	1.86	2.57	
	20	1.32	1.68	2.03	2.73	
	100	2.62				
21-40	0	1.00	1.35	1.71	2.41	8.07
	5	1.07	1.42	1.78	2.48	
	10	1.14	1.49	1.84	2.55	
	20	1.27	1.63	1.98	2.69	
	100	2.37				
41-60	0	1.00	1.35	1.70	2.39	7.96
	5	1.06	1.41	1.76	2.45	
	10	1.12	1.47	1.82	2.51	
	20	1.24	1.59	1.94	2.63	
	100	2.21				

EXAMPLE

GIVEN:

Volume: 480 vehicles/hr
Saturation Flow: 1,600 vehicles/hr
Signal Cycle Time: 60 sec
Effective Green Time: 30 sec
Intersection Approach Speed: 30 mph
5% Single Unit Trucks
5% 3-S2 Combination Trucks

SOLUTION:

λ = 30/60 - 0.5
Capacity of Approach = 0.5 × 1600 - 800
λ = 480/800 = 0.6
(a) Average Stops per Vehicle (per Signal): 0.71
(b) Stopping Delay per Signal: 2.5 hrs
(c) Cost of Stopping: $10.30
Time Cost: 2.5 × $3.00* × 1.35† $10.13
Running Cost: $10.30 × 1.42† 14.63
Total Cost Due to Stopping per 1,000 vehicles per Signal (excludes idling) $24.76

*Assumed hourly value of time per passenger car.
†Adjustment factors for trucks in traffic stream.

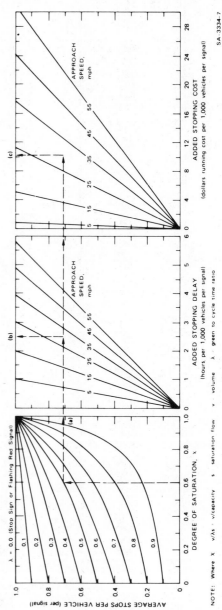

NOTE: Where X = v/λs = v/capacity   v = volume   s = saturation flow   λ = green to cycle time ratio

SA-3334-7

FIGURE 4-3 Costs due to stopping at intersections (excludes idling).

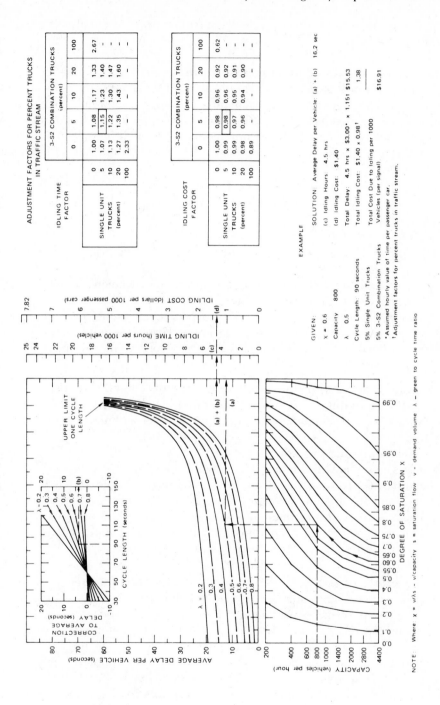

ADJUSTMENT FACTORS FOR PERCENT TRUCKS
IN TRAFFIC STREAM

IDLING TIME FACTOR		3-S2 COMBINATION TRUCKS (percent)				
		0	5	10	20	100
SINGLE UNIT TRUCKS (percent)	0	1.00	1.08	1.17	1.33	2.67
	5	1.07	1.15	1.23	1.40	—
	10	1.13	1.22	1.30	1.47	—
	20	1.27	1.35	1.43	1.60	—
	100	2.33	—	—	—	—

IDLING COST FACTOR		3-S2 COMBINATION TRUCKS (percent)				
		0	5	10	20	100
SINGLE UNIT TRUCKS (percent)	0	1.00	0.98	0.96	0.92	0.62
	5	0.99	0.98	0.96	0.92	—
	10	0.99	0.97	0.95	0.91	—
	20	0.98	0.96	0.94	0.90	—
	100	0.89	—	—	—	—

EXAMPLE

GIVEN:

X = 0.6

Capacity   800

λ   0.5

Cycle Length: 90 seconds

5% Single Unit Trucks

5% 3-S2 Combination Trucks

*Assumed hourly value of time per passenger car.

†Adjustment factors for percent trucks in traffic stream.

SOLUTION: Average Delay per Vehicle: (a) + (b) = 16.2 sec

(c) Idling Hours: 4.5 hrs

(d) Idling Cost: $1.40

Total Delay: 4.5 hrs × $3.00* × 1.151 $15.53

Total Idling Cost: $1.40 × 0.98†   1.38
                                          ───────
Total Cost Due to Idling per 1000
Vehicles (per signal) $16.91

NOTE: Where x = v/λs - v/capacity s = saturation flow v - demand volume λ - green to cycle time ratio typical day, the benefits that are calculated from the one-way analysis should be multiplied by 2.

Effects of induced and diverted traffic

Highway economic analyses should account for changes in traffic volume that are (1) induced by the improvement, that is, trips that would not be made without the improvement, and (2) diverted from other less attractive routes to the new facility. It is therefore recommended that the difference in user costs be multiplied by the average of the traffic volume with and without the improvement.

$$\text{user benefits} = (U_0 - U_1)\left[\frac{(V_0 + V_1)}{2}\right] \tag{4-11}$$

where

U_0 = user cost per unit of traffic (vehicles or trips) without the improvement

U_1 = user cost per unit of traffic with the improvement

V_0 = level of traffic without the improvement

V_1 = level of traffic with the improvement, including any induced or diverted traffic (V_1 may be either larger than V_0, as shown, or smaller, as when traffic is diverted from a link to a new parallel facility)

To properly account for the benefits to both induced and diverted traffic as well as the continuing traffic, the equation should be applied to each link in the highway network being analyzed and summed over all links (6).

Cost updating procedures

It is noted that the nomographs shown as Figures 4-1, 4-2, 4-3, and 4-4 are based on price levels prevailing in January 1975. It is suggested that user costs be reviewed for currentness about once every two years and appropriately updated. Reference 6 recommends a cost updating procedure that employs multipliers based on the commonly available consumer and wholesale price indexes published by the U.S. Bureau of Labor Statistics.

4-6 Example of highway economic analysis calculations

The following example, adapted from Ref. 6, illustrates application of some of the data and procedures presented earlier in the chapter.

A highway agency is considering eliminating (straightening) an S-shaped curve on a multilane highway. Total construction cost for the improvement is estimated at $2 milllion and will be incurred at the inception of the project. Annual maintenance costs are anticipated to be $2000 less after the improvement because the elimination of the curve reduces pavement and guardrail damage.

The S-curve is actually made up on two continuous 8° curves in reverse alignment to each other. Since the curves are the same degree of curvature, they can be analyzed together, and since traffic can be assumed to be uniform over the entire affected roadway, the analyst may view the improvement in terms of a single analysis section, designated *ab* . In analyzing the two curves together, it will be assumed that the average speed over the curves is the same and that there are no vehicular speed changes in passing from one curve to the other. If such assumptions cannot be made, separate analysis sections should be designated for each of the curves.

The characteristics of the existing route (alternative 0) and the proposed route (alternative 1) are summarized in Table 4-4. Note that the length of the section *ab* includes both 8° curves. Note also that after the improvement, the design speed of the straightened section will be 60 mph rather than 50 mph.

The analyst has obtained the traffic estimates with and without the proposed improvement for the 1st and the 20th year (the study years) of the 25-year analysis period. It is assumed that level of service *F* conditions will not occur during the analysis period, and that the hourly variation of traffic is small enough that explicit consideration of peaking characteristics is unnecessary. Hence the analysis can be performed essentially on an *ADT* basis. This entails utilizing traffic volume data for a typical hour of the day in each of the study years. Such traffic data are summarized in Table 4-5, where typical one-way hourly volume is defined as *ADT* divided by 18 (the number of hours assumed, for analysis purposes, in a day). Note that a small increment of traffic is estimated to be induced as a result of the proposed improvement.

Note also that capacity in year 20 is somewhat below that in year 1. This is to reflect normal deterioration of the roadway over the years.

There are so few trucks that the traffic stream will be considered to consist entirely of passenger cars—otherwise, the percentage of trucks by type would also be shown in Table 4-5.

Average running speeds (determined from Figure 4-1 as a function of *v/c*) are also noted in Table 4-5. Since speeds on the analysis section will change as a result of the improvement, transition costs between sections will change; thus speed on the previous and subsequent sections must be specified (as in Table 4-5).

Calculation of user costs

The above traffic and facility data enable calculation of user costs on the

TABLE 4-4
FACILITY DATA

			Basic Section Data			
Alternative	Section	Length (miles)[a]	Lanes (one way)	Grade	Curvature	Design Speed (mph)[a]
0	ab	0.20	2	0%	8°	50
1	ab	0.16	2	0%	0°	60

[a]1 mile = 1.6093 km.

TABLE 4-5
TRAFFIC DATA

| Alternative | Section | Study Year | Period | Number of Vehicles | | v/c Ratio | Average Running Speed (mph)[a] | Speed (mph)[a] | |
				One-Way Volume	One-Way Capacity			Previous Section	N Sec
0	ab	1	Typical hour	1900	3800	0.50	36	41	
1	ab	1	Typical hour	2000	3800	0.53	41	41	
0	ab	20	Typical hour	2900	3500	0.83	30	34	4
1	ab	20	Typical hour	3000	3500	0.86	34	34	4

[a] 1 mile = 1.6093 km.

existing route and the proposed route, as shown in Table 4-6. Unit highway user costs (HU) are calculated from Eq. 4-9 by using basic section costs (B), section length (L), and transition costs (T). Intersection delay costs are not relevant in this example. Accident costs are not considered in this example, but if accident rates were known and changes in accidents were discernible as a result of this improvement, their unit cost would be added to B before multiplication by L. The numbers for B and T are derived from Figures 4-1 and 4-2, using the facility and traffic data in Tables 4-4 and 4-5. A unit value of time of $3 per hour per vehicle is assumed.

Determination of B and T for the existing route (alternative 0) in year one is further described as follows. Entering the lower left-hand (speed-flow) chart of Figure 4-1 with a v/c ratio of 0.50 and a design speed of 50 mph yields an average running speed of approximately 36 mph and a corresponding travel time of 27.8 hr per thousand vehicle-miles. Tangent runing cost at 36 mph on a level (0%) grade is approximately $71 per thousand vehicle-miles (KVM). Added running cost due to an 8° curve is (from Figure 4-1) approximately $20 per KVM. Added running cost due to speed changes at a v/c ratio of 0.50 is approximately $3 per KVM. Total running cost per KVM on the existing route is thus about $94.00 (71 + 20 + 3). Adding this to a total time value of about $83.40 (27.8 × $3 per hour) per KVM yields a total unit basic section cost (B) for the existing route in year 1 of $177.40 per KVM.

Entering Figure 4-2 with a speed of 41 mph on the "faster" previous section and 36 mph on the section being analyzed yields an added one-way transition cost of approximately $1.00 per thousand vehicles (KV). Similarly, for the transition *from* the section, with an assumed speed of 50 mph, the transition cost, from Figure 4-2, is approximately $2.50 per KV. Total one-way section transition cost is thus $3.50 per KV.

Similar utilization of Figures 4-1 and 4-2 applied to the proposed route in

TABLE 4-6
CALCULATION OF HU

Alternative	Section	Year	Period	$(B \times L)^a$	$+T$	$=$	One-Way HU per KV^b
0	ab	1	Typical hour	177.4×0.2	3.5		$38.98
1	ab	1	Typical hour	148.7×0.16	1.7		$25.49
0	ab	20	Typical hour	186.90×0.2	3.0		$40.38
1	ab	20	Typical hour	165.00×0.16	2.0		$28.40

[a]B and T are in dollars; L is in miles; 1 mile = 1.6093 km.
[b]KV = thousand vehicles.

year 1 and both the existing and proposed routes in year 20 yields corresponding values for B and T. The absence of curves under alternative 1 does the most
to reduce the B values relative to alternative 0, while reduction in the magnitude of speed changes between sections reduces the T values for alternative 1
relative to alternative 0. Multiplying basic section costs by section length (L) and
adding section transition costs results in the one-way highway user costs (HU)
per KV in Table 4-6.

Calculation of user benefits
Benefits in the representative 1st and 20th years can be calculated
from the above user cost (HU) estimates by applying Eq. 4-11, as demonstrated in Table 4-7.

Computation of present values and measures of economic desirability
Using an assumption that benefits grow at an equal annual percentage
rate, the benefits for the two representative years may be converted into an
estimate of the present value of all benefits by the application of Eqs. 4-4 and
4-5. Using the ratio α of benefits in the 20th year to those in the 1st year
(232,190/172,827 = 1.34), the rate of growth of annual value $r = \ln(\alpha)/19 =$
0.0155. For an assumed 25-year analysis period and a 4 percent discount rate,
a present value factor PW_g of approximately 18.7 is obtained by Eq. 4-4. The
present value of benefits is then $PW_g \times$ first-year benefits or

$$18.7 \times 172,827 = \$3,231,865$$

Since the improvement affects traffic symmetrically in both directions, this
one-way benefit estimate can be multiplied by 2 to yield the total present value
of the project's benefits, $6.46 million.

TABLE 4-7
USER BENEFIT CALCULATIONS[a]

Section	Year	Period	Days per Year $\times$	Hours per Day in Period $\times$	$(HU_0 - HU_1) \times (V_0 + V_1)/2$ =	One-Way Annual Benefit
ab	1	Typical hour	365	18	$(38.98 - 25.49) \times (1.9 + 2.0)/2$	$172,827
ab	20	Typical hour	365	18	$(40.38 - 28.40) \times (2.9 + 3.0)/2$	$232,190

[a]Unit user costs (HU) are expressed in units of dollars per thousand vehicles and traffic volumes (V) are expressed in units of thousands of vehicles.

A complete project analysis also requires conversion of project costs and related items into present values. Project costs include the capital cost of construction and the change in maintenance expenses that are incurred on the section as a result of the improvement. A present value adjustment for the residual or scrap value of the project at the end of the 25-year analysis period is also necessary. As noted, the total construction cost of $2 million will be incurred at the inception of the project and therefore has a present value of $2 million. Also as noted, annual maintenance costs are anticipated to be $2000 less after the improvement. The project is estimated to have a residual value of roughly $500,000 in its 25th year. Again using a discount rate of 4 percent, the stream of annual maintenance savings is converted to present value by using the appropriate series present worth factor in Table 4-3. The present value of the residual value in the 25 years is also calculated from Table 4-3. The conversion of all project-related costs to present value is summarized in Table 4-8.

The net present value of the project is the difference between the present value of benefits and the present value of costs:

$$\text{net present value} = \text{present value of benefits} - \text{present value of costs}$$
$$= \$6,463,730 - \$1,718,198$$
$$= \$4,745,532$$

The net present value of $4,745,532 indicates that the project is economically desirable. According to Eq. 4-6, the benefit–cost ratio is

$$6,463,730 \div (2,000,000 - 31,244 - 187,558) \text{ or } 3.62$$

Since the benefit–cost ratio is greater than 1.0, the project is economically desirable. However, if there are budget constraints, the decision maker should investigate other projects or project alternatives for their relative economic attractiveness before making a decision to implement the project. In any event, the decision maker should also investigate any significant community or non-costable benefits or costs such as air and noise pollution that might be associated with the project before a final decision is made.

TABLE 4-8
CALCULATION OF PRESENT VALUE OF PROJECT-RELATED COSTS

Item		Present Value
Construction costs		$2,000,000
Changes in annual maintenance expenditures	$-\$2,000 \times 15.62 =$	$-31,244$[a]
Residual value in 25 years	$-\$500,000 \times 0.375 =$	$-187,558$[a]
Present value of all "costs"		$1,781,198

[a]Since incremental maintenance costs are decreased and residual value is a positive amount, these quantities are expressed as negative "costs."

4-7 Cost and benefits of bus transit improvements

In economic analyses involving bus transit, the user benefits should include reductions in highway user costs (as previously described) as well as reductions in bus user costs. Bus user costs consist of the costs of time (including in-vehicle travel time and the time for waiting, walking, and transferring) and the out-of-pocket costs to the user such as bus fares and the costs of getting to and from the bus stops. It is recommended that different costs be specified for the time spent in the transit vehicle and the time spent out of the vehicle. Reference 6 recommends that the costs assigned for walking, waiting, and transferring be 1.5 to 2.0 times that for traveling in the transit vehicle.

The change in transit system costs is the difference in operating the system before and after the improvement adjusted to reflect any change in income from transit fares. If changes in fare revenues are not also accounted for in cost calculations, a simple lowering of the fares might be interpreted to yield a net benefit when in fact the user benefits may be largely offset by losses in bus company revenues (6).

Research has shown that the annual cost of operating a transit system is most closely related to the following variables: (1) annual bus miles produced by the transit system, (2) bus size, (3) bus driver wages, and (4) average service speed. Anderson et al. (6) give an empirical equation by which annual transit operating costs may be estimated.

The present value of user benefits for transit improvements should be compared with the present value of changes in transit system costs. The user benefits must exceed the changes in the system costs (both expressed as present values) if the project is to be judged economically feasible.

The material presented here on economic analyses involving bus transit is admittedly sketchy. For further information on this subject, the reader should consult Ref. 6.

SOCIAL AND ENVIRONMENTAL CONSEQUENCES OF HIGHWAYS

In recent years, there has been increasing public awareness of the impact of highways and other public projects on the environment. Reflecting this concern, Congress passed several laws that brought about fundamental changes in the highway evaluation process. In the rest of this chapter, these laws and certain of the procedures and techniques employed to evaluate social and environmental impacts are described. Generally speaking, the evaluation guidelines described in this chapter are mandatory for federally assisted highway improvements but are not usually required for projects constructed from state or local funds.

4-8 Environmental legislation

The most significant legislative provisions relating to evaluation of social and environmental consequences of highways were contained in the Department of Transportation Act of 1966, the National Environmental Policy Act of 1969, the Clean Air Act of 1963 (and subsequent acts dealing with air pollution), the Uniform Relocation Assistance and Real Property Acquisitions Policies Act of 1970, and the Federal-Aid Highway Act of 1970.

The Department of Transportation Act, passed October 16, 1966, declared it to be the national policy that a special effort should be made to preserve the natural beauty of the countryside. Section 4(f) of that act stated that the Secretary of Transportation shall not approve any program or project that requires any land from a public park, recreation area, wildlife and waterfowl refuge, or historic site unless there is no feasible and prudent alternative to the use of such land. It further stated that transportation programs that require land from such areas must include all possible planning to minimize environmental harm.

The National Environmental Policy Act of 1969 declared it to be a national policy that the federal government would use all practicable means and measures "to create and maintain conditions under which man and nature can exist in productive harmony." It established a three-member Council on Environmental Quality in the Executive Office of the President to develop guidelines for agencies affected by the law. Section 102 of the act, quoted in part below, established the requirement for environmental impact statements.

> Sec. 102. The Congress authorizes and directs that, to the fullest extent possible: (1) the policies, regulations, and public laws of the United States shall be interpreted and administered in accordance with the policies set forth in this Act, and (2) all agencies of the Federal Government shall—
>
> (A) utilize a systematic, interdisciplinary approach which will insure the integrated use of the natural and social sciences and the environmental design arts in planning and in decisionmaking which may have an impact on man's environment;
>
> (B) identify and develop methods and procedures, in consultation with the Council on Environmental Quality established by title II of this Act, which will insure that presently unquantified environmental amenities and values may be given appropri-

ate consideration in decisionmaking along with economic and technical considerations;

(C) include in every recommendation or report on proposals for legislation and other major Federal actions significantly affecting the quality of the human environment, a detailed statement by the responsible official on—

(i) the environmental impact of the proposed action,

(ii) any adverse environmental effects which cannot be avoided should the proposal be implemented,

(iii) alternatives to the proposed action,

(iv) the relationship between local short-term uses of man's environment and the maintenance and enhancement of long-term productivity, and

(v) any irreversible and irretrievable commitments of resources which would be involved in the proposed action should it be implemented.

The Clean Air Act of 1963 encouraged increased state and local programs for the control of air pollution. The act also provided for the development of air quality standards. In 1965, Congress passed the Motor Vehicle Control Act, which initiated controls on motor vehicle manufacturers to require the installation of air pollution control devices on all new vehicles. Since that time, there have been additional laws that further involved both state and federal agencies in controlling air pollution.

In 1970, Congress passed the Uniform Relocation Assistance and Real Property Acquisitions Policies Act in an effort to establish uniform treatment of all people affected by public projects financed with federal funds. The act sets forth standard policies and procedures for the determination of just compensation, for negotiation with property owners, for taking possession of property, and, when required, for instituting formal condemnation proceedings.

As was indicated in Chapter 1, the Federal-Aid Highway Act of 1970 required the Secretary of Transportation to publish guidelines designed to insure that possible adverse economic, social, and environmental effects are properly considered during the planning and development of federal-aid highway projects.

4-9 Federal highway environmental policies

As a means of implementing the various environmental laws, the Federal Highway Administration has issued regulations (*14, 15*) that specify the process by which highway agencies plan and develop highway projects on any federal-aid highway system.

The policy of the FHWA is, to the fullest extent possible, to coordinate all environmental investigations, reviews, and consultations into a single process. The agency further intends that decisions on proposed actions be based on a balanced consideration of the need for safe and efficient transportation and on national, state, and local goals to protect and enhance the environment, conserve energy, and revitalize urban areas.

Other matters of FHWA environmental policy include

1. Involvement of the public in the planning process.
2. Use of a systematic interdisciplinary approach in the planning development process.
3. Incorporation of measures to mitigate adverse impacts from proposed projects.

4-10 Documentation for environmental analyses

The Federal Highway Administration defines three classes of action which prescribe the level of documentation required in the environmental planning process (*14*):

1. *Class I* actions, which require the preparation of an environmental impact statement (EIS). Examples of class I projects include controlled access freeways; projects with four or more lanes on a new location; and major projects whose construction involves a large amount of demolition, extensive displacement of individuals or businesses, or substantial disruption of local traffic patterns.
2. *Class II* actions (categorical exclusions), which do not require an environmental impact statement or environmental assessment. Such projects include planning and technical studies which do not fund the construction of facilities; construction of bicycle and pedestrian lanes, paths, and facilities; and reconstruction of an existing bridge structure on essentially the same alignment or location.
3. *Class III* actions, in which the significance of the impact is not clearly established. Such actions require the preparation of an *environmental assessment* (EA) concisely describing the environmental impacts of the proposed work and its alternatives. This assessment serves as a basis for deciding the nature of further environmental analysis and documentation. If the FHWA determines that a proposed project will not have a significant impact on the environment, a statement to that effect is prepared. This document, which is called a finding of no significant impact (FONSI), must be made available to the public on request.

4-11 Environmental impact statements

An environmental impact statement must be prepared for all federal actions that will have a "significant impact" on the environment. For federally supported highway projects, the FHWA determines the need for an EIS based on preliminary environmental studies or when the review of the environmental assessment and the comments received indicate that the expected impacts may be significant.

The environmental impact statement is a written statement containing an assessment of the anticipated significant effects that the proposed action may have on the quality of the environment. The purpose of the EIS is to assure that careful attention is given to environmental matters and that such matters are appropriately considered in the highway agency's decision. The contents of an environmental impact statement are briefly outlined in Table 4-9.

Environmental impact statements are processed in two stages. The highway agency prepares and circulates a draft statement in a prescribed fashion to public agencies, private groups, and individuals for review and comments. After receiving reactions to the draft EIS and making appropriate modifications, the agency then submits a final EIS to the regional or Washington headquarters of FHWA for review and approval.

A section 4(f) statement is required by Section 4(f) of the Department of Transportation Act. The statement relates to the proposed use of publicly owned land from a park, recreation area, or wildlife refuge. In order to use such lands for highway purposes, the 4(f) statement must declare that

1. There is no feasible and prudent alternative to the use of such land.
2. The proposed program includes all possible planning to minimize harm to the section 4(f) land resulting from such use.

In the planning and development of complex highway projects, the FHWA encourages the preparation of tiered environmental impact statements. The first tier of such statements focuses on broad issues such as mode

TABLE 4-9
CONTENTS OF AN ENVIRONMENTAL IMPACT STATEMENT

1. Description of the proposed action and alternatives considered
   Location, type, and length of facility, termini, number of lanes, right-of-way width
   Other design features such as general horizontal and vertical alignment, structures, etc.
   Deficiencies of existing facilities, anticipated benefits
2. Land use planning
   Description of planning process for the area
3. Probable impact of proposed action on the environment
   Natural, ecological, scenic resource impacts
   Relocation of individuals and families
   Social impacts
   Air quality impacts
   Noise impacts
   Water quality impacts
   Construction impacts
4. Alternatives to the proposed action
5. Probable adverse environmental effects which cannot be avoided
6. The relationship between local short-term uses of man's environment and the maintenance and enhancement of long-term productivity
7. Irreversible and irretrievable commitments of resources
8. The impact on properties and sites of historic and cultural significance
9. Summary of coordination and public and minority involvement

choice, general location, and areawide air quality and land-use implications of alternative transportation improvements. The second tier is site-specific and describes detailed project impacts and measures to mitigate harmful impacts.

4-12 Coordination and public involvement

State highway agencies conduct public hearings to afford interested persons an opportunity to participate in the process of determining the need for and the location of highways. For federal-aid highways, one or more public hearings are normally required for proposed projects

1. If substantial amounts of right-of-way are required.
2. If there is a substantial change in the layout of connecting roadways or the facility being improved.
3. If there is a substantial adverse effect on abutting property.
4. If the property otherwise has significant social, economic, or environmental effects.

At such hearings, a representative of the state highway agency explains the project's purpose and need, its major design features, the social, economic, and environmental impacts, and measures planned to mitigate such effects. The agency must also accept oral and written statements from the public and consult with citizens at various stages of federally supported projects.

In the early stages of the development of all important federal-aid highway projects, the responsible state highway agency must send notice of the project and available plans to a state or metropolitan area clearinghouse. The clearinghouse then informs other interested agencies and invites their comments on the proposed work. The same process is used for annual programs of proposed future projects. The coordination process was formerly known as the "A-95 notification," named after a U.S. Office of Management and Budget circular that required it. It is now called the "Intergovernmental Review of Federal Programs." Various federal agencies have published rules for intergovernmental review in the *Federal Register*, implementing Executive Order 12372 (*16*).

HIGHWAY NOISE PLANNING

In the next several sections, some of the common environmental impacts of highways will be described. The first impact to be considered is perhaps the most troublesome to control: highway noise. First, noise characteristics and measurement will be described, and then some remedial programs to reduce noise levels and mitigate their harmful effects will be examined.

4-13 Noise characteristics and measurement

Noise is defined as excessive or unwanted sound. It is unwanted because it annoys, interferes with conversation, disturbs sleep, and, in the exteme, is a danger to public health. Sound, whether noisy or noiseless, is produced by vibrations in air, water, steel, or other substance. When an object vibrates, it produces rapid small-scale variations in the normal atmospheric pressure. This disturbance is propagated from the source in a repetitive spherical pattern at a speed (in air) of approximately 1100 ft/sec (340 m/sec). It may be reflected, partially absorbed, or attenuated before reaching an eardrum to produce a sensation of sound.

Noise is characterized by its *sound level*, its frequency spectrum, and its variation over time. The term sound level refers to a physical measure that corresponds to the hearer's subjective conception of loudness. It is a function of the magnitude of the pressure fluctuations about the ambient barometric air pressure. One can speak of the strength of these fluctuations in terms of several variables, the most common being *sound intensity* and *sound pressure*.

Sound intensity (also called sound power density) is the average rate of sound energy transmitted through a unit area perpendicular to the direction of sound propagation, typically measured in picowatts (10^{-12} watt) per square meter. The human ear can detect sound intensities as weak as 1 picowatt and tolerate intensities as high as 10^{13} picowatts. Because of the difficulties of dealing with such a large range of numbers, a logarithmic measure called the decibel (dB) is used to describe sound level. The sound intensity, expressed in decibels, is

$$\text{sound intensity} = 10 \log_{10}\left(\frac{I}{I_0}\right) \tag{4-12}$$

where

I = sound intensity (picowatts/m^2)
I_0 = 1 picowatt/m^2, a standard reference intensity representing approximately the weakest audible sound

Since no instrument is available for directly measuring the power level of a source, sound pressure, which is usually proportional to the square root of sound power, is used as a measure of the magnitude of a sound disturbance (*17*). The sound pressure, in decibels, is

$$\text{sound pressure} = 10 \log_{10}\left(\frac{P}{P_0}\right)^2$$

$$= 20 \log_{10}\left(\frac{P}{P_0}\right) \tag{4-13}$$

where

P = root-mean-square sound pressure, typically expressed in units of newtons per square meter
P_0 = 20 micronewtons/m^2 or 0.0002 dyne/cm^2, a standard reference pressure corresponding to the weakest audible sound

Sound level is measured by a sound level meter, which consists essentially of a microphone that converts the pattern of sound pressure fluctuations into a similar pattern of electrical voltage, amplifiers, and a voltage meter that is normally calibrated to read in decibels. For practical purposes, the decibel scale ranges from zero, the threshold of hearing, to about 140 dB, the onset of pain. For every increase of about 10 dB, there is a doubling of the sound's apparent loudness.

The apparent loudness of a sound also depends on the *frequency* of the sound. Frequency is the rate of occurrence of the sound pressure fluctuations, commonly expressed in cycles per second or hertz (Hz). The frequency determines the pitch of the sound; the higher the frequency, the higher the pitch. The normal human ear can hear sounds with frequencies from about 20 to 20,000 Hz but is more sensitive to sounds in the middle- to high-frequency range.

> Most noises are made up of a mixture of components having different frequencies: the sound of a diesel tractor/trailer at high speed on the freeway combines the high-pitched singing of the tires and the low-pitched roar of the engine and the exhaust, both of which the ear readily distinguishes. . . . A flute, on the other hand, if played softly, makes an almost pure tone containing only a single prominent frequency. Depending on how the components of a noise are distributed in frequency, our ears make a subjective judgment of "quality." (*17*)

A great many scales have been used to express noise levels. One of the simplest and most straightforward noise measurement techniques consists of measuring the *overall sound pressure level* that is related to the total sound energy over the audible frequency range. However, the unweighted overall sound pressure level is not strongly correlated with a hearer's subjective response to the noise. As indicated earlier, the human ear tends to be more sensitive to sounds with relatively high frequencies.

The *A-weighted sound level* was devised to more closely represent a person's subjective response to sounds. In the *A*-weighted filter network, the lower frequencies are deemphasized in a manner similar to human hearing. The *A*-weighted sound level, measured in decibels (dBA), is the generally accepted scale for measuring highway transportation noises.

The extent to which a person is annoyed by noise also depends to a considerable degree on its variation over time. The temporal effect is manifested in the duration of a single noise event as well as its repetitiveness and time of occurrence. Clearly, the longer a noise lasts and the more frequently it is repeated, the greater the interruption of human activity and the more pronounced the annoyance.

One factor that gives a good representation of both the magnitude and the fluctuation pattern of noise is the 10th percentile noise level, L_{10}, that is, the noise level exceeeded 10 percent of the time (*18*). With reference to Figure 4-5, the L_{10} noise level is that value for which the sum of the deltas equals 10 percent of the total time being considered, T_T.

For obvious reasons, noise tends to be more annoying during evening and nighttime hours than during the day. There is a greater need for quiet during sleeping hours, and during those periods when there is less noise from other

$$\Delta_1 + \Delta_2 + \Delta_3 + \Delta_4 = .10 \times T_T$$

FIGURE 4-5 Statistical measures of noise. (Courtesy American Association of State Highway and Transportation Officials.)

sources, peak traffic noises tend to be more disturbing than even higher levels during daylight hours.

4-14 Noise studies and standards

AASHTO (*18*) recommends that a noise study be prepared during the location stage of all alternative lines under consideration for each highway project. Early in the design process, the present and future uses of land adjacent to the highway should be determined and noise-sensitive areas should be identified.

Existing noise levels should be determined by field measurements at representative times and locations, with special emphasis on potentially noise-sensitive times and areas. There is considerable variation in types of equipment and levels of sophistication and effort employed by the various agencies for noise studies. Detailed procedures for making such measuements are given in Ref. *18*.

The prediction of future noise levels is a complex task in which the effects of a large number of parameters should be considered, including traffic volumes and composition, accelerations and speeds, roadway gradients, topography, the presence and types of barriers, and the distance between the highway and the affected area. Some guidance on the prediction of future noise levels is given by Refs. *19* and *20*.

AASHTO (*18*) has established standards for evaluation of the impact of highway-generated noise on nearby land uses or activities. The standards provide a basis for the selection of appropriate noise attenuation measures. The following L_{10} exterior design noise levels are recommended for the land uses and activities indicated:

Category	Description	Noise Level
A	Lands on which serenity and quiet are of extraordinary significance and serve an important public need and where the preservation of those qualities is essential if the area is to continue to serve its intended purpose	60 dBA (exterior)
B	Picnic areas, recreation areas, playgrounds, active sports areas, parks, residences, motels, hotels, schools, churches, libraries, and hospitals	70 dBA (exterior)
C	Developed lands, properties, or activities not included in categories A and B above	75 dBA (exterior)
D	Undeveloped lands	—
E	Residences, motels, hotels, public meeting rooms, schools, churches, libraries, hospitals, and auditoriums	55 dBA (interior)

Unusual conditions may, of course, require special analyses that could lead to the establishment of design noise levels different from those indicated above.

4-15 Noise attenuation

Highway engineers can lessen the impact of highway noise on nearby human activities in three general ways:

1. By providing a buffer zone between the highway and adjacent land activities.
2. By modification of the horizontal or vertical alignment of the highway.
3. By providing noise shielding.

Since noise tends to spread out uniformly as it travels away from the source, the sound level decreases at a rate of about 6 dB for each doubling of distance. One approach to noise attenuation, therefore, is to provide a buffer zone between the highway and the users of adjacent land. The acquisition of additional right-of-way width may, of course, be expensive and not be justified for the purpose of noise attenuation alone. However, it may be possible to use such buffer areas for other purposes such as for bicycle paths, pedestrian walkways, traffic safety recovery zones, and so forth.

Designers may greatly decrease the noise impact of new highways by locating the alignment so as to avoid noise-sensitive areas. It may also be possible to lower noise levels along adjacent land areas by utilizing independent alignment for opposing traffic lanes, and taking advantage of natural features for noise attentuation (18). Other possible desirable modifications to the highway alignment include depressing or elevating the roadway section and reducing the highway grade.

A variety of noise shields or barriers have been used to lessen the impact of noise on adjacent land, including (1) earth berms, (2) timber walls, (3) concrete and masonry walls, (4) metal walls, (5) vegetative screens, and (6) stucco and chain link fences. Experience indicates that a well-designed barrier will provide a noise attenuation of about 10 dBA.

Some general principles for noise barrier design are outlined below.

1. Barriers should be high enough and long enough to intercept the line of sight between the source and the observer. It is especially important that the barrier be high enough to shield truck exhaust noise. To accomplish this without aesthetic sacrifice, some agencies have constructed combination barriers consisting of an earth berm which supports some type of vertical wall.
2. To be most effective, barriers should be constructed so as to cast a big sound "shadow." This generally means that the barrier should be placed close to either the source or the receiver. It should be remembered that noise barriers are fixed objects that, if placed close to the traffic lanes, may be unsafe. For this and other reasons, noise barriers are frequently located near the right-of-way line.
3. Barriers should be constructed of a dense material, and there should be no air paths through or under the barrier that could constitute noise leaks.
4. In the interests of community acceptance, barriers should be made as attractive as possible. They should be compatible with local architecture and blend into the forms of the landscape.

Reference 21 gives additional information on the selection, design, and construction of highway noise barriers.

HIGHWAY AIR QUALITY

Air pollution may constitute the most serious environmental impact caused by highway transportation. It is a complex problem, and its evaluation and control may require the involvement of highly trained environmental specialists. A detailed treatment of the subject is well beyond the scope of this textbook. This section will include a brief description of the air pollution problem and ways in which highway engineers may lessen its impact. For additional information, it is suggested that the reader consult the FHWA publication *Fundamentals of Air Quality (22)*, which served as the principal source of reference for the following material.

4-16 Air pollution

Air pollutants are natural and man-made contaminants in the atmosphere. The Environmental Protection Agency estimated that 198 million tons of air pollutants were emitted in the United States during 1974. Transportation sources* accounted for 50 percent of the total emissions.

Air pollutants may be grouped into five major classes:

1. Carbon monoxide (CO).
2. Hydrocarbons (HC) and photochemical oxidants.

---

*Transportation sources include all highway motor vehicles plus off-highway mobile sources such as aircraft, railroads, vessels, and agricultural, industrial, and construction machinery.

3. Nitrogen oxides (NO_x).
4. Particulate matter.
5. Sulfur dioxide (SO_2).

Carbon monoxide, the most widely distributed and the most commonly occurring air pollutant, is a colorless, odorless, highly poisonous gas that results from the incomplete combustion of carbonaceous fuels. Gaseous organic compounds of carbon and hydrogen (hydrocarbons) and oxides of nitrogen are also emitted during the combustion process. While prevailing concentrations of hydrocarbons do not appear to be detrimental to human health, certain of these substances may react with nitrogen oxides to produce harmful pollutants. Ozone and other oxidizing agents are formed when hydrocarbons and nitrogen oxides are exposed to sunlight. These photochemical oxidants can cause irritation of the respiratory and alimentary systems and damage vegetation, metals, and other materials. There is also some evidence that long-term exposure of humans to nitrogen dioxide, even in low concentrations, may contribute to chronic respiratory diseases.

Particulate matter is any solid or liquid matter dispersed in air under standard conditions of temperature and pressure. Particulate matter discharged into the atmosphere may be in the form of dust, soot, fly ash, and so on. Generally, particles released into the atmosphere from man-made sources are 1 to 10 μm in size.

Sulfur dioxide is generated by the combustion of sulfur-bearing fuels and by certain industrial processes that use sulfur-bearing raw materials. Motor vehicles are not a major source of sulfur dioxide pollution.

Highway vehicles are the most important source of air pollution in the United States. They are the principal source of carbon monoxide and hydrocarbons and account for nearly half of the total nitrogen oxides emitted each year. In addition, motor vehicles are the chief source of particulate lead in the atmosphere.

Highway air pollution results principally from emissions from internal combustion and diesel engines. Internal combustion engines produce emissions that originate from three basic sources within the vehicle system (22):

1. *Exhaust emissions.* Pollutants that are present in the exhaust gas as it is discharged into the atmosphere.
2. *Evaporative emissions.* Vapors lost directly to the atmosphere from the fuel tank, carburetor, or any other part of the system.
3. *Crankcase blow-by.* Gases and vapors that under pressure escape from the combustion chamber, past the engine pistons, and enter the crankcase.

The diesel engine emissions problem has not yet been well defined or quantified. The most apparent source of atmospheric pollution from a diesel engine is the smoke and odor from its exhaust. In addition, diesel engines are known to emit significant concentrations of unburned hydrocarbons, nitric oxides, and oxygenated compounds (22). Studies have shown that the concentration of emissions from diesel engines varies widely depending on engine speed and load.

4-17 Air quality standards and pollution modeling

The U.S. Environmental Protection Agency (EPA) has the overall responsibility for the establishment of national ambient air quality standards. Such standards have been established and published in the *Federal Register*. Enforcement of these standards is the responsibility of state air pollution control agencies approved by the EPA.

To thoroughly assess the impact of a proposed highway improvement, it is necessary to estimate the quantities of pollutant concentrations within the study area both before and after development. Such an assessment depends heavily on a thorough statistical analysis of past meteorological data and the evaluation of the effects of existing and forecast traffic volumes. The evaluation process is a complex one for which two types of computer programs or models have been developed: (1) emission models and (2) dispersion models.

Emission models provide estimates of quantities of various types of pollutants based on the amount of travel, driving patterns, and emission rates for various classes of vehicles. Dispersion models use data on source emissions from emission modeling, and, with additional data on meteorological conditions, geographic boundaries, and so forth, compute the dispersion of pollutants over the area of interest for specified periods of time.

4-18 Control of highway air pollution

Programs to control air pollution from motor vehicles have concentrated primarily on modification of engine designs to meet federally prescribed maximum levels of emissions. However, motor vehicle emissions are influenced not only be engine operating characteristics but also by a wide variety of transportation system characteristics. The following paragraphs, which are quoted directly from Ref. 22, describe ways of lessening the impact of air pollution by appropriate planning, design, and operation of the highway facilities.

> *Planning.* Air quality is affected by both mobile emission sources such as automobiles and aircraft, and stationary emission sources such as factories and power plants. In theory, there should be an optimum location pattern for these various activities that would minimize ground level concentrations of pollutants. Different urban forms, such as urban sprawl, a centralized core, or satellite cities, affect air quality by requiring different transportation systems and energy expenditure. Actual development of these urban forms, however, which approximates theoretical forms, is generally limited to new towns. This may allow incorporation of air quality planning considerations into the basic layout of a proposed new community.
>
> More commonly, planning studies are conducted within the context of an existing pattern of development. Planning for improvements to highway and transit networks should be coordinated to

permit opportunities for transfer between public transit facilities and private automobiles. This planning strategy is the most direct method for reducing highway-related air pollution in cities, since transit vehicles can serve high-density urban core areas more effectively than automobiles. This method of reducing private vehicle travel, however, is probably the most difficult to implement since it reduces an individual's travel flexibility.

Planning for highway improvements can provide for increased capacity to reduce congestion, higher speed links in the road network, and shorter travel distances. The first two planning strategies tend to reduce emissions of CO and HC due to smoother traffic flow and higher and more uniform average speeds. The third item reduces all emissions. It should be recognized that these are long-term approaches to reducing emissions, with implementation periods on the order of five to twenty years.

Design. The physical layout of highway facilities also affects vehicle emissions. In general, highway design features which reduce congestion and increase operating speeds will also contribute to reduced emisssions of air pollutants.

Number of travel lanes. The number of lanes is an obvious factor that affects the potential of total pollutants which may be emitted in a given corridor. With a greater number of lanes, the roadway capacity is increased, resulting in a potential for higher traffic densities and therefore higher pollutant emissions.

Lane capacity. All roadway lanes are limited in their ability to accommodate vehicular traffic, and emissions are dependent on the number of vehicles utilizing each lane. One of the many variables that enters into the evaluation of lane capacity is driver behavior. Lane width affects capacity on two-lane and multi-lane highways and at intersections, because drivers tend to encroach on adjacent lanes (or totally avoid a lane) when lane width is narrow or conditions make a lane appear narrow. Such behavior causes inefficient lane usage and lower operating speeds, and therefore, a possible reduction in capacity. Closely related to lane width is the presence of lateral obstructions, such as median barriers, curbs, and bridge abutments. The presence of a single lateral restriction can cause a bottleneck, leading to alternate deceleration and acceleration, and the products of increased emissions. The absence of adequate shoulders along the highway or on structures can also create bottlenecks at times of accidents or breakdowns, even under moderate traffic loadings.

Horizontal and vertical alignment. Emission rates are also affected by features such as horizontal and vertical curves, due to varying vehicle power requirements. Emissions from all types of vehicles vary whenever there is a deviation from steady-state speed or power. Network capacity can be decreased by long, steep grades, though truck climbing lanes can help alleviate this problem.

Consideration must also be given to the varying effects of at-grade, elevated, or depressed roadway sections. Though these configurations do not affect emission rates, they do influence the dispersion capabilities of a given roadway section. In general, ground-level concentrations downwind of an elevated roadway will be less than those from an at-grade section. The dispersion capabilities are more limited in a cut section, where the potential exists for pollutants becoming trapped and accumulating.

Operating Speeds. As noted previously, vehicle pollutant emissions are speed-dependent. As operating speeds for a given roadway segment increase, emissions of carbon monoxide and hydrocarbons decrease, while those for oxides of nitrogen increase. Changes in speed contribute to higher emissions, since emissions of HC and NO_x in the acceleration mode are significantly higher than those at a steady-state speed.

Interchanges and Intersections. The existence of interchanges and ramps affects operating speeds. Directional freeway interchanges generally allow smoother flow and require fewer speed changes, and therefore cause lower emissions than cloverleaf or diamond interchanges.

At-grade intersections are significant generators of increased emissions, due chiefly to changes in operating mode brought about by vehicles decelerating, idling, and accelerating. Generally, any strategy used to improve traffic flow also contributes to a decrease in the total emission burden.

Operation. Transportation system management strategies can contribute to reductions in pollutant emissions by ensuring a smooth flow of traffic on arterials and freeways. For example, a system of synchronized traffic signals contributes to lower emissions by decreasing vehicle time delays and increasing travel speeds. By replacing stop signs with yield signs, or allowing right turns during the red phase of the cycle, average vehicle speeds are increased, resulting in lower emissions of CO and HC. On freeways, ramp metering and variable message information signs can reduce congested conditions, and thereby also reduce emissions.

OTHER CONSEQUENCES OF HIGHWAYS

There are, of course, many other social and environmental impacts from highway improvements that should be considered as a part of the evaluation process, including

1. Land use and developmental impacts.

2. Hydrologic impacts including flooding and disruption of water movements.

3. Water pollution including soil erosion.

4. Ecological impacts.

It is regretted that space limitations preclude an extended discussion of these highway impacts.

PROBLEMS

4-1   Determine the basic section costs for passenger cars using a multi-lane highway under the following conditions:

design speed = 60 mph

volume/capacity ratio = 0.7

level of service = B

grade = 2%

curvature = 6°

4-2   Estimate the one-way section transition costs for the highway described in Problem 4-1 given that the average running speed on an adjacent section is 35 mph.

4-3   The volume of traffic along a certain intersection approach is 600 passenger cars per hour. Determine the costs due to stopping (excluding idling) at the intersection under the following conditions:

saturation flow = 1650 vehicles/hr

signal cycle time = 60 sec

effective green time = 33 sec

approach speed = 45 mph

4-4   For the conditions described in Problem 4-3, determine the costs due to idling.

4-5   A certain highway project is planned that would have an initial investment cost of $1.2 million. The user benefits for the facility (in excess of maintenance costs) are estimated to be $87,000 per year over its useful life of 20 years. At the end of the 20-year period, the residual (salvage) value would be $200,000. On the basis of present worth concepts, should the project be built? Assume an interest (discount) rate of 8 percent. What is the benefit–cost ratio?

4-6   A highway agency is considering two alternative energy attenuation systems, described below, to lessen the severity of crashes into bridge columns and other fixed object hazards. System A has low initial cost but must be repaired extensively after each "hit." System B has high initial cost but needs little maintenance. It is expected that such a device will be hit once every two years. On the basis of present worth concepts, which system is preferred? Use an annual interest rate of 10 percent.

	System A	System B
First cost	$3500	$8500
Routine annual maintenance	300	200
Repair cost after each hit	1200	None

Life	20 yrs	20 yrs
Salvage value	1000	2500

4-7 The benefits from a certain highway project are estimated to be $12,000 in the first year and $30,000 in the 21st year. Assuming a uniform growth rate and an interest rate of 9 percent, calculate the present value of benefits over a 20-year analysis period.

REFERENCES

1. American Association of State Highway Officials. *Road User Benefit Analysis for Highway Improvements.* Washington, D.C. (1952).

2. Winfrey, Robley. *Economic Analysis for Highways.* International Textbook, Scranton, Pa. (1968).

3. Claffey, P. J. *Running Costs of Motor Vehicles as Affected by Road Design and Traffic.* National Cooperative Highway Research Program (NCHRP) Report No. 111, Washington, D.C. (1971).

4. Winfrey, R., and Zellner, C. *Summary and Evaluation of Economic Consequences of Highway Improvements.* NCHRP Report No. 122, Washington, D.C. (1971).

5. Curry, D. A., and Anderson, D. G. *Procedures for Estimating Highway User Costs, Air Pollution, and Noise Effects.* NCHRP Report No. 133, Washington, D.C. (1972).

6. Anderson, D. G., Curry, D. A., and Pozdena, R. J. *User Benefit Analysis for Highway and Bus Transit Improvements.* Final Report NCHRP Project 2-12, Stanford Research Institute, Menlo Park, Calif. (published as *A Manual on User Benefit Analysis of Highway and Bus—Transit Improvements, 1977* by the American Association of State Highway and Transportation Officials).

7. *Accident Facts, 1985 Edition.* National Safety Council, Chicago (1985).

8. *Estimating the Cost of Accidents 1983.* National Safety Council Bulletin, Chicago (1983).

9. Cribbins, P. D., Arey, J. M., and Donaldson, J. K. Effects of Selected Roadway and Operational Characteristics on Accidents on Multi-Lane Highways. *Highway Research Record No. 188* (1967).

10. Dart, Olin K., and Mann, Jr., Lawrence. Relationships of Rural Highway Geometry to Accident Rates in Louisiana. *Highway Research Record No. 312* (1970).

11. Kihlberg, J. K., and Tharp, K. J. Accident Rates as Related to Design Elements of Rural Highways. NCHRP Report No. 47, Washington, D. C. (1968).

12. Grant, Eugene L., and Ireson, W. Grant. *Principles of Engineering Economy,* 7th ed. Wiley, New York (1982).

13. *Highway Capacity Manual.* Transportation Research Board Special Report No. 209, (1985).

14. *Code of Federal Regulations.* Title 23, Part 771.

15. *Federal-Aid Highway Program Manual,* Vol. 7, Chap. 7. Federal Highway Administration, Washington, D.C.

16. *Intergovernmental Review of Federal Programs.* Executive Order 12372, *Federal Register,* Vol. 48, No. 16, January 24, 1983.

17. Schultz, T. J., Bolt Benanek and Newman, Inc. *Noise Assessment Guidelines: Technical Background.* Prepared for U.S. Department of Housing and Urban Development, Washington, D.C. (1972).

18. *Guide on Evaluation and Attenuation of Traffic Noise.* American Association of State Highway and Transportation Officials, Washington, D.C. (1974).

19. *Highway Noise—A Design Guide for Highway Engineers.* NCHRP Report No. 117, Highway Research Board, Washington, D.C. (1971).

20. *Fundamentals and Abatement of Highway Traffic Noise.* Federal Highway Administration, National Highway Institute, Washington, D.C. (3 vols.): Vol. 1 (1973), Vol. 2 (1974), and Vol. 3 (1974).

21. Snow, C. H. *Highway Noise Barrier Selection, Design, and Construction Experiences, A State of the Art Report—1975.* Implementation Package 76-8, Federal Highway Administration, Washington, D.C. (1976).

22. *Fundamentals of Air Quality.* Implementation Package 76-5, prepared for the Federal Highway Administration by Greiner Engineering Sciences, Inc., Baltimore, Md. (1976).

chapter five

TRAFFIC CHARACTERISTICS

A knowledge of traffic characteristics is useful to the highway engineer in developing highway and transportation plans, performing economic analyses, establishing geometric design criteria, selecting and implementing traffic control measures, and evaluating the performance of transportation facilities. Dozens of measures have been employed to describe the quality and quantity of traffic flow. In this chapter, information is presented on those flow characteristics that fundamentally bear on the planning, design, and operation of highway and transport facilities: traffic speed, travel time, volume, and density. In a section on highway capacity, we will consider ways of estimating the ability of various highway facilities to accommodate traffic flow. Finally, we will describe the nature and severity of the highway accident problem and examine the causes of traffic crashes.

TRAFFIC FLOW CHARACTERISTICS

5-1 Speed and travel time

Speed of travel is a simple and widely used measure of the quality of traffic flow. Basically, speed is the total distance traversed divided by the time of travel. Speed is commonly expressed in miles per hour or feet per second. Its reciprocal, travel time, is usually expressed in units of minutes per mile.

There are three basic classes or measures of speed of travel:

1. Spot speed.
2. Overall speed.
3. Running speed.

Spot speed is the "instantaneous" speed of a vehicle as it passes a specified point along a street or highway. There are, of course, practical difficulties in measuring instantaneous speeds since, by definition, speed is the distance traveled divided by the travel time. Spot speeds may be determined by manually measuring the time required for a vehicle to traverse a relatively short specified distance. A variety of electromechanical and electronic devices are commonly employed to measure spot speeds. Such devices typically involve some sort of vehicle detectors (e.g., pneumatic tubes) that actuate and stop a timing mechanism, the time of travel or speed being printed on a tape or recorded on a graph. Radar meters have also been widely used by traffic engineers and enforcement officers to measure spot speeds.

The average of a series of measures of spot speeds can be expressed in two ways, as a *time-mean speed* and a *space-mean speed*. Time-mean speed is the arithmetic mean of speeds of all vehicles passing a point during a specified interval of time. The time-mean speed is

$$\bar{u}_t = \frac{\sum_{i=1}^{n} u_i}{n}$$

(5-1)

where
u_i = observed speed of ith vehicle
n = number of vehicles observed

The space-mean speed is the arithmetic mean of speeds of vehicles occupying a relatively long section of street or highway at a given instant. It is the average of vehicle speeds weighted according to how long they remain on the section of road. The space-mean speed is

$$\bar{u}_s = \frac{nd}{\sum_{i=1}^{n} t_i}$$

(5-2)

where
d = length of roadway section
t_i = observed time for the ith vehicle to travel distance d

Space-mean speed and time-mean speed are not equal. In fact, Wardrop (*1*) has shown that

$$\bar{u}_t = \bar{u}_s + \frac{\sigma_s^2}{\bar{u}_s}$$

(5-3)

where
σ_s^2 = variance of the space distribution of speeds

For general-purpose usage, no distinction is normally made between time-mean and space-mean speeds. For theoretical and research purposes, the type of mean should be specified.

Overall speed and *running speed* are speeds over a relatively long section of street or highway between an origin and a destination. These measures are used in travel time studies to compare the quality of service between alternative routes. Overall speed is defined as the total distance traveled divided by the total time required, including traffic delays. Running speed is defined as the total distance traveled divided by the running time. The running time is the time the vehicle is in motion; time for stop-delays is excluded.

Overall and running speeds are normally measured by means of a test vehicle that is driven over the test section of roadway. The driver attempts to travel at the average speed of the traffic stream or to "float" in the traffic stream, passing as many vehicles as pass the test vehicle. A passenger uses a stopwatch to record time of travel to various previously chosen points along the course. Distances are usually recorded by the vehicle's odometer. The test drive is repeated several times and the average travel time is used to compute the overall and running speeds.

Spot speeds vary with time, location, and environmental and traffic conditions. Since 1942, the average speed on main rural highways has generally increased, rising from about 40 mph in 1944 to 50 mph in 1951 and 60 mph in 1968. Following a petroleum embargo and the subsequent imposition of a nationwide 55 mph speed limit, the average speed on main rural highways decreased to 55.7 mph in 1983.

Speeds vary with the quality of traffic service, being generally higher along expressways and other well-designed facilities and during times when traffic congestion is not a factor. Oppenlander (2) found that mean spot speeds along two-lane rural highways were positively related to lane width and minimum sight distance and negatively related to degree of curvature, gradient, and the number of roadside establishments per mile of highway.

At a given time and location, speeds are widely dispersed and can generally be represented by a normal probability distribution. As Figure 5-1 illustrates, the range of speeds decreases with increase in traffic volume.

5-2 Traffic volume and rate of flow

Traffic volume is defined as the number of vehicles that pass a point along a roadway or traffic lane per unit of time. A measure of the quantity of traffic flow, volume is commonly measured in units of vehicles per day, vehicles per hour, vehicles per minute, and so forth.

Two measures of traffic volume are of special significance to the highway engineer: average daily traffic (*ADT*) and design hourly volume (*DHV*). The average daily traffic is the number of vehicles that pass a particular point on a

FIGURE 5-1 Typical distribution of passenger car speeds in one direction of travel under ideal uninterrupted flow conditions on freeways and expressways. (Courtesy Transportation Research Board.)

roadway during a period of 24 consecutive hours averaged over a period of 365 days.

> *ADT* is a fundamental traffic measurement needed for the determination of vehicle-miles of travel on the various categories of rural and urban highway systems. *ADT* values for specified road sections provide the highway engineer, planner, and administrator with essential information needed for the determination of design standards, the systematic classification of highways, and the development of programs for improvement and maintenance. Vehicle-miles values are important for the development of highway financing and taxation schedules, the appraisal of safety programs, and as a measure of the service provided by highway transportation.(*3*)

It is not feasible to make continuous counts 365 days a year along every section of a highway system. Average daily traffic values for many road sections are therefore based on a statistical sampling procedure described in Chapter 6.

The design hourly volume is a future hourly volume that is used for design. It is usually the 30th highest hourly volume of the design year. Traffic volumes are much heavier during certain hours of the day or year, and it is for these peak hours that the highway is designed.

It has been found that, for the United States as a whole, traffic on the maximum day is approximately 233 percent of the annual average daily traffic and traffic volume during the maximum hour is approximately 25 percent of the annual average daily traffic. In order to design a highway properly, it is necessary to know the capacity that must be provided in order to accommodate the known traffic volume.

The relation between peak hourly flows and the annual average daily traffic on rural highways is shown in Figure 5-2. Experience has indicated that it would be uneconomical to design the average highway for an hourly volume greater than that which will be exceeded during only 29 hr in a year. The hourly traffic volume chosen for design purposes, then, is that occurring during the "30th highest hour."

An approximate value of the 30th highest hour can be obtained by applying an empirically based percentage to the future *ADT*. The thirtieth highest hour, as a percentage of the average daily traffic, ranges from 8 to 38 percent, with an average for the United States of 15 percent for rural locations and 12 percent for urban locations.

Early studies of U.S. traffic indicated that the relationship between the thirtieth highest hour and the annual average daily traffic remained unchanged from year to year. However, later studies suggest that the thirtieth highest hour factor has a tendency to decline slightly with the passing of time. If this trend continues, appropriate adjustments will have to be made in the design hourly volume for any future year.

On a given roadway, the volume of traffic fluctuates widely with time.

FIGURE 5-2 Relation between peak hourly flows and annual average daily traffic on rural highways. (Courtesy Federal Highway Administration.)

Figures 5-3, 5-4, and 5-5 illustrate the variations in volume that occur with time of day, day of the week, and season of the year. These variations tend to be cyclical and to some extent predictable. The nature of the pattern of variation depends on the type of highway facility. Urban arterial flow is characterized by pronounced peaks during the early morning and late afternoon hours, due primarily to commuter traffic. The peaking pattern is not generally evident on weekends, and such facilities experience lowest flows on Sundays. Rural highways tend to experience less pronounced daily peaks, but they may accommodate heaviest traffic flows on weekends and holidays because of recreational travel. Highway facilities generally must accommodate heaviest flows during the summer months. Peaks typically occur during July or August. As might be expected, the seasonal fluctuations are most pronounced for rural recreational routes.

The term *rate of flow* accounts for the variability or the peaking that may occur during periods of less than 1 hr. The term is used to express an equivalent hourly rate for vehicles passing a point along a roadway or for traffic during an interval less than 1 hr, usually 15 min (*4*).

FIGURE 5-3 Hourly variations of volume of traffic on an average weekday. (Courtesy Georgia Department of Transportation.)

FIGURE 5-4 Traffic volume fluctuation by day of week. (Courtesy Georgia Department of Transportation.)

The distinction between volume and rate of flow may be illustrated by an example. Suppose the following traffic counts were made during a study period of 1 hr:

Time Period	Number of Vehicles	Rate of Flow (vehicles/hr)
8:00–8:15	1000	4000
8:15–8:30	1100	4400
8:30–8:45	1000	4000
8:45–9:00	900	3600
Total	4000	

The total volume is the sum of these counts or 4000 vehicles/hr. The rate of flow varies for each 15-min period and during the peak period is 4400 vehicles/hr. Note that 4400 vehicles did not actually pass the observation point during the study hour, but they did pass *at that rate* for one 15-min period.

FIGURE 5-5 Monthly variation in traffic volume. (Courtesy Georgia Department of Transportation.)

Consideration of peak rates of flow is of extreme importance in highway capacity analyses.* Suppose the example roadway section is capable of handling a maximum rate of only 4200 vehicles/hr. In other words, its capacity is 4200 vehicles/hr. Since the peak rate of flow is 4400 vehicles/hr, an extended breakdown in the flow would likely occur even though the volume, averaged over the full hour, is less than the capacity.

The Highway Capacity Manual (4) uses a *peak hour factor* to relate peak rates of flow to hourly volume. The peak hour factor is defined as the ratio of total hourly volume to the maximum rate of flow within the hour. If there was no variability in flow rate during the hour, the peak hour factor would be 1.00. Typical peak hour factors for two-lane roadways range from about 0.83 to 0.96.

5-3 Traffic density

Traffic density, also referred to as traffic concentration, is defined as the average number of vehicles occupying a unit length of roadway at a given instant. It is generally expressed in units of vehicles per mile. As Section 5-5

*Highway capacity analyses are discussed in more detail in Section 5-6.

indicates, traffic density bears a functional relationship to speed and volume. Density has not been extensively employed in the past by highway and traffic engineers to describe traffic flow; however, it is now recommended as the basic parameter for describing the quality of flow along freeways and other multilane highways. It has also been the focus of a number of theoretical and analytical studies.

5-4 Spacings and headways

There are many situations that engineers encounter for which it is necessary to consider the behavior of individual vehicles in the traffic stream rather than the average traffic stream characteristics. Such situations include calculating the probability of delay and average delay for vehicles or pedestrians crossing a traffic stream and predicting the length of waiting lines at toll booths, traffic signals, and entrances to parking facilities. Two measures are of fundamental importance in such calculations: spacing and time–headway between successive vehicles. The spacing is simply the distance between successive vehicles, typically measured from front bumper to front bumper. It is the reciprocal of density. Time–headway is the time between the arrival of successive vehicles at a specified point and it is the reciprocal of volume.

For many light traffic situations, traffic can be described by the Poisson probability distribution. The equation for the Poisson distribution is

$$P(x) = \frac{m^x e^{-m}}{x!}$$

(5-4)

where

$P(x)$ = the probability that exactly x randomly arranged vehicles will be observed in a unit length of road, or the probability of arrival of exactly x vehicles in a unit length of time

$m = Vt/3600$ = the average number of vehicles arriving in an interval of length t

V = traffic volume (vehicles/hr)

t = length of time interval (sec)

EXAMPLE 5-1. *The Poisson Distribution*   Consider the following example, which has been taken from Ref. 5. The number of vehicles arriving along a Los Angeles street was recorded for each of 120 30-sec intervals. During 9 of the intervals, no vehicles arrived; during 16 intervals exactly one vehicle arrived; and so forth. See Table 5-1. The average number of arrivals is

$$m = \frac{Vt}{3600} = \frac{368(30)}{3600} = 3.067$$

$$P(0) = \frac{(3.067)^0 e^{-3.067}}{0!} = 0.047$$

$$P(1) = \frac{(3.067)^1 e^{-3.067}}{1!} = 0.143$$

TABLE 5-1
EXAMPLE DATA FOR THE POISSON DISTRIBUTION

Traffic Arrivals: 30-sec Intervals
(Durfee Avenue, Northbound)

Number of cars per interval X_i	Observed frequency[a] f_i	Total cars observed f_iX_i	Probability of X_i, $P(X_i)$	Theoretical frequency F_i
0	9	0	0.047	5.6
1	16	16	0.143	17.2
2	30	60	0.219	26.3
3	22	66	0.224	26.9
4	19	76	0.172	20.6
5	10	50	0.105	12.6
6	3	18	0.054	6.5
7	7	49	0.023	2.8
8	3	24	0.009	1.1
≥9	1	9	0.002	0.4
Total	120	368	1.000	120.0

$$m = \frac{368}{120} = 3.067 \qquad e^{-m} = 0.047$$

$$F_i = 120 \, (P(X_i))$$

[a]Data by courtesy of Los Angeles County Road Department.

SOURCE: *Poisson and Other Distributions in Traffic.* Eno Foundation for Transportation, Saugatuck, Conn. (1971).

Other probabilities are shown in Table 5-1. The theoretical frequencies corresponding to the probability $P(x_i)$ are

$$F_i = 120 \, P(x_i)$$
$$F_0 = 120(0.047) = 5.6 \text{ intervals etc.}$$

Corresponding to the Poisson counting distribution for free-flow conditions is the negative exponential gap or headway distribution. The equation is

$$P(g \geq T) = e^{-VT/3600} \tag{5-5}$$

where
$(P(g \geq T)$ = the probability of a headway equal to or greater than T sec

The value e is the Naperian logarithmic base and V and T are as previously defined.

EXAMPLE 5-2. ***The Negative Exponential Distribution***   Given the conditions described in Example 5-1, estimate the percentage of a randomly chosen group of gaps that will be greater than 10 sec.

According to the negative exponential distribution,

$$P(g \geq T) = e^{-368(10)/3600} = 0.36$$

Thirty-six percent of the gaps will be greater than 10 sec.

It should be emphasized that the Poisson and negative exponential distributions are applicable to random or free-flow traffic situations. More complex distributions have been proposed to describe a wider variety of traffic flow conditions. References 5 and 6 give detailed descriptions of many counting and headway distributions.

5-5 Speed–volume–density relationships

The relationship between traffic speed, volume, and density is shown by the fundamental equation

$$q = k\bar{u}_s \tag{5-6}$$

where
q = average volume or flow (vehicles/hr)
k = average density or concentration (vehicles/mile)
$\bar{u}_s$ = space-mean speed (mph)

Although a number of theoretical and analytical speed–density relationships have been published, the exact shape of the $k-\bar{u}_s$ curve has not been conclusively established. A model proposed by Greenshields (7) assumed a linear relationship between speed and density. With that assumption, a parabolic volume–density model results. The Greenshields model is illustrated by Figure 5-6 along with the corresponding relationships between speed and volume and volume and density. The curves represent a hypothetical situation with a maximum (or mean-free) speed of 50 mph and a maximum density of 175 vehicles/mile.

Let us consider three key points on the curves shown in Figure 5-6. Point A represents a low-density, high-speed situation. Low volumes exist because few vehicles are on the road. Point B is the point of maximum flow. For this condition, intermediate levels of speed and density occur. Point C represents the worst possible type of flow situation, with maximum density occurring and speeds (and volume) approximating zero.

Empirical research has indicated that speed decreases exponentially with increases in density. Figure 5-7 shows empirical data collected in the Lincoln Tunnel to which a logarithmic flow–concentration curve has been fitted. The equation is

$$q = u_m k \ln\left(\frac{k_j}{k}\right) \tag{5-7}$$

where
u_m = speed at maximum flow
k_j = maximum density

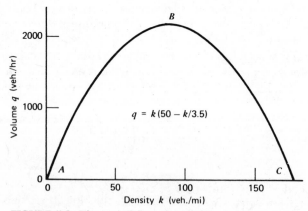

FIGURE 5-6 Theoretical flow, density, speed relationships.

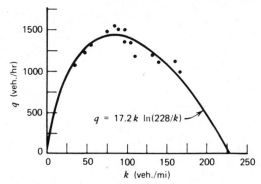

FIGURE 5-7 Logarithmic flow-concentration diagram. (Courtesy Transportation Research Board.)

For the data represented by Figure 5-7, $u_m = 17.2$ vehicles/hr and $k_j = 228$ vehicles/mi.

5-6 Highway capacity

Even with ideal roadway conditions (e.g., a test track), traffic volume tends to reach a maximum point at a relatively low speed. (See Figure 5-8.) This phenomenon, which is puzzling to the casual observer, results from the fact that spacing allowed by the average driver when trailing another vehicle increases nonlinearly with increases in speed (8). Stated another way, traffic density decreases exponentially with increases in speed, as Figure 5-9 illustrates. Referring to Eq. 5-6, increasing speed tends to increase volume, but that effect is offset by concomitant decreases in density.

Another factor limiting the number of vehicles passing a point is interference between vehicles in the traffic stream. This effect is especially noticeable along two-lane roads. With a low traffic volume, the vehicle operator has a wide latitude in selecting the speed at which he or she wishes to travel. As traffic volume increases, the speed of each vehicle is influenced in a large measure by the speed of the slower vehicles. As traffic density increases, a point is finally reached where all vehicles are traveling at the speed of the slower vehicles. This condition indicates that the ultimate capacity has been reached. The capacity of a highway is therefore measured by its ability to accommodate traffic and is usually expressed as the number of vehicles that can pass a given point in a certain period of time at a given speed.

Roadways, of course, are not ideal, and in highway capacity estimates allowances must be made for prevailing roadway and traffic conditions that inhibit the ability of a road to accommodate traffic. In determining highway capacities for uninterrupted flow conditions, the general procedure, described below, is to begin with capacities corresponding to ideal conditions and to apply appropriate empirically based adjustments for prevailing roadway and traffic conditions. The *capacity* of a given section of roadway, either in one direction or in both directions for a two-lane or three-lane roadway, may be defined as the

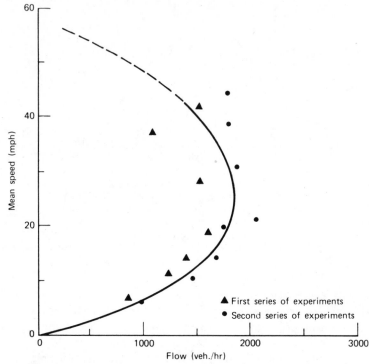

FIGURE 5-8 Experimental speed-flow relationship on test track with straight roadway section. (Courtesy Transportation Research Board)

maximum hourly rate at which vehicles can reasonably be expected to traverse a point or uniform section of a lane or roadway during a given time period under prevailing roadway, traffic, and control conditions (4).

Although the maximum number of vehicles that can be accommodated remains fixed under similar roadway and traffic conditions, there is a range of lesser volumes that can be handled under differing operating conditions. Operation at capacity provides the maximum volume, but as both volume and congestion decrease there is an improvement in the level of service.

Level of service is a qualitative measure that describes operational conditions within a traffic stream and their perception by drivers and/or passengers (4). Six levels of service, A through F, define the full range of driving conditions from best to worst, in that order. These levels of service qualitatively measure the effect of such factors as travel time, speed, cost, and freedom to maneuver, which, in combination with other factors, determine the type of service that any given facility provides to the user under the stated conditions. With each level of service, a *service flow rate* is defined. It is the maximum volume that can pass over a given section of lane or roadway while operating conditions are maintained at the specified level of service.

The levels of service have traditionally been based on relationships involv-

FIGURE 5-9 Speed-concentration curve for straight track at British Road Research Laboratory. (Courtesy Transportation Research Board)

ing operating speed and volume-to-capacity ratio. Figure 5-10, a graphical representation of the level-of-service concept, shows that, as the operating speed on a facility increases, the higher levels of service are attained; on the other hand, as the volume-to-capacity ratio increases, the facility tends to operate at a lower level of service.

Level A represents free flow at low densities with no restrictions due to traffic conditions. Level B, the lower limit of which is often used for the design of rural highways, is the zone of stable flow with some slight restriction of driver freedom. Level C denotes the zone of stable flow with more marked restriction on the driver's selection of speed and with reduced ability to pass. The conditions of Level D reflect little freedom for driver maneuverability, and, while the operating speeds are still tolerable, this region approaches the condition of unstable flow. Low operating speeds and volumes near or at capacity indicate level of service E, which is the area of unstable flow. Level F is the level of service provided by the familiar traffic jam, with frequent interruptions and breakdown of flow, as well as volumes below capacity coupled with low operating speeds.

The operational parameters used to determine the level of service differ with the type of facility. These parameters or "measures of effectiveness"

FIGURE 5-10 General concept of relationship of level of service to operating speed and volume/capacity ratio (not to scale). (Courtesy Transportation Research Board.)

TABLE 5-2
MEASURES OF EFFECTIVENESS FOR LEVEL OF SERVICE DEFINITION

Type of Facility	Measure of Effectiveness	Units
Uninterrupted Flow Facilities		
Freeways		
Basic freeway segments	Density	Passenger cars per mile per lane
Weaving areas	Average travel speed	Miles per hour
Ramp junctions	Flow rate	Passenger cars per hour
Multilane highways	Density	Passenger cars per mile per lane
Two-lane highways	Percent time delay	Percent
	Average travel speed	Miles per hour
Interrupted Flow Facilities		
Signalized intersections	Average individual stopped delay	Seconds per vehicle
Unsignalized intersections	Reserve capacity	Passenger cars per hour
Arterials	Average travel speed	Miles per hour
Transit	Load factor	Persons per seat
Pedestrians	Space	Square feet per pedestrian

SOURCE: *Highway Capacity Manual*, Transportation Research Board Special Report No. 209 (1985).

describe the operating quality for each type of facility. Table 5-2 sets out the measures of effectiveness that best define the level of service for each facility type.

The Highway Capacity Manual

The *Highway Capacity Manual* (4) is the authoritative guide for the performance of highway capacity analyses. The manual reflects over 20 years of comprehensive research by a number of research agencies. This 487-page document was prepared under the guidance of the Transportation Research Board's Committee on Highway Capacity and Quality of Service.

The procedures described in the *Highway Capacity Manual* cover a wide range of facilities, including streets and highways as well as facilities for transit, pedestrians, and bicyclists. As Table 5-2 illustrates, capacity analyses are performed for two general categories of facilities, those with *uninterrupted flow* and those with *interrupted flow*. Uninterrupted flow facilities include two-lane highways, freeways, and other multilane highways. The traffic flow conditions on such facilities result from interactions among vehicles in the traffic stream as well as between vehicles and the physical and ambient characteristics of the roadway.

Interrupted flow facilities generally have traffic control devices that cause periodic interruptions of traffic flow. Examples of such facilities are signalized and two-way and four-way STOP-controlled intersections. The capacity of interrupted flow facilities is limited not only by the physical features of the roadway space but also by the portion of time that is available for various components of traffic flow.

Factors affecting capacity, service flow rate, and level of service

Many of the procedures described in the *Highway Capacity Manual* are based on simple tables or graphs for specified standard conditions, which must be adjusted to account for prevailing conditions different from those specified. The conditions so defined are often descibed in terms of ideal conditions.

Ideal conditions for uninterrupted flow facilities include

1. Twelve-foot lane widths.
2. Six-foot clearance between the edge of travel lanes and the nearest obstructions by objects at the roadside or in the median.
3. A 70 mph design speed for multilane highways and 60 mph design speed for two-lane highways.
4. All passengers cars in the traffic stream (4).

An ideal signalized intersection approach has

1. Twelve-foot lane widths.
2. Level grade.
3. No curb parking allowed on the intersection approaches.
4. All passenger cars in the traffic stream.
5. No turning movements at the intersections.

6. Intersection located outside the central business district.
7. Green signal available at all times (*4*).

Since prevailing conditions are seldom ideal, computations of capacity, service flow rate, or level of service must be adjusted to account for departures from the ideal. Prevailing conditions may be grouped into three categories: roadway, traffic, or control conditions.

Roadway factors include

1. The type of facility and its development environment.
2. Lane widths.
3. Shoulder widths and/or lateral clearances.
4. Design speed.
5. Horizontal and vertical alignments.

Traffic conditions refer to the types of vehicles using the facility and how the traffic flow is distributed by lane use and direction. It is well known that larger and heavier vehicles have an adverse effect on traffic flow in a number of ways. Capacity analyses account for the impacts of trucks, buses, and recreational vehicles, utilizing the results of empirical studies that have quantified these effects.

In addition to the distribution of vehicle types, the effects of two other traffic characteristics on capacity, service flow rates, and level of service must be considered. Directional distribution has a significant impact on the operation of two-lane rural highways. Capacity generally declines as the directional split becomes unbalanced. For multilane highways, capacity analysis procedures focus on a single direction, that of the peak rate of flow. Lane distribution must also be considered for multilane facilities. Normally, the shoulder lane of a multilane facility carries less traffic than other lanes, and analysis procedures must account for these differences.

For interrupted flow facilities, the control of the time available for movement of each traffic flow is of utmost importance in determining capacity, service flow rates, and level of service. Of greatest importance is the control of such facilities by the traffic signal. Signalized operations are affected by the type of signal in use, the phasing, the allocation of green time, and the length of cycle. STOP and YIELD signs also affect capacity, but in a different way. The STOP or YIELD sign assigns the right-of-way permanently to a major street, while a signal assigns designated times when specific movements are permitted.

A complete treatment of highway capacity analyses is beyond the scope of this book. Two examples, one each for uninterrupted and interrupted flow, are given to illustrate the general approach used for such analyses.

General methodology for capacity analysis for freeway segments

With controlled access, limited ramp locations, and no fixed interruptions of traffic flow, freeways operate under the purest form of uninterrupted flow. A capacity analysis of a freeway segment begins with a recognition that a maximum service flow rate per lane exists for each level of service. The various levels of service for freeways are defined by the density, expressed in passenger cars per mile per lane. Suggested maximum service flow rates, MSF_i, are shown in Table 5-3.

These rates, which represent ideal conditions, must be adjusted to account for any prevailing conditions that are not ideal and to reflect the number of lanes in one direction on the freeway segment. The service flow rate for the specified level of service and prevailing conditions is computed by using correction factors, as the following equation demonstrates.

$$SF_i = MSF_i \times N \times f_W \times f_{HV} \times f_P \qquad (5\text{-}8)$$

where

SF_i = service flow rate for level of service i under prevailing roadway and traffic conditions for N lanes in one direction (vehicles/hr)

N = number of lanes in one direction of the freeway

f_W = factor to adjust for the effects of restricted lane widths and/or lateral clearances

f_{HV} = factor to adjust for the effect of heavy vehicles (trucks, buses, and recreational vehicles) in the traffic stream

f_P = factor to adjust for the effect of driver population

The factor f_W is based on the lane width, the number of freeway lanes, the distance to roadside obstructions, and whether obstructions are present on both sides of the freeway. Table 5-4 gives recommended adjustment factors for a range of lane widths and lateral clearances.

TABLE 5-3
LEVELS OF SERVICE (LOS) FOR BASIC FREEWAY SECTIONS WITH DESIGN SPEEDS OF 60 AND 70 MPH

LOS	Density (passenger cars/ mile per lane)	70 mph Design Speed			60 mph Design Speed		
		Speed[a] (mph)	v/c	MSF[b] (passenger cars/ hr per lane)	Speed[a] (mph)	v/c	MSF[b] (passenger cars/ hr per lane)
A	≤ 12	≥ 60	0.35	700	—	—	—
B	≤ 20	≥ 57	0.54	1,100	≥ 50	0.49	1,000
C	≤ 30	≥ 54	0.77	1,550	≥ 47	0.69	1,400
D	≤ 42	≥ 46	0.93	1,850	≥ 42	0.84	1,700
E	≤ 67	≥ 30	1.00	2,000	≥ 30	1.00	2,000
F	> 67	< 30	[c]	[c]	< 30	[c]	[c]

[a]Average travel speed.
[b]Maximum service flow rate per lane under ideal conditions. (Note: All values of MSF are rounded to the nearest 50 passenger cars/hr.)
[c]Highly variable, unstable.

SOURCE: *Highway Capacity Manual.* Transportation Research Board Special Report No. 209 (1985).

TABLE 5-4
ADJUSTMENT FACTOR FOR RESTRICTED LANE WIDTH AND LATERAL CLEARANCE

Distance from Traveled Pavement (ft)	Adjustment Factor f_w, for Lane Width (ft)							
	Obstructions on One Side of Roadway				Obstructions on Both Sides of Roadway			
	12	11	10	9	12	11	10	9
	4-Lane Freeway							
	(2 Lanes Each Direction)							
≥6	1.00	0.97	0.91	0.81	1.00	0.97	0.91	0.81
5	0.99	0.96	0.90	0.80	0.99	0.96	0.90	0.80
4	0.99	0.96	0.90	0.80	0.98	0.95	0.89	0.79
3	0.98	0.95	0.89	0.79	0.96	0.93	0.87	0.77
2	0.97	0.94	0.88	0.79	0.94	0.91	0.86	0.76
1	0.93	0.90	0.85	0.76	0.87	0.85	0.80	0.71
0	0.90	0.87	0.82	0.73	0.81	0.79	0.74	0.66
	6- or 8- Lane Freeway							
	(3 or 4 Lanes Each Direction)							
≥6	1.00	0.96	0.89	0.78	1.00	0.96	0.89	0.78
5	0.99	0.95	0.88	0.77	0.99	0.95	0.88	0.77
4	0.99	0.95	0.88	0.77	0.98	0.94	0.87	0.77
3	0.98	0.94	0.87	0.76	0.97	0.93	0.86	0.76
2	0.97	0.93	0.87	0.76	0.96	0.92	0.85	0.75
1	0.95	0.92	0.86	0.75	0.93	0.89	0.83	0.72
0	0.94	0.91	0.85	0.74	0.91	0.87	0.81	0.70

[a]Certain types of obstructions, high-type median barriers in particular, do not have any deleterious effect on traffic flow. Judgment should be exercised in applying these factors.

SOURCE: *Highway Capacity Manual*. Transportation Research Board Special Report No. 209 (1985).

The factor f_{HV} is used to account for the effects of trucks, buses, and recreational vehicles in the traffic stream. The factor depends on the terrain, specifically the length and magnitude of up grades, and the mix of vehicles in the traffic stream. Heavy vehicles, because of their restricted maneuverability, reduce the number of vehicles that a facility can handle. This reduction is represented by the term *passenger-car equivalent*, which indicates the equivalent number of passenger cars that have been displaced by the presence of each truck, bus, or recreational vehicle. Passenger-car equivalents for heavy trucks and buses are shown in Tables 5-5 and 5-6, respectively. The *Highway Capacity Manual (4)* lists similar values for recreational vehicles.

By using passenger-car equivalents such as those shown in Tables 5-5 and 5-6, the adjustment factor for heavy vehicles can be computed by the following equation:

$$f_{HV} = 1/[1 + P_T(E_T - 1) + P_R(E_R - 1) + P_B(E_B - 1)] \qquad (5\text{-}9)$$

TABLE 5-5
PASSENGER-CAR EQUIVALENTS FOR HEAVY TRUCKS, 300 LB/HP

| | | Passenger-Car Equivalent, E_r, for Percentage of Trucks | | | | | | | | | | | | | | |
| | | 4-Lane Freeways | | | | | | | | 6–8 Lane Freeways | | | | | | | |
Grade (%)	Length[a] (miles)	2	4	5	6	8	10	15	20	2	4	5	6	8	10	15	20
<1	All	2	2	2	2	2	2	2	2	2	2	2	2	2	2	2	2
1	0–¼	2	2	2	2	2	2	2	2	2	2	2	2	2	2	2	2
	¼–½	3	3	3	3	3	3	3	3	3	3	3	3	3	3	3	3
	½–¾	4	4	4	4	3	3	3	3	4	4	4	3	3	3	3	3
	¾–1	5	4	4	4	3	3	3	3	5	4	4	4	3	3	3	3
	1–1½	6	5	5	5	4	4	4	3	6	5	5	4	4	4	3	3
	>1½	7	5	5	5	4	4	4	3	7	5	5	5	4	4	3	3
2	0–¼	4	4	4	3	3	3	3	3	4	4	4	3	3	3	3	3
	¼–½	7	6	6	5	4	4	4	4	7	5	5	5	4	4	4	4
	½–¾	8	6	6	5	5	4	4	4	8	6	6	6	5	5	4	4
	¾–1	8	6	6	6	5	5	5	5	8	6	6	6	5	5	5	5
	1–1½	9	7	7	7	6	6	5	5	9	7	7	6	5	5	5	5
	>1½	10	7	7	7	6	6	5	5	10	7	7	6	5	5	5	5
3	0–¼	6	5	5	5	4	4	4	3	6	5	5	5	4	4	4	3
	¼–½	9	7	7	6	5	5	5	5	8	7	7	6	5	5	5	5
	½–¾	12	8	8	7	6	6	6	6	10	8	7	6	5	5	5	5
	¾–1	13	9	9	8	7	7	7	7	11	8	8	7	6	6	6	6
	>1	14	10	10	9	8	8	7	7	12	9	9	8	7	7	7	7
4	0–¼	7	5	5	5	4	4	4	4	7	5	5	5	4	4	3	3
	¼–½	12	8	8	7	6	6	6	6	10	8	7	6	5	5	5	5
	½–¾	13	9	9	8	7	7	7	7	11	9	9	8	7	6	6	6
	¾–1	15	10	10	9	8	8	8	8	12	10	10	9	8	7	7	7
	>1	17	12	12	10	9	9	9	9	13	10	10	9	8	8	8	8
5	0–¼	8	6	6	6	5	5	5	5	8	6	6	6	5	5	5	5
	¼–½	13	9	9	8	7	7	7	7	11	8	8	7	6	6	6	6
	½–¾	20	15	15	14	11	11	11	11	14	11	11	10	9	9	9	9
	>¾	22	17	17	16	13	13	13	13	17	14	14	13	12	11	11	11
6	0–¼	9	7	7	7	6	6	6	6	9	7	7	6	5	5	5	5
	¼–½	17	12	12	11	9	9	9	9	13	10	10	9	8	8	8	8
	>½	28	22	22	21	18	18	18	18	20	17	17	16	15	14	14	14

[a]If a length of grade falls on a boundary condition, the equivalent from the longer grade category is used. For any grade steeper than the percentage shown, use the next higher grade category.

SOURCE: *Highway Capacity Manual.* Transportation Research Board Special Report No. 209 (1985).

where

f_{HV} = adjustment factor for the combined effect of trucks, recreational vehicles, and buses on the traffic stream

E_T, E_R, E_B = passenger-car equivalents for trucks, recreational vehicles, and buses, respectively

P_T, P_R, P_B = proportion of trucks, recreational vehicles, and buses, respectively, in the traffic stream

TABLE 5-6
PASSENGER-CAR
EQUIVALENTS FOR BUSES

Grade (%)	Passenger-Car Equivalent, E_B
0–3	1.6
4^a	1.6
5^a	3.0
6^a	5.5

[a]Use generally restricted to grades more than ¼ mile long.

SOURCE: *Highway Capacity Manual.* Transportation Research Board Special Report No. 209 (1985).

It is known that certain types of traffic, such as weekend and recreational traffic, use freeways less efficiently than weekday or commuter traffic. Little research has been done on this effect, but it is known that capacities tend to be lower on weekends, particularly in recreational areas. Lacking local data, considerable engineering judgment must be exercised in making an adjustment for the character of the traffic stream. The *Highway Capacity Manual* suggests that an adjustment factor f_P ranging from 0.75 to 0.90 be used to account for this effect.

EXAMPLE 5-3. *Capacity of a Basic Freeway Segment.* (a) Determine the service flow rate with level of service C for a section of a four-lane freeway (two lanes in each direction) with 11-ft lanes and obstructions 5 ft from traveled pavement on one side of the roadway. The section has a 4 percent gradient 0.8 mile long. It is to accommodate 12 percent heavy trucks, 6 percent buses, and no recreational vehicles. Based on local studies, an adjustment factor for the character of the traffic stream, f_P, of 0.90 is indicated. The design speed is 70 mph. (b) What is the capacity of the segment (density = 67 passenger cars per mile per lane)? (c) Estimate the average travel speed that corresponds to the capacity.

Solution to Part (a). From Table 5-3, the maximum service flow rate for LOS C is 1550 passenger cars/hr per lane. The service flow rate for LOS C is computed by Eq. 5-8.

From Table 5-4, the factor to adjust for the effects of lane width and lateral clearances is $f_W = 0.96$.

To adjust for the effect of heavy vehicles, passenger-car equivalents for trucks and buses are obtained from Tables 5-5 and 5-6, respectively:

$$E_T = 8$$
$$E_B = 1.6$$

By Eq. 5-9, the heavy vehicle factor

$$f_{HV} = 1/[1 + 0.12(8 - 1) + 0 + 0.06(1.6 - 1)] = 0.53$$
$$f_P = 0.90 \text{ (given)}$$

By Eq. 5-8, the service flow rate

$$SF_C = 1550 \times 2 \times 0.96 \times 0.53 \times 0.90 = 1420 \text{ vehicles/hr}$$

Solution to Part (b). The capacity corresponds to the critical density of 67 passenger cars/mile and from Table 5-3 is 2000 passenger cars/hr per lane (ideal conditions). For the prevailing conditions, the capacity is

$$SF_E = 2000 \times 2 \times 0.96 \times 0.53 \times 0.90 = 1832 \text{ vehicles/hr}$$

Solution to Part (c). From Table 5-3, the average travel speed is approximately 30 mph.

General methodology for capacity analysis for signalized intersection

The capacity of a signalized intersection is highly dependent on the type of signal control being used. A wide variety of equipment and control schemes may be used for such intersections. The capacity of a signalized intersection is therefore far more variable than that of other types of facilities, where capacity depends primarily on the physical geometry of the roadway.

In intersection analysis, the concepts of capacity and level of service are not as strongly correlated as they are for other types of facilities. The *Highway Capacity Manual* (4) therefore recommends that separate analyses be used to determine the capacity and level of service for a signalized intersection.

The capacity of signalized intersections should be defined for each approach.* Intersection approach capacity is the maximum rate of flow which may pass through the intersection by that approach under prevailing roadway, traffic, and signalization conditions. To account for peaking, the rate of flow is usually measured or projected for a 15-min period, and capacity is expressed in vehicles per hour (4).

Operational analysis of a signalized intersection requires detailed information on the roadway, the signal system, and the traffic at the intersection. Required information on the roadway includes approach grades, the number and width of lanes, parking conditions, and the existence and lengths of exclusive turning lanes.

Complete information is needed on signalization including the phase plan, cycle length, green times, type of signal operation (actuated or pretimed), and existence of push-button pedestrian-actuated phases. (These and other traffic control concepts are discussed in more detail in Chapter 12.)

Capacity analyses of signalized intersections require detailed information on traffic conditions including

---

*It may also be desirable to evaluate separately the capacity of designated lanes or lane groups such as those serving a particular movement or set of movements.

1. Traffic volumes for each movement on each approach.
2. Percentage of heavy vehicles.
3. Volume and pattern of pedestrian traffic.
4. Rate of parking maneuvers within the vicinity of the intersection.
5. Number of local buses picking up or discharging passengers at the intersection.

In addition, information is needed on the shape of the arrival curve distribution and its relationship to the signal operation. This factor describes the platooning effect in arriving flows. It affects the average stopped delay of vehicles passing through the intersection, which defines the level of service. The *Highway Capacity Manual* (4) categorizes arrival distributions by defining five arrival types, defined as follows.

Type 1 is the worst platoon condition, defined as a dense platoon arriving at the intersection at the beginning of the red phase.

Type 2 is a generally unfavorable platoon condition, which may be a dense platoon arriving during the middle of the red phase or a dispersed platoon arriving throughout the red phase.

Type 3 refers to totally random arrivals that are widely dispersed throughout the red and green phases.

Type 4 is a moderately favorable platoon condition, defined as a dense platoon arriving during the middle of the green phase or a dispersed platoon arriving throughout the green phase.

Type 5 is the most favorable platoon condition, defined as a dense platoon arriving at the beginning of the green phase.

The *Highway Capacity Manual* (4) relates the arrival types to a platoon ratio, which is defined by

$$R_p = PVG/PTG$$

$$(5\text{-}10)$$

where
  R_p = platoon ratio
  PVG = percentage of all vehicles in the movement arriving during the green phase
  PTG = percentage of the cycle that is green for the movement

The relationship between arrival type and the platoon ratio is shown in Table 5-7.

The capacity of signalized intersections is based on the concept of a *saturation flow rate*. The *Highway Capacity Manual* (4) defines saturation flow rate as the maximum rate of flow that can pass through an intersection approach or lane group under prevailing roadway and traffic conditions, assuming that the approach or lane group has 100 percent of real time available as effective green time. Saturation flow rate is expressed in units of vehicles per hour of effective green time.

TABLE 5-7
RELATIONSHIP BETWEEN
ARRIVAL TYPE AND
PLATOON RATIO

Arrival Type	Range of Platoon Ratio, R_p
1	0.00 to 0.50
2	0.51 to 0.85
3	0.86 to 1.15
4	1.16 to 1.50
5	≥ 1.51

SOURCE: *Highway Capacity Manual.* Transportation Research Board Special Report No. 209 (1985).

The capacity of a given lane group or approach may be calculated by the equation

$$c_i = s_i \times (g/C)_i \qquad (5\text{-}11)$$

where

c_i = capacity of lane group or approach i (vehicles/hr)

s_i = saturation flow rate for lane group or approach i (vehicle/hr of green)

$(g/C)_i$ = green ratio for lane or approach i

The computation of the saturation flow rate begins with the selection of an "ideal" saturation flow rate, usually taken to be 1800 passenger cars per hour of green time per lane. This value is then adjusted to account for the various prevailing conditions. The equation for estimating saturation flow rate is

$$s = s_o N f_w f_{HV} f_g f_p f_{bb} f_a f_{RT} f_{LT}$$

where

s = saturation flow rate for the subject lane group, expressed as a total for all lanes in the lane group under prevailing conditions (vehicles/hr green)

S_o = ideal saturation flow rate per lane, usually 1800 passenger cars per hour of green time per lane

N = number of lanes in the lane group

f_w = adjustment factor for lane width; 12-ft lanes are standard; given in Table 5-8

f_{HV} = adjustment factor for heavy vehicles in the traffic stream, given in Table 5-9

f_g = adjustment factor for approach grade, given in Table 5-10

f_p = adjustment factor for existence of a parking lane adjacent to the lane group and parking activity in that lane, given in Table 5-11

f_{bb} = adjustment factor for blocking effect of local buses stopping within the intersection area, given in Table 5-12

f_a = adjustment factor for area type, given in Table 5-13

f_{RT} = adjustment factor for right turns in the lane group

f_{LT} = adjustment factor for left turns in the lane group

Adjustment factors for turning movements are not included here but may be found in Ref. 4.

The level of service for signalized intersections is defined in terms of delay. Specifically, level of service is based on the average stopped delay per vehicle for a 15-min. analysis period, as specified in Table 5-14.

TABLE 5-8
ADJUSTMENT FACTOR FOR LANE WIDTH

Lane Width, ft	8	9	10	11	12	13	14	15	≥ 16
Lane Width Factor, f_w	0.87	0.90	0.93	0.97	1.00	1.03	1.07	1.10	Use 2 lanes

SOURCE: *Highway Capacity Manual.* Transportation Research Board Special Report No. 209 (1985).

TABLE 5-9
ADJUSTMENT FACTOR FOR HEAVY VEHICLES

Percent Heavy Vehicles, %HV	0	2	4	6	8	10	15	20	25	30
Heavy Vehicle Factor f_{HV}	1.00	0.99	0.98	0.97	0.96	0.95	0.93	0.91	0.89	0.87

SOURCE: *Highway Capacity Manual.* Transportation Research Board Special Report No. 209 (1985).

TABLE 5-10
ADJUSTMENT FACTOR FOR GRADE

	Downhill			Level		Uphill	
Grade, %	−6	−4	−2	0	+2	+4	+6
Grade, Factor, f_g	1.03	1.02	1.01	1.00	0.99	0.98	0.97

SOURCE: *Highway Capacity Manual.* Transportation Research Board Special Report No. 209 (1985).

TABLE 5-11
ADJUSTMENT FACTOR FOR PARKING

Number of Lanes in Lane Group	No Parking	Number of Parking Maneuvers per Hour, N_m				
		0	10	20	30	40
1	1.00	0.90	0.85	0.80	0.75	0.70
2	1.00	0.95	0.92	0.89	0.87	0.85
3	1.00	0.97	0.95	0.93	0.91	0.89

SOURCE: *Highway Capacity Manual.* Transportation Research Board Special Report No. 209 (1985).

TABLE 5-12
ADJUSTMENT FACTOR FOR BUS BLOCKAGE

Number of Lanes in Lane Group	Number of Buses Stopping per Hour, N_B				
	0	10	20	30	40
1	1.00	0.96	0.92	0.88	0.83
2	1.00	0.98	0.96	0.94	0.92
3	1.00	0.99	0.97	0.96	0.94

SOURCE: *Highway Capacity Manual.* Transportation Research Board Special Report No. 209 (1985).

TABLE 5-13
ADJUSTMENT FACTOR FOR AREA TYPE

Type of Area	Factor f_a
Central business district	0.90
All other areas	1.00

SOURCE: *Highway Capacity Manual.* Transportation Research Board Special Report No. 209 (1985).

TABLE 5-14
LEVEL OF SERVICE CRITERIA FOR SIGNALIZED INTERSECTIONS

Level of Service	Stopped Delay per Vehicle (sec)
A	≤ 5.0
B	5.1 to 15.0
C	15.1 to 25.0
D	25.1 to 40.0
E	40.1 to 60.0
F	> 60.0

The average stopped delay per vehicle can be estimated by the following equation, assuming a random pattern of arrivals for approaching vehicles.

$$d = 0.38C \frac{[1 - g/C]^2}{[1 - (g/C)(X)]} + 173 X^2 \left[(X - 1) + \sqrt{(X - 1)^2 + (16X/c)} \right]$$

$$(5\text{-}13)$$

where
d = average stopped delay per vehicle for the lane group (sec/vehicle)
C = cycle length (sec)
g/C = green ratio for the lane group; ratio of effective green time to cycle length
X = v/c ratio for the lane group
c = capacity of the lane group

Where the arrivals are not random, the delay obtained from Eq. 5-13 can be multiplied by an adjustment factor, which is a function of arrival type, v/c ratio, and type of signalization. Such factors are not included here but can be found in Ref. 4.

EXAMPLE 5-4. *Capacity of a Signalized Intersection.* (a) Determine the capacity of the south approach lanes for a signalized intersection given the following conditions:

2 approach lanes, 11 ft wide

10 percent heavy vehicles

2 percent downgrade along the south approach

location in the central business district

no buses stopping in the intersection and no parking permitted

type 3 (random) arrivals

through traffic only

cycle length = 60 sec

green ratio = 0.45

(b) What is the level of service for a flow rate of 1000 vehicles/hr, assuming equal lane utilization for the two lanes?

Solution to Part a. By Eq. 5-12, and using Tables 5-8 through 5-13,

$$s = 1800 \times 2 \times 0.97 \times 0.95 \times 1.01 \times 0.90$$

$$s = 3015 \text{ vehicles/hr of green}$$

By Eq. 5-11,

$$c_i = 3015 \times 0.45 = 1357 \text{ vehicles/hr}$$

Solution to Part b. For random arrivals, Eq. 5-13 can be used to estimate the average stopped delay per vehicle.

$$d = 0.38(60) \frac{[1 - 0.45]^2}{[1 - (0.45)(0.74)]} + 173(0.74)^2 \left[(0.74 - 1) + \sqrt{(0.74 - 1)^2 + (16 \times 0.74)/1357} \right]$$

$$d = 11.9 \text{ sec per vehicle}$$

From Table 5-14, the approach operates at level of service B.

THE HIGHWAY ACCIDENT PROBLEM

Since 1900, more than 2 million deaths have resulted from motor vehicle accidents, far exceeding the number of Americans killed in all of the nation's wars combined. In 1983 alone, highway accidents in the United States produced 44,600 deaths, 1.9 million disabling injuries, and economic losses of more than 43.3 billion (9). Traffic accidents are the fourth leading cause of death in the country, exceeded only by heart disease, cancer, and stroke. For persons between the ages of 1 and 24, motor vehicle accidents are the leading cause of death.

Despite the long history and pervasiveness of the problem, the nature and causes of traffic accidents are not well understood. While there are differences of opinion among traffic safety professionals concerning why crashes occur, there are greater differences concerning the programs of action required to reduce highway losses.

5-7 Accident loss factors

It is helpful to think of an accident as an event comprised of nine parts, called crash factors. One may visualize an accident in three phases: precrash, at-crash, and postcrash. Within each phase, one or more human, vehicular, and environmental factors may contribute to the initiation of the crash or alter its consequences. Examples of the nine classes of loss factors are shown in Table 5-15.

These factors may be arranged in the form of a nine cell matrix. Countermeasures designed to reduce highway deaths and injuries may be focused on any one or a combination of the cells that comprise the loss factor matrix. Precrash countermeasures are concerned with the prevention of accidents. At-crash countermeasures attempt to reduce injuries by controlling the energy of crashes. Postcrash programs are designed to limit the losses by improving first-aid and other emergency procedures. This part of the chapter is concerned primarily with precrash factors, that is, those aspects of the driver, the vehicle, and the roadway environment that explain the occurrence of the accident event.

5-8 The nature of vehicular crashes

One of the most striking features of scientific accident causation data is the wide variety of human, vehicular, and roadway factors that contribute to vehicular crashes. Many accidents—perhaps one-third to one-half of all crashes—are uncomplicated events that can be explained by the presence of a single human, vehicular, or environmental factor. These factors range from simple lapse of driver attention to the relatively rare case of a driver who uses the automobile as an instrument of suicide.

TABLE 5-15
EXAMPLES OF NINE CLASSES OF ACCIDENT LOSS FACTORS

Designation	Example
Precrash	
Human	Driver fell asleep
Vehicle	Brake failure
Environment	Slippery roadway surface
At-crash	
Human	Seat belts improperly worn
Vehicle	Structural weakness of side of vehicle
Environment	Unyielding signpost near pavement
Postcrash	
Human	Bystanders took improper first-aid action
Vehicle	Vehicle not equipped with a fire extinguisher
Environment	Emergency telephone not available

Many crashes, however, are complex events that are triggered by two or more factors. For example, consider a crash involving a driver driving a vehicle with worn tires at an excessive speed along a wet and slippery roadway. The combined presence of these factors could cause a crash. Removal of any of the factors—the excessive speed, the slippery roadway, or the worn tires—might have prevented the crash from occurring.

For many years, traffic safety specialists have maintained that 90 percent of all accidents are due to improper driving. It appears that this conclusion, which is based on analyses of police accident reports, oversimplifies a complex problem and exaggerates the role of driver behavior. Nevertheless, multidisciplinary research has indicated that human factors are the most prevalent type of contributing factors. It is believed, however, that not more than half of traffic accidents can be attributed to human factors alone.

Research has shown that a substantial percentage of crashes results from a variety of driver malfunctions and conditions. Driver inattention or distraction, excessive speed, and violation of other safe driving practices appear to be recurring contributing factors. Many crashes can be explained by some form of driver impairment such as fatigue, emotional stress, physical illness, or failure to wear needed corrective lenses. By far the most prevalent form of driver impairment is alcohol intoxication.

Vehicular factors that contribute to crashes are more commonly classified as vehicular conditions (e.g., slick tires) rather than malfunctions or catastrophic failures due to poor design. Defective brakes (especially gross brake failure) and tires (especially inadequate tread depth and underinflation) are frequently cited as contributors to crashes. Vehicular factors are rarely solely responsible for an accident but usually occur jointly with human and/or enviromental factors.

Environmental factors commonly interact with some other factor (human or vehicle) to produce an accident. Although a crash may result from a natural disaster or from a malfunction in the highway or its appurtenances, these events are extremely rare. Typical examples of environmental factors that contribute to crashes are wet roadways, missing or improper traffic control devices, and poor geometric design.

5-9 Conceptualization of an accident

Multidisciplinary research has shown that traffic accidents are caused by a diversity of human, vehicular, and environmental factors that often interact in a complex way to trigger the initiation of the event. A fairly wide tolerance is permissible (not accident causative) in any one component of the human/machine/environment system. Even a "drunk" may successfully manipulate an automobile under certain conditions, such as down a wide street with which he is familiar and on which there are no other vehicles or roadside hazards. Conversely, a skilled driver may avoid an unexpected hazard presented either by vehicle failure or by the environment. Blumenthal (*10*) has conceptualized the variability in tolerences in terms of driver performance and system demands, both of which fluctuate unpredictably over a period of time. As Figure

5-11 illustrates, accidents occur at some point in time when the system demands exceed the performance capability of the driver.

5-10 Current accident trends and concerns

In recent years, significant advances have been made in improving highways for safety and in designing more crashworthy vehicles. Numerous and expensive programs aimed at improving the driving function have been sponsored by the federal government and implemented by the state governments. Strong evidence exists that these initiatives have yielded positive results. This is best demonstrated by the fatality rate, which fell to an all-time low of 2.58 fatalities per 100 million vehicle-miles of travel in 1983, approximately half of the rate for the mid-1960s. Nevertheless, vehicular accidents remain one of the nation's most persistent and troublesome public health problems.

By the mid-1980s, traffic safety specialists focused attention on four areas of concern: (1) the influence of alcohol; (2) the nonuse of occupant restraints; (3) crash losses from the growing population of large trucks; and (4) the effects of vehicular crashes on the elderly (*11*).

Researchers have consistently reported that alcohol contributes to more than half of fatal crashes. A nationwide survey in 1983 (*12*) revealed that 57 percent of all fatally injured drivers had been drinking and 46 percent were legally intoxicated. The incidence of alcohol involvement in fatal crashes tends to be greatest in late night and early morning accidents, when 80 to 90 percent of all fatally injured drivers are reported to have been drinking (*12*). It is estimated that alcohol is involved in 29 percent of injury accidents and about 11 percent of all accidents (*11*). Under strong pressure from citizens groups and the general public, legislatures in many states have increased the penalties for driving under the influence of alcohol.

Scientific research has clearly established that vehicle occupants who are

FIGURE 5-11 A concept of a traffic accident.

restrained by safety belts are far less likely to be injured than those who are not restrained. Yet, despite numerous "buckle up" campaigns by public agencies, automobile manufacturers, and private groups, few drivers wear safety belts. Research indicates that about 82 percent of people in cars involved in crashes were not wearing safety belts (*11*). Significant declines in crash-related injuries and deaths are likely to result from greater use of automatic restraint systems. These systems are of two types: automatic belts, which move into position when the door is closed; and air bags, which inflate in a serious crash to cushion the collision of occupants with the vehicle's interior.

For many years, researchers have understood the relationship between vehicle size and injury severity. Clear evidence exists that when two vehicles collide, occupants of the larger vehicle are far less likely to be seriously injured or killed than those traveling in the smaller vehicle. The hazards associated with vehicle size are most prominent in collisions that involve a large truck and an automobile. In such crashes, automobile occupants are 30 times as likely to be killed as occupants of the truck. Deregulation of the trucking industry and the relaxation of restrictions on truck weight and size are likely to increase accident losses. Highway and traffic engineers face the difficult challenge of providing a safe driving environment for drivers of an ever-widening array of vehicle types and sizes.

People over the age of 65 represent a small proportion of those who are involved in crashes; however, when they are involved, they are at greater risk of death or serious, prolonged injury than are young people (*11*). Older drivers are more likely to be driving the struck rather than the striking vehicle in multiple-vehicle accidents and tend to be involved in fewer single-vehicle accidents than younger drivers. Efforts to decrease crash-related injuries and deaths among the elderly should therefore be directed to providing better protection for them in the crash and providing improved emergency medical services and hospital care during the postcrash period.

PROBLEMS

5-1 Three cars travel over a 200-ft section of highway at constant speeds of 30, 35, and 40 ft/sec. Compute the time-mean and space-mean speeds for this condition.

5-2* Four cars travel over a 60-m section of highway at constant speeds of 18, 20 and 22 m/sec. Compute the time-mean and space-mean speeds for this condition.

5-3 For the conditions described in Problem 5-1, estimate the variance about the space-mean period.

5-4* For the conditions described in Problem 5-2, estimate the variance about the space-mean speed in meters per sec.

5-5 The estimated future average daily traffic for a rural highway is 9500. Assume that the relationship between peak hourly flows and *ADT* is as shown in Figure 5-2. Estimate the design hourly flow volume. What would be the effect of using the 10th highest hour for design purposes? The 50th highest hour?

5-6 Arrivals at a parking lot are assumed to follow the Poisson distribution. The average arrival rate is 3.0 per minute. What is the probability that during a given minute no cars will arrive? What percentage of the time will more than three arrive?

5-7 A street has an hourly volume of 300 vehicles. A pedestrian requires a gap of at least 10 sec to cross. Assuming headways can be described by a negative exponential distribution, what is the probability that the pedestrian will have to wait to cross?

5-8 Compute the slope of the volume−density curve shown in Fig. 5-6 at points A, B, and C. What are the units of the slope? Discuss the meaning of the slope of this curve. (See Ref. 5.)

5-9 Compute the density corresponding to the maximum volume for the $q-k$ curve shown in Fig. 5-7.

5-10 Estimate the service flow rate for level of service B for a six-lane freeway (three lanes in each direction) with 12-ft lanes and obstructions 6 ft from the edge of the pavement on one side only. The section is to accommodate 10 percent heavy trucks, 5 percent buses, and no recreational vehicles. Assume the adjustment factor for the character of the traffic stream, f_p, is 0.95. The design speed is 75 mph and the section is level.

5-11 Estimate the capacity of the freeway segment described in Problem 5-10.

5-12 Estimate the capacity of the approach lanes of a signalized intersection given the following conditions:

two approach lanes, 10 ft wide	15 parking manuevers per hour
12 percent heavy vehicles	type 3 (random) arrivals
flat grades	through traffic only
location in central business district	cycle length = 65 sec
20 buses stopping per hour	green ratio = 0.50

5-13 Determine the level of service for a flow rate of 1100 vehicles/hr for the approach lanes described in Problem 5-12. Assume equal lane utilization for the two lanes.

5-14 Consult with the State Office of Highway Safety and prepare a report on the approximate annual costs of the following types of safety countermeasures:
   (a) Driver education.
   (b) Motor vehicle inspection.
   (c) Traffic enforcement.

REFERENCES

   1. Wardrop, J. G. Some Theoretical Aspects of Road Traffic Research. *Proceedings of the Institution of Civil Engineers*, Pt. 2, Vol. I, pp. 325−362 (1952).

   2. Oppenlander, Joseph C. Multivariate Analysis of Vehicular Speeds. *Highway Research Record No. 35*, p. 63 (1963).

   3. *Guide for Traffic Volume Counting Manual. Bureau of Public Roads*, U.S. Department of Commerce, Washington, D.C. (1965).

   4. *Highway Capacity Manual*. Transportation Research Board Special Report No. 209 (1985).

   5. *Poisson and Other Distributions in Traffic*. Eno Foundation for Transportation, Saugatuck, Conn. (1971).

6. *Traffic Flow Theory, A Monograph.* Transportation Research Board Special Report No. 165 (1975).

7. Greenshields, B. D. A. Study of Traffic Capacity. *Proceedings of the Highway Research Board* Vol. 14, pp. 448–477 (1934).

8. *Highway Capacity Manual.* Bureau of Public Roads, U.S. Department of Commerce, Washington, D.C. (1950).

9. *Accident Facts, 1984 Edition.* National Safety Council, Chicago, Ill. (1984).

10. Blumenthal, M. Problem Definition: The Driving Task in the System Context. Presented at the American Psychological Association, San Francisco (1968).

11. *National Accident Sampling System 1983.* National Highway Traffic Safety Administration, Washington, D.C. (1985).

12. *Fatal Accident Reporting System 1983.* National Highway Traffic Safety Administration, Washington, D.C. (1984).

chapter six

TRANSPORTATION PLANNING

Planning is a conscious, deliberate process or activity that is directed toward the achievement of state or community goals (1). It must be capable of dealing with complex subjects and evaluating alternative solutions or courses of action. Meaningful planning is based on the assumption that the power exists to implement the plans that are developed.

The construction of transportation systems has a profound and far-reaching impact on land use and the social and economic well-being of citizens. For this reason, transportation planning is a major component of state and local planning activities. In the United States, the planning of highway systems tends to be the dominant element in multimodal transportation planning.

Prior to 1930, the primary objective of highway agencies was to develop a system of paved rural roads. There was little emphasis on planning, as the highway departments concentrated on getting road users "out of the mud." Sharp increases in motor-vehicle registrations and travel during the early 1930s made it apparent that the highway system would soon become inadequate for future needs. Factual information would be required if sound policies and decisions were to be made regarding the development of an adequate and efficient national system of highways.

An act of Congress adopted in 1934 and known as the Hayden–Cartwright Act provided the impetus for a large number of highway planning surveys starting in 1935. The surveys were designed to determine the condition of existing facilities, the volume and nature of traffic using the highway system, the use and life expectancy of roads comprising the system, and future highway needs.

The Hayden–Cartwright Act authorized the expenditure of funds not to exceed 1.5 percent of the federal-aid apportionment to any state for the making of surveys, plans, and engineering investigations of projects for future construction. By 1940, practically all state highway departments, in cooperation with the Bureau of Public Roads, were doing such surveys. At the time of the passage of the Hayden–Cartwright Act the philosophy of the Federal-Aid Highway Program was to assist the rural areas, but with the increase in motor-vehicle registration and traffic volumes, signs of distress soon became apparent in urban and suburban areas. Thus, it was logical that planning activities should be extended to these areas. With the passage of the Federal-Aid Act of 1944, funds became available for federal-aid highway construction in urban areas.

The Federal-Aid Highway Act of 1962 further defined the requirements for planning processes in urban areas with a population over 50,000, to be

133

effective July 1, 1965. The law required the establishment of a "cooperative, comprehensive, and continuing" transportation planning process in each of these areas as a prerequisite for federal aid. By the deadline, nearly all of the more than 200 such areas in the country had met the requirement.

By the mid-1970s, multimodal transportation planning had become a key activity of practically all state highway or transportation agencies and of most urban planning agencies. In this chapter, the planning policies and procedures employed by such agencies are described.

6-1 Definition

As the term suggests, transportation planning is concerned with the development of a transportation plan for an urban area or for an entire state. This involves generating and comparing alternative plans, evaluating the social, economic, and environmental impacts of proposed transportation actions, and, with appropriate participation of citizens, political representatives, and public agencies, selecting the preferred plan.

Transportation planning should be viewed as a continuing process that takes into account changes in the needs and wishes of the citizens and their political leaders and that makes use of advances in planning procedures and techniques. For this reason, some planners believe that the publication of a single long-range master transportation plan should be avoided (1). Rather, transportation planning agencies should consider the practice of updating and publishing transportation plans or reports on a periodic basis, for example, every two years.

Transportation planning consists of three levels of planning: (1) policy planning, (2) systems planning, and (3) project or facility planning. Policy planning is concerned with the establishment of transportation policy and the development of goals and objectives. It is "involved with questions of resource allocation . . ., both in terms of allocation by geographic area and by transportation mode. It is not involved with the review or recommendation of specific facilities or corridors or even transportation networks" (2).

Systems planning is "a process under which transportation networks and corridors are defined . . ., starting from forecasts of population and economic growth and continuing through estimates of person and goods movement to a physical description of the systems required to meet those real or implied needs" (2).

Project or facility planning is concerned primarily with determining the scale and layout of individual facilities to accommodate the anticipated demand. In a state highway agency this level of planning is normally performed by modal divisions (aviation, highways, etc.) or, in a functional organization, by the design division. Aspects of project planning are covered in other chapters of this text and will not be discussed further here.

Planning may be subdivided into three categories to account for differences in the planning period:

1. Long-range planning.

2. Intermediate-range planning.
3. Short-range planning.

Long range planning is typically conducted for a period of about 15 to 30 years into the future (3). The level of detail for such planning is generalized, and long-range plans should be periodically reappraised and changed to account for unexpected shifts in population, development, and so forth. Intermediate-range planning covers a period of about 5 to 15 years in the future. "In this planning phase, the facility characteristics and performance, as well as route location, would be more specific, with evaluation of alternative designs and route locations in terms of a performance analysis, including traditional benefit/cost analysis, as well as environmental impacts and developmental considerations" (3). Short-range planning is typically conducted for a period of about five years. In short-range plans, the physical features of transport facilities and the traffic and land developments that these facilities are expected to serve can be specified in considerable detail and with reasonable precision.

6-2 Organization

As indicated in Chapter 2, state transportation department organizational structures may be classified as modal, functional, or mixed modal and functional. (See Figure 2-2). Because of differing needs and capabilities among the various states, there is no single ideal organizational structure that would be suitable for all states.

Some planners recommend that policy planning and systems planning be performed by a single planning unit that reports directly to the chief executive of the transportation department (3). In a department with a functional or mixed organizational structure, the planning unit might be one of the principal operating (or line) divisions. In a modal organization, the planning unit is more likely to be a staff-level organization with the planning director reporting directly to the chief executive officer. Another recommendation (4) calls for a staff-level unit for policy planning and a line-level unit for systems planning.

Because of the nature of the planning function, a multidisciplinary staff is preferred. Typically, a statewide transportation planning unit will be comprised of persons from a variety of disciplines including transportation planners/engineers, environmentalists, transportation economists, public finance specialists, behavioral psychologists, and sociologists.

It is also important that policies and plans be developed with broad involvement of other agencies, interest groups, and individual citizens. Generally, the following types of agencies, groups, and individuals should contribute to the planning process (2):

1. The state legislature.
2. General policy and budgetary groups in the executive branch of the state government.
3. Other state agencies such as the industrial development department, public service commission, recreation commission, and travel commission.

4. Regional and local planning commissions.
5. Special interest groups.
6. Private transportation companies.
7. Multistate coordinating groups.
8. Citizens and citizens groups.

Planners of federally assisted projects in the United States must now coordinate the planning activity through an areawide or state clearinghouse (5). In metropolitan areas, the clearinghouse is an areawide agency (e.g., a regional planning commission) that has been recognized by the Office of Management and Budget as an appropriate agency to evaluate, review, and coordinate federally assisted programs and projects. In nonmetropolitan areas, the clearinghouse is a comprehensive planning agency designated by the governor or by state law to review and coordinate such activities.

Typically, coordination with public agencies, private groups, and individuals is accomplished by the establishment of two or more committees. For example, a technical review committee might be formed consisting of individuals representing governmental agencies, commissions, and so forth. A citizens advisory committee could be formed with representatives of civic organizations, special interest groups, and citizens. It will be recalled from Chapter 4 that current regulations require that citizens be given the opportunity to participate in public hearings regarding the construction of certain federally assisted highway projects. Such hearings give the citizens the opportunity to comment on such projects at the corridor or design stage.

POLICY PLANNING

6-3 The Policy Planning Process

Policy planning is a deliberate process that leads to a set of coordinated policy decisions designed to achieve a defined set of goals and objectives (2). It establishes broad direction for a state or local government in transportation matters. It is a means of describing in a general way what is to be done, who is to do it, how it is to be done, and within what constraints.

Policy planning includes the following types of activities (6):

1. Identifying problems, options, and alternative courses of transportation actions.
2. Interpreting information and analytical data for policymakers and citizens.
3. Monitoring and analyzing federal legislation and evaluating its impact on state or local policy.
4. Providing resource materials for management and public information personnel.

6-4 Goals and standards of transportation planning

An important function of policy planning is the establishment and articulation of goals or objectives. To planners, goals are expressions of purposes toward which plans are directed. Transportation plans are based on such goals as reducing traffic congestion, providing safe travel, conserving energy, fostering economic growth, and preserving and enhancing the environment. To the extent possible, goals or objectives should be stated in specific and measurable terms and should be accompanied by standards or criteria by which goal performance is to be measured. For example, one of the objectives listed by the state of Wisconsin was "to provide a plan which is both economical and efficient, satisfying all other objectives at the lowest possible cost" (3). The standards used to measure that objective were as follows:

1. The sum of the highway transportation system's capital costs and users' operating costs should be minimized.
2. The fullest economic use should be made of existing and committed major highway facilities.
3. Additional highway facilities should be provided to serve or induce future travel demand at the desired level of service.

SYSTEMS PLANNING

Great differences exist among the various highway agencies at the systems level of planning in terms of both needs and level of sophistication employed. At this level of planning, transportation links and networks are defined, forecasts of future traffic are made, and the physical facilities required to serve future needs are described.

Traditional planning procedures, which are still employed in a number of states, establish needs for improvements on a link-to-link basis rather than systemwide. In those states, highway needs are established to some extent subjectively on the basis of various surveys including highway inventories, traffic surveys, road-life studies, and highway sufficiency studies. Data from such surveys provide a basis for projections of trends and the identification and ranking of highway improvement needs.

Increasingly, state agencies are performing statewide transportation planning by employing planning procedures that have been widely used to develop urban transportation plans. Typically, such procedures consider transport needs on a network or systemwide basis and include the following sequential activities or steps:

1. Data collection.
2. Data analysis.
3. Plan generation and evaluation.
4. Implementation of the plan.

6-5 Collection of data to measure existing transportation conditions

To plan intelligently, planners must have a great deal of data on the transportation system and its use. They obtain such data by making studies on the extent, use, and condition of existing facilities; the life expectancy of the system; and the nature and extent of traffic served. Some of these studies are described in the following paragraphs.

Road inventory

For many years, various highway agencies have conducted road inventories, recording information on the mileages, physical characteristics, and condition of all roads in the system. The following types of information should be collected in a road inventory:

1. The functional roadway class.
2. The classification by type of pavement.
3. The type, dimensions, and condition of all structures.
4. The locations of farms, rural dwellings, schools, churches, and other cultural features.
5. The physical design features including pavement widths, grades, curvature, sight distances, location of roadwide obstacles, and other hazards.

Many highway agencies display road inventory information on maps similar to that shown as Figure 6-1. They make other maps by reproducing the base maps and adding other desired information. For example, the agency may make maps showing (1) surface conditions and type of surface, (2) common carrier, bus, and truck routes, (3) postal and school bus routes, and (4) average daily traffic.

Highway agencies should maintain an up-to-date inventory of roads under their jurisdiction. If this has not been done, a road inventory should be made and maintained.

Road-use studies

Highway planners perform road-use studies to determine the relative use of various parts of a highway system. Information may be obtained by personal interviews with a representative sample of the registered motor vehicle operators within the highway jurisdiction. Drivers are asked to state the total mileage driven in one year and the proportion of that mileage that was driven on various classifications of streets and highways. The planners can use such information to determine the proportionate use made of the various roads by urban, suburban, and rural residents. By correlation with financial data, planners may then compare the benefits received with the revenues contributed by various groups of citizens. Many states use such information to more equitably distribute the costs of highway transportation among the various benefiting groups in accordance with the concepts presented in Chapter 2.

Road-life studies

State highway agencies have been conducting road-life studies since the mid-1930s. The primary objectives of such studies are to determine the average rate of retirement and the estimated average service life for each type of highway surface. Roads are usually retired by resurfacing, reconstructing, or

FIGURE 6-1 A general highway map. (Courtesy Georgia Department of Transportation.)

abandoning them. Information is also assembled on costs of construction, maintenance, and depreciation.

With this information, planners may estimate the amount and cost of replacements that will be required in the future. They can program construction and reconstruction operations on the basis of anticipated future revenue receipts. Highway engineers use road-life data in economic calculations, as the previous chapter indicated.

Traffic volume studies

Engineers and planners have been conducting traffic volume studies for more than 40 years. The procedures used in such studies are well documented elsewhere (7,8) and will be only briefly described in this section. Traffic volume studies provide highway officials with essential information on the amount of usage of the roads in the highway system. Such information is needed for "the determination of design standards, the systematic classification of highways and the development of programs for improvement and maintenance" (9). From traffic volume studies, planners may estimate the vehicle-miles of travel on the various classes of rural and urban roadways. Such travel data indicate a measure of service provided by the system and facilitate the appraisal of safety programs and the development of equitable highway finance and taxation programs.

The fundamental measure for traffic volume studies is annual average daily traffic (*ADT*). By using empirically-based relationships, *ADT* values can easily be converted to peak hourly volumes.

Because of the wide variability of traffic flows during different times and at different locations, it is practically impossible to arrive at a true accounting of the traffic volumes on each section within a highway system. Even if comprehensive and continuous monitoring of traffic were possible, the costs of maintaining such a counting program would be prohibitive. Traffic volume studies are therefore carried out on a sample basis.

Many highway agencies have established statewide or areawide traffic counting programs. In such programs, traffic volumes are measured by means of a large number of short counts covering an entire state or area. The locations for the counts are called *coverage stations*. Coverage stations are typically spaced at intervals of 2 to 5 miles along rural highways and about 1 mile apart along urban highways and streets. Measurements at coverage stations are typically recorded on portable counters actuated by pneumatic detectors. At coverage stations, the traffic volume is not monitored continuously but is sampled over time. The duration and frequency of such counts generally depend on the class of the street or highway and the amount of traffic it serves. Typically, coverage counts consist of a single nondirectional 24- or 48-hr count repeated at four-year intervals.

In order to convert coverage counts to estimates of average daily traffic, traffic volumes are measured at a small number of strategically selected *control stations*. At those locations, traffic counts are usually made over a 7-day period to measure hourly and daily traffic patterns of variation. In addition, 24-hr nondirectional counts are normally made at control stations on a monthly basis to measure the seasonal variation. On the basis of temporal patterns of variation measured at the control stations, the sample coverage counts can be adjusted by correction factors to provide estimates of *ADT* throughout the system.

In a number of highway jurisdictions, permanent counting stations have been established at a small number of selected locations. At those locations, traffic volumes are measured continuously, providing an accurate measure of the hourly, daily, and seasonal patterns of variation. Generally, automatic traffic detection and recording equipment is employed at continuous counting stations, and in some states traffic volume data are transmitted to a central computer over telephone lines for processing and analysis (*10, 11*).

Traffic counts made with automatic devices generally must be supplemented by manual surveys to record information on the character of traffic. Such surveys are often undertaken at weighing or loadometer stations, There, information is collected on the type of vehicle, its length, width, and height, the rated capacity, and the axle and gross loads. Additional information on the type of commodity carried, the origin, the destination, and other pertinent facts may also be noted. When all the information of the traffic survey has been assembled, the expansion factors have been determined, and the annual average 24-hr traffic has been computed, traffic maps and flow maps can be prepared. A base map previously prepared by the road inventory survey is used for the traffic map. The numerical traffic volumes for each segment of the road are usually placed on this map. For more rapid visualization, the volume of traffic may be shown on the traffic flow map by bands proportional in width to the volume of traffic passing through the various points at which base station counts were taken. Figure 6-2 shows a typical traffic flow map.

Travel surveys

Travel surveys identify where and when trips begin and end, the trip purpose, the mode of travel, as well as certain social and economic characteristics of the trip maker. Such surveys may also determine the types of land use at trip termini, automobile occupancy, and, for freight, the type of commodity transported. Travel surveys serve as a fundamental source of data for urban transportation studies and are being increasingly used in statewide transportation studies (*10, 11*).

By means of travel surveys, planners attempt to obtain a sample cross section of all travel on a typical day within a specified area. By expanding the data, they are able to produce an estimate of the average travel demand on the transportation system at the time the survey was conducted.

There are four general classifications of travel surveys:

1. Household travel surveys.
2. Roadside surveys.
3. Modal surveys.
4. Goods movement surveys.

FIGURE 6-2 A portion of a traffic flow map. (Courtesy Georgia Department of Transportation.)

Household travel surveys These surveys are especially useful for determining the number and characteristics of person-trips or auto-driver trips made by residents within a specified study area. By such surveys, planners may collect extensive data on socioeconomic characteristics of households and relate such information to household travel. Such relationships are extremely valuable for making forecasts of future travel by persons who reside in the study area (*11*). Household travel surveys include home interview, telephone, and mail surveys.

The home interview method involves identifying a sample of all of the households within the survey area, arranging for interviews, and questioning respondents about trips made by persons in their respective households the previous day. The sample size varies with the population of the survey area, ranging from 20 percent in areas with a population less than 50,000 to 4 percent in areas with over 1 million. Home interviews will usually be preferred over other household survey methods if a long or complicated survey instrument is to be used. Higher response rates may be expected in home interview surveys than in telephone or mail surveys. On the other hand, home interview surveys are usually more expensive than other methods. DiRenzo (*11*) reported that the costs of interviewing and field work for three home interview surveys ranged, in 1974 dollars, from $10.09 to $23.79 per completed interview. He reported that the costs of telephone and mail surveys were $11.07 and $4.25 (1974 dollars) per completed interview, respectively. Such surveys typically have lower response rates and require greater care in their design and supervision to avoid underreporting of trips and the introduction of biases.

The roadside survey method This method consists of stopping vehicles and asking the drivers for information on trip origin, destination, and purpose. Typically, all of the vehicles passing the roadside survey station are counted, but only a sample of the drivers (e.g., 20 percent) are interviewed. Alternative roadside survey methods involve handing out voluntary return postcard questionnaires to drivers or recording license numbers and mailing questionnaires to the vehicle owners. The stations where the counts are made are usually located on all the main approach highways to the survey area. If the stations are located along an imaginary line encircling a survey area, they are referred to as "cordon stations," and the counts are known as "cordon counts." Roadside surveys are also used to intercept and interview drivers of vehicles crossing selected boundary lines (i.e., screen lines) within a state or region.

Roadside surveys are especially useful in recording relatively long and infrequent trips, including those made by persons who reside outside the survey area. Such surveys provide only limited information about the nature of the trips surveyed and practically nothing about the social and economic characteristics of the trip maker. Roadside surveys must be carefully planned and executed to prevent intolerable congestion, especially along heavily traveled routes.

Modal surveys In a large number of states where multimodal transportation agencies have been established, special surveys may have to be made of travel by bus, rail, and air transportation. Typically, modal surveys are made by interviewing passengers at terminals or while onboard public carriers. To a lesser extent, telephone interviews and surveys of license plates at terminals have been used to collect modal travel data. The U.S. Department of Transportation has published detailed guidelines on statewide modal surveys (*11*) and urban mass transportation travel surveys (*12*) to which the reader should refer for additional information.

Goods transportation surveys   The transportation of goods within a state or urban area is complex, involving several transport modes, a combination of line-haul and distribution traffic, and a variety of commodity types and users. The impact of goods movement on the functioning of a transportation system is significant, and planners should endeavor to incorporate goods transportation planning into the existing planning process. Unfortunately, planning procedures for goods transport analyses have not kept pace with those for passenger transportation.

Because of the competitive nature of the freight industry, freight carriers may be reluctant to respond to mailed questionnaires. For this reason, experience suggests that the personal visit–interview technique is the most effective means of gathering confidential information from transportation industry sources.

In certain instances, goods transportation surveys may be limited to the collection of origin–destination information for trucks by truckload type and weight. On the other hand, comprehensive multimodal goods transportation surveys may be desired in which detailed shipping information is sought for specific commodities, including consignee, shipper, carrier, commodity type, origin, destination, weight, value, time of origin and destination, and shipping rate. It may be possible to collect a great deal of terminal facility data on a confidential basis from terminal operators, including the number of movements per day (by mode type) and the class and volume of cargo handled during average and peak periods. Additional information on goods transportation surveys is given by Ref. 13.

Urban travel surveys   Before describing the analysis of systems planning data, let us look at how the various data collection procedures described above might be employed in an urban origin–destination (O–D) survey. One of the first steps in performing an urban O–D survey is to divide the geographic area under study into convenient planning units. It is necessary to establish the location of the external cordon boundary, an imaginary line encircling the survey area that separates the *external area* from the *internal area*. The external cordon line will normally encompass entire political jurisdictions, that is, towns, cities, census tracts, and so forth, and the "urbanized area" as defined by agencies such as the Bureau of the Census and the Federal Highway Administration.

The internal area is subdivided into analysis zones. The boundaries of these zones should, whenever feasible, coincide with census tract boundaries and will generally follow existing streets and highways. The zones should be small enough to avoid inclusion of a large number of intrazonal trips. The

FHWA (*14*) suggests that intrazonal trips should comprise not more than 10 to 15 percent of the total trips. The zones are generally systematically numbered to facilitate coding and analysis of travel data during the analysis phase.

Comprehensive urban O−D studies usually involve the use of a combination of the survey procedures described previously: household travel surveys, roadside surveys, modal surveys, and goods movement surveys.

Such surveys consist of two parts:

1. An external survey involving the collection of travel data on vehicles crossing the external cordon line.
2. An internal survey that focuses on the travel habits of persons who live within the area bounded by the external cordon line.

The external survey data are collected along the external cordon line. Directional traffic counts are normally obtained for every street and highway that crosses the cordon line. Origin−destination data are collected by roadside surveys performed on all major routes that cross the cordon line.

The internal survey consists of the following subdivisions: (1) home interviews (trips by all individuals), (2) truck studies, (3) taxi studies, and (4) public transit information.

Interviews are made on a predetermined schedule and forms are provided for tabulating the necessary information. The information required is for the day preceding the home interview. This information includes the number of occupants in a household, occupation, number making the trip, use of private or public transportation, origin and destination, purpose of the trip, and so forth.

In selecting the sample, use is made of certain available basic records. These may include (1) Sanborn maps, a series of copyrighted maps prepared for the use of insurance companies, assessors, and so forth, which show all structures together with other information; (2) city directories; (3) Bureau of the Census "Block Statistics," which show blocks and block numbers; and (4) land-use maps that may be available.

Truck and taxi information is generally available from the records of the companies that operate them. Usually 20 percent of the trucks and 50 to 100 percent of the taxis registered in an area are sampled. Public transit information is usually obtained directly from the transit companies. This should include location of existing routes, schedule of operations, and the total number of passengers carried on an average weekday during the period of the survey.

The completeness with which trips are reported in the internal survey may be checked by three methods: (1) use of control points for comparison, (2) a screen-line comparison, and (3) a cordon-line comparison. Not more than three or four control points should be selected for use in any one study. Control points are usually viaducts, bridges, underpasses, or other points of constriction through which large volumes of traffic pass. The average traffic volume and classification at each control point must be determined by manual counts made during the survey so that the expanded counts obtained from home interviews can be compared with the actual ground counts at these points.

Screen lines are natural barriers such as rivers or railroads; they usually have a limited number of crossings at which ground counts can be made at a minimum expense. The purpose of the screen line is to divide the area of the internal survey into two parts in order to determine the number of vehicles moving from one part to the other. A comparison can then be made between the number of trips having their origin on one side and destination on the other, as actually counted, and the number of trips as determined from the expanded interviews.

A cordon-line comparison deals with passenger-car trips by residents of the internal area and truck trips by trucks registered in the area, but only those trips which cross the cordon line. The total number of such trips recorded in the external survey can be compared with the total number recorded in the internal survey.

In many recent surveys conducted in various municipalities by the method just described, about 90 percent of the traffic passing control points and screen lines during 16 hr of operation of external stations has been accounted for by the expanded interview data. The remaining 10 percent has been assumed to be made up of cars circulating in search of a parking place, of unimportant short trips not reported in interviews, and of trips by persons living outside the area and not intercepted at the external cordon stations. These trips are considered to be of minor importance as far as the principal purposes of the survey are concerned.

Figure 6-3 shows a comparison of the number of vehicles counted at control points with the number passing those points as determined from the interviews.

FIGURE 6-3 Baltimore travel habit survey. Comparison of traffic passing three control points as reproduced from interviews and as actually counted at those points. (Courtesy Federal Highway Administration.)

For cities with a population of less than 5000 it may not be necessary to conduct an internal survey. It may be possible to collect sufficient O−D information by means of roadside interviews at the external cordon. In slightly larger cities (with populations in the range of 5,000 to 50,000, in lieu of an internal survey, roadwise surveys may be performed along two cordon lines, one at the edge of the city and one at the fringe of the central business district.

Origin−destination data have traditionally been displayed in the form of desire line maps. Desire lines indicate the desires of vehicle users as direct lines of travel from one point to another within a given area, assuming that direct routes are available. The information that these lines represent is especially helpful to planning and design engineers when routes for new arterial improvements are to be selected.

Plans for improvement should be keyed to the major demands of traffic, as indicated by the desire lines having the greater widths in a plot of the type shown in Figure 6-4. Assuming that this can be done, the routes thus improved should best meet the desires of the largest number of vehicle users.

Parking surveys

The purpose of a parking survey is to determine the parking habits and requirements of motorists and the relation of these factors to other uses of existing parking facilities.

A parking survey should be designed so that the information collected will provide needed data for the evaluation of the several factors in the parking problem. The information secured should include (1) the location, kind, and capacity of existing parking facilities; (2) the amount of parking space needed to serve present demands; (3) the approximate location of possible additional parking facilities; and (4) the legal, administrative, financial, and economic aspects of parking facilities.

The first phase of the parking survey is to make an inventory of all available parking facilities. This includes curb and off-street parking areas such as parking lots, garages, and service stations. Any physical and legal restrictions are also to be noted. The theoretical capacity of a parking lot or garage is usually determined by dividing the gross area by the area needed by one car, which varies from about 250 to 400 ft^2, depending on the type of facility and the design. (See Chapter 9.) Twenty-two lineal feet is usually used in determining theoretical curb capacities. Calculation of the theoretical capacity does not imply more cars cannot be accommodated. Some cars are parked for short periods of time, and their space then becomes available to others. Cooperation of the owners of off-street parking facilities must be obtained in order to determine the actual capacities or numbers of cars that are accommodated for periods of the survey so that the results can be correlated with capacities of other facilities. Additional information on the duration of parking is also to be obtained.

The determination of actual capacities of curb parking facilities is more readily done. Parking, when permitted, is usually for 1- or 2- hr periods. Therefore, the number of space-hours available is determined from the inventory of

FIGURE 6-4 Comparison of traffic on actual routes (above) and desire lines (below) of travel for Rome, Georgia. (Courtesy Georgia Department of Transportation.)

FIGURE 6-4 *(continued)*

the curb space available. It is assumed that parking is for the legal time specified. Studies are also made to ascertain the extent of illegal parking and its effect on the total effective capacity of curb parking facilities.

In order to determine the number of cars that may utilize available parking spaces in a central area, a cordon line is usually drawn around the area to be studied and cordon counts are made of all vehicles entering and leaving it. The counts are made manually so that a motor-vehicle classification can be made. Automatic recorders are often used to supplement the manual counts and to serve as a basis of control for abnormal conditions.

Motorists who are interviewed at all parking facilities within the cordon area are asked for their home address, destination, and the purpose of the trip. Figure 6-5 shows the results of a parking survey in the city of Baltimore.

A parking survey should also include a study of the area for further off-street parking facilities. This should include the investigation of vacant lots, obsolete buildings, and blighted residential or business areas within the area of the parking survey. Mention may be made of other parking areas outside the limits of the central shopping area that may also be used. These areas are usually referred to as fringe parking areas, the idea being that motorists can park near the central business district and use available mass transportation facilities as a means of arriving at their destination. Many municipalities have tried this type of facility, but as a general rule it has not proved as successful as desired.

In order to be properly evaluated, a parking survey should also include an inventory of the state laws, local ordinances, and judicial decisions pertaining to parking facilities. This should include public regulations for the licensing of commercial off-street parking facilities, curb restrictions, and police regulations.

An investigation into the administration of parking facilities should be made in order to determine which state and local agencies are responsible for the planning, location, financing, construction, operation, and maintenance of parking facilities within the municipality that is being surveyed. For example, in a given municipality the police may have the responsibility for enforcing curb parking restrictions, while a traffic commission may have the authority to establish such parking regulations, including meter parking. The zoning commission may require parking provisions under zoning laws for various property uses. All administrative facts should be included so that in the final analysis recommendations may be made to remedy any deficiencies that may exist.

In matters pertaining to the financing of parking services, studies should include the aggregate revenue from parking meters and its disposition as well as the assignment of costs of parking facilities for off-street parking according to the benefits received. Benefits may be apportioned to property owners, business establishments, the motorists, the general community, and the municipality itself.

The information received in a parking survey may point to a number of recommendations for the relief of congestion and for general improvement in traffic facilities. Such recommendations might, for example, include the provi-

FIGURE 6-5 Parking survey of number of passenger cars and trucks entering and leaving the downtown area of Baltimore on a weekday between 10:00 A.M. and 6:00 P.M.

sion of additional parking facilities, either publicly or privately owned, better enforcement of parking regulations to make available space at the curb now being used by motorists parking illegally, changes in zoning regulations, and so forth.

6-6 Data analysis

Because of the magnitude and changing nature of the problem, it is difficult to prepare an accurate up-to-date inventory of an existing transportation system and the traffic it serves. It is even more difficult to accurately forecast future traffic that will use a proposed transportation system. Transportation planners face the challenge of making reliable forecasts of traffic demand that reflect the effects of changes in population and social and economic conditions as well as changes in the physical transport system. Unless reliable forecasts of future traffic are made, transportation officials face the risk of building facilities that will either receive little use or be prematurely overloaded.

Transportation planners have developed methodologies designed to forecast traffic demand for a given route or corridor as well as for an entire transportation system (network).

The traditional approach, employed principally for rural facilities, has been to forecast traffic for a specific highway section by subdividing the traffic into its various components and to make separate projections for each component. The recognized components of future traffic for a new or improved facility include:

1. *Existing traffic.* Traffic currently using an existing highway that is to be improved.
2. *Normal traffic growth.* Traffic that can be explained by anticipated growth in state or regional population or by areawide changes in land use.
3. *Diverted traffic.* Traffic that switches to a new facility from nearby roadways.
4. *Converted traffic.* Traffic changes resulting from change of mode.
5. *Change of destination traffic.* Traffic that has changed to different destinations, where such change is attributable to the attractiveness of the improved transportation and not to changes in land use.
6. *Development traffic.* Traffic due to improvements on adjacent land in addition to the development that would have taken place had the new or improved highway not been constructed.
7. *Induced traffic.* Traffic that did not previously exist in any form, but results when new or improved transportation facilities are provided.

In recent years, planners have developed methodologies for estimating the distribution of future traffic over an entire transportation network. These procedures, which have been used for both urban and statewide (3) systems, involve the use of computer simulation programs, comprised typically of five types of models:

1. Land use.
2. Trip generation.

3. Trip distribution.
4. Traffic assignment.
5. Modal split.

The models are mathematical equations and procedures that collectively relate travel patterns to land use, demographic characteristics, and parameters of the transportation system. The models are developed and "calibrated" for a given study area so as to reproduce existing travel patterns as determined from actual counts. Assuming the basic relationships between travel, land use, and socio-economic characteristics remain constant over time, planners use the models to evaluate future alternative land use and transportation systems.

Land-use models

"The land use model is a procedure which estimates future development by analysis zone. These estimates include not only land use *per se*, but also estimates of the socioeconomic variables which are used in the trip generation models, such as population, dwelling units, auto ownership, income, employ-ment, retail sales, etc." (*15*). Such estimates are normally made by economic or demographic planners rather than by highway or transportation specialists.

Trip generation models

Trip generation models provide a measure of the rate of trip-making for each analysis zone. Trip generation rates, which vary with trip purpose, are normally expressed as a function of land-use and demographic parameters. Studies have shown that, within an urban area, trip generation values are most closely related to three characteristics of land use: (1) intensity of land use (dwelling units per acre, employees per acre, etc.); (2) character of land use (e.g., average family income, car ownership); and (3) location relative to the city center.

Trip generation relationships "usually take the form of mathematical equations of several independent variables, or tables that classify each zone or dwelling unit according to its characteristics and give the number of trips which may be expected to begin and end at the zone or dwelling unit (trip ends)" (*15*).

An example of the procedure used in estimating trips produced by a residential zone is shown as Figure 6-6.

Trip distribution models

Trip distribution models begin with the number of trip ends generated by each zone and answer the question, "What zone are the trips going to and coming from?" Several trip distribution models have been described in the literature, but only two of the most prominent models will be described here—the gravity model and the Fratar method.

The gravity model The gravity model is so named because of its similarity to Newton's law of gravitation. Employed first for sociological and marketing research, the gravity model began to be used for transportation studies in the early 1950s. Since that time, the model has been slightly modified and has

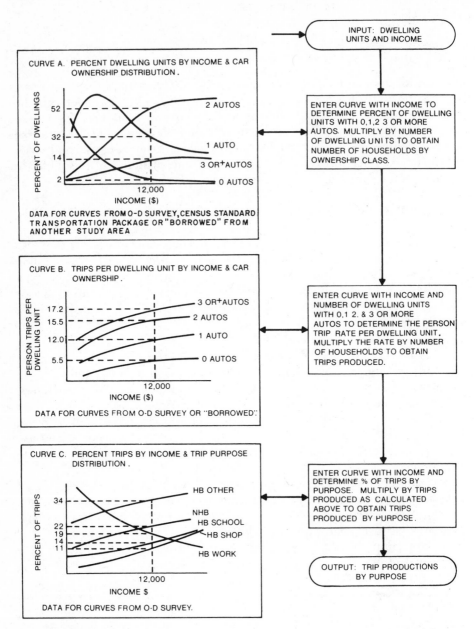

FIGURE 6-6 Example of urban trip production procedure. (Courtesy Federal Highway Administration.)

become the predominant technique for trip distribution. The original version of the model, which was introduced by Voorhees (*16*), was of the form:

$$T_{ij} = \left[\frac{\dfrac{A_j}{(D_{ij})^n}}{\dfrac{A_1}{(D_{i1})^n} + \dfrac{A_2}{(D_{i2})^n} + \cdots + \dfrac{A_m}{(D_{im})^n}} \right] P_i \qquad (6\text{-}1)$$

where

T_{ij} = trips from zone i to zone j for a specified purpose
P_i = total trips produced at zone i for the specified purpose
A_j = a measure of attraction of the jth zone for trips of this purpose
D_{ij} = distance from zone i to zone j
n = some exponent that varies with trip purpose

Consider the following numerical example. Given a residential zone that produces a total of 110 shopping trips per day, distribute these trips to shopping centers 1, 2, 3 in accordance with the gravity model. Distances between zones are shown on the sketch. The value of n in the gravity model is 2. Use the amount of commercial floor space within the destination zone as the measure of attractiveness:

Shopping Center	Floor Space (thousand ft²)
1	184
2	215
3	86

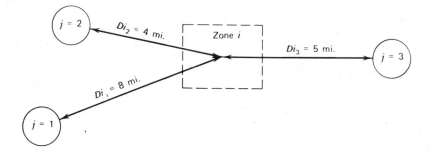

$$\text{Trips from zone } i \text{ to zone } 1 = \frac{\dfrac{184}{(8)^2}}{\dfrac{184}{(8)^2} + \dfrac{215}{(4)^2} + \dfrac{86}{(5)^2}} \times 110 = 16$$

$$\text{Trips from zone } i \text{ to zone 2} = \frac{\dfrac{215}{(4)^2}}{\dfrac{184}{(8)^2} + \dfrac{215}{(4)^2} + \dfrac{86}{(5)^2}} \times 110 = 75$$

Similarly, the trips form zone i to zone 3 $\qquad = \underline{\quad 19 \quad}$

$\qquad\qquad\qquad\qquad\qquad\qquad$ Total trips $= 110$

The gravity model has been modified in recent years to reflect research and experience with the model. It has been found that decreases in travel propensity are more closely related to travel time than to distance. In addition, it has been established that the exponent of travel time, n, varies not only with trip purpose but also with trip length. Trip distribution analyses are therefore usually stratified according to travel time t with different calibrated values of the exponent being determined for a given city and trip length. Furthermore, to facilitate efficient computer use of gravity models, it is now the practice to represent the effect of spatial separation on travel between zones in the form of travel time factors

$$F_t = \frac{C}{t^n}$$

where C is a constant. Instead of a surrogate measure of attractiveness such as commercial floor space use or number of employees, actual zonal total trip attractions are used in the equation. Current gravity models permit an analyst to make adjustment to allow for special social or economic conditions by choice of socioeconomic adjustment factor.

Currently, the recommended formulation of the gravity model is

$$T_{ij} = \frac{A_j F_{t_{ij}} K_{ij}}{\displaystyle\sum_{\text{all zones}} A_x F_{tij} K_{ix}} \times P_i \tag{6-2}$$

where

$F_{t_{ij}}$ = travel time factor for travel time between zones i and j = $\dfrac{C}{t_{ij}{}^n}$

K_{ij} = socioeconomic adjustment factor between zones i and j

A_j = total attractions at zone j

The Fratar method   Proposed by T. J. Fratar in 1954, this method is designed to compute trip interchanges where there is nonuniform growth within various sections of a study area. This method and variations on it are called "growth factor" methods. A growth factor for a particular zone is simply the ratio of expected future traffic to the existing traffic emanating from the zone. According to the Fratar method, future travel patterns between zones are determined by the present travel patterns and growth factors at the destination zone. The method is an iterative factoring process in which the number of future origins at each zone is held constant. It is analogous to the Hardy Cross method of successive approximations for moment distribution in indeterminate structures. In recent years, the method has been principally used to predict trip interchanges between external sections of a study area. It has also been employed for statewide transportation studies. For further information on the Fratar method, the reader should consult Ref. 17.

Traffic assignment models

Traffic assignment is defined as the process of determining the routes of travel and allocating the zone-to-zone trips to these routes. The process is one of the most important and complex phases of transportation planning. It is a systematic and reproducible technique that enables the planner to predict the probable traffic loads on each segment of a transportation network. The costs and operational efficiency of various system designs can be compared and evaluated and, after proper analysis, the results may be utilized to identify changes that would improve the system.

Traffic assignment is a computerized process, and the planner must first describe the street system in terms that will facilitate computer processing. The street or highway network is defined by "nodes" and "links." A node is a point at which two or more route sections meet, allowing for a change in travel direction. A link is a one-way part of a route that lies between two intersections or nodes. Contingency checks are made to ensure that the coded system is free of anomalies and errors. Then for each link the following information is usually stored in the computer: (1) the length of the link, (2) the travel time or speed, and possibly (3) the capacity and existing volume.

The computer selects the minimum-time paths by systematically searching travel time information stored in its memory. All the minimum-time path routes from one loading node to all others constitute a "tree," and the process is called "tree building." In this process, the minimum-time path and travel time is recorded. After the trips have been assigned from one zone, the computer then selects the next zone and repeats the process.

In assigning traffic to various routes, it is sometimes assumed that all drivers would choose the route with the least travel time, and assignments are made on an "all-or-nothing" basis. That is, all the trips between a given pair of nodes will be assigned to the minimum-time path and none to the next shortest time path. More commonly, trips are proportionately assigned to the two best routes on the basis of travel time or distance or both. Empirical diversion curves have been developed to determine what proportion of a movement should be assigned to the shortest path.

Some of the traffic loads on the individual links may exceed the capacity of the transportation facilities. This would affect the travel time and possible change the minimum-time paths. Travel times are therefore adjusted and new minimum paths are selected by using the adjusted times. When this is done automatically, it is called "capacity restraint" (*18*).

Modal split models

Modal split models are used to estimate the proportion of future (person) trips that will be made by transit and by private automobile. Such models usually classify trips or trip-ends by type and economic status of the tripmaker and provide an estimate of percentage of travel by transit. The estimate is based on some characteristic of the transportation system such as relative travel times, distances, or costs. The models take the form of mathematical equations or empirical tables and curves.

There are two general classes of modal split models: (1) trip-end models, in which the *origins and destinations* are divided by mode, that is, before trip distribution; and (2) trip-interchange models, in which the trip *movements* forecast by a trip distribution are divided by mode.

Empirical evidence indicates that the following factors influence modal choice (*19*):

1. Type of trip (e.g., trip purpose, time of day).
2. Characteristics of the tripmaker (e.g., income, age, auto ownership, residential density).
3. Characteristics of the transportation system (e.g., ratio of transit travel time to auto travel time).

Interestingly, an opinion survey (*20*) of 90 professional planners and engineers revealed that the most important mass transportation attributes are

1. Reliability (arriving on time).
2. Safety (crime, accident, etc.).
3. Door-to-door travel time.
4. Total time spent waiting.
5. Rider's attitude toward public transit.
6. Parking availability at suburban terminals.
7. Terminal access and location.

Many attributes related to comfort and convenience, such as adjustable seats, attractiveness of the vehicle, and providing music, were ranked relatively low by the respondents.

Many planners believe that current modal split analysis techniques are inadequate and unreliable, especially in predicting trips by "choice" riders (*20*). Further research is needed to develop modal split models that will take into consideration the effects of technological innovation, public policy, and sociological changes.

6-7 Plan generation and evaluation

The next step in the transportation planning process involves the generation and evaluation of alternative transportation plans. A transportation plan consists of a set of proposed actions to improve the transportation system as well as policies to protect and control the construction and improvement of the system (*3*).

Planners begin with the existing transportation system plus any "committed" facilities. Committed facilities are those that have not been built but have progressed through the planning, design, and property acquisition stages to such an extent that a policy decision has been made to build them regardless of the outcome of the current study (*15*). To the existing and committed system, future traffic demand is applied by traffic assignments. This will likely reveal deficiencies in the existing and committed network (e.g., high volume/capacity ratios). A new transportation plan is formulated and evaluated on the basis of goals, objectives, constraints, and evaluation criteria. See Table 6-1. Such a process is continued iteratively, employing planning models previously described. To the extent possible, the consequences of each alternative plan should be established, preferably in quantitative terms (hours of delay, numbers of persons killed in traffic crashes, numbers of families displaced, pounds of emissions, etc.). The impacts from each alternative plan should be evaluated in terms of the goals that were previously established. From such analyses, several preferred alternative plans may evolve, from which the "best" plan must be chosen. It may then be desirable to rank the preferred plans quantitatively by means of attitude surveys or a rating panel similar to the procedure described by Jessiman et al. (*21*).

6-8 Implementation of the plan

Implementation involves those activities that are necessary to put the transportation plan into effect in an orderly manner. Planners must accomplish this step cooperatively with others as the planning process is followed by other phases of development (location, design, property acquisition, etc.). At this point, local rather than regional goals and objectives must be defined and interpreted in light of the proposed transportation improvements.

It is vitally important that the improvements called for in the transportation plan be made with a minimum of disruption and annoyance to the public. Priorities should be established, and a capital improvement program should be developed in which construction projects are staged in order of those priorities. Traditionally, priorities have been "assigned to improvement projects largely on the basis of subjective judgments developed from past experience. Priorities that are established subjectively run the risk of personal engineering bias, lack of comprehensiveness, and political bias. Furthermore, the increasing number, magnitude, and complexity of the programs will soon make subjective priority analysis unmanageable" (*22*).

> Frequently, the (programming) process is based on an examination of base year (current) congested facilities, and target year (future) volumes on facilities. Improvements that are expected to relieve current congestion receive highest priority, then the other improvements are given importance related to their future volumes. Considerations of continuity of the system while being constructed and the distribution of capital costs over the years are employed to convert the priorities into five-year construction programs.(*15*)

A number of states have performed highway sufficiency analyses as a means of identifying where the highway needs are greatest. Such studies

TABLE 6-1
EXAMPLES OF GOALS STATEMENTS FOR THE EVALUATION OF TRANSPORTATION SYSTEMS[a]

Goals	Objectives	Evaluation Criteria	Constraints
Provide economical transportation	Maximize ratio of benefits to costs	Benefit–cost ratio	B/C ratio < 1.0
Provide efficient transportation	Minimize highway congestion	Travel volume in peak period	Vol./capacity ratio ≦ 0.75
Provide safe transportation	Minimize losses from highway crashes	Number of accidents per vehicle-mile	Accident rate in target year < acc. rate in base year
Reduce environmental pollution	Minimize vehicular emissions	Pounds of emissions	Emissions in target year < emissions in base year

[a]After *A Policy on Design of Urban Highways and Arterial Streets*. American Association of State Highway and Transportation Officials (1973).

employ a rating scheme that attempts to classify the segments of an existing highway system in a manner that is unprejudiced, objective, and uniform as possible. Various elements of highway sections are evaluated to obtain a measure of condition or structural adequacy, safety, and service. A composite score is calculated for each project and the projects are then ranked according to their scores.

Other states have performed economic analyses and ranked various projects in accordance with their economic importance. Typically, when this approach is used, benefit–cost ratio or rates of return, computed as described in Chapter 4, serve as the basis for priority assignment.

Mak and Jones (22) proposed the use of a priority analysis procedure that incorporates a number of intangible parameters such as continuity considerations, state and local political reactions, and social, economic, and environmental consequences.

6-9 Maintaining the plan

Finally, it should be emphasized that transportation planning is a continuing and dynamic process. Plans should be constantly reappraised and modified to reflect changes in levels of funding, land use, social and economic conditions, and community, state, and national goals. Surveillance and updating of inventory data and periodic reappraisal of the transportation plan are required if it is to be responsive to public wishes and needs.

6-10 Transporation planning software

A great deal of research effort has focused on the development of standard computer programs for transportation planning. For example, the Urban Transportation Planning System (UTPS) has been made available by the Urban Mass Transportation Administration and the Federal Highway Administration for the support of systems planning (23). Designed for mainframe computers, UTPS consists of computer programs, documentation, users' guides, and manuals. It is suitable for use where the alternatives to be analyzed are few in number, and the level of detail is sufficient to permit estimates of system costs, levels of service, major facility and corridor volumes, and social and environmental impacts.

Beginning in the 1980s, transportation planners turned increasingly to the use of microcomputers in transportation planning. Extensive software has been developed for microcomputers and is described in published reports of the Department of Transportation (DOT), for example, Ref. 24. The Federal Highway Administration funds a microcomputer user support group in transportation which provides software and technical assistance and publishes a user group bulletin called *Microcomputers in Transportation* (25).

Microcomputers are especially suitable for short-term planning and for the analysis of traffic impacts of proposed highway system changes along corridors or in other small to medium-sized planning subareas. The smaller computers hold the promise of providing analytical capabilities in transportation planning to many more agencies at low cost and with fast response times.

PROBLEMS

6-1 A household travel survey is to be made for a city with a population of 42,000. Estimate (in today's dollars) the cost of collecting data by home interviews, by telephone survey, and by mail survey. Assume an inflation rate of 6 percent per year and an average family size of 3.5.

6-2 Obtain a highway map and plan a travel survey for a nearby county or city using the roadside survey technique. Show the locations of survey stations on the map and prepare a report describing the planned survey times and durations and sample size. Indicate the size of survey crews and a description of duties for each crew member. Describe how the survey data could be used to estimate average daily traffic and design hourly volumes for the roads surveyed.

6-3 From current transportation literature, prepare a report briefly describing at least four trip distribution models.

6-4 Referring to the sketch shown in Section 6-6, calculate the interzonal trips due to 450 work trips produced at zone i. There are 750 attractions at zone 1, 400 attractions at zone 2, and 300 attractions at zone 3. The exponent of travel time is 0.6, and the travel times are: to zone 1, 9 min; to zone 2, 5 min; and to zone 3, 7 min. Assume all socioeconomic adjustment factors = 1.0.

6-5 Learn how the local transit agency measures the level of service provided to various areas of the city. Make a list of various measures that can be used to describe the amount of bus service provided to a given area or region.

REFERENCES

1. Creighton, Roger L. State of the Art in Statewide Transportation Planning. *Issues in Statewide Transportation Planning.* Transportation Research Board Special Report No. 146 (1974).

2. Pyers, Clyde E. Workshop I: Organization and Administration for Statewide Transportation Planning. Transportation Research Board Special Report No. 146 (1974).

3. *Statewide Transportation Planning Needs and Requirements.* National Cooperative Highway Research Program Synthesis of Highway Practice 15, Highway Research Board, Washington, D.C. (1972).

4. Sturm, Byron, D. Resource Paper. Transportation Research Board Special Report No. 146 (1974).

5. *Intergovernmental Review of Federal Programs.* Executive Order 12372, *Federal Register,* Vol. 48, No. 16, January 24, 1983.

6. Campbell, E. Wilson. Workshop II: Policy Planning. Transportation Research Board Special Report No. 146 (1974).

7. *Transportation and Traffic Engineering Handbook.* Institute of Transportation Engineers, Washington, D.C. (1976).

8. *Manual of Traffic Engineering Studies,* 4th ed. Institute of Transportation Engineers, Washington, D.C. (1976).

9. *Guide for Traffic Volume Counting Manual.* Bureau of Public Roads, Washington, D.C. (1965).

10. Hazen, P. I. A Comparative Analysis of Statewide Transportation Studies. *Highway Research Record No. 401* (1972).

11. DiRenzo, John F. *Travel Survey Procedures for Statewide Transportation Planning.* Federal Highway Administration, Washington, D.C. (1976).

12. Urban Transportation Systems Associates, Inc. *Urban Mass Transportation Travel Surveys.* U.S. Department of Transportation, Washington, D.C. (1972).

13. *Goods Transportation in Urban Areas.* Informational Report, Institute of Traffic Engineers, Washington, D.C.

14. *Urban Origin Destination Surveys.* Federal Highway Administration, U. S. Department of Transportation, Washington, D.C.

15. *A Policy on Design of Urban Highways and Arterial Streets.* American Association of State Highway and Transportation Officials, Washington, D.C. (1973).

16. Voorhees, A. M. A General Theory of Traffic Movement. *Proceedings of the Institute of Traffic Engineers,* Arlington, Va. (1955).

17. Fratar, T. J. Vehicular Trip Distribution by Successive Approximations. *Traffic Quarterly.* Vol. VIII, No. 1, pp. 53–65 (1954).

18. *Traffic Assignment Manual.* Bureau of Public Roads, Washington, D.C. (1964).

19. Fertal, Martin J., Weiner, Edward, Balek, Arthur J., and Sevin, Ali F. *Modal Split.* Federal Highway Administration, Washington, D.C. (1970).

20. Wallin, Rex J., and Wright, Paul H. Factors Which Influence Modal Choice. *Traffic Quarterly,* (April 1974).

21. Jessiman, William, et al. A Rational Decision-Making Technique for Transportation Planning. *Highway Research Record No. 180* (1967).

22. Mak, K. K., and Jones, Paul. Priority Analysis Procedure for Ranking Highway Improvement Projects. *Transportation Research Record 585* (1976).

23. *Urban Transportation Planning System: Introduction.* Urban Mass Transportation Administration and Federal Highway Administration, Washington, D.C. (December 1982).

24. *Software and Source Book: Microcomputers in Transportation.* Urban Mass Transportation Administration, U.S. Department of Transportation, Washington, D.C. (February 1986).

25. *Microcomputers in Transportation.* User Group Bulletin, published periodically by the Federal Highway Administration, Washington, D.C.

chapter seven

COMPUTER APPLICATIONS IN HIGHWAY ENGINEERING*

One of the most challenging aspects of an engineering career, including highway engineering, is trying to keep up with the "state of the art." One of the major changes in this state of the art is the use of computers in the highway design process. The "computer revolution" has come to the point where the typical day of a highway engineer usually includes some work on a computer.

Most of the highway design presented in this textbook used to be done just as you have learned it—by hand, with the help of a hand-held calculator. The computer is also a tool, but with much greater capabilities. Given the right combination of computer equipment (hardware) and programs that tell the computer what to do (software), *all* of the highway design taught in this text can be performed by a computer.

Gone are the days when the highway designer did nothing but calculate and draw (or tell the draftsman what to draw). The engineer (or draftsman) is now able to communicate his or her ideas directly with a computer and see the results, as shown in Figure 7-1. The Intergraph Drafting Systems Station shown in the photograph is not in some experimental research laboratory, but instead is in the offices of many state departments of transportation. One change in any aspect of a design can be made and the ramifications of that change for other aspects of the project can be seen quickly and in many different ways, including three-dimensional simulation with resolution rivaling that of a photograph.

This is what the highway designer works with daily, and it shows the importance of giving you an introduction to what you will see after graduation.

GENERAL COMPONENTS OF COMPUTERS

Any computer, regardless of size, has several different parts that make up what is known as "the computer." Each part is unique in function and is discussed separately below.

7-1 Input/output devices

The input/output (I/O) devices are what the user comes into contact with most of all. They exist mainly for the benefit of the human user and usually are not necessary for the computer to carry out its main task, that is, handling information.

*By John L. Hibbard and Peter S. Parsonson.

164

FIGURE 7-1 An Intergraph drafting systems station. (Courtesy Texas State Department of Highways and Public Transportation.)

Video display terminal (VDT)

The VDT, or cathode ray tube (CRT), is a very popular I/O device. The VDT most closely resembles a small television screen with a keyboard attached to it, and it is often mistakenly referred to as "the computer." It is a very versatile I/O device and is able to display both text and graphics (pictures or graphs generated by computer programs). The VDT's major drawback is that it does not have the ability to give the user a "hard copy" or "printout" of its contents without using some other sort of I/O device, like a printer.

A terminal, which is often a VDT, is an I/O device that has no computing ability on its own, but can show a computer's output and can transfer input to a computer. Figure 7-2 shows a terminal being used to review part of a topographic map. The VDT's ability to show both graphics and text at the same time is evidenced by the "K1 Command:" at the bottom of the screen.

Printing terminal

This is also an I/O device, like the VDT, but instead of displaying the input and output on a monitor, it displays the I/O to the user on paper, like a typewriter. In fact, many printing terminals look very much like typewriters. The printing terminal is not as popular as the VDT for several reasons: it usually prints slowly; it is more expensive than a VDT; and it is usually not able

FIGURE 7-2 A terminal being used to review a portion of a topographic map.

to print out graphics. The printing terminal also gives a hard copy of everything, whether the user wants it or not, since printing on paper is the only way it
can communicate with the user. Operating costs. consequently, are higher for a
printing terminal than for a combination of VDT and selectable printer.

Punched cards

This I/O medium has been rendered obsolete by the advent of low-cost
VDTs, printers, and magnetic media (tapes and disks). Even so, it is still
available for use on many mainframe computers. To use punched cards it is
necessary first to use a keypunch machine, where the user types commands
onto the punched cards, one command to a card.

The stack of cards, with punched holes on each card conveying the user's
command to the computer, are then stacked in a card reader, which looks at the
holes on the card and interprets them to the computer. The computer itself
may have a keypunch that can give the output of a program on punched cards,
if desired. Punched card systems have never been well liked, since a run can be
ruined if the cards are out of order of if a card is worn or missing.

Mass storage media

Disks and tapes will be discussed later as storage media, but they are also
important I/O devices. A program might tell the computer to look to a disk or
tape for some data, and the same program might tell the computer to write
output to a disk or tape. Using storage media like this to provide input and store

output from a computer program can enable the computer to run programs even when no user is actively giving the computer information—for example at night.

Modem

Modem is short for modulator–demodulator, and it is a device that enables a computer or a terminal to "talk" to another computer. A modem takes output from its computer (or terminal), translates it into audible tones (modulates the output), and sends it along telephone lines to another computer's modem, which translates the audible tones into electric pulses that its computer can understand (demodulates the data). A modem can, for example, connect a person using a small computer or terminal in New York with a large computer in San Francisco. Modems are very popular with microcomputer users as they enable the users to gain access to the vast memory and capabilities of a mainframe.

Digitizer

A digitizer is a specialized type of input device which is very useful to highway engineers. A digitizer enables the computer to "read in" data from a graph, map, or chart. To do this the document is placed on the surface of the digitizer, and then its boundaries are "marked" by placing a hand-held cursor on the corners of the document and pressing a button on the cursor. The user can then use the cursor to read in the desired data points within the predetermined boundaries. Although digitizers are quite expensive, their worth is easily seen when one is dealing with large irregular documents.

Printer

The printer is a popular (and potentially inexpensive) output-only device. Most printers fall into one of three main categories: dot matrix, letter-quality (daisy-wheel), and laser printers.

Dot matrix printers are cheap, fast, and usually able to print graphic output. Their main disadvantage is that when they are operated at top speed their print quality may not be acceptable for formal business correspondence. At lower speeds, however, even inexpensive dot matrix printers are able to print nearly as cleanly as a letter-quality printer.

Letter-quality printers work in much the same way as a typewriter, giving very clear text but at a low speed. They are usually unable to print graphics but are preferred for text printing. Most letter-quality printers in use today are of the daisy-wheel variety. A daisy-wheel printer has a printing element shaped like a daisy with the various characters located at the ends of the petals. To print a character the printer rotates the daisy-wheel to the proper position, and a small hammer strikes the end of the petal on its back, pushing the character against a ribbon and then against the paper (*1*).

Laser printers are a recent printing development and are rapidly gaining popularity. Although they are very expensive, advantages such as tremendous speed, quality, and quietness are worth the cost in many users' eyes. Laser printers work like some popular copying machines, even using the same "laser engine" to guide the print head to the prescribed point.

Plotters

A plotter is an output device that has the ability to draw intricate shapes and designs on paper, vellum, mylar, or transparent film. Maps, graphs, blueprints, and other charts can be generated on the plotter easily and with very high quality. Plotters are expensive, but they perform printing/drawing tasks very well and can be used on everything from microcomputers to mainframes. Plotters draw by using small ink pens, which have up to 14 different colors on some plotters. The computer tells the plotter to pick up a certain pen, move it to a certain point identified with *X-Y* coordinates, put the pen down, and draw according to instructions from the computer.

Figure 7-3 shows a "flat-bed" plotter, which operates by holding the drawing surface fixed on a flat bed and moving the print head (Figure 7-4) on the surface to create the desired picture. Figure 7-5 shows the output from this plotter, which is the geometrics of a freeway interchange.

7-2 Central processing unit

The central processing unit (CPU) is often referred to as the "brain" of the computer and is divided into two major units and several minor units. The first major unit is the arithmetic logic unit (ALU), which performs the arithme-

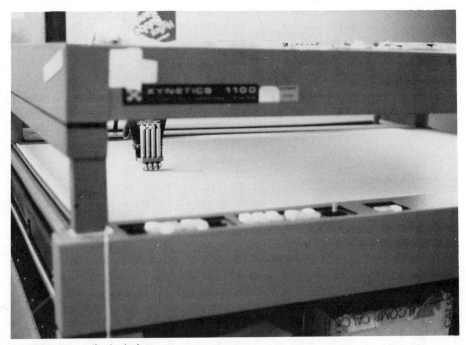

FIGURE 7-3 A flat bed plotter.

FIGURE 7-4 The print head on a flat-bed plotter.

tic computing tasks. The second major unit is the control unit, which organizes the processes of the ALU and consequently the functions of the entire computer. Smaller registers, also part of the CPU, provide the temporary data storage necessary to the functioning of the computer as well as advancing the program counter, which tells the computer to execute the next program line (2).

Microcomputers are often described on the basis of their CPU (microprocessor) "word size." The three main types of microprocessors popular as of the mid-1980s are referred to as 8 bit, 16 bit, and 32 bit. The "words" referred to here are not words such as those you are reading, but instead are groups of bits (0's and 1's) which might make up words or numbers or any type of character. Any microprocessor picks up the largest amount of bits that it can handle, manipulates them, and returns them to the computer's memory. Simply put, the larger the word size, the fewer cycles it takes for the microprocessor to perform a certain task, which means a shorter execution time (3).

Another characteristic often used to describe the rate at which a microprocessor can perform functions is its "clock speed." A microprocessor's clock is much faster than the clock you use to tell time; it functions at 5 to 10 MHz (million cycles per second). Obviously, the higher the clock speed, the faster the microprocessor does its job. Consequently, in terms of speed, the larger the word size and the higher the clock speed, the faster the computer.

FIGURE 7-5 Typical output from a flat-bed plotter. (Courtesy Georgia Department of Transportation.)

7-3 Internal memory

The next pieces of computer hardware to be discussed are the parts that store programs and data which the user plans to use in the near future. Also included here is the part of the internal memory which cannot be altered either by the computer or by the user, since it is essential for the proper functioning of the computer.

Read-only memory

Read-only memory (ROM) is the permanent memory, which resides within the computer on a series of silicon chips. "Written" on the chips are very basic instructions essential to the operation of the computer. Without the instructions contained in ROM, the computer would not know what to do when first switched on and would be useless. Most ROM information tells the computer to look to a certain input device or a certain place in memory, where the computer will obtain additional information. Consequently, this memory can be read only by the computer and can never be written over by the computer.

Random access memory

Random access memory (RAM), or read/write memory, is the memory space within the computer where programs and data are manipulated. RAM is like ROM in that it is contained in chips. The program or data can be loaded into RAM by using any of the various input devices discussed earlier. This memory is called random access memory because the computer has the ability to look at any one part of it (memory address) without looking at any of the other memory. RAM is volatile memory; that is, its contents are not retained when the computer is turned off. In addition, information in RAM can be written over if desired.

7-4 Mass storage

Mass storage is a relatively permanent medium where programs and data are kept. Information is stored on a mass storage device by changing the polarity of a very small section of magnetic film which has been applied to either a plastic or metal base. This means that any device which generates a magnetic field has the potential to erase information in mass storage. Also, over the life of the storage device, the polarity of the particles of metallic oxide will wear off and the stored data will gradually be lost. Consequently, it is a good idea to make a backup disk or tape, itself renewed every few years, so that information will not be lost.

Tape storage

Tape has always been a popular medium for storing very large amounts of information. The tapes range in size from large reels (such as those in Figure 7-6) to microcassettes. Large computers typically use tape storage, and the tape used with some of these computers is kept on reels 1 ft in diameter. Early microcomputers used tape storage, but problems developed since loading and storing times are slow and tape breakage occurred all too frequently.

FIGURE 7-6 Large tape reels.

Aside from the problems mentioned above, tape storage is a reliable method of keeping information and is used today to back up data for other storage media because of its large capacity.

Disk storage

There are two types of disks: "hard" disks, which are aluminum disks covered with microscopic magnetic particles, and "floppy" disks, which are thin plastic disks covered with magnetic particles. Many microcomputers use both types.

Hard disks are becoming increasingly popular for microcomputers as their cost drops. They are preferred because of their ability to store large quantities of data (10 to 30 megabytes was common in 1985) and recall that information very quickly. From time to time the data on a hard disk can be lost (due often to user error), so hard disks are often backed up by copying their information onto floppy disks or tape cassettes.

Floppy disks are very popular among microcomputer users and come in three popular sizes: 8, 5¼, and 3½ in. The 8-in. size is declining in popularity, partly because of the large size of the disk and also because of the increasing efficiency of the smaller disks. The middle size is the most popular today and is often used as a double-sided, double-density (DS/DD) disk, with a typical capacity of 360K bytes (kilobytes). The small, 3½ in. size is in little use today, but it is gaining popularity since it can store up to 400K bytes of information.

7-5 Monitor programs and operating systems

Every CPU chip has had built into it an instruction set, which is a list of commands to which the chip can respond. These instructions are quite technical and would be most tedious for a user to program (particularly in machine language, which is a series of 0's and 1's). Therefore one of the first programs that must be written for a new CPU chip is a monitor (not to be confused with the display screen, which resembles a television set). The monitor program watches for input, such as characters types in from a keyboard, and routes it to a desired destination such as the display screen or a RAM memory location. Similarly, the monitor program handles output, such as to a diskette or printer.

The monitor program is a major part of an operating system (OS) developed for a CPU chip. The OS is a file manager and is the main link between the user and the computer. It provides user access to the computer's capabilities. The OS of a microcomputer is usually based on storage on floppy disks, so it is called a DOS (rhymes with floss). An example of a DOS command is DIR, which instructs the microcomputer to read a disk's library of files and to send that list to the display screen (*4*).

A widely used DOS in the early to middle 1980s was MS-DOS, developed by Microsoft Corporation for Intel's 8088 CPU chip. This is a 16-bit chip used in microcomputers marketed by a number of companies.

COMPUTER PROGRAMMING LANGUAGES

The OS of a computer permits the user to perform only limited, file-managing types of tasks. Computer programming languages enable the user to tell the computer specifically what is to be done. As seen below, there are many different types of languages, each with its own specialty.

7-6 Types and characteristics

There are two major types of computer programming languages: general purpose and special purpose. General-purpose languages are designed so that the user can create a program to perform almost any task he or she wants done by the computer. Special-purpose languages, on the other hand, permit programming on a limited, specific purpose basis. Special-purpose languages perform their limited number of functions more efficiently than a general-purpose language performs the same functions, but they have many fewer functions in their vocabulary. An example of a special-purpose language is dBASE III, developed by the Ashton—Tate company to manage data base systems.

There are three major characteristics of computer programming languages that will be used to describe the most popular languages. A language is "threaded" if the user can define the computer functions and "nonthreaded" if the user must use a predefined list of commands.

A programming language can also be "structured" or "unstructured." The most extreme degree of an structured language is one which has no

GOTO statement at all, because all the "branching" of the program is implicitly embodied in the program A GOTO statement is a statement that transfers the execution of a program to another part in the program.

A language can also be "compiled" or "interpreted." A program line in an interpreted language is translated from the language to the computer's machine language and executed, and the computer then goes on to the next line in the program. In a compiled language, the computer translates the entire program into machine language (compiles it) and stores its machine language version of the program in memory. When the user tells the computer to run a certain program in a compiled language, it runs this "compiled" version. Compiled languages tend to run programs much faster than interpreted languages, because they have less work to do on the program when they run it. They are suitable for large, complex programs. For simple programs interpreted languages are easier to use; if a program fails to run, most interpreters will provide helpful error messages and the user can rerun the corrected program without going to the trouble of recompiling it (5).

7-7 General-purpose languages
General-purpose languages enable the user to do whatever he or she wants to, within limits. These are very popular with students, since a course in a general-purpose language is often a student's first introduction to computers.

BASIC
BASIC, the Beginner's All-purpose Symbolic Instruction Code, has been a very popular language ever since its introduction at Dartmouth College in the mid-1960s. Reading a BASIC program is easy, since most of its commands are single-word commands taken from everyday English. It is a nonthreaded, fairly unstructured language, and is usually interpreted, although compiled versions of BASIC exist. Although it is a relatively slow language, it remains very popular because of its simplicity. Most microcomputers can be programmed in BASIC, and the most popular version of BASIC is MBASIC, written by the Microsoft Corporation. Although there is not a standardized BASIC, MBASIC is close to being a standard.

FORTRAN
FORTRAN stands for FORmula TRANslation and is a language developed for use by scientists and engineers. Its commands are very similar to those of BASIC, with more emphasis on numerical work, and it is always compiled. FORTRAN is nonthreaded and relatively unstructured. It is popular on large computers and is also used on microcomputers. FORTRAN is an older computer language and was one of the first "English-like" languages. Standardized FORTRAN versions include FORTRAN V and FORTRAN 77.

Pascal
Pascal is a slightly newer computer language which is beginning to receive wide use in all fields. Named for the French mathematician Blaise Pascal and

developed at the University of California at San Diego, Pascal is a very readable language, like BASIC, but is highly structured. It is a nonthreaded, compiled language used mainly on microcomputers. Pascal is widely used in business applications; Pascal programs are beginning to replace COBOL programs, since they are more efficient. UCSD Pascal is recognized as a standard.

COBOL

COBOL, the COmmon Business-Oriented Language, is an older language, like FORTRAN. It also resembles FORTRAN in that it is compiled, unstructured, and nonthreaded. It is much more business-oriented than FORTRAN, though. Most COBOL programs are long, but they execute very quickly. Because of the program length, COBOL is very popular on large computers, where there is no shortage of memory space. ANSI COBOL is recognized as the standard version.

Forth

Forth is a new language, designed to appeal to BASIC users, but it is more efficient than BASIC. It is a structured, threaded language which is usually compiled. There is no standardized form of Forth, so many different "dialects" exist. It is also the only major threaded language, permitting the user to define terms and function if desired.

C

The C programming language is popular among professional computer programmers. C is compiled, nonthreaded, and relatively structured. Its increasing popularity is due in large part to its close association with the UNIX operating system developed by Bell Laboratories.

7-8 Special-purpose languages

Special-purpose languages are becoming increasingly popular as the languages are able to accomplish more and more with greater ease. Discussed below are two of the most popular special-purpose languages, as of 1985.

dBASE III

dBASE III is short for data-Base, 3rd edition. It is a data base manager, which means that it can take data input by the user, mathematically manipulate it (if necessary), and put the data into a blank form designed by the user and kept in the computer memory. dBASE III uses a very specific language and is called a "command-driven" application. This means that the user must know the commands or have an instruction manual handy in order to use it, in contrast to a "menu-driven" program, where choices are presented to the user one screenful at a time. This type of package has many highway engineering applications, such as inventorying a city's road or traffic signal system or keeping an inventory of accidents in a county or city.

Lotus 1-2-3

Lotus 1-2-3 is a popular spreadsheet program, designed to be flexible to meet the needs of many different needs of many different types of users. Spreadsheets are large blank grids of rows and columns. They allow the user to fill the individual "cells" with column headings, row titles, the data, and the results of calculations which can be performed on the data. Spreadsheets are excellent for manipulating and reducing large quantities of data quickly and easily with little error, in contrast to the tedious and error-prone method of giving the data manipulation and reduction job to a technician. Highway applications are mainly in the traffic analysis area, where the field data (volume counts, spot speed data, highway capacity data, etc.) are entered into the cells on the spreadsheet and the sheet is "recalculated," giving new results based on these new values.

Word processors

Word processors are very popular special-purpose languages, as they allow the user to create text files for many different needs and alter them as necessary. The computer can become a very advanced kind of typewriter; they user can make any number of manipulations to a file and store it when done. Popular word procrressors include WordStar, Volkswriter and Word Perfect.

Other Special-Purpose Languages

Many other special-purpose languages exist, and some of these languages designed especially for highway applications will be discussed later in this chapter.

TYPES OF COMPUTERS

Most computers fall into one of three major classes: mainframes, minicomputers, and microcomputers. Mainframes are the largest, with entire rooms full of computing equipment, which is usually organized by installing it in frameworks. Minicomputers do not take up nearly as much room as mainframes, but they can be as large as a small room. Microcomputers are the most inexpensive and the smallest, usually able to sit on a desktop.

7-9 Mainframes

The first computers were mainframes, but not very efficient ones. Today's mainframes share one major characteristic with the early computers: they take up a lot of space. The similarities end there, since modern mainframes have huge memory capacities and can serve many users at once. Large corporations and universities with heavy computer needs use mainframes, since they need a computer than can serve many users very efficiently. Time sharing, using modems and terminals, is a popular way for small users who cannot afford or do not need a mainframe to make use of a mainframe's vast memory space while providing some income to the mainframe owner. Popular mainframe makers include IBM, Honeywell, CDC, and Sperry.

7-10 Minicomputers

Minicomputers take up about the same space as a full-sized desk, and their memory space is much more limited than that of a mainframe. Minicomputers can have more than one user. Minicomputers are faster than microcomputers but not as fast as mainframes, and today they are able to do many of the jobs that mainframes used to perform. Today's microcomputers do just as much as the minicomputers of a few years ago did, and they are much cheaper. Minicomputers are made by IBM, Hewlett-Packard, Honeywell, and CDC, to name a few. Figure 7-7 shows a CDC/VAX minicomputer used by the Georgia Department of Transportation for its computerized highway design work.

7-11 Microcomputers

The microcomputer field has grown by leaps and bounds since about 1980. Microcomputers are popular with smaller offices because of their low cost and the large amount of software available for them. Microcomputers can now be connected together to share such devices as hard disks, printers and plotters, thereby lowering the cost per workstation. Microcomputers are becoming

FIGURE 7-7 A CDC/VAX minicomputer used by the Georgia Department of Transportation.

increasingly able to perform the same tasks as a minicomputer (just as minis can now do mainframe type tasks) and are able to do them more cheaply.

One of the first popular microcomputers was the Apple II, introduced in 1977. As of 1985 the Apple II series included the Apple IIC, primarily for the home user, and the Apple IIe, directed to the business or education user. Both use the same 8-bit CPU chip. In 1981 IBM entered the microcomputer field with its IBM PC, which was an instant success. A year later a model with a hard disk was introduced—the IBM PC/XT. Both are based on the Intel 8088 CPU chip.

As Intel continues to refine the 8088 into the 80286 and the 80386, these major personal computer manufacturers will continue to come out with up-dated versions of their products, as IBM has with its 80286-based PC/AT. The major advantage of Intel's 80_86 chips is that much of the software for the older CPU chips can be used on the new ones, eliminating the problem of available software for a new computer.

Much of the success of the Apple and IBM computers is attributable to the companies' policies of encouraging vendors of software and hardware to develop and sell products that are compatible with their computers.

7-12 Lap-top and hand-held computers

A lap-top computer is not much bigger than a typical microcomputer keyboard, but it usually comes complete with a small liquid crystal screen, printer, and tape storage device. A popular lap-top is the Epson HX-20, which has 16K bytes of memory and built-in BASIC programming as well as the peripheral devices mentioned above. This small computer, weighing about 4 lb, makes computerized field data collection both practicable and affordable. Another very popular type of small computer is an outgrowth of the hand-held calculator. Casio and Sharp make small computers which are about the size of a thick business envelope but have capabilities similar to those of the Epson mentioned above.

Closer yet to the hand-held calculator is the Hewlett-Packard HP-41 series of hand-held computers. The HP-41 is a very small computer which can function as a calculator when needed but also has enough internal memory to hold useful application programs. Some digital surveying equipment is now made with output jacks to take advantage of the four ports on the HP-41.

Many application programs exist for the HP-41; they are stored on magnetic cards, magnetic tapes, or as bar code on paper (read into the HP-41 by using an optical wand) or are kept in the HP-41's constant memory. Because of its portability, the HP-41 is very useful for both the field engineer and the highway designer.

7-13 Highway design software packages

These packages are the ones that highway engineers are most interested in, since they make the engineer's life so much easier. Some are languages, and some are programs written in one of the previously mentioned languages. Regardless of which category the packages fall into, each has its own "vocabulary" of commands used to access its computing capabilities.

COGO

COGO (short for COordinate GeOmetry) was the first major highway design software package. Developed by Charles L. Miller at the Massachusetts Institute of Technology in the early 1960s, it was, and is, a valuable time-saving tool for highway design. COGO was written with the aid of a grant from the Puerto Rico Department of Transportation and was originally run on the IBM 1620, an early personal computer.

COGO was originally a program run in BATCH mode; that is, all the information about the highway location and the necessary commands were first put into a file, and the program ran by using that file as input. Most of today's COGO programs are interactive ones, meaning that the program handles the data as completely as possible as it is entered into the computer. The ICS (Integrated Cogo System) can be run by using Intergraph stations, which will be discussed later (6).

After the highway engineer does the preliminary location work, he or she then inputs the coordinates of the beginning and end points of the tangents and the basic information about the horizontal curves. The computer outputs stationing, dimensions, and all the curve data.

Although many programs exist today which do more than COGO, it is still a very popular computer design tool, and many different "COGO" exist. These different versions of COGO all do essentially the same thing; they differ in terms of the specific commands used, the computers they will run on, and their compatability with other design-oriented programs such as CAD (Computer-Aided Design) programs.

Charles L. Miller's current work

Professor Miller continues to develop highway design software packages for mini- or microcomputers. Some of his programs can stand alone, working independently of any other programs, while others are designed to be used as an integrated package with other CLM/Systems software. Figure 7-8 shows one combination of hardware that supports CLM/Systems software: two IBM PC/ATs, a digitizer (left foreground), a drum plotter (right), and a video monitor (center, back).

CLM COGO is an updated version of the original COGO developed by Professor Miller. It is a command-driven program for micro- and minicomputers, like the IBM PC/AT microcomputer and the DEC/VAX minicomputer. CLM COGO is intended for civil and site engineering, surveying and mapping, and transportation engineering, mainly in the highway design area. CLM COGO is a stand-alone program, but it can be used in combination with other CLM /Systems software.

CLM TOPO is another stand-alone program and is very useful to surveyors, highway designers, geologists, and hydrologists, all of whom often need to know what the land looks like three-dimensionally. It enables the user to produce surface and terrain models, contour models, and topographic maps. CLM TOPO can also produce terrain profiles, cross sections, digital maps, and overlays. As well, CLM TOPO can compute volumes and surfaces. Although this program has many non-highway design applications, in conjunction with

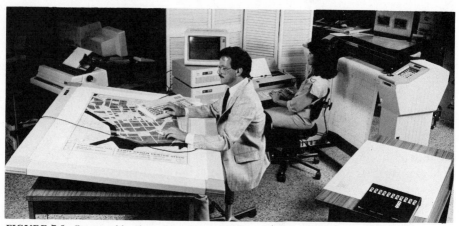

FIGURE 7-8 One combination of hardware which supports CLM/Systems software.

the preceding and following programs it makes a very powerful highway design package which can be used on remarkably small computers.

CLM ROADS is different from the two preceding programs in that it can run only in conjunction with COGO and TOPO. When used, however, it provides a very powerful software package for the highway designer. CLM ROADS provides for the interactive design of horizontal alignments, vertical alignments, the roadway proper, cross sections, spirals, plan and profiles, and earthwork. In addition, it has a library of templates, which are already-designed roadway sections for many different types of highways which the designer can call up and later to fit specific needs. This feature can be of tremendous use when the designer is confronted with similar situations frequently, such as the design of certain intersections or exit-entrance ramp situations. User-designed templates can be added to the library of existing templates, if desired.

COGO-CAD is another integrated program which works in conjunction with the previous programs and provides computer-aided design capabilities for those programs. This permits much more flexibility in terms of the output, since the user can now add more to the output than the program itself might give.

COGO-LINK is an integrated program which provides for file transfer between any of the preceding programs and other computers, total stations (computerized surveying equipment), and data gatherers (like the lap-top computers or the HP-41s). This program can save much time in transferring information gathered in the field to the CLM/System software package and can avoid errors inherent in hand transfer (7).

RDS

RDS, the Roadway Design System, is a very popular software package used by 28 state departments of transportation for their design work. Developed by the Texas State Department of Highways and Public Transportation, it is a large program and is usually run on mainframes or minicomputers. RDS is a comprehensive design package and can be used to establish roadway location, alignments, grades, geometrics, earthwork quantities, and right-of-way requirements. In addition, RDS provides geodetic control and bridge geometry and design.

The bridge design capability is especially needed today because curved beams and box girders, requiring many complex calculations, are often necessary to achieve the desired bridge geometrics.

The input for RDS is gathered in two different ways. The first method uses total station surveying equipment, which has the capability of storing survey data (distances, elevations, and angles) on magnetic tape. These data are transmitted through telephone lines to a processor at the central office. The second method involves stereophotography from airplanes and stereoplotters to determine distances and elevations. The information derived from the stereophotographs is used to add detail to the field survey information. By combining these two techniques, field survey cost is reduced to a minimum while the necessary detail is provided by the stereophotographs.

These topographic data are transformed into digital input for the computer and used as the major input for RDS. Text-oriented details such as street names, river names, and descriptions of unique objects (radio towers, etc.) are then input by an operator at a video graphics terminal.

RDS is run with two types of terminals: alphanumeric terminals, used to review the text portion of the input and output; and interactive graphics drafting system (Intergraph) stations, like the one shown in Figure 7-9, used for viewing geographic data, obtaining input graphically, and displaying calculated design elements. Since RDS is often run on large minicomputers or mainframes, it can have several users are once.

The main drawback of RDS is that it can be run only in batch mode, limiting the abilities of the Intergraph stations, Any changes in the design must be input through the alphanumeric terminals and then viewed on the Intergraph stations. AASHTO is now completing work on an interactive RDS (IGRDS). IGRDS was was begun by the Texas Highway Department and was financed by the DOTs of 20 states. When interactive RDS is released the interactive graphics stations will become much more versatile, since it will allow the user to make changes in the video output at the Intergraph stations and immediately see the results of those changes. This will make the entire RDS system even more attractive to large users.

Prior to the Intergraph stations, the computerized highway design process involved inputting all the topogrpahic data, then inputting all the design elements, and routing the output to a "drum" plotter (Figure 7-10) to view the results such as those in Figure 7-11. If any aspect of the design was not correct, the changes had to be input at the alphanumeric terminals (or through a keypunch and card reader) and another plot of the design generated. The Intergraph stations, even in batch mode, help make this design process much easier (8).

The Intergraph stations have been around since the early 1970s. It is only

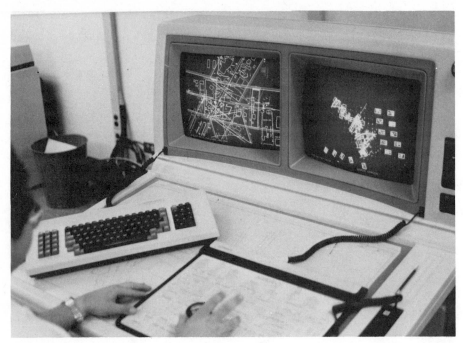

FIGURE 7-9 An interactive graphics drafting system station.

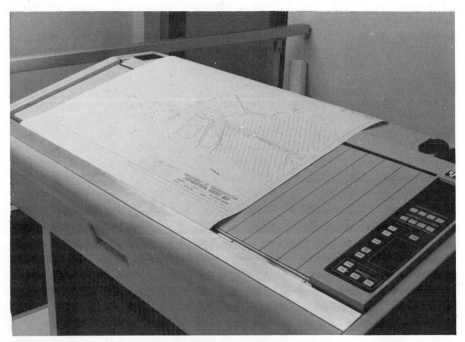

FIGURE 7-10 Plotter for a computerized highway design system.

FIGURE 7-11 A portion of output from a computerized highway design system. (Courtesy Georgia Department of Transportation.)

with the advent of the new interactive RDS that these stations can be used to their full potential. The Intergraph station is able to show three-dimensional renderings of the finished roadway, complete with landscaping and vehicles, in color. If desired, the perspective can be changed by pushing only a few buttons.

The new IGRDS will be a large step forward in the field of highway design. The interactive capability will permit the designer to carry on a virtual conversation with the computer. Alternative designs, which would not have been explored before the advent of IGRDS, can now be considered by virtue of the speed and flexibility of the system. The end result will be better design, since more alternatives can be investigated and changes can be made more easily. While there is no substitute for experience, the advantages of IGRDS will allow a younger, more inexperienced engineer to make a good, safe design, since revisions are made quickly and alternative designs are easily tried (9).

RDS requires much computer space to operate—so much, in fact, that many agencies run RDS on a minicomputer of its own, with phone lines used to connect remote users with the main system. This enables an engineer in a branch office to communicate with an engineer in the home office over the phone while both of them see the same drawing on their Intergraph stations. The Georgia Department of Transportation's organizational plan for their computerized highway design is shown in Figure 7-12 as an example of the way remote and main office users both use the same system.

ROADCALC

ROADCALC is an interactive, menu-driven highway design program available commercially for use on microcomputers. ROADCALC performs all COGO tasks plus layout, earthwork calculations, and geometric design. ROADCALC is also able to work with AutoCAD, a general design package discussed below. This compatability increases the power of ROADCALC, since AutoCAD gives it additional graphics capabilities (10).

AutoCAD

AutoCAD was not developed specifically for highway engineering at all but instead serves all engineering fields, since it can augment or replace everyday drafting tasks performed by draftsmen. AutoCAD is a commercially available program that provides computer-aided design capabilities for microcomputers equipped with a digitizer and a plotter. This CAD package is menu-driven and can be used by those with a minimum of computer experience. Until the advent of AutoCAD, CAD capabilities were available only on mainframes and minicomputers. AutoCAD can be used with many programs for highway design, including those mentioned above, and can be used separately if needed. It is particularly suited to structural and architectural work.

7-15 What does the future hold?

Two general trends are predicted for the computing field. The first major trend is that the price of computer hardware will continue to fall, even though the present cost of hardware is much less than it was as recently as three years

FIGURE 7-12 Organizational plan for a computerized highway design system. (Courtesy Georgia Department of Transportation.)

ago. This will allow even the smallest companies to computerize their design and drafting procedures.

Falling prices will be accompanied by continuing improvements in capability and reliability. The boundaries between the various sizes of computers will become even more fuzzy than they are at present. As more companies produce "supermicros" with capabilities rivaling those of minicomputers or "superminis" which seem just like mainframes, it will be very difficult to determine what category a given computer will fall into.

The class of computer might well be tied to the size of the hardware, as it seems to be now: microcomputers can sit on a desk, minicomputers can be anywhere from desk-sized up to small room-sized, and mainframes are room-sized and larger.

Software will be more capable and "user-friendly," and this may serve to attract customers who were previously afraid of computers and the potential for making "fatal" mistakes. This is evidenced in the large amount of time and money being invested in the development of an interactive version of RDS.

"User-friendly" often implies that more programs will become menu-driven, in an effort to accommodate the unfamiliar user by leading the user by the hand through a program. Experienced users often have no need for menus, preferring command-driven programs which let them work through the program more rapidly.

REFERENCES

1. Toong, Hoo-min D., and Gupta, Amar. Personal Computers. *Scientific American* (December 1982).

2. *Microcomputer Literacy Program for Executives, Managers and Professionals.* McGraw-Hill, Inc., Washington, D.C. (1982).

3. Sippl, Charles J. *Microcomputer Dictionary.* Howard W. Sams & Co., Indianapolis, Ind. (1981).

4. Pournelle, Jerry. The Operating System Jungle. *Popular Computing.* (June 1984).

5. *Getting a Running Start: A Quick Introduction to Micros.* Microcomputer Support Centers, U.S. Department of Transportation, Cambridge, Mass. (1984).

6. Baumgaertner, W.E., et al. Traffic Signal Design by Computer. *1985 Compendium of Technical Papers.* Institute of Transportation Engineers, Washington, D.C. (1985).

7. CLM/Systems, Inc. Material transmitted by letter of November 4, 1985.

8. Henry, H.A. *Interactive Graphics in Highway Design.* Transportation Research Record No. 923, (1983).

9. AASHTO Computerized Design Process to Revolutionize Highway Design Throughout the Country. *AASHTO Quarterly* (July 1985).

10. Engineering Data Systems Corporation. Material transmitted by letter of October 25, 1985.

chapter eight

GEOMETRIC DESIGN
OF HIGHWAYS

This chapter describes the design standards and procedures for establishing the horizontal and vertical alignment and the dimensions and slopes of a highway cross section. It concerns the "geometric design" or the proportioning of the visible elements of a highway or street. The first and major part of the chapter deals with the design of motor-vehicle facilities. The chapter includes brief treatments of the design of facilities for bicyclists and pedestrians.

DESIGN OF MOTOR-VEHICLE FACILITIES

Geometric design practices of the state highway and other designing agencies are not entirely uniform. A considerable variation exists in the laws of the various states limiting the size and weight of motor vehicles. Differences also exist in the financing ability of the various governmental agencies, and these influence the designers' decisions and modify their design standards. There are also differences in experience and the interpretation of research, which are reflected in a variety of design practices.

These differences are tolerated by the Federal Highway Administration as unavoidable and are accepted when approval is requested on plans developed by the various states for federal-aid improvements. The strongest force tending to standardize these differences lies in the various technical committees of the American Association of State Highway and Transportation Officials. All state highway agencies and the Federal Highway Administration hold membership in this association and join in the deliberations of its technical committees. Upon approval by a required majority, a standard submitted is declared adopted and becomes, in effect, a guide for all members of the association.

8-1 Design controls and criteria

The elements of highway design are influenced by a wide variety of design controls and criteria. Such factors include

1. Functional classification of the roadway being designed.
2. Traffic volume and composition.
3. Design speed.
4. Topography.
5. Cost and available funds.
6. Human sensory capacities of drivers, bikers, and pedestrians.

7. Size and performance characteristics of the vehicles that will use the facility.
8. Safety considerations.
9. Social and environmental concerns.

These considerations are not, of course, independent. The functional class of a proposed facility is largely determined by the volume and composition of traffic to be served. It may also depend on availability of funds and social and environmental considerations. For a given class of highway, the choice of design speed is governed primarily by topography, which in turn is a determinant of facility cost. Once a design speed is chosen, many of the elements of design (e.g., horizontal and vertical alignment) may be established on the basis of fundamental human sensory capabilities and vehicular sizes and performance characteristics described in Chapter 3.

The design features of a highway influence its capacity and efficiency, its safety performance, and its social acceptability to highway users, owners of abutting property, and the general public.

The principal design criteria for highways are traffic volume, design speed, the physical characteristics of the vehicles, and the proportions of vehicles of various sizes that use the highways. These criteria are discussed in more detail in the following paragraphs.

8-2 Relationship of traffic to highway design

The major traffic elements that influence highway design are average daily traffic *(ADT)*, design hour volume *(DHV)*, directional distribution *(D)*, percentage of trucks *(T)*, and design speed *(V)*. As indicated in Chapter 5, *ADT* is a fundamental measure of traffic flow. To the designer, the most significant measure of traffic volume is the design hour volume, a two-way value, which may be determined by multiplying the *ADT* by a representative percentage. This percentage, *K*, is typically 8 to 12 percent for urban facilities and 12 to 18 percent for rural facilities. The directional distribution *(D)* is the one-way volume in the predominant direction of travel, expressed as a percentage of the two-way design hour volume. For rural roads, *D* ranges from 55 to 80 percent and typically is about 67 percent. "As a general rule, traffic tends to be more equally divided by direction near the city center, where a value of 55 is common, than in the outer fringe of the city where values of 67 or even 70 occur with considerable frequency" *(1)*.

Composition of traffic *(T)* is usually expressed as the percentage of trucks (exclusive of light delivery trucks) during the design hour. That percentage typically varies from about 5 to 10 percent. In urban areas, the percentage of trucks during peak hours tends to be considerably less than percentages on the daily basis *(1)*.

The objective in highway design is to create a highway of appropriate type with dimensional values and alignment characteristics such that the resulting design service volume will be at least as great as the design volume, but not enough greater as to represent extravagance or waste. Where this objective is accomplished, a well-balanced, economical highway system will result *(2)*.

It will be recalled from Chapter 5 that the traffic volumes that can be served at each level of service are termed "service volumes." Once a level of service has been chosen for design, the corresponding service volume logically becomes the design service volume. This implies that if the traffic volume using the facility exceeds that value, the operating conditions will be inferior to the level of service for which the roadway was designed *(2)*. The level of service appropriate for the design of various types of highways is shown by Table 8-1.

Table 8-2 generally describes the operating characteristics of various types of highways operating at each level of service. Such information is useful in determining the number of lanes required to serve the estimated design year traffic. A more extensive capacity analysis, based on data and procedures given in Ref. 3, may be required to identify potential bottlenecks and to facilitate the design of intersections, ramp terminals, weaving sections, and the like.

Design speed

The assumed design speed for a highway may be considered as "the maximum safe speed that can be maintained over a specified section of highway when conditions are so favorable that the design features govern" *(2)*. The choice of design speed will depend primarily on the terrain and the functional class of the highway. Other factors determining the selection of design speed include traffic volume and composition, costs of rights-of-way and construction, and aesthetic considerations.

Design speeds typically range from 20 to 70 mph and intermediate values are chosen in increments of 10 mph: 30, 40, 50 mph, and so forth. Where

TABLE 8-1
GUIDE FOR SELECTION OF DESIGN LEVELS OF SERVICE

Highway Type	*Type of Area and Appropriate Level of Service*[a]			
	Rural Level	*Rural Rolling*	*Rural Mountainous*	*Urban and Suburban*
Freeway	B	B	B	C
Arterial	B	B	C	C
Collector	C	C	D	D
Local	D	D	D	D

[a]A, free flow, with low volumes and high speeds; B, stable flow, but speeds beginning to be restricted by traffic conditions; C, in stable flow zone, but most drivers restricted in freedom to select their own speed; D, approaching unstable flow, drivers have little freedom to maneuver; E, unstable flow, may be short stoppages.

SOURCE: *A Policy on Geometric Design of Highways and Streets.* American Association of State Highway and Transportation Officials, Washington, D.C. (1984).

TABLE 8-2
LEVEL-OF-SERVICE CHARACTERISTICS BY HIGHWAY TYPE

Level of Service	Controlled-Access Highways	Multilane Rural without Access Control	Two Lanes	Urban and Suburban Arterials
A	Free flow. Operating speeds at or greater than 60 mph. Service volume of 1400 passenger cars per hour on two lanes, one direction. Each additional lane serves volume of 1000-vph lane.	Operating speed 60 mph or greater. Under ideal conditions, volume is limited to 600 passenger cars per lane per hour or 30 percent of capacity. Average speeds are likely to be influenced by speed limits.	Operating speeds of 60 mph or higher. 75 percent of passing maneuvers can be made with little or no delay. Under ideal conditions, a service volume of 400 passenger vph, total two-way, can be achieved.	Average overall travel speed of 30 mph or more. Free flowing with volume/capacity ratio of 0.60. Load factor at intersections near the limit of the 0.0 range. Peak-hour factor at about 0.70.
B	Higher speed range of stable flow. Operating speed at greater than 55 mph. Service volume on two lanes in one direction not greater than 2000 passenger vph. Each additional lane above two in one direction can serve 1500 vph.	Beginning of stable flow area. Volume at which actions of preceding vehicle will have some influence on following vehicles. Volume will not exceed 50 percent of capacity of 1000 passenger vehicles per lane per hour at a 55 mph operating speed under ideal conditions	Operating speeds of 50 mph or higher. Volumes may reach 45 percent of capacity with continuous passing sight distance. Volumes of 900 passenger cars per hour, total two-way, can be carried under ideal conditions.	Average overall speeds drop due to intersection delay and intervehicular conflicts, but remain at 25 mph or above. Delay is not unreasonable. Volumes at 70 percent of capacity and peak-hour factor approximately 0.80. Load factor at intersections approximately 0.1.
C	Operation still stable, but becoming more critical. Operating speed of 50 mph. Service flow on two lanes in one direction at 75 percent of capacity or not more than 5-min flow rate of 3000 passenger cars per hour. Under ideal conditions each additional lane above two in one *[text cut off]*	Stable flow to a volume not exceeding 75 percent of capacity or 1500 passenger cars per lane per hour, under ideal conditions, maintaining at least a 45-mph operating speed.	Flow still stable. Operating speeds of 40 mph or above with total volume under ideal conditions equal to 70 percent of capacity with continuous passing sight distance, or 1400 passenger vph total two-way.	Service volumes about 0.80 of capacity. Average overall travel speeds of 20 mph. Operating conditions at most intersections approximate load factor of 0.3. Peak-hour factor approximately 0.85. Traffic flow still stable with acceptable delays.

D	Lower speed range of stable flow. Operation approaches instability and is susceptible to changing conditions. Operating speeds approx. 40 mph. Service flow rates at 90 percent of capacity. Peak 5-min flow under ideal conditions cannot exceed 3600 vph for two-lanes, one direction; 1800 vph for each added lane.	Approaching unstable flow at volume up to 90 percent of capacity or 1800 passenger cars per hour at an operating speed of about 35 mph under ideal conditions.	Approaching unstable flow. Operating speeds approximately 35 mph. Volumes, two-direction, at 85 percent of capacity with continuous passing opportunity, or 1700 passenger cars per hour total two-way under ideal conditions.	Beginning to tax capabilities of street section. Approaching unstable flow. Service volumes approach 90 percent of capacity. Average overall speeds down to 15 mph. Delays at intersections may become extensive with some cars waiting two or more cycles. Peak-hour factor approximately 0.90; load factor of 0.7.
E	Unstable flow. Overall operating speeds of 30–35 mph. Volumes at capacity or 2000 vph lane under ideal conditions. Traffic flow metered by design constructions and bottlenecks, but long backups do not normally develop upstream.	Flow at 100 percent of capacity or 2000 passenger cars per lane per hour under ideal conditions. Operating speeds of about 30 mph or less.	Operating speeds in neighborhood of 30 mph but may vary considerably. Volumes under ideal conditions, total two-way, equal to 2000 passenger vph. Level E may never be attained. Operation may go directly from Level D to Level F.	Service volumes at capacity. Average overall traffic variable but in area of 15 mph. Unstable flow. Continuous backup on approaches to intersections. Load factor at intersections in range 0.7 to 1.0. Peak-hour factor likely to be 0.95.
F	Forced flow. Freeway acts as a storage for vehicles backed up from downstream bottleneck. Operating speeds range from near 30 mph to stop-and-go operation.	Forced flow, congested condition with widely varying volume characteristics. Operating speeds of less than 30 mph.	Forced, congested flow with unpredictable characteristics. Operating speeds less than 30 mph. Volumes under 2000 passenger cars per hour, total two-way.	Forced flow. Average overall traffic speed below 15 mph. All intersections handling traffic in excess of capacity with storage distributed throughout the section. Vehicular backups extend back from signalized intersections, through unsignalized intersections.

SOURCE: *A Policy on Geometric Design of Highways and Streets*. American Association of State Highway and Transportation Officials, Washington, D.C. (1984)

feasible, a constant design speed should be used in the design of a highway of substantial length. Where changes in terrain or other conditions dictate a change in design speed, such change should be made over a sufficient distance to permit drivers to change speed gradually. The changes in design speed should, of course, be indicated by appropriate traffic control devices.

Design designation

A tabulation of the major traffic controls for which a highway is designed is called the "design designation." Examples of such designations are given below. The tabulation on the left is for a two-lane highway; that on the right is for a multilane highway.

		Control of access = full	
ADT (1985) = 2,500		ADT (1985) = 10,200	
ADT (2005) = 5,200		ADT (2005) = 2,950	
DHV	= 720	DHV	= 2,950
D	= 65%	D	= 60%
T	= 12%	T	= 8%
V	= 60 mph	V	= 70 mph

The highway design designation is normally placed on the plans for proposed improvements in order to indicate to all those handling and reviewing the plans the traffic controls used in the design of the project.

8-3 Design vehicle

The dimensions of the motor vehicle also influence design practice. The width of the vehicle naturally affects the width of the traffic lane; length has a bearing on roadway capacity and affects the turning radius; the height of the vehicle affects the clearance of the various structures. Weight, as mentioned previously, affects the structural design of the roadway.

The American Association of State Highway and Transportation Officials recommends *(1)* that 10 design vehicles be used for determining the controls for the geometric design of a highway. These vehicles range in size from the passenger car to the large semitrailer–full trailer combination. Normally, the design engineer will select for design the largest vehicle that is expected in significant numbers in the design year. The dimensions of these design vehicles are given in Table 8-3 and exemplified by Figure 8-1.

8-4 The highway cross section

The width of the surfaced road should be adequate to accommodate the type and volume of traffic anticipated and the assumed design speed of vehicles. Roads presently in use include two-lane, three-lane, multilane undivided, and multilane divided.

TABLE 8-3
DESIGN VEHICLE DIMENSIONS

Design Vehicle Type	Symbol	Overall			Overhang		WB_1	WB_2	S	T	WB_3
		Height	Width	Length	Front	Rear					
Passenger car	P	4.25	7	19	3	5	11				
Single unit truck	SU	13.5	8.5	30	4	6	20				
Single unit bus	BUS	13.5	8.5	40	7	8	25				
Articulated bus	A-BUS	10.5	8.5	60	8.5	9.5	18		4[c]	20[c]	
Combination trucks											
Intermediate semitrailer	WB-40	13.5	8.5	50	4	6	13	27			
Large semitrailer	WB-50	13.5	8.5	55	3	2	20	30			
"Double Bottom" semi-trailer—full-trailer	WB-60	13.5	8.5	65	2	3	9.7	20	4[d]	5.4[d]	20.9
Recreation vehicles											
Motor home	MH		8	30	4	6	20				
Car and camper trailer	P/T		8	49	3	10	11	5	18		
Car and boat trailer	P/B		8	42	3	8	11	5	15		

Heading: *Dimension (ft)*^{a,b}

[a] WB_1, WG_2, WB_3, are effective vehicle wheelbases; S is the distance from the rear effective axle to the hitch point; T is the distance from the hitch point to the lead effective axle of the following unit.
[b] 1 ft = 0.3048 m.
[c] Combined dimension 24, split is estimated.
[d] Combined dimension 9, 4, split is estimated.

SOURCE: *A Policy on Geometric Design of Highways and Streets.* American Association of State Highway and Transportation Officials, Washington, D.C. (1984).

FIGURE 8-1 Minimum turning path for WB-50 design vehicle. (Courtesy American Association of State Highway and Transportation Officials.)

Two-lane highways

The bulk of our highway system is composed of two-lane highways, and over 90 percent of the rural roads in the United States are of this type. These two-lane roads vary from low-type roads, which follow the natural ground surface, to high-speed primary highways with paved surfaces and stabilized shoulders. As traffic density, vehicle speeds, and truck widths have increased, two-lane highways have also increased in width from 16 ft to the current recommended value of 24 ft, with 10-ft shoulders on either side for primary routes. Where lighter traffic volumes are expected, the American Association of State Highway and Transportation Officials recommends minimum surfacing widths based on design speed and traffic volume.

Three-lane highways

Many three-lane roads were built in previous years and are still in use. Their great advantage seemed to stem from the large improvement in capacity over the two-lane road, with only a moderate increase in construction and right-of-way costs. The three-lane road, however, does appear to have a high accident rate, and the construction of these roads is no longer recommended.

Four-lane highways

On four-lane highways, traffic flows in opposite directions on each pair of lanes, and passing is accomplished within the lanes of forward movement of traffic and not in the lanes of opposing traffic. This feeling of freedom from opposing traffic results generally in smoother operation and increase in the capacity per lane over that of two- and three-lane pavements. Four-lane highways have at least four times the capacity of a two-lane highway for the same assumed design speed.

Although the four-lane highway is the basic multilane type,, traffic volumes sometimes warrant the use of highways, having six, eight, or more lanes, particularly in urban areas.

The undivided multilane highway does, however, appear to have an accident rate higher than that of the two-lane highway. This would appear to be due to higher traffic volumes with traffic frictions from cars traveling both in the same direction and in opposing directions.

It is generally agreed that for rural roads and where design speeds are high, where traffic volumes are sufficient to require multilane construction, traffic separation is desirable.

8-5 Divided highways

In order to provide positive protection against conflict of opposing traffic, highways are frequently divided by a median strip (see Figure 8-2). On such highways, lane widths should be a minimum of 12 ft, with 13-ft lanes provided where many large truck combinations are likely to occur. It is highly desirable that all multilane highways should be divided. The width of these median strips varies from 4 to 60 ft or more. A median strip less than 4 to 6 ft in width is considered to be little more than a centerline stripe, and its use, except for special conditions, should be discouraged. The narrower the median, the longer must be the opening in the median be to give protection to vehicles making left turns at points other than intersections. Where narrow medians must be used, many agencies install median barriers to physically separate opposing streams of traffic and minimize the number of head-on collisions. Medians of 14 to 16 ft have been used and are sufficient to provide most of the separation advantages of opposing traffic while permitting the inclusion of a median lane at crossroads; however, medians 16 to 60 ft wide and more are now recommended. The median should also be of sufficient width to maintain vegetation and to support low-growing shrubs that reduce the headlight glare of opposing traffic. Median strips at intersections should receive careful consideration and should be designed to permit necessary turning movements.

Divided highways need not be of constant cross section. The median strip may vary in width; the roads may be at different elevations; and the superelevation may be applied separately on each pavement. In rolling terrain, substantial savings may be effected in construction and maintenance costs by this variation

FIGURE 8-2 A divided highway in Connecticut. (Courtesy Connecticut Department of Transportation.)

in design. This type of design also tends to eliminate the monotony of a constant width and equal grade.

Where it is necessary to narrow the median strip, or where intersections make it desirable to widen the median strip in tangent alignment, the change may be effected by reverse curves of 1° or less which may be without superelevations or transitions. Where such changes in width on curves are desirable, they should be accomplished if possible by changing the curvature of one or both roads.

8-6 Limited-access highways

A very important feature of the design of multilane highways is the control of access from adjacent property.

A limited-access highway may be defined as a highway or street especially designed for through traffic, to which motorists and owners of abutting property have only a restricted right of access. Limited- or controlled-access highways may consists of (1) freeways that are open to all types of traffic and (2) parkways from which all commercial traffic is excluded. Most of our present expressway systems have been developed as freeways.

Limited-access highways may be elevated, depressed, or at grade. Many examples of the various types may be found in the United States in both rural and urban areas.

The control of access is attained by limiting the number of connections to and from the highway, facilitating the flow of traffic by separating cross traffic with overpasses or underpasses, and eliminating or restricting direct access by owners of abutting property.

The design of limited-access routes should provide for adequate width of right-of-way, adequate landscaping, prohibition of outdoor advertising on the controlled access proper, and provisions for controlling abutting service facilities such as gas stations, parking areas, and other roadside appurtenances.

In urban areas, the design of a limited-access facility is usually accompanied by the design of frontage roads, parallel to the facility, which serve local traffic and provide access to adjacent land. Such roads may be designed for either one-way or two-way operation. Reasonably convenient connections should be provided between through-traffic lanes and frontage roads.

8-7 Pavement crowns

Another element of the highway cross section is the pavement crown, which is the raising of the centerline of the roadway above its edges. Pavement crowns have varied greatly throughout the years. On the early low-type roads, high crowns were necessary for good drainage and were ½ in. or more per foot. With the improvement of construction materials, techniques, and equipment which permit closer control, crowns have been decreased. Present-day high-type pavements with good control of drainage now have crowns as low as ⅛ in. per foot. Low crowns are satisfactory when little or no settlement of the pavement is to be expected and when the drainage system is of sufficient capacity to quickly remove the water from the traffic lane. When four or more traffic lanes are used, it is desirable to provide a higher rate of crown on the outer lanes in order to expedite the flow of water.

Crowns may be formed by intersecting tangent lines or by curved lines that emanate from the road centerline. In the latter case, circular arcs of long radii are used, as well as parabolic arcs. It makes little difference which is employed, but the parabolic arc lends itself better to making computations for the initial offsets or ordinates in constructing templates or in setting grades.

8-8 Shoulders

Closely related to the lane width is the width of the shoulders. It is necessary to provide shoulders for safe operation and to develop full traffic capacity. Well-maintained, smooth, firm shoulders increase the effective width of the traffic lane as much as 2 ft, as most vehicle operators drive closer to the edge of the pavement in the presence of adequate shoulders. To accomplish their purpose, shoulders should be wide enough to permit and encourage vehicles to leave the pavement when stopping. The greater the traffic volume, the greater is the likelihood of the shoulders being put to emergency use. A usable outside shoulder width of at least 10 and preferably 12 ft clear of all obstructions is desirable for all heavily traveled and high-speed highways. Inside shoulders often are not as wide. In mountainous areas, where the extra cost of providing shoulders of this width may be prohibitive, or on low-type highways, a minimum width of 4 ft may be provided; a width of 6 to 8 ft is preferable. Under these conditions, however, emergency parking strips should be provided at proper intervals. In terrain where guardrails or other vertical elements (such as retaining walls) are required, an additional 2 ft of shoulder should be provided.

The slope of the shoulder should be greater than that of the pavement. A shoulder with a high-type surfacing should have a slope of at least ⅜ in. per foot. Sodded shoulders may have a slope as high as 1 in. per foot in order to carry water away from the pavement.

8-9 Guardrails

Guardrails should be provided at points where fills are over 8 ft in height, when shoulder slopes are greater than 1 on 4, and at locations where there is a sudden change in alignment and where a great reduction in speed is necessary. Deep roadside ditches with steep banks or right-of-way limitations, often make it necessary to steepen the side slopes and require the use of guardrails. Where guardrails are used, the width of the shoulders is increased approximately 2 ft to allow space for placing the posts.

Various types of guardrail are in use at the present time. The most important of these are the "W" beam guardrail, the cable guardrail, and the box beam guardrail. These may be installed on wood, steel, or concrete posts. Some agencies have successfully used the so-called weak post system. In this system, the posts collapse when struck, and the rail deflects and absorbs the energy of impact. When strong posts are used, the rail should be blocked out or supported away from the post to minimize "pocketing" of the vehicle. Design features of guardrails that have performed satisfactorily are given in Table 8-4.

8-10 Curbs, curb and gutter, and drainage ditches

The use of curbs is generally confined to urban and suburban roadways. The design of curbs varies from a low, flat, lip-type to a nearly vertical barrier-type curb. Curbs adjacent to traffic lanes, where sidewalks are not used, should be low and very flat. The face of the curb should be no steeper than 45° so that vehicles may mount them without difficulty. Curbs at parking areas and adjacent to sidewalks should be 6 to 8 in. in height, with faces nearly vertical. Clearance should be sufficient to clear fenders and bumpers and to permit the opening of car doors. From the pedestrian viewpoint, curbs should be one step in height. When a barrier curb is used, it should be offset a minimum of 10 ft from the edge of the traffic lane. Figure 8-3 shows some typical curb sections (2).

Drainage ditches should be located and shaped to avoid creating a hazard to traffic. Under normal conditions, ditches should be low enough to drain the water from under the pavement. A broad, flat, rounded ditch section has been found to be safer than a V-type ditch, which also may be subject to severe washing action. The profile gradient of the ditch may vary greatly from that of the adjacent pavement.

8-11 Slopes

Side slopes and back slopes may vary a great deal, depending on the type of material and the geographic location. Well-rounded flat slopes present a pleas-

TABLE 8-4
DESIGN FEATURES OF GUARDRAILS THAT HAVE PERFORMED SATISFACTORILY

System	Cable	"W" Beam (Steel Weak Post)	Box Beam	Blocked-out "W" Beam (Wood Post)
Deflection	12 ft.[a]	8 ft	4 ft 2 ft	2 ft
Post spacing	16 ft-0 in.	12 ft-6 in. nominal	6 ft-0 in. 4 ft-0 in.	6 ft-3 in.
Post	S3×5.7	S3×5.7		8×8 in. Douglas fir
Beam	Three $\frac{3}{4}$ in. diam. steel cables	Steel "W" section	6×6×0.180 in. steel tube	Steel "W" section
Offset brackets			L5×3$\frac{1}{2}$×$\frac{1}{4}$ in. steel angle 4$\frac{1}{2}$ in. lg	8×8×14 in. Douglas fir blocks.
Mounting	$\frac{5}{16}$ in. diam. steel hook bolts	$\frac{5}{16}$ in. diam. steel bolt	$\frac{3}{8}$ in. diam. steel bolt (beam to angle)	$\frac{5}{8}$ in. carriage bolts
Footing	$\frac{1}{4}$ in. steel plate welded to post	$\frac{1}{4}$ in. steel plate welded to post	$\frac{1}{4}$ in. steel plate welded to post	None
Developed by	New York	New York	New York	California
Remarks	Revised 1971	Revised 1971	Increase height of rail from 30 to 35 in. on the outside of superelevated curve. Revised 1971	Southern yellow pine is acceptable alternative to Douglas fir.

[a] 1 ft = 0.3048 m.

Source: *A Handbook of Highway Safety Design and Operating Practices.* Federal Highway Administration, Washington, D.C., 1973 and 1978 Editions.

FIGURE 8-3 Typical curb sections. (Courtesy American Association of State Highway and Transportation Officials.)

ing appearance and are most economical to build and maintain. Side slopes of 1 on 4 are used a great deal in both cut and fill sections up to about 10 ft in depth or height, but where the height of cut or fill does not exceed 6 ft a maximum side slope of 6:1 is recommended. Extremely flat slopes are sometimes used in swamp sections. As previously mentioned, where guardrail is used, slopes may be as high as 1 on 1.5. Slopes as high as 1 on 1 are generally not satisfactory and present an expensive maintenance problem. In certain fill sections, special slopes may be built with riprap, dry masonry rubble, reinforced concrete cribbing, and various types of retaining wall.

The back slopes in cut areas may vary from 6 on 1 to vertical in rock sections or loess formations, and to 1 on 1.5 in normal soil. It is sometimes advisable to have back slopes as flat as 1 on 4 when side borrow is needed. Slope transitions from cuts to fills should be gradual and should extend over a considerable length of the roadway.

8-12 Right-of-way

The right-of-way width for a two-lane highway on secondary roads with an annual average daily traffic volume of 400 to 1000 vehicles, recommended by the American Association of State Highway and Transportation Officials, is 66 ft minimum and 80 ft desirable. On the interstate system minimum widths vary, depending on conditions, from 150 ft (without frontage roads) and 250 ft (with frontage roads) to 200 and 300 ft for an eight-lane divided highway without and with frontage roads, respectively. On high-type two-lane highways in rural areas a minimum width of 100 ft, with 120 ft desirable, is recommended. A minimum width of 150 ft and a desirable width of 250 ft are recommended for divided highways.

Sufficient right-of-way should be acquired in order to avoid the expense of purchasing developed property or the removal of other encroachments from the highway right-of-way. A wide section of right-of-way provides a safer highway, permits gentle rounded slopes, and, in general, lowers maintenance and snow removal costs.

In summing up the various elements of the cross section, it can readily be seen that there is an opportunity for wide variation in design in the selection of crowns, pavement and shoulder widths, slopes, and other features. Every feature, from the centerline to the extremity of the right-of-way, must be given careful consideration for a balanced design. Typical designs are illustrated in Figures 8-4 and 8-5.

FIGURE 8-4 Typical cross-section dimensions for arterial streets. (Courtesy American Association of State Highway and Transportation Officials.)

FIGURE 8-5 Typical cross sections.

8-13 Alignment

An ideal and most interesting roadway is one that generally follows the existing natural topography of the country. This is the most economical to construct, but there are certain aspects of design that must be adhered to which may prevent the designer from following this undulating surface without making certain adjustments in vertical and horizontal directions.

The designer must produce an alignment in which conditions are consistent. Sudden changes in alignment should be avoided as much as possible. For example, long tangents should be connected with long sweeping curves, and short sharp curves should not be interspersed with long curves of small curvature. The ideal location is one with consistent alignment where both grade and curvature receive consideration and satisfy limiting criteria. The final alignment will be that in which the best balance between grade and curvature is achieved.

Terrain has considerable influence on the final choice of alignment. Generally, the topography of an area is fitted into one of the following three classifications: level, rolling, or mountainous.

In level country, the alignment is in general, limited by considerations other than grade, that is, cost of right-of-way, land use, waterways requiring expensive bridging, existing roads, railroads, canals and power lines, and subgrade conditions or the availability of suitable borrow.

In rolling country, grade and curvature must be considered carefully. Depths of cut and heights of fill, drainage structures, and number of bridges will depend on whether the route follows the ridges, the valleys, or a cross-drainage alignment.

In mountainous country, grades provide the greatest problem, and, in general, the horizontal alignment (curvature) is conditioned by maximum grade criteria.

8-14 Circular curves

Circular curves may be described by giving either the radius or the "degree of curve." In highway design, the degree of curve is defined as the central angle subtended by an arc of 100 ft. This is known as the "arc definition." According to this definition, the circumference of a circle bears the same relationship to 360° as a 100-ft arc does to the degree of curve D. Therefore,

$$\frac{2\pi R}{360°} = \frac{100 \text{ ft}}{D} \text{ and } R = \frac{5729.58}{D} \tag{8-1}$$

Books on railroad location define the degree of curve as the central angle subtended by a chord of 100 ft, and some highway departments follow this procedure. Arc or chord measurements can be considered alike for all curves less than 4° without appreciable error. An examination of tables will show that the following chords may be assumed to be equal to the arc without appreciable error:

100-ft chords up to 4°

P.C. = Point of curvature

P.I. = Point of intersection

P.T. = Point of tangency

E = External distance

M = Middle ordinate distance

R = Length of radius of curve

T = Length of tangent (P.C. to P.I. and P.I. to P.T.)

D = Degree of curve (angle subtended at the center of curve by an arc of 100 feet)

L = Length of curve in feet

L.C. = Long chord

Δ = External angle

$$T = R \text{ tangent } \frac{\Delta}{2}$$

$$E = R \text{ exsecant } \frac{\Delta}{2}$$

$$L = 100 \frac{\Delta}{D}$$

$$M = R \text{ versine } \frac{\Delta}{2}$$

$$L.C. = 2 R \sin \frac{\Delta}{2}$$

FIGURE 8-6 A simple highway curve.

50-ft chords up to 10°

25-ft chords up to 25°

10-ft chords up to 100°

Figure 8-6 shows a simple highway curve and its component parts, with the necessary formulas for finding the values of the various elements. A combination of simple curves can be arranged to produce compound or reverse curves.

In general, the maximum desirable degree of curvature is from 5 to 7° in open country and not over 10° in mountainous areas. Many states limit curvature to 3° on principal highways. However, exceptions to these limits will occur.

Two curves in the same direction connected with a short tangent, known as "broken back" curves, should be combined into one continuous curve. This may be accomplished by compounding the curve, provided that the difference between the two branches of the curve does not exceed 5°.

Where reversed curves are necessary, they should be separated by a tangent distance of at least 200 ft in order to allow proper easement from one curve to another.

For small deflections a curve should be at least 500 ft long for a 5° central angle, with an increase of 100 ft for each decrease of 1° in the central angle.

Other factors that must be considered with a curve alignment, and that materially affect the design, are superelevations, transitions, and widening.

8-15 Superelevation of curves

On rural highways, most drivers adopt a more or less uniform speed when traffic conditions permit them to do so. When making a transition from a tangent section to a curved section, if the sections are not properly designed, the vehicle must be driven at reduced speed for safety as well as for the comfort of the occupants. This is due to the fact that a force is acting on the vehicle that tends to cause an outward skidding away from the center of the curve. Most highways have a slight crowned surface to take care of drainage. It can readily be seen that when these crowns are carried along the curve, the tendency to slip is retarded on the inside of the curve because of the banking effect of the crown. The hazard of slipping is increased on the outside of the curve, however, due to the outward sloping of the crown. In order to overcome this tendency to slip and to maintain average speeds, it is necessary to "superelevate" the roadway section, that is, raise the outside edge or "bank" the curve.

Analysis of the forces acting on the vehicle as it moves around a curve of constant radius indicates that the theoretical superelevation is equal to

$$e + f = \frac{v^2}{gR} = \frac{V^2}{15R} \tag{8-2}$$

where
e = rate of superelevation (ft per foot)
f = side-friction factor
v = velocity (ft/sec)
V = velocity (mph)
R = radius of curve (ft)
g = acceleration due to gravity (32.2 ft/sec^2)

Research and experience have established limiting values for e and f. Use of the maximum e with the safe f value in the formula permits determination of minimum curve radii for various design speeds. Safe side-friction factors to be used in design range from 0.17 at 20 mph to 0.10 at 70 mph. Present practice suggests a maximum superelevation rate of 0.12 ft per foot. Where snow and ice conditions prevail, the maximum superelevation should not exceed 0.08 ft per foot. Some states have adopted a maximum superelevation rate of 0.10; however, other rates may have application on some types of highways in certain areas.

The maximum safe degree of curvature for a given design speed can be determined from the rate of superelevation and the side-friction factor. The minimum safe radius R can be calculated from the formula given above:

$$R = \frac{V^2}{15(e + f)}$$

using D as the degree of curve (arc definition)

$$D = \frac{5729.6}{R} \text{ or } D = \frac{85,900(e + f)}{V^2} \tag{8-3}$$

The relationship between superelevation and degree of curve is illustrated in Figure 8-7 and Table 8-5.

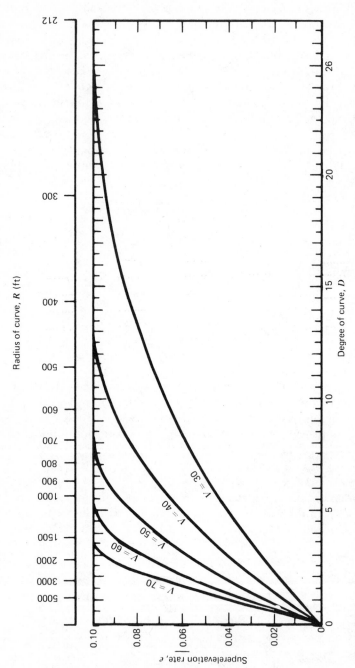

FIGURE 8-7 Design superelevation rates, $e_{max} = 0.10$. (Courtesy American Association of State Highway and Transportation Officials.)

TABLE 8-5
**MAXIMUM DEGREE OF CURVE AND MINIMUM RADIUS FOR LIMITING VALUES OF
E AND F, RURAL HIGHWAYS AND HIGH-SPEED URBAN STREETS**

Design Speed (mph)[a]	Maximum e	Maximum f	Total (e + f)	Maximum Degree of Curve	Rounded Maximum Degree of Curve	Minimum Radius (ft)[b]
20	0.04[c]	0.17	0.21	44.97	45.0	127
30	0.04	0.16	0.20	19.04	19.0	302
40	0.04	0.15	0.19	10.17	10.0	573
50	0.04	0.14	0.18	6.17	6.0	955
60	0.04	0.12	0.16	3.81	3.75	1528
20	0.06	0.17	0.23	49.25	49.25	116
30	0.06	0.16	0.22	20.94	21.0	273
40	0.06	0.15	0.21	11.24	11.25	509
50	0.06	0.14	0.20	6.85	6.75	849
60	0.06	0.12	0.18	4.28	4.25	1348
65	0.06	0.11	0.17	3.45	3.5	1637
70	0.06	0.10	0.16	2.80	2.75	2083
20	0.08	0.17	0.25	53.54	53.5	107
30	0.08	0.16	0.24	22.84	22.75	252
40	0.08	0.15	0.23	12.31	12.25	468
50	0.08	0.14	0.22	7.54	7.5	764
60	0.08	0.12	0.20	4.76	4.75	1206
65	0.08	0.11	0.19	3.85	3.75	1528
70	0.08	0.10	0.18	3.15	3.0	1910
20	0.10	0.17	0.27	57.82	58.0	99
30	0.10	0.16	0.26	24.75	24.75	231
40	0.10	0.15	0.25	13.38	13.25	432
50	0.10	0.14	0.24	8.22	8.25	694
60	0.10	0.12	0.22	5.23	5.25	1091
65	0.10	0.11	0.21	4.26	4.25	1348
70	0.10	0.10	0.20	3.50	3.5	1637

[a] 1 mph = 1.6093 km/hr.
[b] 1 ft = 0.3048 m.
[c] In recognition of safety considerations, use of e_{max} = 0.04 should be limited to urban conditions.

SOURCE: *A Policy on Geometric Design of Highways and Streets*. American Association of State Highway and Transportation Officials, Washington, D.C. (1984).

8-16 Spirals or transition curves

Transition curves serve the purpose of providing a gradual change from the tangent section to the circular curve, and vice versa. Spiral curves give more comfort and safety at high operating speeds than other methods and should be used on all curves except those that are very flat; when used in combination with superelevated sections, superelevation should be attained within the limits of the transition.

The minimum length of the transition curve is given as (*4*)

$$L_s = 1.6 \frac{V^3}{R} \qquad (8\text{-}4)$$

where
L_s = length of the transition (ft)
V = speed (mph)
R = radius (ft)

8-17 Attainment of superelevation

The transition from the tangent section to a curved superelevated section must be accomplished without any appreciable reduction in speed and in such a manner as to ensure safety and comfort to the occupants of a vehicle.

In order to effect this change, it will readily be seen that the normal road cross section will have to be tilted to the superelevated cross section. This tilting usually is accomplished by rotating the section about the centerline axis. The effect of this rotation is to lower the inside edge of the pavement and raise the outside edge without changing the centerline grade. Another method is to rotate about the inner edge of the pavement as an axis so that the inner edge retains its normal grade but the centerline grade is varied, or rotation may be about the outside edge. Rotation about the centerline is used by a majority of the states, but for flat grades too much sag is created in the ditch grades by this method. On grades below 2 percent, rotation about the inside edge is preferred. Regardless of which method is used, care should be exercised to provide for drainage in the ditch sections of the superelevated areas.

The roadway on full superelevated sections should be a straight inclined section. When a crowned surface is rotated to the desired superelevation, the change from a crowned section to a straight inclined section should be accomplished gradually. This may be done by first changing the section from the centerline to the outside edge to a level section. Second, raise the outside edge to an amount equal to one-half the desired superelevation, and at the same time change the inner section to a straight section and put the whole section in one inclined plane. Third, continue rotation about the centerline until full superelevation is reached.

The distance required for accomplishing the transition from a normal to a superelevated section, sometimes called the "transition runoff," is a function of the design speed and the rate of superelevation. Recommended minimum lengths of superelevation runoff for two-lane pavements, based primarily on appearance and comfort, are shown in Table 8-6. AASHTO (*2*) recommends that the superelevation runoff lengths for wider pavements be computed as follows:

three-lane pavements—1.2 times length for two-lane highway.

four-lane undivided pavements—1.5 times length for two-lane highway.

six-lane undivided pavements—2.0 times length for two-lane highway.

TABLE 8-6
LENGTH REQUIRED FOR SUPERELEVATION RUNOFF—TWO-LANE PAVEMENTS

Superelevation Rate, e	L—Length of Runoff (ft)[a] for Design Speed (mph)[b] of						
	20	30	40	50	60	65	70
12-ft lanes							
0.02	30	35	40	50	55	60	60
0.04	60	70	85	95	110	115	120
0.06	95	110	125	145	160	170	180
0.08	125	145	170	190	215	230	240
0.10	160	180	210	240	270	290	300
10-ft lanes							
0.02	25	30	35	40	45	50	50
0.04	50	60	70	80	90	95	100
0.06	80	90	105	120	135	145	150
0.08	105	120	140	160	180	190	200
0.10	130	150	175	200	225	240	250
Design minimum length regardless of superelevation	50	100	125	150	175	190	200

[a] 1 ft = 0.3048 m.
[b] 1 mph = 1.6093 km/hr.

SOURCE: *A Policy on Geometric Design of Highways and Streets.* American Association of State Highway and Transportation Officials, Washington, D.C. (1984).

Superelevation is usually started on the tangent at some distance before the curve starts, and full superelevation is generally reached beyond the point of curvature (P.C.) of the curve. In curves with transitions, the superelevation can be attained within the limits of the spiral. In curves of small degree where no horizontal transition is used, between 60 and 80 percent of the superelevation runoff is put into the tangent. In order to obtain smooth profiles for the pavement edges, it is recommended that the breaks at cross sections be replaced by smooth curves (see Figure 8-8). Different states adopt different methods to attain smooth edge profiles. Figure 8-9 indicates the methods adopted by one state.

8-18 Widening of curves
Extra width of pavement may be necessary on curves. As a vehicle turns the rear wheels follow the front wheels on a shorter radius, and this has the effect of

A-PAVEMENT REVOLVED ABOUT CENTER LINE

B-PAVEMENT REVOLVED ABOUT INSIDE EDGE

C-PAVEMENT REVOLVED ABOUT OUTSIDE EDGE

Note: Angular breaks to be appropriately rounded as shown by dotted line.

FIGURE 8-8 Diagrammatic profiles showing methods of attaining superelevation. (Courtesy American Association of State Highway and Transportation Officials.)

FIGURE 8-9 Superelevation using reversed parabolic curves. (Courtesy Michigan Department of State Highways and Transportation.)

increasing the width of the vehicle in relation to the lane width of the roadway. Studies of drivers traversing curves have shown that they tend to drive a curved path longer than the actual curve, shifting the vehicle laterally to the right on right-turning curves and to the left on left-turning curves. Thus, on right-turning curves the vehicle shifts toward the inside edge of the pavement, creating a need for additional pavement width. The amount of widening needed varies with the width of the pavement on tangent, the design speed, and the degree of curve.

Present practice requires no widening when the degree of curvature is less than 9° on a two-lane pavement 24 ft wide. For greater curvatures and/or narrower pavements, widening normally is from 2 to 4 ft, depending on design speed and pavement width. It is suggested that no pavement be widened less than 2 ft.

8-19 Grades and grade control

The vertical alignment of the roadway and its effect on the safe and economical operation of the motor vehicle constitute one of the most important features of road design. The vertical alignment, which consists of a series of straight lines connected by vertical parabolic or circular curves, is known as the "grade line." When the grade line is increasing from the horizontal it is known as a "plus grade," and when it is decreasing from the horizontal it is known as a "minus grade." In analyzing grade and grade controls, the designer usually studies the effect of change in grade on the centerline profile.

In the establishment of a grade, an ideal situation is one in which the cut is balanced against the fill without a great deal of borrow or an excess of cut to be wasted. All hauls should be downhill if possible and not too long. Ideal grades have long distances between points of intersection, with long vertical curves between grade tangents to provide smooth riding qualities and good visibility. The grade should follow the general terrain and rise and fall in the direction of the existing drainage. In rock cuts and in flat swampy areas it is necessary to maintain higher grades. Future possible construction and the presence of grade separations and bridge structures also control grades.

Changes of grade from plus to minus should be placed in cuts, and changes from a minus grade to a plus grade should be placed in fills. This will generally give a good design, and many times it will avoid the appearance of building hills and producing depressions contrary to the general existing contours of the land. Other considerations for determining the grade line may be of more importance than the balancing of cuts and fills.

Urban projects usually require a more detailed study of the controls and a finer adjustment of elevations than do rural projects. It is often best to adjust the grade to meet existing conditions because of the additional expense of doing otherwise.

In the analysis of grade and grade control, one of the most important considerations is the effect of grades on the operating costs of the motor vehicle. An increase in gasoline consumption and a reduction in speed are apparent when grades are increased. An economical approach would be to balance the added annual cost of grade reduction against the added annual cost of vehicle operation without grade reduction. An accurate solution to the problem depends on the knowledge of traffic volume and type, which can be obtained only by means of a traffic survey.

While maximum grades vary a great deal in various states, AASHTO recommendations make maximum grades dependent on design speed and topography (2). Present practice limits grades to 5 percent for a design speed of 70 mph. For a design speed of 30 mph, maximum grades typically range from 7 to 12 percent, depending on topography (2).

Wherever long sustained grades are used, the designer should not substantially exceed the critical length of grade without the provision of climbing lanes for slow-moving vehicles. Critical grade lengths vary from 1700 ft for a 3 percent grade to 500 ft for an 8 percent grade.

Long sustained grades should be less than the maximum grade used on any particular section of a highway. It is often preferred to break the long sustained uniform grade by placing steeper grades at the bottom and lightening the grades near the top of the ascent. Dips in the profile grade in which vehicles may be hidden from view should also be avoided.

Minimum grades are governed by drainage conditions. Level grades may be used in fill sections in rural areas when crowned pavements and sloping shoulders can take care of the pavement surface drainage. It is preferred, however, to have a minimum grade of at least 0.3 percent under most conditions in order to secure adequate drainage.

8-20 Vertical curves

The parabolic curve is used almost exclusively in connecting grade tangents because of the convenient manner in which the vertical offsets can be computed. A typical vertical curve is shown in Figure 8-10.

Offsets for vertical curves may be computed from the formulas given in Figure 8-10. It is usually necessary to make calculations at 50-ft stations, while

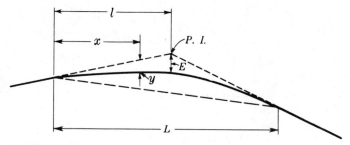

FIGURE 8-10 A typical vertical curve.

some paving operations require elevations at 25-ft intervals or less. It is often necessary to compute other critical points on the vertical curve in order to ensure proper drainage or clearances, such as at sags and crests. The low point or high point of a parabolic curve is not usually vertically above or below the vertex of the intersecting grade tangents, but it is to the left or right of this point. The distance from the P.C. of the curve to the low or high point is given as

$$x = \frac{Lg_1}{g_1 - g_2}$$

(8-5)

and the difference in elevation is given as

$$y = \frac{Lg_1{}^2}{2(g_1 - g_2)}$$

(8-6)

where
  x = distance from the P.C. to the turning point in stations
  L = length of the curve in stations
  g_1 = grade tangent from the P.C. (percent)
  g_2 = grade tangent from the point of tangency (P.T.) (percent)
  y = difference in elevation between the P.C. and the turning point (ft)

The above formulas apply only for the symmetrical curve, that is, one in which the tangents are of equal length. The unequal tangent or unsymmetrical vertical curve is a compound parabolic curve. Its use is generally warranted only where a symmetrical curve cannot meet imposed alignment conditions.
Consider the following numerical example.

EXAMPLE 8-1. *Calculation of Elevations along Vertical Curve*   A plus 3.0 percent grade intersects a plus 0.9 percent grade at station 50 + 00 and at an elevation of 630.00 ft. Calculate the vertical curve elevations at stations 49 + 00 and 51 + 50 for a 500-ft vertical curve. A sketch of the given conditions is shown below.

The external distance

$$E = \frac{AL}{8}$$

$$E = \frac{(3.0 - 0.9)5}{8} = 1.31 \text{ ft}$$

The elevation of the tangent line at station 49 + 00 is 630.00 − 3.0 = 627.00 ft. The offset

$$y = \left(\frac{x}{l}\right)^2 E = \left(\frac{150}{250}\right)^2 1.31 = 0.47 \text{ ft}$$

The curve elevation at station 49 + 00 = 627.00 − 0.47 = 626.53 ft.

The elevation of the tangent line at station 51 + 50 is 630.00 + 0.9(1.5) = 631.35 ft. The offset

$$y' = \left(\frac{100}{250}\right)^2 1.31 = 0.21$$

The curve elevation at station 51 + 50 = 631.35 − 0.21 = 631.14 ft.

8-21 Sight distance

Safe highways must be designed to give drivers a sufficient distance of clear vision ahead so that they can avoid hitting unexpected obstacles and can pass slower vehicles without danger.

Sight distance is the length of highway visible ahead to the driver of a vehicle. When this distance is not long enough to permit passing an overtaken vehicle, it is termed "stopping (or nonpassing) sight distance," The stopping distance is the minimum distance required to stop a vehicle traveling at or near the design speed before it reaches an object in its path. This stationary object may be a vehicle or some other object on the roadway. When the sight distance is long enough to enable a vehicle to overtake and pass another vehicle on a two-lane highway without interference from an oncoming vehicle, it is termed "passing sight distance."

8-22 Minimum stopping sight distance

Sight distance at every point should be as long as possible but never less than the minimum stopping sight distance. The minimum stopping sight distance is based on the sum of two distances: (1) the distance traveled from the time the object is sighted to the instant the brakes are applied, and (2) the distance required for stopping the vehicle after the brakes are applied. The first of these two distances is dependent on the speed of the vehicle and the perception time and brake-reaction time of the operator. The second distance depends on the speed of the vehicle; the condition of brakes, tires, and roadway surface; and the alignment and grade of the highway.

Perception–reaction distance

As indicated in Chapter 3, a certain amount of time is required for a driver to perceive and react and for the brakes to be actuated. For purposes of design, a combined perception–reaction and brake–reaction time of 2.5 sec is recommended. The corresponding distance traveled during this time is 2.5 times the speed in feet per second.

*reaction distance = 2.5 * speed* ⟵ *time*

Braking distance

The approximate braking distance of a vehicle on a level highway is determined by

$$d = \frac{v^2}{2fg} \tag{8-7}$$

where
d = braking distance (ft)
v = velocity of the vehicle in feet per second when the brakes are applied
g = acceleration due to gravity
f = coefficient of friction between tires and roadway

Changing v in feet per second to V in miles per hour and substituting 32.2 for g, we have

$$d = \frac{V^2}{30f} \tag{8-8}$$

⟵ mph

It is assumed that the friction force is uniform throughout the braking period. This is not strictly true; it varies as some power of the velocity. Other physical factors affecting the coefficient of friction are the condition and pressure of the tires, type and condition of the surface, and climatic conditions such as rain, snow, and ice. Friction factors for skidding are assumed to vary from 0.60 at 30 mph to 0.55 at 70 mph for dry pavements. For wet pavements these values are much lower. Recommended minimum stopping sight distances are shown in Table 8-7. In this table, perception and brake reaction times are combined.

Effect of Grade on Stopping Distance When a highway is on a grade, the formula for braking distance is modified to

$$d = \frac{V^2}{30(f \pm g)} \tag{8-9}$$

in which g is the percentage of grade divided by 100. The safe stopping distances on upgrades are shorter and on downgrades are longer than horizontal stopping distances. Table 8-8 shows the effect of grade on stopping distance.

TABLE 8-7
STOPPING SIGHT DISTANCE ON WET PAVEMENTS

Design Speed (mph)	Assumed Speed for Condition (mph)[b]	Brake Reaction Time (sec)	Brake Reaction Distance (ft)[c]	Coefficient of Friction f	Braking Distance On Level[a] (ft)[c]	Stopping Sight Distance Computed[a] (ft)[c]	Stopping Sight Distance Rounded for Design (ft)[c]
20	20–20	2.5	73.3–73.3	0.40	33.3–33.3	106.7–106.7	125–125
25	24–25	2.5	88.0–91.7	0.38	50.5–54.8	138.5–146.5	150–150
30	28–30	2.5	102.7–110.0	0.35	74.7–85.7	177.3–195.7	200–200
35	32–35	2.5	117.3–128.3	0.34	100.4–120.1	217.7–248.4	225–250
40	36–40	2.5	132.0–146.7	0.32	135.0–166.7	267.0–313.3	275–325
45	40–45	2.5	146.7–165.0	0.31	172.0–217.7	318.7–382.7	325–400
50	44–50	2.5	161.3–183.3	0.30	215.1–277.8	376.4–461.1	400–475
55	48–55	2.5	176.0–201.7	0.30	256.0–336.1	432.0–537.8	450–550
60	52–60	2.5	190.7–220.0	0.29	310.8–413.8	501.5–633.8	525–650
65	55–65	2.5	201.7–238.3	0.29	347.7–485.6	549.4–724.0	550–725
70	58–70	2.5	212.7–256.7	0.28	400.5–583.3	613.1–840.0	625–850

[a]Different values for the same speed result from using unequal coefficients of friction.
[b]1 mile = 1.6093 km.
[c]1 ft = 0.3048 m.

SOURCE: A *Policy on Geometric Design of Highways and Streets.* American Association of State Highway and Transportation Officials, Washington, D.C. (1984).

TABLE 8-8
EFFECT OF GRADE ON STOPPING SIGHT DISTANCE
IN WET CONDITIONS

	Increase for Downgrades				*Decrease for Upgrades*			
Design Speed (mph)[a]	*Correction in Stopping Distance (ft)[b]*			*Assumed Speed for Condition (mph)[a]*	*Correction in Stopping Distance (ft)[b]*		
	3%	*6%*	*9%*		*3%*	*6%*	*9%*
30	10	20	30	28	—	10	20
40	20	40	70	36	10	20	30
50	30	70	—	44	20	30	—
60	50	110	—	52	30	50	—
65	60	130	—	55	30	60	—
70	70	160	—	58	40	70	—

[a] 1 mile = 1.6093 km.
[b] 1 ft = 0.3048 m.

SOURCE: *A Policy on Geometric Design of Highways and Streets.* American Association of State Highway and Transportation Officials, Washington, D.C. (1984).

Where an unusual combination of steep grades and high speed occurs, the minimum stopping sight distance should be adjusted to provide for this factor.

8-23 Measuring Stopping Sight Distance

The sight distance available over a crest vertical curve depends on the fundamental characteristics of the curve, namely the algebraic difference in grades and the length of the curve. The basic relationships are

$$\text{when } S < L, L = \frac{AS^2}{100(\sqrt{2h_1} + \sqrt{2h_2})^2} \tag{8-10}$$

$$\text{when } S > L, L = 2S - \frac{100(\sqrt{2h_1} + \sqrt{2h_2})^2}{A} \tag{8-11}$$

where
S = sight distance available over crest vertical curve (ft)
L = length of vertical curve (ft)
A = algebraic difference in grades (percent)
h_1 = height of eye of average driver (ft)
h_2 = height of object sighted (ft)

For the purposes of design, it is assumed that the height of eye of the average driver $h_1 = 3.5$ ft and the height of a stationary object $h_2 = 0.5$ ft. For these conditions, the above equations become

$$\text{when } S < L, S = 36.46\sqrt{\frac{L}{A}} \tag{8-12}$$

$$\text{when } S > L,\ S = \frac{L}{2} + \frac{664.6}{A}$$

(8-13)

These equations give the stopping sight distances provided by a given or assumed crest vertical alignment. Combining these equations and the requirements shown in Table 8-7, AASHTO *(2)* has published a chart (Figure 8-11*a*) that can be used to determine the minimum length of vertical curve to provide stopping sight distance for a given design speed. A similar chart for sag vertical curves based on headlight sight distance has also been published. (See Figure 8-11*b*) The latter chart shows the minimum length of sag vertical curve to provide a light beam distance sufficient to stop for different design speeds.

8-24 Minimum passing sight distance

The majority of U.S. highways carry two lanes of traffic moving in opposite directions. In order to pass slower-moving vehicles, it is necessary to use the lane of opposing traffic. If passing is to be accomplished safely, the vehicle driver must be able to see enough of the highway ahead in the opposing traffic lane to have sufficient time to pass and then return to the right traffic lane without cutting off the passed vehicle and before meeting the oncoming traffic. The total distance required for completing this maneuver is the passing sight distance.

When computing minimum passing sight distances, various assumptions must be made about traffic behavior. On two-lane highways it may be assumed that the vehicle being passed travels at a uniform speed and that the passing vehicle is required to travel at this same speed when the sight distance is unsafe for passing. When a safe passing section is reached, a certain period of time elapses in which the driver decides whether or not it is safe to pass. When he decides to pass, it is assumed that he accelerates his speed during the entire passing operation. It is also assumed that the opposing traffic appears the instant the passing maneuver starts and arrives alongside the passing vehicle when the maneuver is completed.

Accepting these assumptions, it can be shown that the passing minimum sight distance for a two-lane highway is the sum of four distances: d_1, the distance traveled during perception and reaction time and during the initial acceleration to the point where the vehicle will turn into the opposite lane; d_2, the distance traveled while the passing vehicle occupies the left lane; d_3, the distance between the passing vehicle at the end of its maneuver and the opposing vehicle; and d_4, the distance traveled by the oncoming vehicle for two-thirds of the time the passing vehicle occupies the left lane.

The preliminary delay distance d_1 is computed from the formula

$$d_1 = 1.47 t_1 \left(v - m + \frac{a t_1}{2} \right)$$

(8-14)

where
t_1 = time of preliminary delay (sec)
a = average acceleration rate (mphps)
v = average speed of passing vehicle (mph)
m = difference in speed of passed vehicle and passing vehicle (mph)

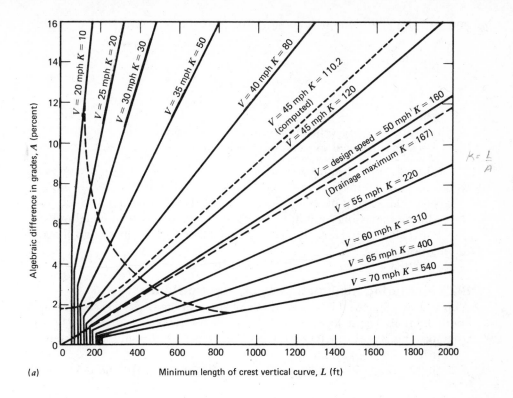

$K = \dfrac{L}{A}$

(a)   Minimum length of crest vertical curve, L (ft)

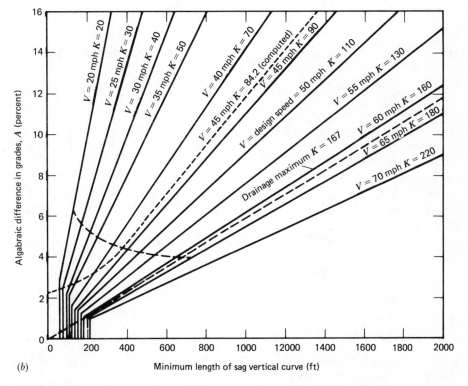

(b)   Minimum length of sag vertical curve (ft)

FIGURE 8-11 Design controls for vertical curves. (*a*) Crest vertical curves. (*b*) Sag vertical curves. (Courtesy American Association of State Highway and Transportation Officials.)

The distance

$$d_2 = 1.47vt_2$$

where

t_2 = time passing vehicle occupies the left lane (sec)
v = average speed of passing vehicle (mph)

The clearance distance d_3 varies from 110 to 300 ft. These distances are adjusted as shown in Table 8-9 and Figure 8-12. The distance $d_4 = 2d_2/3$.

The relationship of d_1, d_2, d_3, and d_4 is illustrated in Figure 8-12.

Minimum passing sight distances for the design of two-lane highways are shown in Table 8-10. Here the passing speed is related to the design speed of the highway.

8-25 Measuring minimum passing sight distance

As previously stated, the height of eye of the average driver is assumed to be about 3.5 ft above the pavement. As vehicles are the objects that must be seen when passing, it is assumed that the height of object for passing sight distance is 4.25 ft high on the pavement.

Substituting $h_1 = 3.5$ ft and $h_2 = 4.25$ ft into Eqs. 8-10 and 8-11, we ob-

TABLE 8-9
ELEMENTS OF SAFE PASSING SIGHT DISTANCE—TWO-LANE HIGHWAYS

Speed Group (mph)[a] *Average Passing Speed (mph)*	*30–40* *34.9*	*40–50* *43.8*	*50–60* *52.6*	*60–70* *62.0*
Initial maneuver:				
a = average acceleration (mphps)[b]	1.40	1.43	1.47	1.50
t_1 = time (sec)[b]	3.6	4.0	4.3	4.5
d_1 = distance traveled (ft)[c]	145	215	290	370
Occupation of left lane:				
t_2 = time (sec)[b]	9.3	10.0	10.7	11.3
d_2 = distance traveled (ft)	475	640	825	1030
Clearance length:				
d_3 = distance traveled (ft)[b]	100	180	250	300
Opposing vehicle:				
d_4 = distance traveled (ft)	315	425	550	680
Total distance, $d_1 + d_2 + d_3 + d_4$ (ft)	1035	1460	1915	2380

[a] 1 mile = 1.6093 km.
[b] For consistent speed relation, observed values adjusted slightly.
[c] 1 ft = 0.3048 m.

Source: *A Policy on Geometric Design of Highways and Streets.* American Association of State Highway and Transportation Officials (1984).

FIGURE 8-12 Passing sight distance relationships. (Courtesy American Association of State Highway and Transportation Officials.)

tain equations that give the passing sight distance provided by a given crest vertical alignment:

$$\text{when } S < L, \ S = 55.61\sqrt{\frac{L}{A}} \tag{8-15}$$

$$\text{when } S > L, \ S = \frac{L}{2} + \frac{1546.4}{A} \tag{8-16}$$

TABLE 8-10
MINIMUM PASSING SIGHT DISTANCE FOR
DESIGN OF TWO-LANE HIGHWAYS

Design Speed (mph)[a]	Assumed Speeds		Minimum Passing Sight Distance (ft)[b]	
	Passed Vehicle (mph)	Passing Vehicle (mph)	Figure 8-12	Rounded
30	26	36	1090	1100
40	34	44	1480	1500
50	41	51	1840	1800
60	47	57	2140	2100
65	50	60	2310	2300
70	54	64	2490	2500

[a] 1 mile = 1.6093 km.
[b] 1 ft = 0.3048 m.

SOURCE: *A Policy on Geometric Design of Highways and Streets.* American Association of State Highway and Transportation Officials (1984).

8-26 Horizontal sight distances

Horizontal sight distances may be scaled directly from the plans with a fair degree of accuracy. However, where the distance from the highway centerline to the obstruction is known, it is possible to determine the sight distance for a known radius or degree of curve from the relations

$$R = \frac{m}{1 - \cos\dfrac{SD}{200}} \tag{8-17}$$

$$S = \frac{200}{D} \cos^{-1}\left(\frac{R - m}{R}\right) \tag{8-18}$$

where

m = distance from center of roadway to edge of obstruction (ft)
S = sight distance along the center of the inside lane (ft)
R = radius of curvature (ft)
D = degree of curvature

The formulas do not apply when the length of the circular curve is less than the sight distance or when the radius of curvature is not constant throughout.

8-27 Design standards

Because of functional differences, a wide array of design standards must be used for the many types of facilities comprising a highway system. Design standards vary widely for different functional classes. For example, freeways are designed predominantly for traffic movement, and freeway standards are

TABLE 8-11
LOCAL STREET DESIGN GUIDELINES

Terrain classification	Level			Rolling			Hilly		
Development density	Low	Medium	High	Low	Medium	High	Low	Medium	High
Right-of-way width (ft)[a]	50	60	60	50	60	60	50	60	60
Pavement width (ft)	22–27	28–34	36	22–27	28–34	36	28	28–34	36
Type of curb (V = vertical face; R = roll-type; 0 = none)	0/R	V	V	0	V	V	V	V	V
Sidewalks and bicycle paths (ft)	0	4–6	→	0	←	4–6	0	←	4–6
Sidewalk distance from curb face (ft)	—	6	6	—	6	6	—	6	6
Minimum sight distance (ft)	←	200	→	←	150	→	←	110	→
Maximum grade (%)	←	4	→	←	8	→	←	15	→
Maximum cul-de-sac length (ft)	1000	700	700	1000	700	700	1000	700	700
Minimum cul-de-sac radius (right-of-way) (ft)	←	50	→	←	50	→	←	50	→
Design speed (mph)[b]	←	30	→	←	25	→	←	20	→
Minimum centerline radius of curves (ft)	←	250	→	←	175	→	←	110	→
Minimum tangent between reverse curves (ft)	←	50	→	←	50	→	←	50	→

[a] 1 ft = 0.3048 m.
[b] 1 mph = 1.6093 km/hr.
SOURCE: *Recommended Guidelines for Subdivision Streets: A Recommended Practice.* Institute of Transportation Engineers, Washington, D.C. (1984).

TABLE 8-12
COLLECTOR STREET DESIGN GUIDELINES

Terrain classification	Level			Rolling			Hilly		
Development density	Low	Medium	High	Low	Medium	High	Low	Medium	High
Right-of-way width (ft)[a]	70	70	70	70	70	70	70	70	70
Pavement width (ft)	36	36	40	36	36	40	36	36	40
Type of curb (v = vertical face)	V	V	V	V	V	V	V	V	V
Sidewalk width (ft)		4–6			4–6			4–6	
Sidewalk distance from curb face (ft)		10			10			10	
Minimum sight distance (Ft)		250			200			150	
Maximum grade		4			8			12	
Minimum spacing along major traffic route (ft)		1300			1300			1300	
Design speed (mph)[b]		35			30			25	
Minimum centerline radius (ft)[c]		350			250			175	
Minimum tangent between Reverse curves (ft)		100			100			100	

[a] 1 ft = 0.3048 m.
[b] 1 mph = 1.6093 km/hr.
[c] Assumes superelevation.
SOURCE: *Recommended Guidelines for Subdivision Streets: A Recommended Practice.* Institute of Transportation Engineers, Washington, D.C. (1984).

characterized by high design speeds, wide lanes, and straight horizontal and vertical alignments. At the other end of the spectrum, standards for subdivision streets reflect the emphasis on land access function. Such standards, exemplified by Tables 8-11 and 8-12, are based on design speeds of 20 to 35 mph. In fact, local streets should be designed to discourage excessive speeds through the use of curvilinear alignment and discontinuities in the street system (5).

Within a functional class, design standards may vary with the type of terrain, anticipated traffic to be served, and whether the highway is to be in an urban or rural area. Table 8-13 illustrates the importance of these variables for highway standards. It shows the minimum and desirable design speeds used for rural roads in North Carolina.

PLANNING AND DESIGN OF BICYCLE FACILITIES

Since 1970 there has been a phenomenal growth in bicycling activity in the United States. Between 1970 and 1973 annual bicycle sales more than doubled, the greatest growth occurring in sales of bicycles for adults. By 1972 the rate of bicycle sales exceeded that of automobiles, and the estimated number of bicycle users in the United States was 85 million.

The remarkable growth in bicycling was accompanied by demands for more and better bicycle facilities. In recent years, many highway agencies have undertaken studies of demand for bicycling facilities, bikeway design and location criteria, and bicycle safety. A number of agencies have made provision for bicycling an integral part of their planning and design activities.

In the following paragraphs, criteria and standards for the planning and design of bicycle facilities will be described briefly. It should be remembered that variance exists among jurisdictions in recommended bikeway planning and design practices.

TABLE 8-13
MINIMUM AND DESIRABLE DESIGN SPEEDS (mph) FOR RURAL ROADS IN NORTH CAROLINA

	Type of Terrain		
Traffic Volume	Level	Rolling	Mountainous
Current *ADT*<50	40(55)[a]	30(55)	20(55)
Current *ADT* 50−250	40(55)	30(55)	20(55)
Current *ADT* 250−400	50(60)	40(60)	20(60)
Current *ADT* 400−750 and *DHV* 100−200	50(60)	40(60)	30(60)
DHV 200−400	50(60)	40(60)	30(60)
DHV≥400	50(60)	40(60)	30(60)

[a]Desirable design speeds are shown in parentheses.

SOURCE: *Design Manual*, Vol. I, *Roadway*. North Carolina Department of Transportation, Raleigh (1985).

8-28 Bicycle facility planning

Bicycle planning procedures are similar to those for other transportation modes. The process should include

1. Establishing goals and objectives.
2. Collecting information on characteristics of current bicycle riding, generators of bicycle traffic, as well as inventories of existing and potential bike facilities.
3. Establishing planning principles and design standards.
4. Forecasting demand for bicycling facilities.
5. Development of a bicycle facility plan.

Warrants for bicycle facilities

Warrants have not generally been established for bicycle facility improvements. The most important factors to consider in setting warrants are

1. Automobile volume or volume/capacity ratio.
2. Volume of bicycle traffic.
3. Difference in auto and bicycle speeds.

When determining the need for a separate bicycle facility, other factors that might be considered include available space for a bicycle path, cost, and other possible means of providing for bicycle traffic.

AASHTO *(6)* gives the following general warrants for separate bicycle facilities:

bicycle volumes ≥200/day and motor vehicle volumes ≥2000 ADT

or

bicycle volumes ≥200/day and motor-vehicle speeds −40 mph.

Forecasting demand

A forecast of bicycling demand is needed to see if a bike facility is required, to determine the type of facility, and to help establish design criteria. Unfortunately, there are no generally accepted methods and little experience in forecasting demand, especially latent demand. Forecasts of demand can be based on special surveys (e.g., origin−destination studies), experience with other nearby facilities, and experience in other cities or countries. Caution should be exercised when basing demand estimates on the experience at other locations, being sure to allow for differences in conditions.

Major factors responsible for large bicycling demand are

1. Good weather.
2. Level terrain.
3. Favorable demographic characteristics (e.g., large numbers of young adults).
4. Lack of alternative transportation modes.

Development of bicycle facility plan

A bicycle facility plan should be developed within a framework of possible recreational and utilitarian networks. This recognizes that there are two general classes of bicycle trips: recreational trips and all others (termed utilitarian trips). Reference 7 suggests the consideration of eight bikeway network concepts, shown as Table 8-14.

Potentially suitable bicycle facility corridors should be identified, preferably following closely the desire lines between origins and destinations. The lengths of designated routes vary widely, but routes serving mainly recreational trips should be over 5 miles in length. Generally, commuting trips will be shorter than 5 miles. Bike routes should take advantage of good views, historic sights, and so forth and should be compatible with existing land use.

8-29 Design of bicycle facilities

As Figure 8-13 illustrates, bicycle facilities may be grouped into three classes:

Class I—Bike path or trail. A completely separate roadway designated for the exclusive use of bicycles; typically separated from motor-vehicle roadways by an open space or barrier.

Class II—Bike lane. A portion of a roadway which has been designated for exclusive or preferential use by bicycles; normally distinguished by a paint stripe, curb, or barrier.

Class III—Shared roadway or bike route. A roadway that has been officially designated and marked as a bicycle route but which is used by both motor vehicle and bicycle traffic.

TABLE 8-14
BIKEWAY NETWORK CONCEPTS

Concept	Examples
A. Recreational	
1. Linear networks	Abandoned railroads, utility corridors
2. Paired linear networks	Paired pathways on either side of a stream
3. Off-street looped networks	Bikeway around a lake
4. On-street looped networks	Bikeways along residential streets
B. Utilitarian	
1. Grid networks	Selected improvements along existing arterial and collector streets
2. Penetrator networks	Bike lanes giving access to central business district or other major generators
3. Selected improvements	Grade separations, channelization
4. Midblock networks	Off-street facilities constructed during original area development

SOURCE: *Bicycling in Tennessee, Planning and Design Manual.* Barton-Aschman Associates, Minneapolis, Minn. (1975).

BIKE ROUTE

(CLASS III)

PED.   STREET OR HIGHWAY   PED.
WALK   WALK

BIKE LANE

(CLASS II)

PED.   BIKE   STREET OR HIGHWAY   BIKE
WALK   LANE   LANE

BIKE PATH

(CLASS I)

STREET OR HIGHWAY   PED.   BIKE
                             WALK   PATH

FIGURE 8-13 Principal alternative bikeway facilities. (Courtesy Barton-Aschman Associates, Inc.)

Width of bicycle facilities

The minimum width of a one-lane bikeway is 3.5 ft. This allows 2.0 ft for the bike width and 0.75 ft for weaving space on each side. Since an additional shy distance from obstructions is needed, a desirable surface width of 4.0 ft is recommended (6). For a two-lane bikeway, the surface width should be at least 7.0 ft and preferably 8.0 ft. An additional width of 0.5 to 1.0 ft should be provided if the bikeway is bordered by a raised curb, and 1.0 to 2.0 ft additional width is suggested if there are raised curbs on both sides. A minimum horizontal clearance of 0.8 ft and preferably 2.0 ft is needed between the edge of the bikeway and trees, poles, and other obstacles. At least 2.0 ft lateral clearance should be allowed between the edge of bikeways and parked cars.

Bikeway capacity

Little research has been performed to establish precisely the capacity of bikeways. Available evidence suggests that "once basic operating space require-ments are met, the bikeway would have ample capacity for almost all situations" *(8)*. Empirical studies indicate that a one-way bicycle lane with an effective width* of 4 ft has a capacity of approximately 1275 bikes per hour. A two-way lane with an effective width of 8 ft has a capacity of about 1900 bikes per hour *(8)*.

Bikeway design speed

Studies show that nearly all bikers travel within a range of 7 to 15 mph, with an average of 10 to 11 mph. Higher speeds are noted on downgrades. AASHTO *(6)* recommends the use of a design speed of 15 mph normally and 20 mph for long downgrades.

Curvature and superelevation

If one analyzes the forces on a biker traveling around a curve, it can be shown that

$$e + f = \frac{V^2}{15R} (1 - ef)$$

(8-19)

where
  e = superelevation rate (ft per foot)
  f = coefficient of friction; for dry pavement $f = 0.4$ is recommended *(9)*
  V = design speed (mph)
  R = radius of curvature (ft)

This equation, a precise version of Eq. (8-2), also applies to motor-vehicle facilities. When e and f are both small, Eq. (8-2) gives satisfactory results.

Engineers are not in agreement as to what constitutes a suitable maximum superelevation rate for bicycle facilities. AASHTO *(6)* recommends a maxi-mum value of 0.05 ft per foot. The state of Oregon *(9)* uses a maximum rate of 0.06 where pedestrians constitute more than half of the traffic. Superelevation rates should never exceed 0.12 ft per foot.

Once a maximum value is chosen, a minimum radius of curvature can be computed with Eq. (8-2). For example, using $f = 0.4$ and $e_{max} = 0.05$, we obtain:

Design Speed (mph)	Minimum Radius (ft)
15	33
20	60

*Effective width = width minus shy distance (10 to 20 in.) from path side obstructions.

Because of a biker's tendency to lean to the inside of a curve, pavements should be widened around curves. AASHTO (6) recommends that bikeway pavements be widened up to 2 ft along curves where the radius is less than 50 ft and the design speed is 10 mph or greater. The Oregon State Highway Division (9) specifies that highway pavements along curves be widened as much as 4 ft and that tapers be provided at each end of the widened section.

Grades

The maximum grade that a bicyclist will be able to negotiate depends on the capability of the individual biker, the length of grade, as well as the condition of the bicycle and road surface, weather conditions, and so forth. Where the grade exceeds 5 percent, the length should be not more than 300 ft and preferably not more than 100 ft. AASHTO (6) recommends a maximum grade of 10 percent. Where a 10 percent gradient is used, its length should not exceed about 50 ft.

Sight distance

Sight distance criteria are similar to those for the design of motor-vehicle facilities. Minimum lengths of vertical curves may be determined by Eqs. 8-10 and 8-11, using a height of eye of the cyclist $h_1 = 4.5$ ft and a height of object $h_2 = 0.33$ ft.

As described in Section 8-22, minimum stopping distance is the sum of the perception−reaction distance and the braking distance. Design stopping distances are shown in Table 8-15 based on a perception−reaction time of 2.5 sec and a coefficient of skid resistance $f = 0.25$.

Checks should be made to ensure that adequate sight distance is provided around curves and at intersections with other bikeways or with motor-vehicle

TABLE 8-15
DESIGN STOPPING SIGHT DISTANCES
FOR BICYCLES

Design Speed (mph)[a]	Stopping Sight Distances (ft)[b] for Downhill Gradients of			
	0%	5%	10%	15%
10	50	50	60	70
15	85	90	100	130
20	130	140	160	200

[a] 1 mile = 1.6093 km.
[b] 1 ft = 0.3048 m.

SOURCE: Guide for Bicycle Routes. American Association of State Highway and Transportation Officials, Washington, D.C. (1974).

roadways. The arc distance available for stopping along a bikeway centerline may be computed by the equation

$$S = \frac{200R}{5730} \cos^{-1}\left(\frac{R - m}{R}\right) = \frac{R}{28.65} \cos^{-1}\left(\frac{R - m}{R}\right) \qquad (8\text{-}20)$$

The symbols used in this equation are defined in Section 8-26. The distances computed by the equation should in each instance be greater than the safe stopping sight distances shown in Table 8-15. Reference 8 describes the procedures for checking sight distances at intersections.

PLANNING AND DESIGN FOR PEDESTRIANS

Each year approximately 10,000 pedestrians are killed after being struck by motor vehicles and about 120,000 are injured. Many pedestrians who do not have access to an automobile or public transit, such as the young, the elderly, and the handicapped, face special difficulties in moving about streets and sidewalks congested with motor vehicles and other pedestrians. Increasing traffic volumes and pedestrian accident losses have been accompanied by a growing public awareness of the need for safe and convenient pedestrian facilities. In the following paragraphs we will briefly describe standards and criteria for the planning and design of pedestrian facilities.

8-30 Sidewalks

The need for and width of sidewalks depend primarily on the type and density of land development and the volume of pedestrian and vehicular traffic. In rural areas, sidewalks are seldom provided since pedestrian traffic is usually very light. Along residential subdivision streets, 5-ft-wide sidewalks are recommended where the development density exceeds two dwelling units per acre *(5)*. Sidewalks should generally be provided along all arterial streets that are not provided with shoulders, even though pedestrian traffic is light *(1)*. A minimum width of 6 ft is recommended for commercial areas and major school routes.

Sidewalk widths in central business districts and other high-density areas may be determined objectively on the basis of capacity analyses. The quality and rate of pedestrian flow depend on the amount of space available per pedestrian. According to Pushkarev and Zupan *(10)*, pedestrian flow can be grouped into seven categories ranging from "open flow" to "jammed." The seven groups are listed in Table 8-16 along with the corresponding rates of flow and the amount of space per person.

Fruin *(11)* described six levels of service for pedestrian flows analogous to those for highway flows (Chapter 5). These are described in Table 8-17 and illustrated in Figure 8-14.

In estimating pedestrian flow along sidewalks, the effective width should be reduced by an additional 2 ft or more to account for the constricting effects of mailboxes, fire hydrants and other street furniture.

TABLE 8-16

CHARACTERISTICS OF PEDESTRIAN FLOW IN A HOMOGENEOUS STREAM

Type of Flow	Flow Rate, (persons per min per ft)[a]	Space per Person (ft²)	Description of Flow
Open	<0.5	>530	Complete freedom to select speed and direction of movement.
Unimpeded	0.5−2.0	530−130	Frequent indirect interaction with others.
Impeded	2.0−6.0	130−40	Constant indirect interaction with others.
Constrained	6.0−10.0	40−24	Crossing and passing movements are possible, but with interference and likely conflicts.
Crowded	10.0−14.0	24−16	Probability of conflicts is fairly high, and passing is difficult.
Congested	14.0−18.0	16−11	Frequent body contacts; difficult to walk at a normal pace.
Jammed	0−25.0	2−11	Only shuffling movement is possible

[a]1 ft = 0.3048 m.

Source: *Urban Space for Pedestrians*. Boris S. Pushkarev with Jeffrey M. Zupan, MIT Press, Cambridge, Mass. (1975).

EXAMPLE 8-2. *Sidewalk Flow Evaluation*   A new office building is expected to add 1000 pedestrians to an 18-ft sidewalk during the peak 10-min period. The sidewalk already has a flow of 1600 pedestrians during the peak period. Because of parking meters, light standards, and other obstructions, subtract 3 ft from the actual sidewalk width to obtain the effective width. Characterize the quality of flow by using Tables 8-16 and 8-17.

The pedestrian volume P per foot of sidewalk width per minute is

$$P = \frac{2600 \text{ pedestrians}}{10 \text{ min} \times 15 \text{ ft.}} = 17.3$$

According to Table 8-16, the sidewalk flow would be termed "congested." By Table 18-17 and Figure 8-14, a level of service D would be expected.

The numbers of pedestrians within a given area can be determined by manual counts, analysis of time-lapse movies made from a high vantage point, and study of a series of timed, sequentially overlapping photographs taken from a helicopter or light aircraft (sky counts).

The forecasting of pedestrian traffic generated by proposed buildings is a complex process, the details of which are beyond the scope of this text. Generally, for a given building, estimates of such traffic can be made by reference to

TABLE 8-17
LEVEL OF SERVICE DESCRIPTION FOR WALKWAYS

Level of Service	Flow Rate, (persons per min per ft)[a]	Space per Person, (ft²)	Description of Flow
A	7 or less	35	Pedestrians freely select walking speed and are able to bypass slower pedestrians and avoid crossing conflicts.
B	7–10	25–35	Where reverse direction or crossing movements exist, minor conflicts occur, slightly lowering mean speeds and volumes.
C	10–15	15–25	Freedom to select individual walking speed and freely pass others is restricted.
D	15–20	10–15	The majority of persons have normal walking speeds restricted and reduced. Reverse flow and crossing movements severely restricted.
E	20–25	5–10	Virtually all persons have normal walking speeds restricted requiring frequent adjusting of gait. At lower end of range, forward progress made only by shuffling.
F	Variable up to 25	5 or less	All pedestrian walking speeds are extremely restricted. Frequent, unavoidable contact with others. Forward progress made only by shuffling.

[a]1 ft = 0.3048 m.

SOURCE: *Pedestrian Planning and Design*. John J. Fruin, Metropolitan Association of Urban Designers and Environmental Planners, Inc., New York (1971).

empirical trip generation rates. See Table 8-18. In aggregating pedestrian volumes for a given location, it is necessary to account for variations in peaking patterns (which depend on the type of building use) and in trip lengths (which depend on trip purpose). For further information on forecasting pedestrian traffic, see Refs. 10 and 11.

Generally, sidewalks should be located well away from the curb face. Such a location provides increased safety, facilitates driveway gradient design, and

FIGURE 8-14 Level of service standards for walkways. (Source: John J. Fruin, *Pedestrian Planning and Design*, Metropolitan Association of Urban Designers and Environmental Planners, Inc., New York, (1971).)

TABLE 8-18
DAILY AND PEAK-PERIOD TRIP GENERATION RATES
FOR VARIOUS BUILDING USES

| | *In and Out Trips per* | |
| | *1000 ft² (93 m²) of Floor Space* | |
Building Use	*Daily*	*Peak 15 Min*	*Time of Peak Period*
Cafeteria	492[a]	22.36	12:00 noon
Dept. store	252	15.64	12:45 P.M.
Supermarket	285	11.97	5:00 P.M.
Restaurant	173	10.20	1:15 P.M.
Office, hdqrs.	14.2	0.98	8:45 A.M.
Office, hdqrs.	13.2	0.97	5:00 P.M.
Office, mixed use	17.3	0.89	4:45 P.M.
Residence	8.3	0.25	5:45 P.M.

[a]10-hr rate only.

SOURCE: *Urban Space for Pedestrians*. Boris S. Pushkarev with Jeffrey M. Zupan, MIT Press, Cambridge, Mass. (1975).

provides an area for street furniture and temporary storage of snow. Many agencies place the sidewalks 1 ft from the right-of-way line *(5)*.

8-31 Provisions for elderly and handicapped pedestrians

Special consideration must be given to elderly and handicapped pedestrians. Curb heights should generally not exceed 7 in., and at locations where significant wheelchair traffic is to be expected, ramps should be provided. Recommended ramp designs are shown by Figure 8-15. The suggested 1:12 ramp slope should also be satisfactory for bicycle traffic.

Sidewalk pavements should be free of holes, bumps, and other irregularities, and the surface should be skid-resistant. Gratings and manhole covers, which tend to become slippery and create tripping hazards, should be removed from crosswalks and other walkway areas whenever feasible. Special barriers and surface textures that can be felt may be needed in areas where there are significant numbers of blind pedestrians.

FIGURE 8-15 Typical curb ramp design. (Courtesy California Department of Rehabilitation.)

PROBLEMS

8-1   Given $\Delta = 32.4°$ right, $D = 4.25°$ and the P.I. station $75 + 28.10$. Compute the curve data and the station of the P.T. Compute the deflection angles at even 100-ft stations. Use the arc definition formulas.

8-2*   Solve Problem 8-1 in SI (metric) units. The "metric station"[+] of the P.I. would become M22 + 94.593. Compute the curve data and metric station of the P.T. and the deflection angles at metric stations and half metric station.

8-3   What superelevation rate would you recommend for a roadway with a design speed of 75 mph and a degree of curve $D = 4.0$. Assume $f = 0.11$.

8-4*   Develop an equation similar to Eq. 8-2 in SI (metric) units. What superelevation rate would you recommend for a roadway with a design speed of 100 km/hr and a radius of 450 m?

8-5   A 1-mile-long racetrack is to be designed with ¼-mile turns ($L = 1320$ ft) at each end. Determine the superelevation rate for a design speed of 70 mph, assuming $f = 0.11$.

8-6*   A 1-km-long racetrack is to be designed with turns 250 m in length at each end. Determine the superelevation rate you would recommend for a design speed of 140 km/hr, assuming $f = 0.2$.

8-7   A plus 5.1 percent grade intersects a minus 2.8 percent grade at station $365 + 85$ at an elevation of 427.50 ft. Calculate the centerline elevations for every even 100-ft station for a 600-ft vertical curve.

8-8*   A plus 3.9 percent grade intersects a plus 1.1 percent grade at a metric station of $M50 + 25$ and an elevation of 622.40m. Calculate the centerline elevations at metric stations $49 + 50$, $50 + 00$, and $51 + 00$ for a 200-m curve.

8-9   A vertical parabolic curve is to be used under a grade-separation structure. The curve is 775 ft long and the minus grade of 3.8 percent intersects the plus grade of 6.3 percent at station $250 + 75$. Calculate the low point of the curve if the intersection at station $250 + 75$ is at elevation 426.83 ft.

8-10*   Develop stopping sight distance equations similar to Eqs. 8-12 and 8-13 using SI (metric) units.

8-11   Calculate the stopping sight distance over a crest of a 1500-ft vertical curve with a plus grade of 4.4 percent and a minus grade of 2.3 percent.

8-12*   Develop passing sight distance equations similar to Eqs. 8-15 and 8-16 using SI (metric) units.

8-13   Calculate the passing sight distance over the crest described in Problem 8-11.

8-14*   Develop an equation to determine the minimum length of transition curve similar to Eq. 8-4 using SI (metric) units.

8-15   Given a horizontal curve with a 1360-ft radius, estimate the minimum length of spiral necessary for a smooth transition from tangent alignment to the circular curve. The design speed is 60 mph.

[+] As used here, metric station = 100m.

8-16* Develop an equation for horizontal sight distance similar to Eq. 8-18 using SI (metric) units.

8-17 The superelevation rate for a certain highway is 0.084. Prepare a plot showing the ordinates required for the transition from a tangent section to the full superelevated section, assuming that the pavement is 24 ft wide. (Use Figure 8-9 as a guide.)

8-18 A corner of a building is situated adjacent to a horizontal curve ($R = 60$ ft) 10 ft from the edge of a two-lane, two-way bikeway. The bikeway width is 8 ft. Determine if stopping sight distance is provided for the curve. Assume a design speed of 15 mph and zero gradient.

8-19 A total of 1000 pedestrians use a sidewalk during the peak 10-min period. If the effective width is 9 ft, characterize the quality of flow using Tables 8-18 and 8-19.

REFERENCES

1. *A Policy on Design of Urban Highways and Arterial Streets.* American Association of State Highway Officials, Washington, D.C. (1973).

2. *A Policy on Geometric Design of Highways and Streets.* American Association of State Highway and Transportation Officials, Washington, D.C. (1984).

3. *Highway Capacity Manual.* Transportation Research Board Special Report No. 209 (1985).

4. Barnett, Joseph. *Transition Curves for Highways.* U.S. Government Printing Office, Washington, D.C. (1940).

5. *Recommended Guidelines for Subdivision Streets.* Institute of Transportation Engineers, Washington, D.C. (1984).

6. *Guide for Bicycle Routes.* American Association of State Highway and Transportation Officials, Washington, D.C. (1974).

7. *Bicycling in Tennessee, Planning and Design Manual.* Barton-Aschman Associates, Minneapolis, Minn. (1975).

8. *Bikeways—State of the Art, 1974.* Prepared for the Federal Highway Administration by DeLeuw Cather and Company, San Francisco (1974).

9. *Bikeway Design.* Oregon State Highway Division, Salem (1974).

10. Pushkarev, Boris S., with Zupan, Jeffrey M. *Urban Space for Pedestrians.* MIT Press, Cambridge, Mass. (1975).

11. Fruin, John J. *Pedestrian Planning and Design.* Metropolitan Association of Urban Designers and Environmental Planners, New York (1971).

chapter nine

INTERSECTIONS, INTERCHANGES, TERMINALS

In the previous chapter, we presented standards and procedures for the design of highway links, the facilities that serve "line-haul" transportation. In this chapter, we will discuss the design of facilities to accommodate trips at nodes and termini. We will discuss the design of (1) intersections and interchanges that provide for change in travel direction and (2) parking facilities and terminals that provide for interaction with passengers and the storage of vehicles at the termini.

INTERSECTIONS AT GRADE, GRADE SEPARATIONS, AND INTERCHANGES

9-1 Intersections at grade

Most highways intersect at grade and the intersection area should be designed to provide adequately for turning and crossing movements, with due consideration given to sight distance, signs, grades, and alignment.

Simple intersections at grade consist of a junction of three, four or more road approaches. A junction of three approaches is indicated as a "branch," **T**, or **Y**. A branch may be defined as an offshoot of a main-traveled highway, and it usually has a small deflection angle. A **T** intersection is one in which two approaches intersect to form a continuous highway and the third approach intersects at, or nearly at, right angles. A **Y** intersection is one in which three approaches intersect at nearly equal angles. In addition to the above types of intersection, the flared intersection may be used. This consists of additional pavement width or additional traffic lanes at the intersection area. A few examples of these types of intersections are given in Figure 9-1.

The design of the edge of the pavement for a simple intersection should provide sufficient clearance between the vehicle and the other traffic lanes. It is frequently assumed that all turning movements at intersections are accomplished at speeds of less than 20 mph, and the design is based on the physical characteristics of the assumed design vehicle.

The design of the curb line at intersections depends on the composition of the traffic, the angle of intersection, the right-of-way available, the width and number of lanes on the intersecting streets, and the needs of pedestrians. To facilitate traffic movements, the curb line should closely conform to the turning path of the largest class of vehicles that, in significant numbers, are to turn at

ALL-PAVED CROSSING   CROSSING WITH SEPARATE TURNING LANES

ALL-PAVED T   T WITH SEPARATE TURNING LANES
(NOT A Y)

ALL-PAVED BRANCH   BRANCH WITH SEPARATE TURNING LANE
(NOT A Y)

CHANNELIZED Y

ALL-PAVED Y   Y WITH SEPARATE TURNING LANES

FLARED CROSSING

FIGURE 9-1   Types of intersections at grade. (Courtesy Federal Highway Administration.)

the intersection. Where traffic predominantly consists of passenger vehicles, arcs of circles with radii of 15 to 25 ft may be sufficient. Whenever feasible, corner radii of 25 and 30 ft should be provided, respectively, for minor and major cross streets on new construction and reconstruction (1). Tables 9-1 and 9-2 give recommended minimum edge of pavement designs at intersections. It will be noted that three-centered compound curves are recommended, especially for sharp turns and for the larger classes of vehicles. Alternatively, simple

TABLE 9-1

MINIMUM EDGE OF PAVEMENT DESIGNS FOR TURNS AT INTERSECTIONS—SIMPLE CURVES AND TAPERS

Angle of Turn (degrees)	Design Vehicle	Simple Curve Radius (ft)	Simple Curve Radius with Taper		
			Radius (ft)[a]	Offset (ft)	Taper (ft : ft)
45	P	50	—	—	—
	SU	75	—	—	—
	WB-40	120	—	—	—
	WB-50	—	120	2.0	15 : 1
	WB-55	—	140	2.0	15 : 1
60	P	40	—	—	—
	SU	60	—	—	—
	WB-40	90	—	—	—
	WB-50	—	95	3.0	15 : 1
	WB-55	—	110	3.0	15 : 1
75	P	35	25	2.0	10 : 1
	SU	55	45	2.0	10 : 1
	WB-40	—	60	2.0	15 : 1
	WB-50	—	65	3.0	15 : 1
	WB-55	—	90	3.0	15 : 1
90	P	30	20	2.5	10 : 1
	SU	50	40	2.0	10 : 1
	WB-40	—	45	4.0	10 : 1
	WB-50	—	60	4.0	15 : 1
	WB-55	—	75	4.0	15 : 1
105	P	—	20	2.5	8 : 1
	SU	—	35	3.0	10 : 1
	WB-40	—	40	4.0	10 : 1
	WB-50	—	55	4.0	15 : 1
	WB-55	—	65	5.0	15 : 1
120	P	—	20	2.0	10 : 1
	SU	—	30	3.0	10 : 1
	WB-40	—	35	5.0	8 : 1
	WB-50	—	45	4.0	15 : 1
	WB-55	—	50	6.0	15 : 1
135	P	—	20	1.5	15 : 1
	SU	—	30	4.0	8 : 1
	WB-40	—	30	8.0	6 : 1
	WB-50	—	40	6.0	10 : 1
	WB-55	—	45	7.0	10 : 1

[a]1 ft = 0.3048 m.

SOURCE: *A Policy on Geometric Design of Highways and Streets*. American Association of State Highway and Transportation Officials, Washington, D.C. (1984).

TABLE 9-2

MINIMUM EDGE OF PAVEMENT DESIGNS FOR TURNS AT INTERSECTIONS—THREE-CENTERED COMPOUND CURVES

Angle of Turn (degrees)	Design Vehicle	3-Centered Compound Curve Radii (ft)[a]	Symmetric Offset (ft)	3-Centered Compound Curve Radii (ft)	Asymmetric Offset (ft)
45	P	—	—	—	—
	SU	—	—	—	—
	WB-40	—	—	—	—
	WB-50	200-100-200	3.0	—	—
	WB-55	200-120-200	3.0	140-100-300	1.0- 1.5
60	P	—	—	—	—
	SU	—	—	—	—
	WB-40	—	—	—	—
	WB-50	200- 75-200	5.5	200- 75-275	2.0- 6.0
	WB-55	200- 85-200	5.0	120- 80-240	1.0- 7.5
75	P	100- 25-100	2.0	—	—
	SU	120- 45-120	2.0	—	—
	WB-40	120- 45-120	5.0	120- 45-200	2.0- 6.5
	WB-50	150- 50-150	6.0	150- 50-225	2.0-10.0
	WB-55	200- 70-200	7.0	120- 60-200	2.0-10.0
90	P	100- 20-100	2.5	—	—
	SU	120- 40-120	2.0	—	—
	WB-40	120- 40-120	5.0	120- 40-200	2.0- 6.0
	WB-50	180- 40-180	6.0	120- 40-200	2.0-10.0
	WB-55	200- 65-200	7.0	100- 55-260	2.0-10.0
105	P	100- 20-100	2.5	—	—
	SU	100- 35-100	3.0	—	—
	WB-40	100- 35-100	5.0	100- 55-200	2.0- 8.0
	WB-50	180- 45-180	8.0	150- 40-210	2.0-10.0
	WB-55	240- 50-240	8.0	100- 45-500	4.0-10.0
120	P	100- 20-100	2.0	—	—
	SU	100- 30-100	3.0	—	—
	WB-40	120- 30-120	6.0	100- 30-180	2.0- 9.0
	WB-50	180- 40-180	8.5	150- 35-220	2.0-12.0
	WB-55	400- 40-400	10.0	100- 35-500	6.0-15.0
135	P	100- 20-100	1.5	—	—
	SU	100- 30-100	4.0	—	—
	WB-40	120- 30-120	6.5	100- 25-180	3.0-13.0
	WB-50	160- 35-160	9.0	130- 30-185	3.0-14.0
	WB-55	300- 35-300	10.0	100- 35-400	3.0-15.0

[a]1 ft = 0.3048 m.

SOURCE: *A Policy on Geometric Design of Highways and Streets*. American Association of State Highway and Transportation Officials, Washington, D.C. (1984).

curves combined with spirals or tapers to conform to the paths of vehicles may be used.

9-2 Islands and channels

Traffic islands or pavement markings may be used at intersections to facilitate the safe and orderly flow of vehicles and pedestrians. Properly designed channelization systems increase intersection capacity and decrease conflicts and accidents.

Islands in an intersection serve one or more of the following purposes *(1):*
1. Separation of conflicts.
2. Control of angle of conflict.
3. Reduction of excessive pavement areas.
4. Regulation of traffic flow in the intersection area.
5. Arrangements to favor a predominant turning movement.
6. Protection of pedestrians.
7. Protection and storage of turning and crossing vehicles.
8. Location of traffic control devices.

Islands are generally grouped into three major classes: directional or channelized, divisional, and refuge. General types and shapes of islands are shown in Figure 9-2.

FIGURE 9-2 General types and shapes of islands. (Courtesy American Association of State Highway and Transportation Officials.)

Directional islands are designed primarily to guide the motorist through the intersection by indicating the intended route. Where spacious area exists at an intersection and leaves much to the discretion of the driver, islands may be used to channel the motorist into the desired lane by placing a channeling island in the little-used portion of the intersection.

The placing of directional islands should be such that the proper course of travel is immediately evident and easy to follow. A complicated system of islands where the desired course of travel is not immediately evident may result in confusion of the motorist and may be of more hindrance than help in maintaining a steady traffic flow. Islands should be so placed that crossing streams of traffic will pass at approximately right angles and merging streams of traffic will converge at flat angles. By use of such angles there will be less hindrance to traffic on the thoroughfare, and the possibility of accidents in the intersection will be decreased.

Divisional islands are most frequently used on undivided highways approaching intersections. They serve to alert the driver to the intersection and regulate the flow of traffic into and out of the intersection. Their use is particularly advantageous for controlling left-turning traffic at skewed intersections.

A refuge island is located at or near crosswalks to aid and protect pedestrians crossing the roadway. Refuge islands are most generally used on wide streets in urban areas for loading and unloading of transit riders.

Islands should be large enough to command attention. AASHTO *(1)* recommends that islands have an area of at least 75 ft^2 and preferably 100 ft^2. This means that the sides of triangular islands should be not less than about 12 ft but preferably 15 ft after rounding of corners. It is recommended that elongated or divisional islands be at least 4 ft wide and 20 to 25 ft long *(1)*.

Islands should be visible to drivers for a sufficient distance to allow appropriate response. Generally speaking, islands should not be used on or near the crest of vertical curves or where sight distance is restricted due to horizontal curvature.

> For median curbed islands, the face of the curb at the approach island nose should be offset at least 2 ft, and preferably 4 ft from the normal (median) edge of pavement. The median curbed island should then be gradually widened to its full width. For other curbed islands, the total nose offset should be 4 to 6 ft from the normal edge of through pavement, and 2 to 3 ft from the pavement edge of a turning roadway. *(1)*

Steep-sided island curbs approximately 7 in. in height are recommended in congested areas to provide a greater degree of protection for pedestrians. Mountable curbs are appropriate in outlying areas where pedestrian volumes are low. Painted flush islands may be used in sparsely developed areas where approach speeds are relatively high, pedestrian traffic is light, and signs, signals, or lighting standards are not needed on the island area.

9-3 Rotary intersections

A rotary intersection is one in which all traffic merges into and emerges from a one-way road around a central island. Advantages of a rotary intersection include

1. Continuous traffic movement from all legs at reduced speed when operating at low volumes.
2. Since crossing movements are eliminated, accidents are likely to be less serious.
3. Where more than four intersection legs are involved, the design layout may be simplified. In such a case, a channelized layout may become extremely complicated.
4. The cost of an at-grade intersection may be considerably less than that of grade separation structures.

Disadvantages include

1. A rotary can handle no more traffic than an adequately designed channelized layout.
2. It has been found that unsatisfactory functioning occurs when two or more legs approach design capacity.
3. The area involved to satisfy proper geometric design of a rotary is extensive. This requirement for large flat areas may render the cost prohibitive in urban areas or areas of difficult topography.
4. Channelization will often prove more acceptable where large pedestrian traffic is expected.
5. Most rotaries are designed to function at low speeds. Where high speeds are anticipated the large lengths of required weaving sections may cause prohibitive land costs.
6. For proper functioning, it has been found that adequate lighting must be provided and controlled access is required on the intersecting legs. Failing these provisions, the intersection may become a serious safety hazard.

Because of these disadvantages, highway engineers in the United States seldom recommend the use of rotary intersections for new construction.

9-4 Grade separations and interchanges

Intersections at grade can be eliminated by the use of grade-separation structures that permit the cross flow of traffic at different levels without interruption. The advantage of such separation is the freedom from cross interference with resultant saving of time and increase in safety for traffic movements.

Grade separations and interchanges may be warranted (1) as part of an express highway system designed to carry heavy volumes of traffic, (2) to eliminate bottlenecks, (3) to prevent accidents, (4) where the topography is such that other types of design are not feasible, (5) where the volumes to be catered for would require the design of an intersection, at grade, of unreasonable size, and (6) where the road user benefit of reducing delays at an at-grade intersection exceeds the cost of the improvement *(1)*.

An interchange is a grade separation in which vehicles moving in one direction of flow may transfer by the use of connecting roadways. These connecting roadways at interchanges are called ramps.

Many types and forms of interchanges and ramp layouts are used in the United States. These general forms may be classified into four main types:

1. **T-** and **Y-**interchanges.
2. Diamond interchanges.
3. Partial and full cloverleafs.
4. Directional interchanges.

T- and Y-interchanges

Figure 9-3 shows typical layouts of interchanges at three-legged junctions. The geometry of the interchange can be altered to favor certain movements by the provision of large turning radii and to suit the topography of the site. The trumpet interchange has been found suitable for orthogonal or skewed intersections. Figure 9-3*a* favors the left turn on the freeway by the provision of a semidirect connecting ramp. Figure 9-3*c* indicates an intersection where all turning movements are facilitated in this way.

Diamond interchanges

The diamond interchange is adaptable to both urban and rural use. The major flow is grade-separated, with turning movements to and from the minor flow achieved by diverging and merging movements with through traffic on the minor flow. Only the minor flow directions have intersections at grade. In rural areas, this is generally acceptable owing to the light traffic on the minor flow. In urban areas, the at-grade intersections generally will require signalized control to prevent serious interference of ramp traffic and the crossing arterial street. The design of the intersection should be such that the signalization required does not impair the capacity of the arterial street. To achieve this, widening of the arterial may be necessary in the area of the interchange. Care must also be taken in the design of the ramps so that traffic waiting to leave the ramps will not back up into through lanes of the major flow.

One disadvantage of the diamond interchange is the possibility of illegal wrong-way turns, which can cause severe accidents. Where the geometry of the intersection may lead to these turns, the designer can use channelization devices and additional signing and pavement marking. Wrong-way movements are, in general, precluded by the use of cloverleaf designs.

Figure 9-3*d* shows the conventional diamond interchange. Increased capacity of the minor flows can be attained by means of the arrangement shown in Figure 9-3*e* or Figure 9-3*f*. The arrangement shown in Figure 9-3*g* is suitable where two diamond ramps are in close proximity. Weaving movements which would, in this case, inhibit the flows of the major route, are transferred to the parallel collector–distributor roads.

Partial and full cloverleafs

The partial cloverleafs shown in Figure 9-4 are sometimes adopted in place of the diamond interchange. Traffic can leave the major flow either

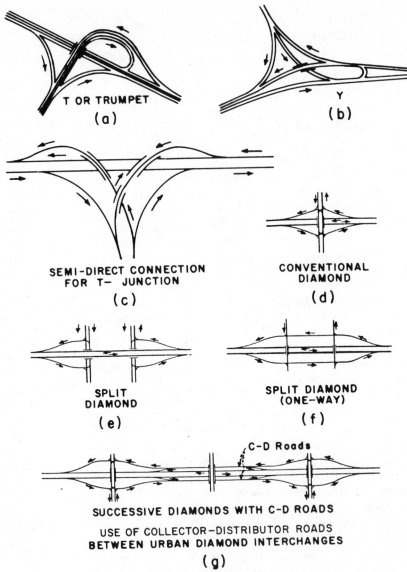

T OR TRUMPET
(a)

Y
(b)

SEMI-DIRECT CONNECTION
FOR T- JUNCTION
(c)

CONVENTIONAL
DIAMOND
(d)

SPLIT
DIAMOND
(e)

SPLIT DIAMOND
(ONE-WAY)
(f)

C-D Roads

SUCCESSIVE DIAMONDS WITH C-D ROADS

USE OF COLLECTOR-DISTRIBUTOR ROADS
BETWEEN URBAN DIAMOND INTERCHANGES
(g)

FIGURE 9-3 Highway interchanges. (Partially adapted from Adaptability of Interchanges to Interstate Highways, *Transactions of the American Society of Civil Engineers*, Vol. 124, p. 558.)

before or after the grade-separation structure, depending on the quadrant layout. The intersections at grade for the minor road are present as for the diamond interchange, but the probability of illegal turning movements can be reduced. By the provision of two on-ramps for each direction of the major route as in Figure 9-4*c*, left-turn traffic on the minor route can be eliminated.

The more conventional arrangement of the full cloverleaf, which can be adapted to nonorthogonal layouts, eliminates at-grade crossings of all traffic

FIGURE 9-4 Cloverleaf interchanges. (Adapted from *A Policy on Geometric Design of Rural Highways—1965*, American Association of State Highway and Transportation Officials.)

streams for both major and minor roads. The ramps may be one-way, two-way, two-way separated, or two-way unseparated roads. Although all crossing movements are eliminated, the cloverleaf design has some disadvantages: (1) the layout requires large land areas and (2) decelerating traffic wishing to leave the through lanes must weave with accelerating traffic entering the through lanes. Figure 9-4*e* is a layout using collector–distributor roads to overcome this second disadvantage. The use of collector–distributor roads also has certain operational and signing advantages. An exiting driver makes only one decision (i.e., to transfer to the cross road) while traveling on the high-speed, high-volume facility. The second decision—for example, to proceed north or to proceed south—is made off the high-speed facility (2).

Figure 9-5 shows a full cloverleaf interchange in California.

Directional interchanges

Directional interchanges are used whenever one freeway joins or intersects another freeway. The outstanding design characteristic of this type of interchange is the use of a high design speed throughout, with curved ramps and roadways of large radius. The land requirements for a directional interchange are therefore very large. In cases where volumes for certain turning

FIGURE 9-5 A full cloverleaf interchange in California. (Courtesy Federal Highway Administration.)

movements are small, design speeds for these movements are reduced and the turnoff is effected within a loop. Figures 9-6 and 9-7 show examples of directional interchanges.

Recently, experience with directional interchanges has revealed operational problems associated with left-hand entrance and exit ramps. Most drivers expect to exit freeways to the right and to enter from the right. When those expectations are violated, confusion, erratic maneuvers, and accidents sometime result.

Consider the alternative designs shown in Figure 9-8. Design A is a directional interchange in which a driver wishing to go right bears right and a driver wishing to go left bears left. A northbound driver must make two decisions as he approaches the first exit: (1) that he is going to leave the freeway at this exit, and (2) whether his destination is to the right (east) or left (west). Confronted with such interchange configuration, a driver may be in the right-hand lane, decide that he must go west to reach his destination, and be faced with a need to cross several lanes of high-speed traffic in order to exit to the left.

Design B (Figure 9-8) is a preferred single exit configuration on which a northbound driver makes decisions one at a time. The driver exits to the right and then decides whether to proceed east or west while on the lower-speed, less congested exit ramp.

FIGURE 9-6 A four-level directional interchange. (Courtesy Federal Highway Administration.)

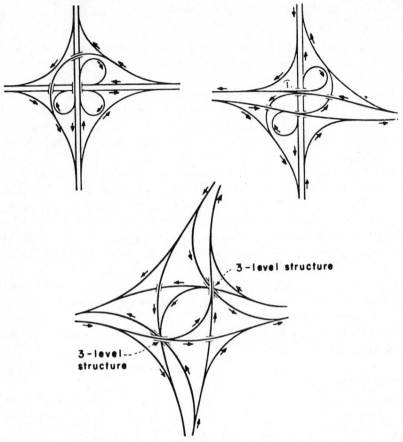

FIGURE 9-7 Examples of directional interchanges. (Courtesy American Association of State Highway and Transportation Officials.)

FIGURE 9-8 One-exit versus two-exit interchanges. (Courtesy Federal Highway Administration.)

250

9-5 Design of ramps and turning roadways

A "turning roadway" is a connecting roadway for traffic turning between two legs of an intersection. The term "ramp" is used to refer to all types of turning roadways that connect two or more legs at an interchange (1). The principles for the design of open highways are generally applicable to the design of ramps and turning roadways. There are, of course, differences. The design speeds for turning roadways are considerably lower than for the open highway but should be as high as practicable.

Research has shown that drivers will operate on turning roadways at higher speeds in relation to curvature than on the open highway. Thus, larger side-friction factors are used for design. AASHTO (1) recommends friction factors of 0.32, 0.23, and 0.18 for design speeds of 15, 25, and 35 mph, respectively. On the other hand, limitations in space preclude the development of superelevation rates on intersection curves as large as those on the open highway. With appropriate selection of values for a friction factor and a maximum rate of superelevation, the minimum radius of curvature can be calculated by Eq. 8-2.

Compound and transition curves are recommended for turning road-ways to more closely fit the natural paths of vehicles. Where compound curves are used, it is preferred that the ratio of the flatter radius to the sharper radius not exceed 1.75. Where that ratio exceeds 2, a spiral transition should be placed between the two curves (1).

Gradients for turning roadways should be as flat as feasible and generally should not exceed 6 percent. Stopping sight distance should be provided for turning roadways.

The recommended widths of turning roadways depend on the radius of curvature of the inner edge of the pavement, the class of design vehicle, and the type of operation (one-way, two-way, etc.). The recommended widths vary from 12 to 50 ft depending on the conditions.

9-6 Railroad grade intersections

In the design of a highway that intersects a railroad at grade, consideration must be given to approach grades, sight distance, drainage, volume of vehicu-lar traffic, and the frequency of regular train movements at the particular intersection. The particular type of surfacing and kind of construction at railroad crossings at grade will depend on the class of railroad and kind of roadway improvement.

All railroad intersections at grade require proper advance warning signs. At crossings on heavily traveled highways where conditions justify, automatic devices should be installed. Recommended standards for railroad–highway grade-crossing protection have been adopted by the Association of American Railroads.

The use of grade separations at railroad crossings is recommended at all main-line railroads that consist of two or more tracks and at all single-line tracks when regular train movements consist of six or more trains per day. Other considerations for separating railroad and highway traffic are the elements of delay and safety.

Railroad grade-separation structures may consist of an overpass on which the highway is carried over the railroad or an underpass that carries the highway under the railroad. The selection of the type of structure will depend in large part on the topographic conditions and a consideration of initial cost. Drainage problems at underpasses can be serious. Pumping of surface and subsurface water may have to be carried on a large part of the time, and failure of power facilities sometimes causes flood conditions at the underpass with the resultant stoppage of traffic.

DESIGN OF PARKING FACILITIES

Widespread parking problems exist in business districts, on college campuses, at airports, and in industrial areas and other highly developed areas. Shortages of parking spaces in close proximity to major traffic generators have worsened as automobile registrations have increased and more and more people have opted to travel by automobile. Parking problems have been aggravated in many cities as curb parking has been eliminated to improve flow.

The procedures for conducting parking surveys are described in Chapter 6. In the following paragraphs, we will discuss principles for the location, layout, and design of off-street and on-street parking facilities.

9-7 Location of off-street parking facilities

Perhaps the most important aspect of parking facility planning and design is the choice of a site. Experience has shown that an improperly located lot or garage is likely to fail or have limited use, even if located only a few blocks away from the center of parking demand.

Most parkers are reluctant to walk even short distances from the parking location to their ultimate destination. The maximum walking distance a parker will tolerate depends on trip purpose, city size, and the cost of parking. Office workers may be willing to walk a maximum distance of about 2000 ft (five or six blocks) if the rates are attractively low (3). Short-term parkers such as shoppers will usually resist walking more than one or two blocks. In general terms, the smaller the city and the higher the parking cost, the less distance a parker is willing to walk.

Preferably, parking lots and garages should be located on or near major arterials; garages should have access to two or more streets. It is also desirable that both lots and garages be accessed by right-hand turning movements.

Safe and convenient pedestrian access to parking facilities should be provided. To alleviate parkers' fear for their personal security, pedestrian walkways should be well lighted, suitably marked, and free of blind corners. Most pedestrians prefer overhead bridges to tunnels. To the extent feasible, pedestrian conflicts with vehicular traffic should be avoided.

In summary, parkers prefer parking facilities near their destination, that can be accessed with ease, that can be used without fear for one's personal safety, and that cost little or no money.

9-8 Layout of parking lots and garages

Preferably, a parking lot or garage should be rectangular with cars parked on both sides of access aisles. Parking sites should be wide enough to provide two or more bays approximately 60 ft wide. Relatively long aisles, preferably 250 ft or longer, function well. Aisles running with the long dimension of the lot or garage provide a better search pattern and yield more spaces per unit area.

Ninety degree parking is generally used for two-way traffic. Right-angled parking tends to require slightly less area per parked car than other configurations.

Parking stalls oriented at angles of 45 to 75° to the aisles are often used with one-way circulation. Where such an arrangement is used to serve a building along the end of a parking lot, drivers must circulate at least once next to the building during entry or exit, causing more vehicle−pedestrian conflict than the two-way 90° arrangement.

Where space permits, it is desirable to provide pedestrian sidewalks between adjacent lines of parked cars. When sidewalks are provided, it is usually necessary to install wheel stops to prevent vehicle encroachment. For 90° pull-in parking, the wheel stop setback from the edge of the curb should be about 2.5 ft. For back-in parking, a setback of at least 4.0 ft is recommended (3).

Stall widths of 8.5 to 9.0 ft are commonly used for parking facilities in the United States. Widths of 9.5 or even 10.0 ft are preferred for supermarket lots and other areas where packages are customarily being placed in cars (3). Stall depths of 18.0 to 18.5 ft are most often used.

The recent downscaling of the size of U.S. automobiles and the need to minimize the amount of parking space allocated to each car have resulted in two design strategies to accommodate the smaller cars more efficiently. These strategies have involved either a general reduction in the standard dimensions of automobile stalls or the provision of spaces designed for the exclusive use of small cars.

The trend has been to a general downscaling of parking spaces. Many zoning agencies now specify a stall width of 8 ft 4 in.; however, some cities allowing the narrower width limit their use to low-turnover parking such as employee parking (4). Most agencies have retained the same stall depth as for full-size cars.

Where it is feasible to provide separate parking spaces for the exclusive use of small cars, the stall dimensions commonly used are 7.5 by 15 ft or 8 by 16 ft (4).

The most common and preferred layout pattern is the bumper-to-bumper interlocked configuration illustrated by Figure 9-9. The herringbone or nested interlock pattern is sometimes used with 45° parking. In that configuration, the bumper of one car faces the fender of another car, necessitating the installation of wheel stops and increasing the probability of vehicular damage. Parking layout dimensions for 9 by 18.5 ft stalls arranged at various angles are given by the table which accompanies Figure 9-9. The best parking layout for a given site, usually determined by a process of trial, will depend primarily on

1. The size and shape of the available area.

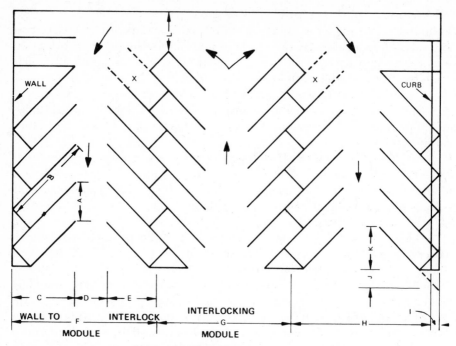

X = STALL NOT ACCESSIBLE IN CERTAIN LAYOUTS

PARKING LAYOUT DIMENSIONS (in feet) FOR 9-FT STALLS
AT VARIOUS ANGLES

Dimension	On Diagram	45°	60°	75°	90°
Stall width, parallel to aisle	A	12.7	10.4	9.3	9.0
Stall length of line	B	25.0	22.0	20.0	18.5
Stall depth to wall	C	17.5	19.0	19.5	18.5
Aisle width between stall lines	D	12.0	16.0	23.0	26.0
Stall depth, interlock	E	15.3	17.5	18.8	18.5
Module, wall to interlock	F	44.8	52.5	61.3	63.0
Module, interlocking	G	42.6	51.0	61.0	63.0
Module, interlock to curb face	H	42.8	50.2	58.8	60.5
Bumper overhang (typical)	I	2.0	2.3	2.5	2.5
Offset	J	6.3	2.7	0.5	0.0
Setback	K	11.0	8.3	5.0	0.0
Cross aisle, one-way	L	14.0	14.0	14.0	14.0
Cross aisle, two-way	–	24.0	24.0	24.0	24.0

FIGURE 9-9 Parking stall layout elements. (Courtesy Transportation Research Board.)

2. Type of facility (attendant, self-park).
3. Type of parker (especially long-term versus short-term).
4. Type of operation (pull-in, back-in, one-way, two-way, etc.).

9-9 Parking garage design criteria

Special design criteria for parking garages, taken largely from Ref. 5, are given in the following paragraphs.

Single entrances and exits, with multiple lanes, are preferable to several openings. Entrances and exits should be located away from street intersections to prevent congestion on the street. Lane widths of 12 to 14 ft are typically used for entrances and exits. Lanes should be tapered to a width of 9 to 10 ft for approaches to ticket dispensers and cashiers' booths.

When self-parking is used, there is little need for reservoir space at the entrance. Studies indicate that the capacity of an entrance lane with an automatic ticket dispenser is approximately 400 vehicles per hour. One entrance lane for every 300 to 500 spaces is typically provided.

If attendant parking is used, a storage reservoir will be needed. The reservoir space required will depend on the passenger unloading time and the time required for parking each vehicle. Table 9-3 gives the number of reservoir spaces required for various vehicle arrival and storage rates. Use of the recommended values should result in overloading less than 1 percent of the time.

If possible, clear span construction should be provided with column spacing equal to the unit parking depth (module width). A clear ceiling height of 7 ft and preferably 7.5 ft should be provided.

Vehicular access between floors can be provided by sloped floors or ramps. For self-park facilities, the floor slopes should not exceed 3 to 4 percent. Floor slopes up to 10 percent can be used for attendant park facilities. Ramp slopes should preferably not exceed 10 percent. Straight ramps should be at least 9 ft wide. Curved ramps are usually at least 12 to 13 ft in width. A minimum radius of ramp curvature of 30 ft is recommended, measured at the face of the outer curb of the inside lane.

Counterclockwise circulation is preferred for parking garages. When helical ramps are used, the down ramp should be placed inside and the up ramp outside.

9-10 Curb parking

Curb parking tends to seriously impede traffic flow and contribute to conflicts and accidents. Most highway and traffic engineers therefore recommend that

TABLE 9-3
RESERVOIR SPACES REQUIRED FOR
VARIOUS VEHICLE ARRIVAL AND STORAGE RATES[a]

	Average Number of Cars Arriving during Peak Hour					
Rate of Storage	*20*	*60*	*100*	*140*	*180*	*220*
0.9 × rate of arrival	14	26	35	43	50	56
0.95 × rate of arrival	13	23	30	36	42	47
1.00 × rate of arrival	12	20	25	29	33	36
1.05 × rate of arrival	11	17	20	22	23	24
1.10 × rate of arrival	10	13	15	15	15	15

[a]Adapted from *Traffic Design of Parking Garages.* The Eno Foundation, Saugatuck, Conn. (1957).

curb parking be prohibited along major streets. It should also be prohibited adjacent to fire plugs, within pedestrian crosswalks and bus stops, and in the vicinity of intersections, alleys, and driveways. Where permitted, curb parking should be regulated to minimize its effect on accidents and congestion and to ensure that available parking spaces are used appropriately and efficiently.

Properly marked stalls tend to make it easier for drivers to park, lessening the likelihood of congestion and crashes. One study (3) found a 43 percent reduction in the average parking time after suitably marked parking stalls were painted. Stalls are especially recommended for locations where there is great demand for curb parking spaces and where parking meters are used.

Three basic types of curb stalls are used (3): (1) end stalls, (2) interior stalls, and (3) paired parking stalls. End stalls are situated adjacent to intersections, alleys, driveways, and other restricted areas. End stalls are typically 20 ft in length. Interior stalls are usually 22 ft long, providing approximately 4.5 ft between adjacent cars for maneuvering. Paired parking consists of pairs of contiguous 18-ft stalls separated by an 8-ft open space that may be used for maneuvering. Paired parking is frequently used in conjunction with double parking meters, that is, two meters installed on a single post.

Curb parking stalls are designated by white lines extending out from the curb a distance, typically, of 7 ft. The end row of curb parking spaces is usually marked with an inverted L-shaped line, and interior stalls are designated by a T-shaped line.

Curb parking stalls should not be placed closer than 20 ft to the nearest sidewalk edge at nonsignalized intersections. At signalized intersections, a clearance to the nearest sidewalk edge of 50 and preferably 100 ft should be observed. Parking stalls should be placed not closer than 15 ft away from fire hydrants and driveways.

TRUCK TERMINAL PLANNING AND DESIGN

The planning and design of facilities to accommodate the loading, unloading, and parking of trucks can have a significant impact on the operational efficiency of the street and highway system. Although truck loading docks and berths are usually located off the rights-of-way of public roadways, inadequacies in such facilities will cause bottlenecks and congestion on nearby streets and highways. Furthermore, trucks making pickups and deliveries for commercial establishments often make use of street and alley space for maneuvering into truck loading docks and berths. It is apparent that, while public highway agencies do not usually become involved in the planning and design of truck loading/ unloading facilities, it is in the public interest that such facilities be properly planned and designed.

9-11 Truck terminal planning

Truck terminal planning involves the determination of the overall size and general layout of facilities required to handle the forecast quantity of freight. This includes estimating the number of loading berths (doors) as well as the amount of space required for the handling and temporary storage of the freight. To make reliable estimates of the number of loading berths, it is necessary to make a thorough analysis of the truck movements and cargo-handling procedures. The design year workload must also be forecast, taking into account daily and seasonal variations in the rate of truck arrivals and the effect of loading and unloading rates and procedures. Detailed procedures for the planning of shipper–motor carrier facilities, which are beyond the scope of this text, are given in Ref. 6.

9-12 Truck terminal design criteria

To allow proper clearances on each side of 8.5-ft-wide vehicles, a minimum width of 11 ft, and preferably 12 ft, is recommended for each spot (6). The dock approach length is the length of the largest tractor-trailer combination using the dock plus the apron length necessary to maneuver the vehicle in and out of the parking spot. As a rule, the dock approach should be at least twice the length of the longest tractor–trailer combination. Minimum recommended apron and dock approach lengths are given in Table 9-4. If 11-ft widths are used for dock spots, the apron lengths shown in the table should be increased about 2 ft.

The apron dock approach area should be nearly level. Although most trucks are designed to negotiate a 15 percent grade, the startup grade for pulling away from a dock should not exceed 3 percent (6).

The dock height should be as close as possible to the bed height of the vehicles using the dock. Generally, trailer bed heights vary from about 48 to 52 in., while pickup and delivery vehicle bed heights range from about 44 to 48 in. In order to ensure that vehicle doors can be opened and closed, it is best to select a dock height that is too low rather than too high.

Vehicle clearances of at least 15 ft should be provided to accommodate the 13.5-ft-high trailers.

TABLE 9-4
MINIMUM RECOMMENDED APRON LENGTHS AND DOCK APPROACH LENGTHS FOR TRUCK TERMINALS

Overall Length of Tractor-Trailer (ft)[a]	Apron Length (ft)	Dock Approach Length (ft)
40	43	83
45	49	94
50	57	107
55	62	117
60	69	129
65	75	140

[a] 1 ft = 0.3048 m.
SOURCE: *Shipper–Motor Carrier Dock Planning Manual*. The Operations Council, The American Trucking Associations, Washington, D.C. (1970).

Traffic circulation plans should encourage vehicles to circulate in a counterclockwise direction. For two-way traffic, service roads should be a minimum of 23 ft and preferably 24 ft wide. For one-way roads, the minimum widths on straight sections should be 12 ft. On curves, the pavement must be widened to conform to the wheel paths of the turning vehicles. (See Figure 8-1.)

The width of parking spaces for trucks should be about 12 ft, and the length should be the vehicle length plus 20 percent.

PROBLEMS

9-1   Sketch a right-angled T-intersection that is to accommodate passenger cars and WB-55 trucks. Show all dimensions of the curb line. Assume that both streets will have two-way operation.

9-2*   Work Problem 9-1 using SI (metric) units.

9-3   Sketch a right-angled T-intersection that is to accommodate passenger cars, buses, and WB-50 trucks. The minor or intersecting road is one way, allowing right turns only. Show all dimensions of the curb line.

9-4*   Work Problem 9-3 using SI (metric) units

9-5   How many cars could be placed in a commercial parking lot 130 ft wide (street frontage) and 280 ft deep. Make a sketch showing the arrangement. Assume two driveways and one-way circulation.

9-6*   From Figure 9-9, prepare a table in SI (metric) units showing dimensions A, B., C, . . . , M for a 90° parking layout.

9-7   Seventy-five cars arrive at a parking garage during the peak hour. If cars can be stored at a maximum rate of 80 per hour, estimate the number of required reservoir spaces.

9-8   Prepare a report on the experience in your city on the location and lengths of bus stops. Indicate whether bus stops are located on the near side or far side of intersections or at midblock. Discuss the advantages and disadvantages of the system and indicate steps that could be taken to alleviate any traffic or safety problems related to bus movements.

REFERENCES

1. *A Policy on Geometric Design of Highways and Streets.* American Association of State Highway and Transportation Officials, Washington, D.C. (1984).

2. *Dynamic Design for Safety Seminar Notes.* Prepared by the Institute of Traffic Engineers for the Federal Highway Administration, Arlington, Va. (1975).

3. *Parking Principles.* Highway Research Board Special Report No. 125, (1971).

4. Weant, Robert A. The Influence of Smaller Cars on Parking Geometry. *Transportation Quarterly*, Vol. XXXIX, No. 3, Eno Foundation for Transportation (July 1985).

5. *Traffic Design of Parking Garages.* The Eno Foundation for Highway Traffic Control, Saugatuck, Conn. (1957).

6. *Shipper–Motor Carrier Dock Planning Manual.* The Operations Council, The American Trucking Associations, Washington, D.C. (1970).

chapter ten

HIGHWAY MASS TRANSIT FACILITIES

Highway engineers have long recognized that the facilities they plan and design must accommodate a wide range of vehicle types and sizes. Events of the 1970s, especially increasing urban congestion and scarcity of fuel, heightened the awareness of engineers that steps could be taken to cause the highway system to be used more efficiently. Deliberate actions could be taken to encourage more extensive use of streets and highways by buses, vans, and other high-occupancy vehicles (HOVs).* The highway design would not only accommodate mass transit vehicles but also facilitate the movement of people rather than vehicles and give priority to HOV transportation.

In this chapter, we describe the planning and design of special facilities for buses and other high-occupancy vehicles and examine some traffic control measures and institutional programs designed to encourage more efficient use of streets and highways, especially in heavily congested urban areas.

10-1 Bus stops on streets

Bus stops that are properly located, adequately designed, and effectively enforced can improve bus service and expedite traffic flow (2). Preferably, the frequency of stops should not exceed 8 to 10 per mile. Practically speaking, this means that stops should be provided every block where city blocks are greater than about 500 ft long and every other block where blocks are shorter.

Bus stops may be located on the near side or far side of intersections or at midblock. The choice of location will depend on the geometric design of the intersection, the proximity of large generators of passengers, whether turning movements are permitted at the intersection, and a number of other factors.

Some of the advantages of *far-side* bus stops include

1. They make the curb lane available for traffic, increasing the intersection capacity.
2. They reduce conflicts between stopped buses and right-turning vehicles.
3. They lessen the problem of buses blocking the view of drivers approaching the intersection.
4. They reduce pedestrian accidents. (3)

---

*The definition of HOV varies. Commonly, it includes all vehicles with two or more occupants; however, some jurisdictions have restricted HOV lanes to vehicles with three or more occupants (1).

On the other hand, *near-side* bus stops are preferred where transit flows are heavy but traffic and parking problems are not extensive *(2)*. Near-side stops are usually near a crosswalk, making it easier for passengers to board the buses. Also, with near-side stops there is less interference with traffic turning into the bus route from a side street. Furthermore, bus drivers tend to prefer near-side stops because it is easier for the buses to reenter the traffic stream, especially where curb parking is allowed.

Midblock bus stops are sometimes provided in downtown areas where several bus routes converge and require long boarding areas. At midblock areas, buses create less serious sight restrictions to drivers and pedestrians. In addition, passengers are able to assemble at less congested areas of the sidewalk.

Sufficient space should be provided at bus stops to permit bus drivers to easily pull in and out of the traffic stream and to park parallel to the curb with the center door not more than 1 ft from the curb *(4)*. Recommendations of the American Transit Association on minimum lengths of curb bus loading zones are given in Table 10-1.

Figure 10-1 shows recommended bus stop design standards for stops at far-side, near-side, and midblock locations. The dimensions shown on the figure are based on an assumed 40-ft bus length and should be adjusted for buses of different length. Where two or more buses are expected to use a bus stop simultaneously, 45 ft should be added to the lengths shown for each additional bus.

10-2 Bus turnouts on streets

Several cities have used bus turnouts by relocating the curb and flaring the street width. Bus turnouts should be considered along heavily traveled high-speed arterial streets, especially where bus dwell times are relatively long. Reference 2 suggests that bus turnouts are warranted whenever the following conditions exist:

TABLE 10-1
CURB BUS LOADING ZONE LENGTHS IN FEET[a] (MINIMUM)

Approx. Bus Seating Capacity	One-Bus Stop			Two-Bus Stop		
	Near Side	Far Side	Midblock	Near Side	Far Side	Midblock
25 and less	60	50	85	90	80	115
30	70	50	95	100	80	125
35	75	55	100	110	90	135
40–45	80	60	105	120	100	145

[a]1 ft = 0.3048 m.

SOURCE: *Parking.* The Eno Foundation for Highway Traffic Control, Saugatuck, Conn. (1957).

FIGURE 10-1 Bus stop design standards. (Courtesy Transportation Research Board.)

1. Right-of-way width is sufficient to permit construction of the turnout without impinging on needed sidewalk space.
2. Peak-hour traffic in the curb lane is at least 500 vehicles per hour.
3. Curb parking is prohibited, at least during periods of peak traffic flow.
4. Average bus dwell time exceeds 10 sec per stop.

5. Bus volumes are insufficient to justify an exclusive bus lane, but bus volumes should be at least 100 per day and consist of 10 to 15 buses transporting 400 to 600 passengers during the peak hour.

Figure 10-2 shows typical layouts for bus turnouts at far-side, near-side, and midblock locations. Figure 10-3 gives recommended dimensions for bus bays to accommodate one standard 40-ft bus.

The capacity of bus turnouts depends on the peak-hour bus volume and the average dwell time. Figure 10-4 provides some guidelines for determining the number of buses that will have to be accommodated at each turnout. The curves assume random (Poisson) arrivals and are based on a confidence level of 95 percent. The criteria are also applicable for determining the lengths of bus stops.

10-3 Bus stops on freeways

In certain instances, special bus stop are needed at interchanges and street crossings to provide the public with the benefit of freeway bus operations. Criteria for the location of bus stops at freeway interchanges include (1) the need for a transit transfer point, (2) the density of population in the area, (3) transit generators within less than one block of the freeway route, and (4) forecasts of future usage of expressway by transit vehicles.

In planning and locating freeway bus stops, there is a need to maximize bus patron convenience but to minimize delays to buses and to other vehicles in the traffic stream. The spacing of the stops should allow the buses to operate at or near the speed of the traffic on the highway. Based on this criterion, AASHTO *(3)* suggests a spacing of 2 miles or more.

Bus stops along freeways may be placed at the street level or the freeway level. Street-level stops are appropriate in or near downtown areas, and either street- or freeway-level stops are suitable for suburban and rural areas. Bus stops at the cross-street level are more convenient for entering and leaving passengers but cause delays for through passengers.

Street-level bus stops are not well suited to cloverleaf or directional interchanges, and AASHTO *(3)* recommends that they not be provided at such interchanges. At diamond interchanges, the preferred location for bus stops is adjacent to on-ramps. There, the bus stop may consist of a widened shoulder area, or it may be on a separate roadway.

Freeway-level bus stops should be isolated from the through freeway lanes. Bus stops generally should be located on a separate roadway at least 20 ft wide to permit buses to pass a standing or stalled bus *(4)*. Acceleration and deceleration lanes at least 100 ft in length should be provided *(2)*. It is usually necessary to provide fencing between the bus stop area and through lanes to deny pedestrians access to the freeway lanes.

Bus stops should be at least 80 ft in length to allow simultaneous loading or unloading by two buses. The pedestrian loading platform should be at least 5 ft and preferably 8 ft in width.

(a) Far side

(b) Near side

(c) Mid-block bus bays

FIGURE 10-2 Typical bus stop turnouts on arterial streets. (Courtesy Transportation Research Board.)

MID- BLOCK LOCATION

NOTES: DESIGNS BASED ON STANDARD 40' BUS.
ADD 45' FOR EACH ADDITIONAL BERTH.

NEAR-SIDE CORNER LOCATION

ADD 40' FOR STOP AFTER
RIGHT TURN.

FAR-SIDE CORNER LOCATION

FIGURE 10-3 Recommended dimensions for bus bays to accommodate one standard 40-ft bus. (Courtesy Transportation Research Board.)

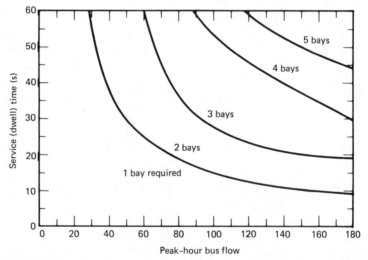

FIGURE 10-4 Recommended criteria for number of bus bays at bus stops and turnouts.

Since bus stops along freeways are usually located at intersecting streets or highways, stairs or ramps for pedestrians are required. Efforts should be made to minimize the total height of stairways, and landings should be provided every 6 or 8 ft (3). The gradients on pedestrian ramps should not exceed 8.33 percent.

Benches and shelters are recommended, and telephones and other amenities may also be warranted.

Figure 10-5a shows a layout for a freeway-level bus stop where the street crosses over the freeway without an interchange. Figure 10-5b illustrates two

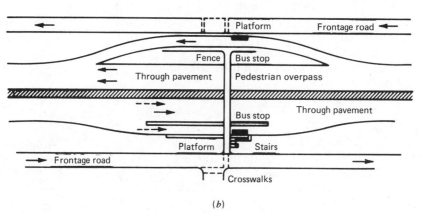

FIGURE 10-5 Layout of a freeway-level bus stop. (*a*) Street crossing over freeway. (*b*) Street crossing under freeway. (Courtesy American Association of State Highway and Transportation Officials.)

layout plans for an undercrossing street without an interchange. In the upper portion of the figure, the turnout to the bus stop is ramped to the level of the frontage road, making it unnecessary for passengers to climb ramps or stairs. In the lower part of Figure 10-5*b*, stairs are shown leading from the bus stop to the undercrossing street. The layout also features a pedestrian walkway built over the freeway.

10-4 Reserved bus lanes

In central business districts and other areas that experience heavy concentrations of bus traffic, lanes may be reserved for exclusive bus use or shared by buses, taxis, and right-turning vehicles. Reserved bus lanes may operate in the same direction as or counter to automobile traffic flow. (Lanes of the latter type are referred to as "contraflow lanes.") Reserved bus lanes are commonly located along curbs but in certain instances may be located in street medians.

Reserved bus lanes may be delineated by painted lines and markings or by some physical separation such as curbs, flexible stanchions, or traffic islands. Normal flow curb lanes are usually separated from other lanes by a solid white line 8 to 12 in. wide. Normal flow median lanes are separated from other lanes by double lines 6 in. wide: white lines facing traffic in the same direction, and yellow lines facing traffic in the opposite direction. Contraflow bus lanes are separated from traffic in the opposite direction by double 6-in. yellow lines, a median divider, or some other physical separation *(2)*.

Large BUS ONLY or BUS LANE messages should be painted on the pavement at least once each block and twice where the block lengths exceed 300 ft. Special signs that conform to the *Manual on Uniform Traffic Control Devices for Streets and Highways (5)* should be installed to facilitate the control of traffic.

Reserved curb lanes are warranted under the following conditions *(2)*:

1. Curb parking should be prohibited whenever the bus lane is in effect.
2. Where the bus lane preempts a lane used by automobiles, there should be sufficient capacity in the remaining lanes or parallel streets to accommodate displaced traffic.
3. Two lanes, and preferably three lanes, should remain for other traffic in the same direction (normal flow reserved lanes) or in the opposite direction (contraflow lanes).
4. Essential services and access to adjacent property should be maintained. It may be possible to provide access to buildings from the rear or during off-peak periods.
5. The number of peak-hour bus passengers transported in the reserve lane should equal or nearly equal the average number of passengers carried by cars in the adjacent lanes.
6. The bus traffic should be at or above the following levels:

	Normal Direction	*Contraflow*
One-way peak-hour traffic	30–40 buses	40–60 buses
One-way peak-hour movement of people	1200–1600 people	1600–2400 people
Daily use of reserved lanes	300 buses	400 buses
Busiest 20-min flow rate	60 buses/hr	60 buses/hr

Ten-foot lane widths may be used for reserved bus lanes used for normal direction of flow. Wider lanes, preferably measuring 12 ft, are desired for contraflow lanes. Typical curb bus lane configurations are shown in Figure 10-6. Typical configurations for contraflow lanes are shown in Figure 10-7.

FIGURE 10-6 Typical curb bus lane configurations. (Courtesy Transportation Research Board.)

FIGURE 10-7 Typical configurations for contraflow lanes. (Courtesy Transportation Research Board.)

Median lanes are warranted under the following conditions *(2)*:

1. The bus lanes are to replace street railway operations in the center of the street, establishing the precedent for center-of-the-street loading.
2. The street should have a wide median that can be paved for buses without harmful environmental effect.
3. The street should be wide enough to provide at least two lanes on each side of the median. (The curb lane may be used for parking.)

4. The number of bus passengers should equal or exceed the average number of passengers carried by automobile in the adjacent lanes throughout a 12-hr base period or, alternatively, the periods the lanes are in effect. Where peak-hour median lanes are provided, buses should transport 1.25 times the number of passengers carried by cars in the adjacent same-direction lanes.

5. The bus traffic should equal or exceed the following amounts:

One-way peak-hour traffic	60−90 buses/lane
One-way peak-hour movement of people	2400−3600 people/lane
Daily use of median lane	600 buses/lane

6. Conflicting left-turning movements should be prohibited or channeled into lanes outside the median.

The effectiveness of reserved bus lanes has not been definitely established, and some consider the success of such measures to be rather limited (3). There are problems of enforcement as well as concerns about safety and increased traffic on the remaining lane street space. Nevertheless, the evidence that is available suggests that, when properly conceived and implemented, reserved bus lanes can result in significant improvements in bus operations that are not outweighed by increases in accidents, nonbus congestion, and detrimental impacts on abutting property.

In a study of contraflow bus lanes in Chicago (6), it was found that there was a significant improvement in bus operations after the installation of the lanes. Average travel speeds of the buses increased 22 percent, resulting in an annual cost saving of approximately $400,000. The study revealed that during peak periods, the contraflow lanes accommodated more people in one lane than were moved on the remaining through lanes and in only one-tenth the number of vehicles.

10-5 Busways

Busways are specially designed roadways for exclusive or predominant use by buses. They are used (2)

1. To provide line-haul express service to the city center.
2. To provide feeder service to rail transit lines.
3. To provide short bypasses of areas of serious congestion.

Busways may be constructed in separate rights-of-way, along the side of a freeway, or in a freeway median. They may (1) be completely grade-separated and provide levels of service comparable to those provided by freeways or rail rapid transit or (2) incorporate some at-grade intersections and provide a level of service comparable to that on arterial streets or light rail transit lines. These two classes of busways have been designated as class A and class B, respectively (2).

Typically, busways have two-lane, two-way operation. The cross section consists of two lanes, each 11 to 13 ft in width, plus space to accommodate breakdowns. A single breakdown lane may be constructed between the two traffic lanes, or breakdown space may be provided as shoulders along the outside of each lane. Typical sections for these two design approaches are shown as Figure 10-8.

Busways may operate with conventional right-hand operations *(normal flow)* or with the buses keeping to the left of the centerline *(contraflow)*. Contraflow operation permits the use of common island station platforms between the two lanes, providing efficiencies in station security, supervision, and maintenance and simplifying access to the platforms. However, contraflow operation is not recommended whenever trucks and/or HOVs might use the bus lanes.

Suggested busway design criteria are given in Table 10-2. The pavement widths shown in the table are for tangent sections. Pavements on two-lane busways should be widened 1.5 to 3.0 ft, depending on normal pavement width, design speed, and sharpness of curve. The ramp widths shown are for one-lane, one-way operation with no provision for passing.

Busway design criteria for horizontal and vertical curvature generally conform to AASHTO highway standards. The table values have been modified somewhat to allow for the special design and operational characteristics of buses and to allow for the later conversion of busways to rail facilities, where appropriate.

Figure 10-9 illustrates a busway in a depressed freeway median with normal flow.

10-6 Other measures to encourage bus and high-occupancy vehicle use

We have described several initiatives that can be taken to foster the use and improve the flow of buses and high-occupancy vehicles on streets and highways. Some of the initiatives described are physical facilities; others are opera-

FIGURE 10-8 Typical sections for busways on special rights-of-way. (*a*) With normal flow. (*b*) With contraflow. (Courtesy Transportation Research Board.)

TABLE 10-2
SUGGESTED BUSWAY DESIGN CRITERIA

Item	Class A Busway	Class B Busway
Loads	H-20-S-16-44	H-20-S-16-44
Design speed (mph)		
Desirable	70	50
Minimum	50	30
Lane width (ft)		
With paved shoulders	12^a	$11-12^a$
Without paved shoulders	13^a	12^a
Paved shoulder width^b (ft)	$8-10$	$6-8$
Total paved width (ft)		
Normal flow	$26-44$	$24-40$
Special flow	$30-36$	
Contraflow	$30-36$	
Minimum viaduct width^c (ft)	28	28
Minimum tunnel width ^{d} (ft)	31	31
Minimum vertical clearance (ft)		
Desirable	$14.5-18^e$	14.5
Absolute minimum		12.5
Min. lat. dist. to fixed obstructions^f (ft)		
Left	3.5	2
Right	6	3
Maximum superelevation (ft/ft)	0.08^g	0.08^g
Min. radius of horiz. curves (ft)		
70 mph	1600	1600
60 mph	1150	1150
50 mph	750	750
40 mph	450	450
30 mph	250	250
Absolute min. radius^h (ft)		
Conv. to conventional rail	250	250
Convertible to light rail	100	100
Nonconvertible	30	30
Maximum gradients (%)		
Desirable		
Convertible to rail	$3-4$	$3-4$
Other	5	6
Ramps, up	6	7
Ramps, down	7	8
Absolute		
Main line	8	8
Ramps	10	10

TABLE 10-2
SUGGESTED BUSWAY DESIGN CRITERIA

(continued)

Ramps		
Design speed (mph)	30−35	15−25
Lane width (ft)		
With paved shoulders	12	12
Without paved shoulders	14	13
Paved shoulder width (ft)	8	8
Total paved width (ft)	14−22	13−20

[a]Increase lane width 1 ft when nonmountable-type curbs are used adjacent to travel lane.
[b]Applies only to normal flow busways.
[c]Curb to curb; excludes pedestrian walks and width required by curbs.
[d]Inside envelope.
[e]Varies according to requirements of the selected rail system.
[f]Distance from edge of traveled lane to vertical face of a noncontinuous obstruction, such as a bridge pier or abutment.
[g]May be reduced to 0.06 in regions where roadway icing is a consideration.
[h]Inner lane edge.

SOURCE: *Bus Use of Highways: Planning and Design Guidelines (2).*

tional changes. There are other measures that will positively affect mass transit and HOV use. Such measures include routine traffic engineering techniques that improve traffic flow generally, for example, planning and effectuation of one-way street systems and implementation of progressive traffic signal systems along arterial streets. In addition, there are measures designed to give priority to bus and HOV movements. Some of these priority measures are described below.

Ramp metering with HOV bypass lanes

A number of traffic engineering agencies use ramp metering to control traffic flow on freeways. Special traffic signals on entrance ramps permit only those vehicles to enter the freeway that can be accommodated without unduly disrupting main-line flow. Vehicles are required to wait at the ramp signal, artificially creating queues. A few agencies have used metered ramps to provide preferential treatment for high-occupancy vehicles. They use bypass ramps such as the one illustrated in Figure 10-10 to allow buses and possibly HOVs to move ahead of vehicles waiting in the queue.

The effectivenes of bypass lanes in inducing greater use of carpools, vanpools, and buses has not been well established. A study of Highway 50 bypass lanes in Sacramento, California, revealed a tendency for automobile occupancy to increase where the carpool bypass lane was implemented; however, the researchers reported that at least some of these changes probably

NORMAL FLOW

FIGURE 10-9 Busway in a depressed freeway with normal flow. (Courtesy Transportation Research Board.)

FIGURE 10-10 Example of a ramp metering system with a bus bypass lane. (Courtesy Transportation Research Board.)

resulted from a shift in route of existing carpools rather than the formation of new carpools *(7)*.

Bus bypass of queues at congestion points

Special bypass facilities may also be suitable for relieving chronic congestion on freeways at bottlenecks that cannot reasonably be remedied by more conventional methods. Such facilities may be (1) short sections of busways built on separate rights-of-way or (2) special bus lanes that are segregated from other lanes by signs and markings or some physical separation. Provision of bus bypass facilities may be appropriate at bridges, toll plazas, and bottlenecks caused by highway construction.

Bus bypass facilities at congestion points are generally warranted where more than 60 design-year peak-hour buses are forecast and it is expected that each bus will save at least 5 min *(2)*. At toll plazas, such facilities may be warranted if buses can avoid an aggregate delay of 30 to 50 min/hr. For example, Ref. 2 recommends a warrant of 15 buses/hr, each saving 2 or 3 min, or 10 buses/hr, each saving 5 min.

It is important that special bus bypass facilities be properly signed and marked and that adequate sight distance be provided at the beginning and at the end of the bypass. Such facilities should not reduce the highway capacity through the bottleneck.

Special traffic control measures

Traffic engineering agencies can give priority to bus movements by actions designed to facilitate bus turning movements at intersections. One action is to allow buses to make left turns that are prohibited to other vehicles (with signs such as NO LEFT TURN, BUSES EXCEPTED). The objective may be to facilitate bus routing or to remove buses from a heavily traveled arterial.

At signalized intersections, a special bus phase may be incorporated in the signal timing plan to allow buses to turn across heavily traveled traffic streams. Signalization of turning movements is considered to be justified wherever buses are required to cross more than two traffic lanes, each accommodating 500 or more vehicles during the peak hour *(2)*.

Priority treatment for buses in mixed flow can also be achieved by signal preemption. This normally involves extending the green phase along the main street in the direction of bus movements. The goal is to improve bus flows by reducing the number of stops due to cross-street traffic. Signal preemption can be achieved by (1) passive systems which are based on signal coordination and improved flow for all the main street traffic or (2) active systems which incorporate special bus detection equipment and signal controllers which adjust the signal phase to give priority to bus movements.

The preemption of traffic signals must take into account its effect on the total signal network. The objective is to reduce total person delay without unduly limiting side-street movements by other vehicles and pedestrians. To justify preemption, there should be at least 10 to 15 buses transporting 400 to 600 people during the peak hour and a daily volume of at least 100 buses *(2)*.

Signal preemption can result in significant improvements in bus flows. Bus delays at traffic signals have been reported to comprise 10 to 20 percent of overall bus trip times and to cause nearly half of all delays *(8)*.

10-7 Park-and-ride facilities

Park-and-ride facilities encourage automobile users to park in outlying areas and complete their journey by express bus or rail rapid transit. Such facilities may be located 1 to 30 miles from a city's central business district (CBD). Change-of-mode lots served by buses are usually closer to the CBD than are lots served by rail.

The layout of park-and-ride facilities should minimize pedestrian travel and give priority to interchanging transit passengers. Facilities should be provided for feeder bus loading and unloading, passenger car loading and unloading (kiss-and-ride), and the short-term and long-term parking. Preferably, the movement of buses and cars should be separated; however, at stations with fewer than 12 to 15 terminating buses per peak hour, buses may share parking area roadways with automobile traffic *(2)*.

Kiss-and-ride facilities should allow for passengers to be dropped off and picked up close to the station entrance and should incorporate short-term parking for 20 to 60 spaces.

The layout and design of the automobile parking lot for park-and-ride facilities conforms to the standards for off-street lots generally, which are described in Chapter 9.

A sketch of a typical park-and-ride facility is shown in Figure 10-11.

10-8 Paratransit

Thus far, we have focused primarily on facilities and programs designed to improve the movement of buses and other high-occupancy vehicles. In this section, we examine more closely the other HOVs and approaches that can be taken to encourage their use.

FIGURE 10-11 A typical park-and-ride facility. (Courtesy Transportation Research Board.)

The term *paratransit* is used to describe a broad range of transportation services falling between conventional fixed-route bus service and the private individually occupied automobile *(8)*. Paratransit is a low-capacity service that includes

1. **Demand-responsive transportation.** This term refers to the family of transportation that usually involves minibuses or vans that are directed from a central dispatching office. Like taxis, they provide on-demand, door-to-door, flexible-route service. Unlike most taxis, ride sharing is accomplished as a matter of operating policy and does not require the passengers' consent. The terms *dial-a-ride* and *shared taxi* have also been used to describe this type of transportation. Demand-responsive transportation is a rapidly growing but very small component of mass transportation. A large part of its market has been groups having special mobility needs: the handicapped, the elderly, and, to some extent, the economically disadvantaged. Demand-responsive transportation has also been used for commuters in low-density areas to provide feeder service to conventional fixed-route transit.

2. *Vanpools and carpools.* This component of paratransit is a high-occupancy mode that supplements conventional transit. It primarily serves commuters, and its major contribution is to help ameliorate highway congestion that occurs during the morning and afternoon rush hours. The vehicles used in vanpools and carpools are usually owned by individuals, and the service is arranged by subscription. Some employers encourage vanpooling and carpooling by administering a ride-matching program, by providing priority parking, and by permitting slight adjustments in work hours.

10-9 Alternative work schedules

Another approach to relieving peak-hour congestion on streets and highways is to manage the travel by (1) shifting demand to less congested times when surplus capacity is available and (2) reducing the need to travel (9). This can be accomplished by means of alternative work schedules.

Three basic types of alternative work schedules may be used (9):

1. Staggered work hours, by which groups of employees are assigned different starting times, for example, at 15-min intervals over a 2-hr period.
2. Flexible work hours, where individual workers are allowed to have some control over their own working hours but all employees are required to be present during a core time.
3. Compressed work weeks, by which workers work fewer than 5 days per week but work more hours per day, for example, 4 days/week, 10 hr/day.

At large places of employment, alternative work schedules can significantly reduce the number of employees arriving and departing during the peak period. Research has shown that staggered and flexible work hours result in lower travel times and improved load factors on transit. Reductions in travel time ranging from 2.5 to 8 min have been reported for commuters who participate in flexible work hours programs (9). Significant but somewhat smaller benefits accrue to all commuters who use the affected highways.

REFERENCES

1. *Enforcement of Priority Treatment for Buses on Urban Streets.* Prepared by John J. Roark, PAWA, Inc., Dallas, Tex. Synthesis of Transit Practice, National Cooperative Transit Research and Development Program, Transportation Research Board, Washington, D.C. (1982).

2. *Bus Use of Highways: Planning and Design Guidelines.* Prepared by Herbert S. Levinson, Crosby L. Adams, and William F. Hoey, Wilbur Smith and Associates, New Haven, Conn. National Highway Cooperative Highway Research Program Report 155, Transportation Research Board, Washington, D.C. (1975).

3. *A Policy on Geometric Design of Highways and Streets.* American Association of State Highway and Transportation Officials, Washington, D.C. (1984).

4. *Bus Stops for Freeway Operations.* ITE Recommended Practice, Institute of Transportation Engineers, Washington, D.C. (1971).

5. *Manual on Uniform Traffic Control Devices for Streets and Highways.* Federal Highway Administration, Washington, D.C. (1978).

6. LaPlante, John, and Harrington, Tim. Contraflow Bus Lanes in Chicago: Safety and Traffic Impacts. *Transportation Research Record 957*, Transportation Research Board, Washington, D.C. (1984).

7. Rogers, Christy A. Effects of Ramp Metering with HOV Bypass Lanes on Vehicle Occupancy. *Transportation Research Record 1021*, Transportation Research Board, Washington, D.C. (1985).

8. Alschuler, David M. Labor Protection, Labor Standards, and the Future of Paratransit. Transportation Research Board *Special Report No. 186* (1979).

9. *Alternative Work Schedules: Impacts on Transportation.* Synthesis of Highway Practice 73, National Cooperative Highway Research Program, Transportation Research Board, Washington, D.C. (1980).

chapter eleven

DRAINAGE AND
DRAINAGE STRUCTURES

One of the most important aspects of the location and design of rural highways and city streets is the necessity for providing adequate drainage. Adequate and economical drainage is absolutely essential for the protection of the investment made in a highway structure and for safeguarding the lives of the persons who use it.

The flow of surface water with which the highway engineer is concerned generally results from precipitation in the form of rain or snow or melting ice. A portion of the surface water infiltrates the soil, while the remainder stays on the surface of the ground and must be carried on, beside, beneath, or away from the traveled way. Artificial drainage resulting from irrigation, street cleaning, and similar operations may also be of consequence in some cases. In certain instances the control of underground water (ground water) may be important, as in the case of an underground flow encountered in a highway cut or in a location where the water table lies close to the surface of the ground.

Measures taken to control the flow of surface water are generally termed "surface drainage," while those dealing with ground water in its various forms are called "subsurface drainage" or, more simply, "subdrainage."

In the following pages, we will discuss some of the fundamental concepts of highway and street drainage. Surface drainage in essentially rural areas is discussed in considerable detail; accompanying this is a discussion of measures for the prevention of erosion of shoulders, side slopes, and side ditches. Considerable space is devoted to the location, design, and construction of culverts. Attention is briefly given to highway bridges. Material is also presented relative to subdrainage, and the chapter concludes with a brief discussion of drainage in municipal areas.

SURFACE DRAINAGE

The portions of the highway structure that provide for surface drainage in rural locations include the roadway crown, shoulder and side slopes, longitudinal ditches (channels), culverts, and bridges. Divided highways in rural areas also have inlets and storm drains (underground pipes) in the median strip to handle a portion of the surface flow.

11-1 Pavement and shoulder cross slopes
Consistent with other design objectives, highway designers should ensure that precipitation is removed from the pavement as expeditiously as possible. As

was indicated in Chapter 8, roadway surfaces are normally crowned (or sloped as in the case of a superelevated section) to facilitate the removal of surface water from the wearing surface. The recommended crown or cross slope tends to be a compromise between the needs of vehicular traffic and those for drainage. Pavement cross slopes should be steep enough to ensure expeditious drainage, but not so steep as to cause driver annoyance, discomfort, or hazard. Recommended ranges of pavement cross slopes are given in Table 11-1. The amount of cross slope varies with the type of surface, being generally small for relatively impervious surfaces such as portland cement concrete and large for pervious surfaces such as gravel or earth.

Shoulders are normally sloped to drain away from the pavement surface. Precipitation that occurs on the shoulder area largely flows to the side ditches or the median swale, as does that which falls on the roadway proper. As Table 11-1 indicates, recommended shoulder cross slopes vary from about ¼ to 1 in. per foot depending on the type of surface and whether or not curbs are provided (1).

11-2 Side slopes and side ditches

Open side ditches are generally provided in cut sections in highway locations in rural areas to provide for surface drainage. Side ditches may also be constructed along embankment sections when needed to supplement natural drainage channels. Both flat-bottomed and **V**-section ditches are used, with preference being given to the former type, with slope changes in the ditch section being rounded to improve appearance and prevent erosion. In either case, side slopes are made as flat as possible consistent with drainage requirements and limiting widths of right-of-ways. Deep, narrow side ditches are to be avoided whenever possible because of the increased hazards presented by such

TABLE 11-1
RECOMMENDED RANGES OF CROSS SLOPES
FOR PAVEMENTS AND SHOULDERS

Roadway Element	Range in Rate of Cross Slope (%)
High-type surface	
2 lanes	1.5−2.0
3 or more lanes in each direction	1.5 minimum; increase 0.5−1.0%/lane; 4.0 maximum
Intermediate surface	1.5−3.0
Low-type surface	2.0−6.0
Urban arterials	1.5−3.0; increase 1.0%/lane
Shoulders	
Bituminous or concrete	2.0−6.0
With curbs	⩾4.0

SOURCE: *Drainage of Highway Pavements.* Highway Engineering Circular No. 12, Federal Highway Administration, Washington, D.C. (March 1984).

construction. Where they must be used, adequate provision must be made for safeguarding traffic through the use of guardrails and similar devices. In Chapter 8, figures have been given for recommended side and ditch slopes, and typical ditch sections have been incorporated in drawings of typical right-of-way cross sections.

Flow of water in side ditches, of course, occurs in a direction that is generally parallel to the roadway centerline. Grades used in open ditches may also be roughly the same as those that are used on the highway centerline; on the other hand, flat roadway grades and steeper ditch grades in the same location are very frequently used. In very flat country, ditch grades as low as 0.1 or 0.2 percent may be used, while in rolling or mountainous terrain the maximum grade may be dictated only by the necessity for preventing erosion.

Side ditches provide open channels for the removal of surface water from within the limits of the highway right-of-way. In certain circumstances, areas adjacent to the right-of-way may also contribute to the flow. The water must be carried to an outlet, in the form of either a natural or an artificial drainage channel. The ditch must be hydraulically capable of handling the anticipated flow of surface water in such fashion that the roadway structure is not endangered or the safety of the motorist threatened. See Section 11-13 for the hydraulic design of side ditches and other open channels.

11-3 Prevention of erosion of shoulders and side slopes

The flow of surface water adjacent to highways is frequently accompanied by detrimental soil erosion that may result in the destruction of productive soils, the creation of areas of unsightly appearance, and the clogging of ditches and drainage structures. Erosion may also endanger the stability of side slopes in embankment and cut sections. A number of construction measures designed to prevent or minimize erosion are discussed in the paragraphs immediately following.

Any decision that is made regarding the installation of any of the special features discussed hereafter with the objective of preventing erosion on shoulders and side slopes must be based largely on judgment and knowledge of local conditions. Many factors enter into such a decision, including such items as intensity and duration of rainfall, soil types and condition, magnitude of slopes, climatic conditions, and many others.

Curbs, gutters, and flumes

Curbs and gutters are sometimes used at the edge of the traveled way on parkways and rural highways to prevent the water that flows from the crowned surface from spilling over and eroding the shoulders and side slopes. On urban highways, curbs and gutters are more commonly used, providing a measure of protection to pedestrians as well as facilitating drainage. Curbs may also be placed on the outer edge of surfaced shoulders when water threatens to erode the embankment slope. At intervals, water that collects along such curbs is channeled into a spillway, paved ditch, or flume and carried into the side ditch or pipe drain. The spacing of the inlets depends on the design discharge, the geometric configuration of the curb and gutter section, and the extent to which water is permitted to spread on the traveled way. More information on the hydraulic design of such facilities is given in Section 11-13 and in Ref.1.

Turf culture

One of the easiest and most effective ways to reduce erosion on side slopes and in side ditches is through the cultivation and development of a firm turf. Every effort is usually made in areas subject to erosion to encourage the growth of native grasses on exposed slopes. Formation of a firm turf may be accomplished by seeding, sprigging, or sodding the slope with suitable native grasses. In recent years, widespread use has been made of mats and linings of various kinds to encourage turf formation. The problems of planting and cultivating these grasses will not be discussed here. Suffice it to say that it is generally possible to establish adequate growths of grass that serve greatly to reduce the erosive effects of a flowing stream or sheet of water in areas of moderate rainfall or on moderate slopes.

Intercepting ditches

Shallow ditches may be placed at the top of outside cut slopes to intercept surface water and thus prevent erosion. Such ditches, which are usually quite shallow and narrow, are termed "intercepting ditches." Water conveyed in such a ditch is collected at suitable intervals along the top of the slope and discharged down the slope in paved or otherwise protected spillways into the longitudinal ditch, or spread out over the adjacent land.

Slope protection under severe erosion conditions

Highways are often subjected to very severe erosive action in which the measures thus far discussed provide only a partial solution of the problem. One example of this situation might be a highway location in a mountainous region subject to heavy rain and snowfalls. In such areas, a large portion of the location will probably consist of side-hill sections with the side slopes being made as steep as possible in the interests of economy. Slopes of 1.5 to 1, 1 to 1, or even steeper are not unusual and are, of course, very susceptible to erosion. The slopes may be protected, for example, by the use of riprap or hand-placed rock.

Still another situation in which a highway embankment may be subjected to very severe erosion is that existing when the location parallels a large or rapidly flowing stream. Such a location may involve a change in alignment of the stream, with resultant scour of the embankment slope. The slope may be protected by riprap, a paved revetment, steel sheet or timber piling, cribbing, or any one of a number of other protective devices.

11-4 Culverts, bridges, and storm drains

Culverts and bridges are visualized as composing the "cross-drainage" system of a highway in a rural location. Their general purpose is to transmit water flowing in natural streams or collected on the high side of the right-of-way from one side of the highway to the other. Additional information on culverts is contained in Sections 11-17 through 11-19.

Where space is restricted, in median swales, in urban areas, or where the natural slope of the ground is unsuited for drainage by open channels, storm drains are provided for the disposal of surface water. Storm drains and appurtenant structures are discussed when consideration is given to the drainage of city streets; see Sections 11-35 through 11-38.

11-5 Design of surface drainage systems

The surface drainage system should be designed to remove all surface water effectively and economically from the roadway area and intercept and dispose of surface water from adjoining areas. Subdrainage systems may be used to supplement surface drainage systems or to solve specific problems associated with the flow of underground water.

Design procedures logically may be divided into three major phases: an estimate of the quantity of water that may be expected to reach any element of the system; the hydraulic design of each element of the system; and the comparison of alternative drainage systems, alternative materials, and other variables in order to select the most economical system that can be devised. In the third phase, attention must be given to selecting the system that has the lowest annual cost, with all variables taken into consideration.

11-6 Streamflow records

When a drainage structure is to handle the flow of an existing stream, as in the case of some culverts and most bridges, peak flow for use in hydraulic design may be based on available records. Flood-frequency studies have been published for a number of states. Some of these studies have been made by the U.S. Geological Survey in cooperation with the various state highway departments, and studies of this type are continuing.

Flood-frequency studies of this type are based on statistical analysis of streamflow records. Usually, they indicate the mean annual flood as a function of the size of the drainage area for each hydrologic region within the state. Factors for computing floods of any frequency (within limits) in terms of the mean annual flood are given on a separate graph. The choice of flood frequency to be used for design is a matter for careful evaluation—frequencies of 10 to 25 years are common for rural highways, and frequencies as low as 2 years and as high as 50 years have been used. The regional flood curves usually cover drainage areas of about 100 to 2000 square miles. Separate flood curves may be given for the larger rivers, which cross regional boundaries and seldom conform to the trend given by the smaller rivers.

Basic considerations in estimating flood frequencies are contained in Ref. 2.

Many of the larger rivers are controlled by dams and other installations. Frequency and magnitude of floods in such cases depend on operation of the system and must be learned from the operating agency.

Where suitable streamflow records are not available, useful information may be gained from observations of existing structures and the natural stream. Drainage installations above and below the proposed location may be studied and a design based on those that have given satisfactory service on other portions of the same stream. Lacking this information, an examination of the natural channel may be made, including the evidences left by flood crests that have occurred in the past, and an estimate made of the quantity of water that has been carried by the stream during flood periods. Measurements may be made and values assigned to the slope, area, wetted perimeter, and roughness coefficient of the flood channel, and the quantity may be estimated by the Chezy formula. The Chezy formula is of the form,

$$Q = Ca\sqrt{RS} \tag{11-1}$$

where
Q = quantity of flow (ft^3/sec)
C = roughness coefficient, varying from 30 to 80 depending on the condition and nature of the channel
a = area of the flow cross section (ft^2)
S = slope of the channel (ft per foot)
R = hydraulic radius (ft) = a/p, where p = wetted perimeter, the length of the boundary of the cross section of the flow channel in contact with the water.

11-7 Hydrologic criteria

In cases where streamflow records and/or examination of the existing stream are inadequate, or where the drainage facility does not carry water that flows in a presently existing stream, the estimate of quantity must be made by some other method. The quantity estimate is important, since its accuracy determines the adequacy of the drainage structure designed.

More than 100 equations involving over 50 variables have been proposed for making peak runoff estimates for ungaged rural watersheds (3). Despite the plethora of available techniques, the estimation of surface runoff is not an exact science. When two or more methods are used to estimate runoff, discrepancies in the estimates of 50 percent or more are not uncommon.

Basic assumptions that must be made in order to obtain even a reasonably accurate estimate of quantity are the following: the rainfall intensity in inches per hour for the design storm, the coefficient of runoff or rate of infiltration, and the time of concentration for critical points in the drainage system.

11-8 Rainfall intensity

Proper design of a highway drainage system is accomplished by consideration of severe storms that occur at intervals and during which the intensity of rainfall and runoff of surface water are far greater than at other periods. Rainfall intensity during the design storm is a function of the occurrence, duration, and intensity.

Selection of the frequency of occurrence of the design storm is largely a matter of experience and judgment, although departmental policy may establish the interval to be used for a given situation. As previously mentioned, frequency of occurrence is a statistical matter. For example, if the system is designed for a frequency of occurrence of 25 years, the statistical assumption is that the system will accommodate runoff from the most severe storm to occur once in 25 years. It is apparent that selection of a frequency of 100 years, instead of 25, would mean designing for a more severe storm and, in general, a more costly system. Conversely, if the frequency is 10 years, the intensity of the design storm will be less and, in general, a less costly drainage system will result, but economic losses from use of the short time might offset the savings in construction costs.

On the Interstate System, the Federal Highway Administration requires that all drainage facilities other than bridges and culverts be designed for storms with a frequency at least as great as 10 years, except that a 50-year frequency is used for underpasses or other depressed roadways where ponded water can be removed only by the storm drainage system.

As Figure 11-1 illustrates, the intensity of rainfall for a particular frequency of occurrence varies greatly with the duration of rainfall. The average rate for a short time—such as 5 min—is much greater than for a longer period, such as 1 hr. In the design of many highway drainage systems, the duration chosen corresponds to the time of concentration. (See Section 11-10).

FIGURE 11-1 Typical rainfall intensity–duration curves. (Courtesy Federal Aviation Administration.)

An accurate estimate of the probable intensity, frequency, and duration of rainfall in a particular location can be made only if sufficient data have been collected over a period of time. If such information is available, standard curves may be developed to express rainfall-intensity relationships with an accuracy sufficient for drainage problems. A number of state highway departments have developed such curves for use in specific areas. Rainfall intensity–duration data have been published by the National Weather Service for various sections of the United States. One such publication is a rainfall-frequency atlas *(4)*, which contains maps of rainfall frequency for durations of 30 min and 1, 2, 3, 6, 12, and 24-hr for periods of 1, 2, 5, 10, 25, 50 and 100 years. Lacking such detailed relationships, recourse may be had to methods based on average relationships, such as the one explained in Section 11-11.

11-9 Surface runoff

When rain falls on a pervious surface, part of it passes into the soil and the remainder disappears over a period of time by either evaporation or runoff or both. In the design of highway drainage systems, the amount of water lost by evaporation is negligible; thus, drainage must be provided for all rainfall that does not infiltrate the soil or is not stored temporarily in surface depressions within the drainage area.

The rate at which water infiltrates the soil is dependent on the following factors: type and gradation of the soil; soil covers; moisture content of the soil; amount of organic material in the soil; temperature of the air, soil, and water; and presence or absence of impervious layers near the surface. Rates of infiltration on bare soil are less than on turfed surfaces. Frozen soil is impervious, and rain infiltrates very little until the frozen layer thaws. The rate of infiltration is assumed to be constant during any specific design storm.

The rate at which runoff occurs depends on the nature, degree of saturation, and slope of the surface. The rate of runoff is greater on smooth surfaces and initially slower where vegetation is present. On pavements and compacted surfaces runoff occurs at a high rate, which varies with the slope and character of the surface at each point For use in connection with the design of a drainage system for a particular area, these variables are considered and a coefficient of runoff selected. Values of the coefficient of runoff for use in one method of estimating the quantity of flow (the rational method) are given in Table 11-2. A more complete list of runoff coefficients has been published in Ref. 5. If the drainage area being considered is composed of several types of surfaces, then a coefficient may be chosen for each surface and the coefficient for the entire area computed as a weighted average of the individual areas.

11-10 Time of concentration

After selecting the design storm frequency, computations are made to determine the duration of the rainfall that produces the maximum rate of runoff. The duration of rainfall required to produce the maximum rate of runoff is known as the "time of concentration." The time of concentration usually consists of the time of overland flow plus the time of flow in the drainage system.

TABLE 11-2
COEFFICIENTS OF RUNOFF TO BE USED IN THE RATIONAL FORMULA

Type of Drainage Area	Coefficients of Runoff, C
Concrete and bituminous pavements	0.70−0.95
Gravel or macadam surfaces	0.40−0.70
Impervious soil	0.40−0.60
Impervious soils, with turf[a]	0.30−0.55
Slightly pervious soils[a]	0.15−0.40
Pervious soils[a]	0.05−0.10
Wooded areas (depending on slope and cover)	0.05−0.20

[a]For slopes from 1 to 2 percent.

The time of overland flow is the time required for a particle of water to flow from the most remote point in any section of the drainage area being considered to the point where it enters the drainage system. To this must be added the time of flow in the drainage system, from the intake to the point being considered. The time of overland flow varies with the slope, type of surface, length, and other factors. A number of empirical studies have been published on the time of overland flow to the defined channels, for example, Ref. 6. When the particular drainage area consists of several types of surfaces, the time of overland flow must be determined by adding together the respective times computed for flow over the lengths of the various surfaces from the most remote point to the inlet. Estimates of time of flow in the drainage system can be made from observed or computed velocities of flow.

11-11 Federal Highway Administration chart

Figure 11-2 shows 30-min rainfall intensities for a 2-year recurrence period at various locations in the 48 contiguous states (7). The map, which was adapted from Ref. 4, is useful for determining approximate rainfall−intensity−duration data when site-specific data are not available.

The 2-year rainfall intensity for other durations may be obtained by multiplying the 30-min intensity for the project location by these factors: 2.22 (5 min); 1.71 (10 min); 1.44 (15 min); 1.25 (20 min); 0.8 (40 min); 0.6 (60 min); 0.5 (90 min); and 0.4 (120 min).

To convert the chart values to other intervals of recurrence, multiply by these factors: 0.75 (1 year); 1.3 (5 years); 1.6 (10 years); 1.9 (25 years); and 2.2 (50 years).

11-12 The rational method

One of the most common methods of estimating runoff from a drainage area is the rational method. Its popularity is due to the fact that is combines engineering judgment with calculations made from analysis, measurement, or estima-

FIGURE 11-2 Map of the contiguous United States, showing 2-year, 30-min rainfall intensity. (Courtesy Federal Highway Administration.)

tion. The method is based on the direct relationship between rainfall and runoff and is expressed by the equation

$$Q = CIA \tag{11-2}$$

where

Q = runoff (ft^3/sec)

C = a coefficient representing the ratio of runoff to rainfall. Typical values of C are given in Table 11-2.

I = intensity of rainfall (in./hr for the estimated time of concentration)

A = drainage area in acres. The area may be determined from field surveys, topographic maps, or aerial photographs.

In view of the preceding discussion of the many variables involved in the rainfall–runoff relationship, shortcomings of this method are apparent . Application of the method should be confined to relatively small drainage areas (up to 200 acres, according to the Federal Highway Administration).

The Federal Highway Administration has published a number of empirical charts that make it possible to develop runoff graphs for large drainage areas. Using such charts, the Georgia Department of Transportation, for example, has developed graphs for estimating runoff for rural watersheds in Georgia with areas up to 1000 square miles *(8)*.

11-13 Design of side ditches and other open channels

With the quantity of water expected to reach any given point in the drainage system known, the design of side ditches, gutters, stream channels, and similar facilities is based on established principles of flow in open channels. The principles also apply to flow in conduits with a free water surface. Most commonly used for design is Manning's formula, which applies to conditions of steady flow in a uniform channel and has the following form:

$$V = \frac{1.486 R^{2/3} S^{1/2}}{n} \tag{11-3}$$

where

V = mean velocity (ft/sec)

R = hydraulic radius (ft); this is equal to the area of the cross section of flow (ft^2) divided by the wetted perimeter (ft)

S = slope of the channel (ft per foot)

n = Manning's roughness coefficient. Typical values of n are given in Table 11-3.

Also applicable is the equation of continuity

$$Q = VA = \frac{1.486}{n} A R^{2/3} S^{1/2} \tag{11-4}$$

where

Q = discharge (ft^3/sec)

A = area of the flow cross section (ft^2)

TABLE 11-3
VALUE OF MANNING'S ROUGHNESS COEFFICIENT

Type of Channel or Structure	Values of n
Open Channels for Type of Lining Shown	
Smooth concrete	0.013
Rough concrete	0.022
Riprap	0.03−0.04
Asphalt, smooth texture	0.013
Good stand, any grass—depth of flow more than 6 in.	0.09−0.30
Good stand, any grass—depth of flow less than 6 in.	0.07−0.20
Earth, uniform section, clean	0.016
Earth, fairly uniform section, no vegetation	0.022
Channels not maintained, dense weeds	0.08
Natural Stream Channels	
(Surface Width at Flood Stage is 100 ft)	
Fairly regular section	
(a) Some grass and weeds, little or no brush	0.030−0.035
(b) Dense growth of weeds, depth of flow materially greater than weed height	0.035−0.05
(c) Some weeds, light brush on banks	0.035−0.05
(d) Some weeds, heavy brush on banks	0.05−0.07
(e) Some weeds, dense willows on banks	0.06−0.08
(f) For trees within channel, with branches submerged at high stage, increase all above values by	0.01−0.02
Irregular sections, with pools, slight channel meander; *increase* values given above about	0.01−0.02
Culverts	
Concrete pipe and boxes	0.012
Corrugated metal	
Unpaved	0.024−0.027
25% paved	0.021−0.026
Fully paved	0.012

A family of charts has been published by the Federal Highway Administration for the solution of Manning's equation for various common channel cross sections *(7, 9)*. One of these charts is shown in Figure 11-3.

At this point it is necessary to introduce the concepts of the "theory of critical flow" as it relates to the flow of water in open channels and culverts. The theory of critical flow gives rise to the definition of "critical depth," which is the depth where the flow changes from "tranquil" to "rapid" or "shooting." For example, water flowing down a relatively flat slope in an open channel is in tranquil flow, while that tumbling down a steep slope is rapid.

FIGURE 11-3 Graphical solution of Manning's equation for one trapezoidal channel section with 2 – 1 side slopes and bottom width 4 ft. (Courtesy Federal Highway Administration.)

In the same channel, with the same quantity of flow, the flow can be changed from tranquil to rapid by an increase in slope. In such a case, the depth of flow decreases from that existing in the section of tranquil flow to a lower value in the section of rapid flow. The decrease in depth begins at some point ahead of the crest of the steep slope and continues gradually over some distance. In the section of rapid flow, since the quantity of flow is the same, the dimensions of the channel the same, and the depth less, it follows from the equation of continuity that the velocity is greater. Critical depth is defined as the depth corresponding to the change from tranquil to rapid flow.

Critical velocity and critical slope are the velocity and slope that correspond to uniform flow at critical depth. Critical depth is independent of channel slope and roughness, but the critical slope is a function of the slope of the channel.

Tranquil flow exists when the normal depth of water in an open channel is greater than the critical depth; conversely, when the depth is less than critical, the flow is rapid.

Rapid flow is difficult to control, because abrupt changes in alignment or cross section produce waves that travel downstream, alternating from side to side and sometimes causing the water to overtop the sides of the channel. Changes in channel slope, shape, or roughness cannot be reflected upstream except for very short distances (this condition is called "upstream control"). This type of flow is common in steep flumes and mountain streams.

Tranquil flow is relatively easy to control. Changes in channel shape, slope, or roughness affect the stream for some distance upstream (downstream control). This kind of flow often occurs in streams in plains and valley regions, where slopes are relatively flat.

Critical depth is important in analysis because it is always a hydraulic control. The flow must pass through critical depth in going from one type of flow to the other.

11-14 Design procedure

Hydraulic design procedures are difficult to simplify because of the wide variety of choices presented to the designer in a typical case and the various assumptions that must be made. Design is usually based on an assumption of uniform flow, since the error involved is relatively slight in most cases.

Basically, the design of a highway drainage channel is done in two parts. The first part involves the selection of a channel section that will carry the given discharge on the available slope. The second part is the determination of the protection required (if any) to prevent erosion of the drainage channel (see Section 11-15).

Use of charts like the one in Figure 11-3 gives a direct solution of the Manning equation for uniform flow in trapezoidal channels with 2:1 side slopes, and with fixed bottom width (in Figure 11-3, bottom width is 4 ft). Depths and velocities shown in the chart apply accurately only to channels in which uniform flow at normal depth has been established by sufficient length of uniform channel on a constant slope when the flow is not affected by backwater.

Depth of uniform flow for a given discharge in a given size of channel on a given slope and with $n = 0.030$ may be determined directly from the chart by entering on the Q-scale and reading normal depth at the appropriate slope line (or an interpolated slope). Normal velocity may be read on the V-scale opposite the same point. This procedure may be reversed to determine discharge at a given depth of flow.

For channel roughness other than $n = 0.030$, compute the quantity Q times n and use the $Q \cdot n$- and $V \cdot n$-scales for all readings, except those which involve values of critical depth or critical velocity. Critical depth for a given value of Q is read by interpolation from the depth lines at the point where the Q-ordinate and the critical curve intersect, regardless of channel roughness. Critical velocity is the reading on the V-scale at the same point. Where $n = 0.030$, the critical slope is read at the critical depth point. Critical slope varies with n; therefore, in order to determine the critical slope for values of n other than 0.030, it is first necessary to determine the critical depth. Critical slope is then read by interpolation from the slope lines at the intersection of this depth with the $Q \cdot n$-ordinate.

As an example, determine the depth and velocity of flow in a trapezoidal channel ($n = 0.030$) with 2:1 side slopes and 4-ft bottom width discharging 150 ft³/sec on a 2 percent slope ($S_0 = 0.02$). Use Figure 11-3. From $Q = 150$ ft³/sec on the Q-scale follow up and read normal depth, $d_n = 2.1$ ft, and normal velocity, $V =$ approximately 8.5 ft/sec.

To find the critical depth, velocity, and slope for these conditions, read upward from $Q = 150$ ft³/sec to the intersection with the critical curve. At this point, read by interpolation critical depth = 2.4 ft; critical velocity = 7.0 ft/sec, and critical slope = 0.013. The normal depth is less than the critical depth, hence the flow is rapid and not affected by backwater conditions.

If $n = 0.020$ and other conditions are the same, the procedure is similar. Use the $Q \cdot n$-scale. $Q \cdot n = 150(0.02) = 3.00$. From 3.00 on the $Q \cdot n$-scale read downward to the intersection with the line $S = 0.02$. At this point, $d_n = 1.8$ ft and $V \cdot n = 0.23$. $V = (0.23/0.02) = 11.5$ ft/sec.

11-15 Prevention of erosion in drainage channels

Erosion of highway drainage channels may create unsightly and hazardous ditches, cause pollution of nearby lakes and streams, and increase maintenance costs of drainage structures because of sedimentation. To prevent erosion, it may be necessary to provide protective linings in the bottom and along the sides of drainage channels. There are two general classes of protective linings: rigid linings and flexible linings. Common types of rigidly lined channels are those paved with portland cement concrete or soil-cement. Examples of flexible linings are rock riprap and vegetation.

Rigid channel linings are more expensive than flexible linings and, being smoother, may cause objectionable high velocities at the end of the lining. On the other hand, rigid linings are capable of preventing erosion under severe service conditions, along steep slopes and in restricted areas where steep sidewall slopes are required.

Flexible linings are generally unsuitable for severe service conditions, for example, where the depth or velocity of flow is great. However, such linings are relatively inexpensive, easy to maintain, and aesthetically pleasing.

Temporary linings

With increasing attention being paid to the environmental impact of highway construction, temporary linings are recommended to protect seeded channels until vegetation has been established. Examples of such linings have been described in Ref. 10.

1. Erosionet 315 is a paper yarn with openings of approximately ⅞ by ½ in. Secured by steel pins, it is normally used to hold other materials such as straw.
2. Jute mesh is a woven mat of jute yarn with openings of about ⅜ by ¾ in. It is normally held in place by steel pins or staples spaced about 3 ft apart in three rows along each strip.
3. Stranded fiberglass roping consists of fine fibers of fiberglass blown into the channel bed with a special nozzle driven by an air compressor. The resulting web of material is normally tacked with asphalt to bind the fibers together and to adhere it to the soil. Alternatively, it may be covered with Erosionet 315.
4. A fiberglass mat similar to a furnace air filter may be used. The mat, which is typically ⅜ by ½ in. in thickness, is held in place by steel pins.
5. Excelsior mat is comprised of dried shredded wood that is held together with a fine paper net. It is also usually secured with steel pins.
6. Straw may also be applied at a rate of about 1.25 lb per square yard and held in place with Erosionet and steel pins.

Design approaches

Since rigid linings are resistant to erosion, the maximum depth of flow which can be accommodated in paved channels generally depends only on the required freeboard. FHWA *(10)* recommends a vertical freeboard height of 1 ft for small drainage channels. For large channels, wave height may have to be considered when establishing freeboard requirements.

Traditionally, the type of lining provided in drainage channels was determined on the basis of maximum allowable velocities. Research *(10)* indicates that for flexible linings, the selection of type of channel lining should be based on the maximum permissible depth of flow. Normann *(10)* explains:

> The maximum permissible depth concept is based on the tractive force theory of channel lining design. Tractive force, basically, is the shear stress exerted by the flow on the channel perimeter. For wide channels of any shape, and for a given channel slope, depth, and lining, the vertical velocity distribution in the central and deepest section, where wall effects are negligible,

should be identical. Also, the first scour occurs at the deepest portion of the channel, since the wall or bottom shear stress is greatest in that portion.

Channel lining design procedures

Selection of channel lining depends on the quantity of flow; the slope, shape, and dimensions of the channel; and the erodibility of the soil. Estimation of the design flow should be based on the hydrologic procedures previously described. A recurrence interval of 5 or 10 years is normally employed for drainage channels. Use of a shorter recurrence period is permitted for the design of temporary linings. The slope, shape, and dimensions of the channel influence the depth and velocity of flow. It is not possible to objectively predict the erodibility of a soil. Rather, the designer classifies a specific soil as "erodible," "erosion resistant," and so forth on the basis of experience and by reference to agricultural soil survey reports or by consultation with soil scientists of the Soil Conservation Service.

The major steps involved in evaluating the erosion resistance of a given channel slope, cross section, and selected lining are given in Ref. 10.

11-16 Stream enclosures

Side ditches paralleling highway locations in rural areas are sometimes eliminated by enclosing the stream in a pipe drain. Conditions that would make this desirable include the elimination of a deep, narrow side ditch made necessary by a narrow right-of-way, the desired widening of an existing pavement built in a narrow right-of-way, and the elimination of erosion. This method has also been employed for the somewhat unusual purpose of enclosing a surface stream so that the highway could be built in the middle of a narrow valley rather than on a side-hill location.

Stream enclosures of the type that have been described are, generally speaking, storm sewers and are designed in a fashion similar to that used in the design of municipal drainage systems. Appurtenant structures such as inlets, catch basins, and manholes are also normally included in the design.

CULVERTS

A culvert is usually, although not always, differentiated from a bridge by virtue of the fact that the top of the culvert does not form part of the traveled roadway. More frequently, culverts are differentiated from bridges on the basis of span length. On an arbitrary basis, structures having a span of 20 ft or less will be called culverts, while those having spans of more than 20 ft will be called bridges. This line of division is by no means standard, and span lengths of 8 to 20 ft are employed by various organizations as limiting culvert lengths. Culverts also differ from bridges in that they are usually designed to flow full under certain conditions, while bridges are designed to pass floating debris or vessels.

Culverts are to be found in three general locations: at the bottom of depressions where no natural watercourse exists; where natural streams intersect the roadway; and at locations required for passing surface drainage carried in side ditches beneath roads and driveways to adjacent property.

11-17 Principles of culvert location

The majority of culverts are installed in natural watercourses that cross the roadway, either at right angles or on a skew. In addition to selecting the proper location or "station number" for the culvert crossing with respect to the centerline of the road, the alignment and grade of the culvert are of importance.

The location of the centerline of the culvert on the centerline of the road may be determined by inspection of the plans or in the field. This location will generally be on the centerline of an existing watercourse or at the bottom of a depression if no natural watercourse exists.

The alignment of the culvert should generally conform to the alignment of the natural stream, and the culvert should, if possible, cross the roadway at right angles in the interests of economy. Skew culverts, located at an angle to the centerline of the road, are needed in many instances. The selection of the natural direction of the stream is somewhat difficult in some areas, where the stream bed is not in a fixed position but shifts with the passage of time. In such a case, judgment must be exercised in selecting the most desirable location for the culvert, and some channel improvements may be necessary to ensure the proper functioning of the culvert after it is built. Where meandering streams are encountered, the water should be carried beneath the roadway at the earliest opportunity. Any changes that are necessary in the direction of the culvert itself should be effected gradually so that excessive head losses and consequent reduction in flow may be avoided.

Similarly, the grade of the culvert should generally conform to the existing grade of the stream. If the grade is reduced through the culvert, the velocity may be reduced, sediment carried in the water will be deposited at the mouth or in the length of the culvert, and the capacity of the structure will thus be further reduced. Culvert grades that are greater than those existing in the natural channel may result in higher velocities through the culvert and at the outlet end. Undesirably high velocities at the outlet will result in scour or erosion of the channel beyond the culvert and may make it necessary to install elaborate and costly protective devices. Changes in grade within the length of the culvert should also be avoided.

11-18 Hydraulic design of culverts

Earlier in this chapter, we discussed concepts and design procedures related to estimation of the quantity of runoff from a drainage basin. In this section, we will briefly discuss principles and techniques for the hydraulic design of culverts. For a more complete treatment of the subject, the reader is advised to consult Refs. 11, 12, and 13.

The purpose of hydraulic design is to provide a drainage facility or system that will adequately and economically provide for the estimated flow throughout the design life without unreasonable risks to the roadway structure or nearby property.

Hydraulic design of culverts involves the following general procedure:

1. Obtain all site data and plot a roadway cross section at the culvert site, including a profile of the stream channel.
2. Establish the culvert invert elevations at the inlet and outlet, and determine the culvert length and slope.
3. Determine the allowable headwater depth and the probable depth of tailwater during the design flood.
4. Select a type and size of culvert and the design features of its appurtenances that will accommodate the design flow under the established conditions. The design of the culvert inlet, discussed in Section 11-19, is especially important to the overall hydraulic efficiency of the structure.
5. Examine the need for energy dissipators, and, where needed, provide appropriate protective devices to prevent destructive channel erosion. (See Section 11-20.)

Whenever a constriction such as a culvert is placed in a natural open channel, there is an increase in the depth of water just upstream of the constriction. The allowable level of the headwater upstream of the culvert entrance is generally the principal control on the culvert size and inlet geometry. The allowable headwater depth depends on the topography and the nature of land use in the culvert vicinity. In establishing the allowable headwater depth, the designer should consider possible harmful effects that flooding may cause, such as damages to the pavement, interruptions to traffic, and inundation of nearby property.

Types of a culvert flow

The type of flow occurring in a culvert depends on the total energy available between the inlet and outlet. The available energy consists primarily of the potential energy or the difference in the headwater and tailwater elevations. (The velocity in the entrance pool is usually small under ponded conditions and the velocity head or kinetic energy can be assumed to be zero.) The flow that occurs naturally is that which will completely expend all of the available energy. Energy is thus expended at entrances, in friction, in velocity head, and in depth.

The flow characteristics and capacity of a culvert are determined by the location of the *control section*. A control section in a culvert is similar to a control valve in a pipeline. The control section may be envisioned as the section of the culvert that operates at maximum flow; the other parts of the system have a greater capacity than is actually used.

Laboratory tests and field studies have shown that highway culverts operate with two major types of control: *inlet control* and *outlet control*. Examples of flow with inlet control and outlet control are shown, respectively, by Figures 11-4 and 11-5.

Culverts flowing with inlet control

Under inlet control, the discharge capacity of a culvert depends primarily on the depth of headwater at the entrance and the entrance geometry (barrel

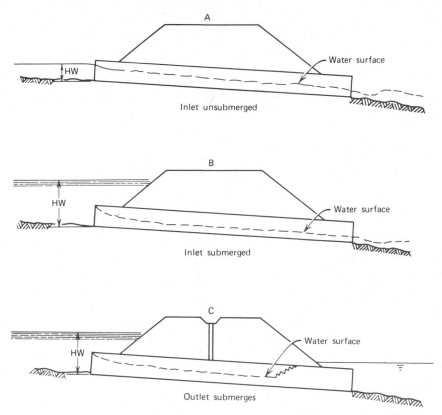

FIGURE 11-4 Inlet controls for culverts. (Courtesy Federal Highway Administration.)

shape, cross-sectional area, and type of inlet edge). Inlet control commonly occurs when the slope of the culvert is steep and the outlet is not submerged.

On the basis of experimental work, analytical relationships have been developed for culverts with inlet control *(13)*. However, these relationships are complex and for a given entrance shape and flood condition (i.e., submerged, unsubmerged) are applicable only within a specified range of discharge factors.* For design expedience, the Federal Highway Administration has published a series of nomographs *(11)* and capacity charts *(12)*. Examples of available nomographs and charts are shown as Figures 11-6, 11-7, and 11-8.

Culverts flowing with outlet control

Maximum flow in a culvert operating with outlet control depends on the depth of headwater and entrance geometry and the additional considerations of the elevation of the tailwater at the outlet and the slope, roughness, and length of the culvert. This type of flow most frequently occurs on flat slopes,

---

*Discharge factor = $Q/D^{5/2}$, where Q is flow and D is culvert diameter.

FIGURE 11-5 Outlet controls for culverts. (Courtesy Federal Highway Administration.)

especially where downstream conditions cause the tailwater depth to be greater than the critical depth.

For culverts flowing full, the difference in head between the upstream and downstream water surface H is equal to the velocity head, plus the energy lost at the entrance and in the culvert:

$$H = \left(1 + K_e + \frac{29n^2L}{R^{4/3}}\right)\frac{V^2}{2g} \tag{11-5}$$

where

H = difference in elevation between the headwater and tailwater surfaces, illustrated in Figure 11-5a, or between the headwater surface and the crown of the culvert at the outlet, as shown in Figure 11-5b.

K_e = entrance loss coefficient; see Table 11-4

n = Manning's roughness coefficient; see Table 11-3

FIGURE 11-6 Nomograph for determining headwater depth for concrete pipe culverts with inlet control. (Courtesy Federal Highway Administration.)

L = length of the culvert (ft)
R = hydraulic radius (ft)
$V^2/2g$ = velocity head (ft)

FIGURE 11-7 Nomograph for determining head for concrete pipes flowing full ($n = 0.012$). (Courtesy Federal Highway Administration.)

Where the critical depth falls below the crown of the culvert at the outlet, as shown in Figures 11-5c and 11-5d, the headwater can be determined analytically only by tedious and time-consuming backwater computations (11). Such computations can generally be avoided by using the design charts published by the FHWA (11, 12). From charts such as these, the headwater depths for both

FIGURE 11-8 Typical culvert capacity charts. (Courtesy Federal Highway Administration.)

inlet and outlet control may be determined for practically all combinations of culvert size, material, entrance geometry, and discharge.

To understand the use of Figures 11-6 and 11-7, the reader should study the examples on the nomographs.

To illustrate the use of capacity charts published in Hydraulic Engineering Circular No. 10 (12), consider the illustrative charts shown as Figure 11-8. Figure 11-8a combines two separate sets of curves for a 36-in. standard corru-

TABLE 11-4
ENTRANCE LOSS COEFFICIENTS K_e; OUTLET CONTROL; FULL OR PARTLY FULL

Type of Structure and Design of Entrance	Coefficient K_e
Pipe—concrete	
Projecting from fill, socket end (groove-end)	0.2
Projecting from fill, sq. cut end	0.5
Headwall or headwall and wingwalls	
Socket end of pipe (groove-end)	0.2
Square-edge	0.5
Rounded (radius = $\frac{1}{12}D$)	0.2
Mitered to conform to fill slope	0.7
End section conforming to fill slope	0.5
Beveled edges, 33.7° or 45° bevels	0.2
Pipe or pipe-arch—corrugated metal	
Projecting from fill (no headwall)	0.9
Headwall or headwall and wingwalls square-edge	0.5
Miltered to conform to fill slope, paved or unpaved slope	0.7
End section conforming to fill slope	0.5
Beveled edges, 33.7° or 45° bevels	0.2
Box—reinforced concrete	
Headwall parallel to embankment (no wingwalls)	
Square-edged on three edges	0.5
Rounded on three edges to radius of $\frac{1}{12}$ barrel dimension, or beveled edges on three sides	0.2
Wingwalls at 30° to 75° to barrel	
Square-edged at crown	0.4
Crown edge rounded to radius of $\frac{1}{12}$ barrel dimension, or beveled top edge	0.2
Wingwall at 10° to 25° to barrel	
Square-edged at crown	0.5
Wingwalls parallel (extension of sides)	
Square-edged at crown	0.7
Side or slope tapered inlet—all three culvert types	0.2

SOURCE: *Hydraulic Design of Improved Inlets for Culverts.* Federal Highway Administration, Washington, D.C. (1972).

gated metal pipe culvert with projecting entrance. Included are the inlet-control curve with $L/100S_0$ of 50 (L is the length of culvert and S_0 is the barrel slope) and two outlet-control curves with $L/100S_0$ of 250 and 450; the additional $L/100S_0$ curves were obtained by interpolation.

Headwater depth is determined by entering the chart at the design Q, then following up the chart to intersect a curve with the $L/100S_0$ value for the particular culvert under study. Headwater is read on the ordinate scale (Figure 11-8a). If $L/100S_0$ is 50 or less, the culvert operates in inlet control, and headwater depth is read on the solid-line curve. No lesser depth of headwater is possible even for smaller $L/100S_0$ values. For any $L/100S_0$ value other than that shown on the curves and within the chart limit, headwater depth must be read by interpolation.

For example, consider a peak discharge rate of 50 cfs. For a 36-in. culvert 78 ft long on a slope of 0.0056 ($L/100S_0$ = 140), headwater depth will be 5.2 ft. Another culvert 166 ft long at 0.5 percent slope ($L/100S_0$ = 330) will require a headwater depth of 6.3 ft for the same discharge.

Factors that govern selection of a culvert size include the design discharge rate, a limiting headwater depth, the culvert length, and invert slope. In the FHWA procedure, the type of culvert must be preselected to locate the capacity chart to be used. An example is Figure 11-8b, which applies to a concrete pipe with a groove-edged entrance.

Assuming a culvert length of 240 ft and a slope of 0.002, a range of headwaters and discharges can be studied for the three sizes of pipe shown in the chart. The $L100S_0$ ratio is 1200. If the situation was such that the design discharge was 100cfs for a limiting headwater depth of 6.0 ft, the designer would select a 48-in. pipe (the smaller, 42-in. pipe gives a headwater of 6.8 ft).

To illustrate further the culvert selection process, assume a design discharge of 120 cfs, and an allowable headwater depth of 6.0 ft. Inspection of Figure 11-8b shows that a 42-in. pipe will not be adequate for any length–slope ratio, since the minimum headwater depth is 7.4 ft for 120 cfs. A 48-in. pipe would be a proper choice for all $L/100S_0$ ratios less than about 900. Some extreme lengths and small slope ratios would require a 54-in. pipe.

Capacity charts (e.g., Figure 11-8) permit the direct selection of culvert size and avoid the trial-and-error approach required by nomographs. Nomographs such as Figures 11-6 and 11-7 are recommended for situations involving high headwater depths and in cases of submerged outlets. In certain ranges of hydraulic operation, the nomographs and charts give identical answers.

It is not possible to describe here all the limitations of, and exceptions to, the use of the charts or the complications that may occur in selecting culverts for particular conditions. The hydraulic design of culverts is a complicated matter, much simplified by the use of the FHWA charts. Suffice it to say that the two FHWA publications provide tools for handling most designs with an accuracy compatible with that of forecasting design discharges and their frequency.

11-19 Improved culvert inlet design

With inlet control, flow in the culvert barrel is very shallow and the potential capacity of the barrel is generally wasted. Since the barrel is usually the most expensive component of the structure, flow under inlet control tends to be uneconomic. Surveys of culvert design practices by highway agencies indicate that millions of dollars could be saved each year by using improved inlet design concepts. An article by Normann *(14)* in *Civil Engineering*, reproduced in part in the following paragraphs, describes fundamental concepts of improved inlet design. For more detailed information on this important subject, the reader should refer to the FHWA publication *Hydraulic Design of Improved Inlets for Culverts (15).*

In 1967, culvert research sponsored by the FHWA was completed by the National Bureau of Standards. Based on this research, a design manual was formulated, in which the best improved inlet designs were selected and compiled, based on several factors including hydraulic efficiency, ease of construction and maintenance, and debris passage capability.

Three basic improved inlet designs are presented: bevel edged inlets (Figure 11-9), side-tapered inlets (Figure 11-10), and slope-tapered inlets (Figure 11-11). These inlets improve hydraulic performance in two basic ways: (1) by reducing the flow contraction at the culvert inlet and more nearly filling the barrel and (2) by lowering the inlet control section and thus increasing the effective head exerted at the control section for a given headwater pool elevation.

Bevel-edged inlet

The bevel-edged inlet shown in Figure 11-9 improves hydraulic performance by reducing the inlet contraction. This is the least sophisticated inlet improvement, and also the least expensive. As a minimum, this degree of improvement should be used on all culverts, both in inlet and outlet control. For concrete pipe culverts, the groove end, left intact and facing upstream, will serve essentially the same purpose as the bevel-edged inlet.

Side-tapered inlet

The side-tapered inlet in Figure 11-10 increases hydraulic efficiency by further reducing the contraction at the inlet control section, which should be the throat section at or near the design discharge. The face section is designed large enough so as to not restrict the flow. The roof and floor of the inlet are straight line extensions of the culvert roof and floor, and the tapered side walls meet the barrel walls at a smaller angle than the beveled edges (9.5 to 14° vs 33 to 35°). Also, the throat section is somewhat lower than the face, thus concentrating more head on the control section for a given headwater elevation.

Slope-tapered inlet

The slope-tapered inlet in Figure 11-11 provides an efficient control section at the throat, similar to that provided by the side-tapered inlet, and further increases the concentration of head on the throat. In this design, the face section remains near the stream bed elevation and the throat is lowered by incorporating a fall within the inlet structure. This fall in the inlet reduces the slope of the barrel and increases the required excavation. The larger the fall required to pass the design discharge, the larger will be the inlet structure.

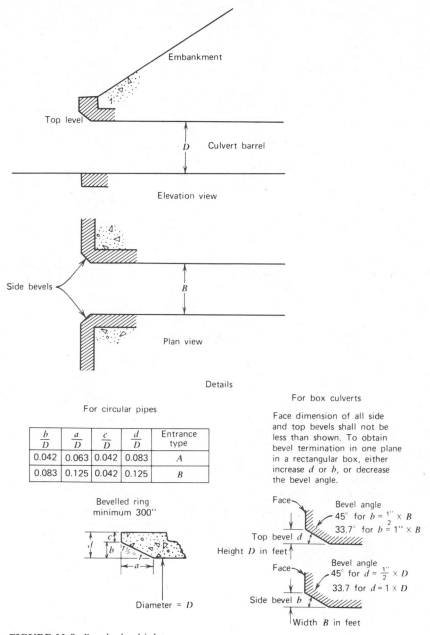

FIGURE 11-9 Bevel-edged inlets.

Details

For circular pipes

$\frac{b}{D}$	$\frac{a}{D}$	$\frac{c}{D}$	$\frac{d}{D}$	Entrance type
0.042	0.063	0.042	0.083	A
0.083	0.125	0.042	0.125	B

Bevelled ring minimum 300″

Diameter = D

For box culverts

Face dimension of all side and top bevels shall not be less than shown. To obtain bevel termination in one plane in a rectangular box, either increase d or b, or decrease the bevel angle.

Face

Bevel angle
45° for $b = \frac{1''}{2} \times B$
33.7° for $b = 1'' \times B$

Top bevel d
Height D in feet

Face

Bevel angle
45° for $d = \frac{1''}{2} \times D$
33.7 for $d = 1 \times D$

Side bevel b
Width B in feet

The slope-tapered inlet incorporates both methods of increasing hydraulic performance: reducing the entrance contraction and lowering the control section. Of course, any of the other inlets, conventional bevel-edged or side-tapered, may be lowered by constructing a sump upstream of the entrance, thus obtaining increased head on the inlet control section.

FIGURE 11-10 Side-tapered inlets.

FIGURE 11-11 Slope-tapered inlets.

In addition to their use on new installations, improved inlets, especially bevel-edged and side-tapered inlets may be added to existing barrels to increase hydraulic performance if the existing inlet is operating in inlet control. Many times, this will preclude the construction of a new barrel when the existing culvert is undersized.

Similar improvements may be applied to pipe culverts. As an example of the resulting economy, savings of $1500 per culvert, or a total of $30,000 were recently reported on an interstate highway project in Georgia from the use of precast side-tapered inlets for pipe culverts.

New culvert design procedure

"Hydraulic Charts for the Selection of Highway Culverts," (U.S. Department of Transportation, Hydraulic Engineering Circular No. 5) is probably the most widely used and accepted culvert design manual in the world. In the conventional design procedure, both inlet and outlet flow control are evaluated. The headwater depth (depth of upstream pool above the inlet invert) is computed for the design discharge assuming that (1) inlet control governs, and (2) that outlet control governs. Then, the highest of the two required headwater depths defines the control type and the culvert is categorized as being in either "inlet control" or "outlet control." Assumptions are simplified to expedite the computations, and by comparing headwaters, the difficult task of defining the actual flow profile through the culvert barrel is avoided. However, in this conventional design method, no attempt is made to modify the unbalanced flow condition that may exist. The inlet control performance curve represents the actual flow capacity of the culvert, while the more favorable outlet performance control curve is the culvert barrel potential. Thus, the full capacity of the barrel is not attained and an uneconomic situation exists.

11-20 Energy dissipators

In the vicinity of culverts, high velocities, abrupt changes in direction of flow, and the presence of waves and eddies may cause severe erosion or "scour" to occur. A common problem at culvert outlets, scour may also occur in the vicinity of culvert inlets and at other locations in roadside channels. Unless preventive measures are taken, scour may cause unsightly and hazardous holes in the channel, damage the embankment, or even cause a culvert structure to fail. It is possible that such problems can be overcome by the use of protective channel linings as discussed in Section 11-15.

Where severe erosion problems are likely to occur, the Federal Highway Administration (16) recommends that consideration be given to the employment of energy dissipators. Such devices provide an effective means of controlling velocities and smoothing the flow.

Examples of energy-dissipating devices include:

1. Drop structures that change the channel slope from steep to mild combined with special stilling basins to dissipate the kinetic energy.

2. Blocks, sills, or other roughness elements installed in the channel to increase the resistance or force and stabilize a hydraulic jump.
3. Stilling wells that dissipate energy by forcing the flow to travel directly upward to reach the downstream channel.

The design of energy-dissipating devices is a complex process that is beyond the scope of this book. The subject is adequately treated in Ref. 16.

11-21 Culvert types and materials

Materials most commonly used in the construction of culverts are reinforced concrete and corrugated metal. Less frequently, culverts are made from timber, cast-iron pipe, vitrified-clay pipe, and, occasionally, stone masonry.

Concrete culverts

Reinforced-concrete pipe intended for use in culverts is made in diameters of 12 to 108 in. and in various lengths, the usual length being 4 to 8 ft. Standard specifications establish five classes of pipe with strengths increasing from class I through class V *(17)*. The specifications show cross-sectional areas of reinforcing steels and concrete strengths for three series of wall thicknesses. Reinforcement may be either circular or elliptical. For special applications, reinforced-concrete culvert pipe may be manufactured with a cross section other than circular—elliptical and "arch" shapes being in quite common use. Concrete culvert pipes have tongue-and-groove or bell-and-spigot joints; the joints are sealed during construction with portland cement mortar, rubber gaskets, or other materials. A greater or lesser amount of care is required in the preparation of the foundation on which the pipes are to be laid. This preparation or "bedding" may vary from simple shaping of the bottom of a trench, or of the ground on which the pipe is laid, to embedment of the pipe in a concrete cradle, depending on foundation conditions, loads on the pipe, and other factors. Pipe culverts are most frequently constructed in what is termed the "projection condition." That is, the culvert is constructed on the surface of the ground, in the open, and the fill is built around it. In such cases, only a nominal amount of attention need be given to bedding of the pipe in normal soils and normal heights of fill.

Concrete box culverts are constructed in place with square or rectangular cross sections; single box culverts vary in size from 2 to 12 ft square, depending on the required area of waterway opening. Most state highway departments use standard designs for various sizes of box culverts; perhaps the most commonly used sizes of concrete box culverts are in the range of 4 to 8 ft square, including such sizes as 4 ×4, 4 × 6, 6 × 6, and 4 × 7 ft. Rectangular cross sections are used where it is desired to reduce the height of the culvert to provide adequate cover between the top of the culvert and the roadway surface. The use of box culverts has declined in recent years, largely because of the time required for their construction.

Both concrete pipe and concrete box culverts are built with more than one opening where additional waterway area is required and when it is desired to avoid the use of excessively large single pipes or boxes. Such installations are called "multiple culverts" and may, for example, be "double" or "triple" concrete pipe or concrete box culverts.

Concrete arches are sometimes used in place of concrete box culverts, although difficulties attendant on their proper design and construction have somewhat restricted their use. Concrete arch bridges are more frequently used.

Corrugated metal culverts (steel)

Corrugated steel is used in various forms in the construction of culverts for use in highway drainage.

Corrugated metal (galvanized steel) pipe is made in diameters of 8 to 96 in. and in lengths of 20 to 40 ft. Various thicknesses of metal are used, generally from 16 to 8 gage. The corrugations that are formed in the sheet metal are $2\frac{2}{3}$ in. from crest to crest and $\frac{1}{2}$ in. deep. Standard pipe is manufactured by bending the corrugated sheet metal into a circular shape and riveting the longitudinal joint. Helically corrugated pipe has a folded seam rather than a riveted longitudinal joint. In the field, lengths of corrugated metal pipe may be joined by a pipe sleeve or by a connecting band that is several corrugations in length—angle irons are riveted to the two ends of the band and connected by bolts. The band is slipped over the ends of the pipes to be connected and the bolts drawn tight to form the connection.

The maximum desirable diameter of standard corrugated metal pipe is 8 ft. This fact led to the development of a method of construction utilizing heavier, curved, corrugated metal plates that are bolted together to form circular pipes or arches. The plates are curved, corrugated, and galvanized at the factory, shipped to the field site, and there bolted together to form the desired structure. The plates are heavier than normal corrugated metal pipe, being available in gages designated as 1, 3, 5, 7, 8, 10, and 12. They are made in various widths and in lengths of 6 and 8 ft. Pipes up to 21 ft in diameter have been fabricated by this method, while arches of almost any desired combination or rise and span are possible; maximum standard size is a rise of 13 ft, 2 in. and a span of 20 ft, 7 in. Development of the "compression ring" method has allowed rational design of multiple-plate structures to fit any combination of circumstances.

In many culvert installations headroom is limited, and a circular pipe that has sufficient hydraulic capacity is not suitable. In such cases, corrugated metal "pipe arches" may be used. Pipe arches made of standard corrugated metal are available in sizes varying from a span of 18 in. and a rise of 11 in. to a span of 72 in. and a rise of 44 in. Figure 11-12 shows a standard drawing for a corrugated metal pipe arch with a flared end section. A typical example of a location where this type of culvert may be used to advantage lies in the culvert opening required to pass water flowing in a side ditch beneath an approach road to a rural highway.

In the interest of increased durability, corrugated metal pipe is sometimes furnished with the invert of the pipe covered with a thick bituminous mixture that completely fills the corrugations in this section of the pipe. Such pipe is called "paved-invert" pipe. The remainder of the pipe may also be coated with a

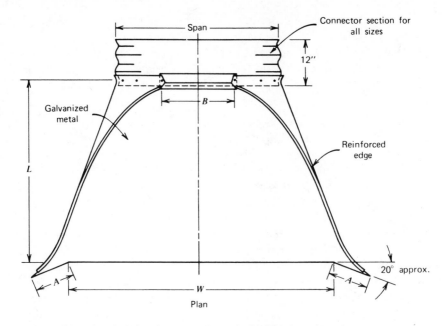

Where flared end—sections are to be used with bituminous coated and paved metal pipe, they are to be galvanized only.

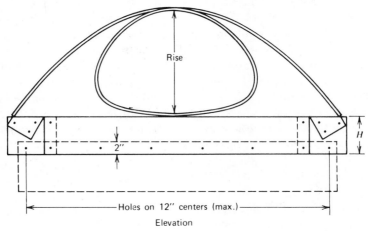

FIGURE 11-12 Flared-end section for metal pipe arches. (Courtesy Virginia Department of Highways and Transportation.)

mixture of bituminous material and asbestos, for very severe exposure conditions, or with bituminous material alone.

Miscellaneous culvert types

Other materials that may be used in the construction of culverts include vitrified-clay pipe, cast-iron pipe, and timber. In addition, masonry arch

Standard coupling band

Approx. 2½ : 1 Slope

L

Typical cross section

8"

FIGURE 11-12 (*Continued*)

culverts are still constructed in some localities where suitable stone is cheaply available, although the use of this type of construction has declined in recent years. Each of these types of culvert has advantages and may be used in areas where the material is economically available and for special-purpose structures.

11-22 Selection of culvert type

The type of culvert selected for use in a given location is dependent on the hydraulic requirements and the strength required to sustain the weight of a fill or moving wheel loads. After these items have been established, the selection is then largely a matter of economics. Consideration must be given to durability and to the cost of the completed structure, including such items as first cost of manufactured units and costs of transportation and installation. Maintenance costs should also be considered in any overall comparison of the cost of different culvert types.

11-23 Culvert headwalls and endwalls

Headwalls and endwalls are provided on culverts principally to protect the sides of the embankment against erosion. Some authorities refer to the upstream wall as a "headwall" and the downstream one as an "endwall." In the following discussion, the term headwall will be used generically.

In addition to their function in erosion control, headwalls may serve to prevent disjointing of section pipe culverts and to retain the fill. Materials most commonly used for culvert headwalls are concrete, masonry (stone or rubble), and metal. Of these, concrete is most widely used because of its adaptability to all types of culverts and because it lends itself to pleasing architectural treatment.

In selecting the size and type of headwall to be used in a given case, matters of economy must again be given consideration. In addition, some weight must be given to aesthetic considerations, as the headwall is the principal portion of the average culvert structure which is visible to the traveler. Headwalls are not always necessary, of course, and their use should be avoided wherever it is feasible to do so because of their cost. Headwalls are an expensive portion of the average culvert installation; the headwall should be made as small as possible consistent with adequate design. Safety of traffic must also be considered in choosing the type of headwall to be used in a given case.

Many different types of headwall are used by different highway agencies; generally, each agency has developed standard designs that are used whenever possible. Several illustrations are given of typical headwall installations as currently used by various organizations.

Figure 11-13 is a drawing that illustrates the details of standard endwalls used for multiple-pipe culverts by the Virginia Department of Highways. The flaring portions of this endwall are generally called "wing-walls." The endwall shown in Figure 11-12 is typical of the prefabricated metal end sections used in connection with small corrugated metal pipe culverts.

In the interest of safety, highway agencies should consider the placement of a traversible grate over the end of culverts if there is danger of vehicles colliding with the headwall. An example of such a structure is shown as Figure 11-14.

11-24 Culvert intakes
In certain circumstances, for example, in easily eroded soils when high velocities occur, unusual precautions must be taken to prevent damage to the inlet of a culvert. Such precautions may consist only of paving ditches that carry water to the culvert entrance. In other cases, special culvert intakes may be provided. An installation of this type appears in Figure 11-15, which shows the details of a culvert intake for a 15-in. to 24-in. pipe culvert as used in Virginia.

As emphasized previously, consideration should also be given to the hydraulic properties of the culvert entrance. That is, the intake should be designed to minimize losses of head due to eddies and turbulent flow. In general, sharp corners or "breaks" in the culvert entrance should be avoided. In certain installations, consideration must be given to prevention of clogging of the culvert entrance by drift and rubbish. Special auxiliary devices are sometimes constructed for this purpose (18).

HIGHWAY BRIDGES

As previously indicated, a bridge may be defined as a drainage structure that has a span of more than 20 ft. As a further distinguishing feature, bridge spans usually, although not always, rest on separate abutments, while culverts are regarded as integral structures. Although the term "bridge" is usually associated with structures that are required to carry the roadbed over an established waterway, it may also be somewhat loosely applied to grade separation structures and elevated highways in urban areas (viaducts).

11-25 Bridge location
In modern practice, bridges of relatively short span are located to conform with the general location of the highway, which has been previously determined.

Note: On shallow fills, where endwalls are 1' or less below shoulder line, the endwalls shall be constructed parallel to the line and grade of the road.

FIGURE 11-13 Standard endwalls for multiple pipe culverts. (Courtesy Virginia Department of Highways and Transportation.)

FIGURE 11-14   Pipe headwall with load-carrying grate. (Courtesy Virginia Department of Highways and Transportation.)

That is, the tentative location for the highway is established after an analysis of all the economic and engineering factors involved, and the bridge engineer is given the problem of providing an economical and adequate bridge design to conform to the roadway location. In some cases the location of a suitable stream crossing may be the most important single factor influencing the location of the highway in a given section; such is usually the case when long bridge spans are involved.

The ideal location for a bridge crossing is, of course, one in which the crossing is made at right angles to the centerline of the stream at the narrowest point, where the alignment of the approach pavement is straight, where the approach grade is slight, and where soil conditions are adequate for the installation of the most economical foundation for the span involved. This ideal combination of circumstances is encountered all too infrequently, except in structures of short span, and many bridges have been located on skew crossings, vertical curves, or with curving alignment. In such cases, considerations related to the general roadway location may still be regarded as controlling factors, and the required adjustments in the location of the bridge are made with these requirements in mind.

FIGURE 11-15 Plans for intake, 15 to 24-in. pipe. (Courtesy Virginia Department of Highways and Transportation.)

Many times, alternative locations of a proposed bridge may seem to offer somewhat similar advantages. A careful comparison must then be made of the several possible locations. The final decision should be based on a complete analysis, including factors related to traffic safety and operating conditions, fulfillment of the purpose of the road (e.g., the direct connection of population centers), and economics. Any complete analysis must include both the bridge and the approaches to it. A comparison of this type will generally result in the selection of one of the possible sites as the most desirable.

Once the general location is determined, the selected site must be subjected to careful scrutiny. This examination may be extremely detailed or somewhat cursory, depending largely on the size and importance of the contemplated structure. A complete survey of the bridge site may include an examination of the channel for some distance above and below the bridge crossing, a complete topographic map of the site, and an extensive soil survey of the area, including securing undisturbed soil samples where required and determining the required waterway opening and the requirements of navigation on large streams.

Although the preceding paragraphs have dealt largely with waterway crossings, similar factors govern the design of grade separation and the bridge portions of complex interchanges. Obviously, a grade separation structure designed to carry the roadway over an existing railroad presents the location engineer and designer with similar problems, as does the design of a highway overpass. Similarly, the general location of an elevated highway in an urban area is usually determined on the basis of maximum serviceability to traffic (and availability of right-of-way) so that the engineer must prepare a design suitable to conditions in a rather limited area.

11-26 Design of waterway opening

In many locations the natural stream channel is somewhat constricted by the bridge structure and roadway approaches. In the interests of economy, the roadway is frequently placed on an embankment on either side of the bridge span, the distance between abutments is reduced as much as possible, and piers may be placed in the stream channel. All these things serve in many cases to reduce severely the area through which the water must pass, particularly when the stream is at flood stage. Two results may immediately be noticed: during flood stage the velocity of the water through the bridge opening may be considerably increased, with resultant danger to the bridge structure through scour at abutments and piers, and the elevation of the water on the upstream side may be increased, with the result that the area subjected to flooding above the bridge site is increased and adjacent property is subjected to overflow beyond the limits of the normal floodplain. It thus seems axiomatic that the bridge must be designed to pass the flow occurring at flood stage without excessive velocity and without damage to property located above the bridge crossing. Estimating flood flows is best done by study of stream-gaging records but is sometimes based on observation of high water marks, the behavior of structures located on the same stream, and hydrological computations.

Bridge openings are also normally designed to pass floating debris carried in the channel at normal and flood states. On navigable streams, requirements of navigation must be evaluated and provided for. Generally speaking, navigable streams are spanned by high-level crossings or movable bridges. Plans for bridges over navigable streams in the United States are subject to approval by the U.S. Corps of Engineers.

11-27 Bridge clearances

Standards related to bridge clearances, both vertical and horizontal, are an important part of the design of highway bridges. Clearances recommended by the American Association of State Highway and Transportation Officials for bridges (and culverts, where applicable) are shown in Figure 11-16 *(19)*. AASHTO specifies that the roadway width at bridges shall be equal to the full shoulder width of the approach section. Along curbed roadways, the full width of the approach section should similarly be carried across the structure. For low-speed, low-volume roads, a minimum horizontal clearance of the width of the approach traveled way plus 8 ft is recommended *(19)*.

Provision of adequate clearances on new bridges is extremely important, as literally thousands of existing bridges that are otherwise adequate have been rendered functionally obsolete by narrow roadways. The provision of adequate width in new construction is much easier and cheaper than modifications that must be made after the design width proves to be inadequate.

FIGURE 11-16 Clearance diagram for bridges. (Courtesy American Association of State Highway and Transportation Officials.)

As Figure 11-16 indicates, the vertical clearance on highway bridges should be at least 14 ft over the entire roadway. An allowance should be added for pavement resurfacing. Along the interstate system and other main highways, a 16-ft vertical clearance is usually provided.

11-28 Bridge live loads

Live loads used in the design of highway bridges are normally those established by AASHTO; two types of loadings are in use—H loadings and H−S loadings *(19)*.

The H loadings consist of a two-axle truck or corresponding "lane loading," the latter being a certain uniformly distributed load and a concentrated load which are equivalent in effect to the specified truck loading. Two H loadings are specified: H20−44 and H15−44. The meaning of these may be explained by examination of the H15−44 loading. The gross weight of this truck is 15 tons (30,000 lb), and this load is distributed 20 percent (6000 lb) on the front axle and 80 percent (24,000 lb) on the rear axle; the spacing between axles is 14 ft. The load has 10-ft clearance and load-lane width; the transverse spacing center-to-center of the wheels is 6 ft. The suffix "44" simply indicates that this standard, as explained, was adopted in 1944. The H20 loading is similarly specified.

The H−S loadings consist of a tractor truck with semitrailer, or the corresponding lane loading. Two H−S loadings are in common use: HS15−44 and HS20−44. These loadings are somewhat similar to the H loadings; the spacing between axles of the truck is fixed at 14 ft, while the axle spacing of the semitrailer is from 14 to 30 ft, this distance being varied to produce maximum stresses. The total load of the HS15−44 loading is 27 tons (54,000 lb), which is distributed as follows: front axle of truck, 6,000 lb; rear axle of truck, 24,000 lb; and axle of semitrailer, 24,000 lb. Total load of the HS20−44 loading is 36 tons (72,000 lb), and it is similarly distributed.

The choice of load to be used in the design of a particular structure depends on the traffic expected to use the bridge during its anticipated life. Heavier loadings are generally used in the design of structures on the various primary systems, and lighter loads are used on other components of the highway system such as county roads. The HS20−44 loading is being used in the design of new bridges on the Interstate System, while many structures on the existing federal-aid system have been designed for H−15 loading. There is a definite trend toward the use of heavier loadings simply because the use of the heavier loads does not greatly increase the cost of the structure and seems to be consistent with the objectives of long-range planning.

11-29 Bridge types

Under the stimulus of a greatly expanded highway construction program, engineers in the United States (and the rest of the free world) have virtually revolutionized bridge design and construction methods in the past decade. The advances apply to short-, medium-, and long-span bridges.

For permanent bridges, the most commonly used materials are steel and concrete. Bridges of many different types are built with these materials, used singly or in combination. Timber may be used for temporary above-water construction, for the elements of a structure that lie below the waterline (particularly timber piles), or for short-span bridges located on secondary roads. A few short-span aluminum bridges have been built in the United States on an experimental basis.

The principal portions of a bridge may be said to be the "substructure" and the "superstructure." This division is used here simply for convenience, since in many bridges there is no clear dividing line between the two.

Common elements of the substructure are abutments (usually at the bridge ends) and piers (between the abutments). Piers and abutments often rest on separately constructed foundations such as concrete spread footings or groups of bearing piles; these foundations are part of the substructure. Occasionally a bridge substructure comprises a series of pile bents in which the piles extend above the waterline and are topped by a pile cap that, in turn, supports the major structural elements of the superstructure. Such bents often are used in a repetitive fashion as part of a long, low, over-water crossing.

In recent years, the dividing lines between short-, medium-, and long-span bridges have blurred somewhat. Currently, spans of 20 to 100 ft are regarded as short by many designers, who have developed many standardized designs to handle these spans economically. Medium spans range up to, perhaps, 400 ft in modern bridge practice, depending on the organization involved and the materials used. Long spans range up to 4000 ft or more, but a clear span above 1000 ft is comparatively rare.

Bridges may also be classed as "deck" or "through" types. In the deck type of bridge, the roadway is above the supporting structure; that is, the load-carrying elements of the superstructure are below the roadway. In the through type of bridge, the roadway passes between the elements of the superstructure, as in a through steel-truss bridge. Deck structures predominate; they have a clean appearance, provide the motorist with a better view of the surrounding area, and are easier to widen if future traffic requires it.

Thousands of short-span bridges have been built in the United States in recent years under the impetus of the accelerated federal-aid highway program and the increasing use of limited access.

Examples of short-span steel bridges include simple-span, wide-flange beam; simple-span, welded girder; continuous, wide-flange beam (the bridge is designed to act as a continuous structure over two or more supports); and continuous, welded girder. Typically, these steel bridges have reinforced-concrete decks designed for composite action with underlying steel members.

Examples of short-span concrete bridges include cast-in-place, reinforced concrete T-beam (and slab); simple-span, prestressed (this type incorporates precast, prestressed I-girders or box girders topped by a cast-in-place deck); and cast-in-place box girder.

The designer of each medium- and long-span bridge tries to devise a structure that is best suited to the conditions encountered at that particular location. The result is an almost bewildering variety of structures that differ either in basic design principles or in design details.

General categories of steel bridges are briefly described in the following paragraphs.

Suspension bridges are used for very long spans or for shorter spans where intermediate piers cannot be built. An example is the Verrazano Narrows Bridge (Figure 11-17), which was completed in 1964. The $305 million, 4260-ft structure spans the entrance to New York Harbor to join Staten Island and Brooklyn.

Girder bridges come in two basic varieties—plate and box girders.

Plate girders are used in the United States for medium spans. They generally are continuous structures with maximum depth of girder over the piers and minimum depth at midspan. The plate girders generally have an I cross section; they are arranged in lines that support stringers, floorbeams, and, generally, a cast-in-place concrete deck. The girders are shop-fabricated by welding; field connections generally are by high-strength bolts.

Welded-steel box girder structures are generally similar to plate girder spans except for the configuration of the bridge cross section.

Rigid frames are used occasionally, most often for spans in the range of 75 to 100 ft and for grade-separation structures.

Arch bridges are used for longer spans at locations where intermediate piers cannot be used and where good rock is available to withstand the thrusts at the arch abutments.

A variation in the arch bridge is the tied arch, in which a horizontal tie that carries the roadway takes much of the horizontal thrust inherent in the arch form.

Truss bridges are built in many forms and in many locations for medium and long spans. Both deck and through trusses are built, with cantilever and continuous trusses being the most common.

Figure 11-18 illustrates the Hale Boggs Bridge, a modern cable-stayed

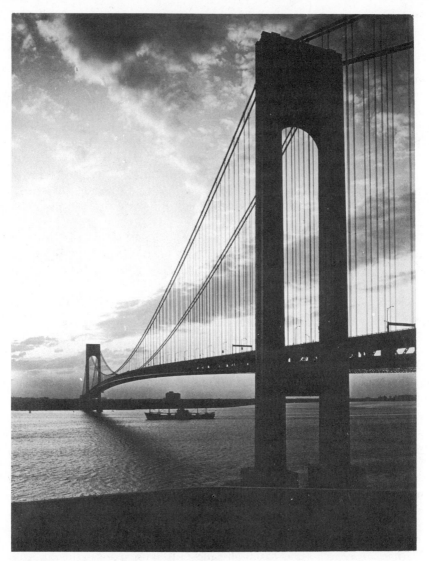

FIGURE 11-17 The Verrazano Narrows Bridge, longest clear span in the world. (Courtesy American Institute of Steel Construction.)

bridge in Louisiana. Its 1222-ft main span is the world's second longest for this type of bridge.

Concrete bridges come in nearly as great a variety as do steel bridges. Listed below are various types now in use and some examples of outstanding concrete structures.

Conventional reinforced concrete is used primarily for short spans, with a variety of innovations from the customary cast-in-place, slab-and-beam design.

Precast, prestressed concrete bridges are popular in the United States for

FIGURE 11-18   A modern cable-stayed bridge in Louisiana. (Courtesy State of Louisana Department of Transportation and Development.)

short, simple spans; they generally comprise prestressed **I**-beams or box girders that support a cast-in-place deck.

One of the world's great water crossings, the 17.6-mile Chesapeake Bay Bridge–Tunnel, contains over 12 miles of repetitive spans that incorporate prestressed concrete cylinder piles, precast pile caps, and 75-ft-long precast, prestressed deck sections. Each span contains four parallel deck sections, each with a double **T** shape.

Cast-in-place prestressed concrete has been used for long-span bridges in Europe and in a few locations in the United States. For example, a 660-ft-long, four-span viaduct in Oakland, California, was prestressed (post-tensioned) with wires longitudinally, transversely, and vertically. In addition, it was stressed by a longitudinal movement of the pier bases produced by jacking after the superstructure was completed.

Arches also are built in concrete. One of the world's longest concrete spans (1000 ft) is an arch at Sydney, Australia. It is made up of 512 huge, hollow concrete boxes, each 20 ft deep.

11-30 Movable bridges

The three basic types of movable bridges—bascule, vertical lift, and swing—are shown in Figure 11-19. They are used where a high-level bridge is uneconomical but provision must be made for navigation.

There is still a fourth type of movable bridge, the floating bridge. Three of these unusual structures are in the Puget Sound area of Washington State.

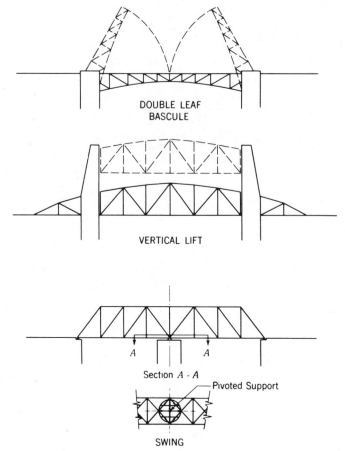

DOUBLE LEAF
BASCULE

VERTICAL LIFT

Section *A - A*

Pivoted Support

SWING

FIGURE 11-19 Movable bridges.

The most recent of these incorporates a four-lane, 1.4-mile-long floating section supported by 35 prestressed concrete pontoons, the largest of which is 360 ft long, 60 ft wide, and nearly 13 ft deep. Each pontoon is secured by cables to precast concrete anchors on the lake bottom in about 200 ft of water. A drawspan at the center provides for passage of ships.

HIGHWAY SUBDRAINAGE

As previously mentioned, the term "subdrainage" is related to the control of ground water encountered in highway locations. Subdrains are a necessary part of the complete drainage system for many highways in rural areas, and they function along with adequate surface drainage facilities to prevent damage caused by water in its various forms.

A highway subdrain usually consists of a circular pipe laid at a suitable depth in a trench, which is then backfilled with porous, granular material.

Materials principally used in subdrains include vitrified clay pipe, porous concrete pipe, and perforated corrugated metal pipe (both galvanized steel and aluminum). Clay pipe is usually laid with open joints, while concrete and perforated corrugated metal pipe are generally laid with sealed joints. Concrete pipe may also be laid with open joints. The size of drain is usually based on previous experience. Hydraulic design of subdrains is difficult but may be carried out on extensive projects. Pipes of 6- and 8-in. diameter are in common use although pipes for subdrains are made as large as 24 in. The slope of the pipe should be sufficient to prevent the deposition or "settling out" of any solid material which may enter the pipe through the joints or perforations. Minimum recommended slope is 0.15 to 0.25 ft per 100 ft.

Subdrains are installed for a number of purposes, most of which may be included in the following classifications:

1. Control of seepage in cuts or sidehill locations—these installations are generally called "intercepting drains."
2. Lowering of ground-water table, as in swampy areas.
3. Base and shallow subgrade drainage.

11-31 Intercepting drains

In rolling or mountainous terrain, cuts made during highway construction frequently expose flowing ground water or "seepage." Seepage that occurs through the cut slope may be a source of damage to the slope and to the roadway itself. Similarly, the seepage zone itself may not be invaded by the construction of a side-hill section, but the roadway and pavement structure may be located only slightly above the zone of flowing underground water and thus be subject to the detrimental effects of capillary action. In such cases the flow of underground water is intercepted by a subdrain located on the uphill side of the section so that the water is prevented from flowing beneath the pavement. An example of such an installation is shown in Figure 11-20. In this example, since there is an impervious soil layer located at a relatively shallow depth, the flow of seepage water is entirely cut off by the subdrain. In cases where the seepage zone is deeper, or where no impervious layer is found, the subdrain may simply be carried to sufficient depth to eliminate the effects of capillarity

FIGURE 11-20 Intercepting subdrain.

on the subgrade or base. Drains of the type that have been described are normally placed parallel to the centerline and are called "longitudinal drains." In some cases, seepage flow may occur in a direction parallel to the centerline. Transverse drains beneath the pavement may then be needed to intercept this longitudinal seepage.

The location of subdrains of this type before construction begins is somewhat difficult unless a very complete study is made of subsurface conditions. Information obtained from a soil survey may permit the approximate location of subdrains. However, many organizations prefer to provide necessary subdrainage on the basis of information gained during construction. Such a procedure demands the exercise of good engineering judgment in the field, but should be more satisfactory than a decision made in the office on the basis of incomplete information regarding ground-water conditions. In any case, water collected in the subdrains must be carried to a suitable outlet.

11-32 Lowering of the water table

In many locations in flat terrain, the roadway may be built on a low embankment in the interests of economy and the base may be only 2 or 3 ft above the water table. If the subgrade soil is one which is subject to capillary action, water will be drawn up into the subgrade and base with resultant loss of stability, possible frost damage, and similar detrimental effects. The answer to such a problem is simply to lower the water table a sufficient amount to prevent harmful capillary action. This is usually accomplished by the installation of parallel lines of subdrains at the edges of the roadway (or shoulders) at the proper elevation. In practice, the solution is frequently complicated by the difficulty of providing satisfactory outlets for the water collected in the subdrains. The depth required beneath the base varies with the type of soil encountered but is generally from 3 to 6 ft.

11-33 Base drainage

Water that falls on the surface of flexible and rigid pavements may enter the base and subgrade through cracks in the surface, joints, and shoulders. If the base is relatively impervious, or if it lies above an impervious subgrade soil, this water may collect in the base and the upper portion of the subgrade. This effect may be noted even where water is not brought up into the base or subgrade from below by capillary action. Again, water thus trapped may cause weakening of the base material, and it is particularly serious in areas that are subject to frost action. Subdrains of the types that have been described may be used to remove water from the base, although experience has not always shown this type of installation to be successful.

11-34 Laying and backfilling

Elements of the construction of a pipe subdrain of the type that has been described include the excavation of a suitable trench, preparation of the trench bottom, laying of the pipe, and backfilling with pervious material.

Excavation may be accomplished by either hand or machine methods. Trenches are usually quite shallow, are excavated with vertical walls whenever possible, and lend themselves to construction through the use of a ditching machine in most localities. When the pipe is being laid in a pervious, water-bearing stratum, very little preparation of the trench bottom is necessary other than a nominal amount of shaping. In the case of an intercepting drain laid at the top of an impervious layer in order to ensure complete removal of the intercepted flow, the pipe is generally placed a short distance down into the impervious layer. If soft, unstable soils are encountered, sufficient granular material must be worked into the upper portion of the soil to ensure uniform support of the pipe.

Perforated circular pipe is laid with the perforations down, when located in pervious soils, so that there is less likelihood of clogging by the entrance of fine-grained soil into the pipe through the perforations. For the same reason, corrugated metal and concrete pipe is generally laid with sealed joints. Open joints in bell-and-spigot vitrified-clay pipe are protected from silting by covering them with tar paper or some similar materials.

Proper backfilling around the pipe and in the trench is of extreme importance if the drain is to function properly without an excessive amount of maintenance. The backfill material must be coarse enough to allow easy passage of water and fine enough to prevent intrusion of fine-grained soil into the pipe.

Criteria which have been developed for adequate filter soils are as follows:

$$\frac{D_{15 \text{ (filter)}}}{D_{85 \text{ (protected soil)}}} \leq 5$$

and

$$\frac{D_{15 \text{ (filter)}}}{D_{15 \text{ (protected soil)}}} \geq 5$$

where D_{15} is the grain diameter which is larger than 15 percent of the soil grains, that is, the 15 percent size on the grain-size distribution curve. (Similarly, for D_{85}.) For most fine-grained soils, standard concrete sand (AASHTO M6 fine aggregate for portland-cement concrete) forms a satisfactory filter material. When such sands contain a sufficient amount of fine gravel, and for pipes with small openings, they are safe against infiltration into the pipe. Otherwise, the relationship between the size of the filter material and the size of the perforations in the pipe is

$$\text{for slots } \frac{D_{85 \text{ (filter)}}}{\text{slot width}} > 1.2$$

$$\text{for circular openings} \frac{D_{85 \text{ (filter)}}}{\text{hole diameter}} > 1.0$$

In some states, material of this approximate gradation is used for the entire depth of the trench with the exception of the top 8 to 12 in., which may be filled with ordinary topsoil. In other states the upper portion of the trench is filled with a coarse material. Under certain circumstances it may be advisable to provide an impervious seal over the top of the trench to prevent washing of silt and other fine material down into the drain. Adequate compaction of the backfill is essential to prevent excessive subsidence of the ground over the subdrain and is especially important if the drain is to be beneath the roadway (as in the case of an interception drain designed to prevent longitudinal seepage) or beneath a shoulder.

DRAINAGE OF CITY STREETS

Precipitation that occurs on city streets and adjacent areas must be rapidly and economically removed before it becomes a hazard to traffic. The removal of surface water in municipal areas is accomplished by methods similar to those employed in the drainage of rural highways, except that the surface water is commonly carried to its eventual disposal point by means of underground pipe drains or "storm drains." The storm drains may be designed to provide for the flow of ground water as well as surface water, or additional subdrains may be required in certain areas No further mention will be made of subdrainage in this discussion, as subdrainage installations in urban areas are very similar to those required in rural areas.

The surface-water drainage system in an average city may be considered to be composed of the following basic elements: pavement crown; curb and gutter; and the storm drains themselves, including such commonly used appurtenances as inlets, catch basins, and manholes.

11-35 Pavement crowns, curbs, and gutters

Water falling on the pavement surface itself is, as in the case of rural highways, removed from the surface and concentrated in the gutters by the provision of an adequate crown; crown requirements previously discussed are applicable to city streets. The crown of any street should be made as small as possible, consistent with proper drainage, in the interest of appearance and safety.

The surface channel formed by the curb and gutter fulfills about the same drainage function as a side ditch on a rural highway, and it must be designed to adequately convey the runoff from the pavement and adjacent areas to a suitable collection point. Typical curb and gutter sections have been illustrated in Chapter 8. The longitudinal grade of the gutter must also be sufficient to facilitate rapid removal of the water; minimum grades are generally recommended as 0.2 or 0.3 percent. The grade of the gutter is usually the same as that of the pavement surface, although it may be different if so required for proper drainage. Hydraulically, gutters are designed as open channels of special shape.

Flow in the gutters generally continues uninterrupted to the intersection unless a low point or "sag" is reached. At such a point an inlet must be provided to carry the water into the storm sewer system. Inlets may also be required between intersections if long blocks are used.

The paved area formed by a street intersection must also be drained, and adjustment of the grades of city streets at intersections to provide for adequate drainage can be a very complex problem. Several analytical methods have been successfully used for determining the elevations required within the intersection to remove water from the intersection area and to adjust any differences in grades that may be present on the streets that approach and leave the intersection. The solution may also be accomplished in the field during the construction process without previous computation, especially where light street grades are involved.

11-36 Inlets

Inlets are generally provided at intersections to intercept the water flowing in the gutters before it can reach the pedestrian crosswalks. It is essential that enough inlets of sufficient size be provided to rapidly remove collected storm water. Improperly drained intersections are a constant source of annoyance and danger to both vehicular and pedestrian traffic.

The selection of the type of inlets to be used and their positioning in the area adjacent to the intersection are matters of individual choice, and considerable variation occurs in practice. As an example of the proper positioning of inlets at the intersection of two streets of normal width, the simple sketch of Figure 11-21 has been prepared. Examples of standard drop inlet designs are shown as Figure 11-22.

On lightly traveled streets it may sometimes be desirable to provide for the flow of surface water across the intersection by means of a shallow paved gutter or "dip." Such a procedure is not generally recommended, but it may be necessary in the interest of economy.

Inlets of various designs are used by different organizations. The type most generally used on new construction may be called the "combined curb-and-gutter inlet." This type of inlet basically consists of a concrete or brick box with an opening in one side, near or at the bottom, into a circular pipe forming one of the laterals of the storm sewer system. Water enters the box through a metal grate placed over a horizontal opening in the gutter and through a relatively narrow opening in the face of the curb. This latter opening in the face of the curb may or may not be protected by a grating of some kind. Inlets of this type are either cast in place or composed of precast units. Inlets are also designed without the opening in the curb, the water flowing in only through a grated opening in the gutter. Some inlets have also been constructed with only curb openings.

While inlets of the types that have been described usually have only an outlet in the form of a circular pipe placed so that the invert of the pipe is flush (or nearly so) with the bottom of the box, they are also constructed with a pipe inlet (in addition to the curb and gutter openings) and a pipe outlet. In other words, the inlet may be a connecting link in the storm sewer system. Still another variation in design occurs in what may be called a "drop inlet." In a

FIGURE 11-21 Drainage at an urban intersection.

drop inlet, the water entering from the gutter must drop some considerable distance to the bottom of the box or may simply be conveyed directly into the storm sewer by means of a vertical pipe.

Inlets are susceptible to clogging with debris and ice, and they must be subjected to continual inspection and maintenance if they are to function properly. The proper hydraulic design of inlets to accommodate a given amount of water under given conditions is beyond the scope of this book. Recent research has lead to increased understanding of the hydraulic principles involved and the development of various design methods. Curb openings for inlets, on the average, are from 2 to 4 ft in length and from 6 to 8 in. high.

11-37 Catch basins

Catch basins are similar to inlets in their function and design. The principal difference between a catch basin and an inlet is that in a catch basin the outlet pipe (and inlet pipe if present) is placed some distance above the bottom of the chamber. The notion behind the installation of a catch basin is that debris flushed from the street into the drainage system is trapped in the bottom of the catch basin and so does not enter the storm sewer itself. This consideration is of less importance now than in former years because of the universal construction of permanent-type street surfaces, more efficient street-cleaning methods, and

FIGURE 11-22 Standard median drop inlet designs. (Courtesy Virginia Department of Highways and Transportation.)

the general design of storm sewer systems to carry water at higher velocities of flow. This latter feature has made many sewer systems practically self-cleaning. Catch basins require good maintenance; if they do silt up and the debris is not removed, they function as inlets.

11-38 Manholes

Storm sewer systems are subject to partial or complete clogging, and facilities must be provided for cleaning them at regular intervals. In modern sewerage practice, manholes are generally placed at points where the sewer changes grade or direction, where junctions are made, and at intermediate points, usually at intervals of 300 to 500 ft. The opening provided must, of course, be large enough to permit a man to enter the manhole chamber and have room in which to work. They are usually about 4 ft in diameter in their main portions and are carried to a depth sufficient to perform the purpose for which they are intended. The pipes generally enter and leave the manhole at the bottom of the chamber, and various arrangements are made for carrying the flow through the manhole with a minimum loss in head. Manholes are built of concrete masonry, concrete block, or brick, and cities generally employ a standard design of manhole to be used in all installations. Manhole covers (and frames) are generally of cast iron; the covers are circular in shape and about 2 ft in diameter. When the entrance to the manhole occurs in the traveled way, as it usually does, special care must be taken in the design and placing of the cover relative to the finished street surface if traffic is not to be impeded by the presence of the manhole.

PROBLEMS

11-1   The cross section of a certain channel has a 10-ft flat bottom with 4:1 side slopes. Evidence at the site indicates that during a flood the depth was 6 ft. The channel slope is 0.9 percent. Assuming a roughness coefficient of 70, estimate the quantity of flow by the Chezy equation.

11-2*   Develop an equivalent expression for the Chezy equation using SI (metric) units.

11-3   Using the Federal Highway Administration chart, estimate the rainfall intensity to be expected once in 25 years from a 1-hour storm on a small drainage area located near Chicago, Ill. What would be the expected rainfall intensity near Mobile, Ala.? Near Denver, Colo.? Make the same estimates for a design frequency of 10 years.

11-4*   Develop an equivalent expression for the rational method, Eq. 10-2, using SI (metric) units.

11-5   Using the rational method, determine the expected runoff from an area of 300 acres located near Chicago and subjected to the 1-hour, 25-year rainfall intensity determined in solving Problem 11-3. The area is 10 percent bituminous pavement, 10 percent gravel, and the remaining 80 percent impervious soil with turf.

11-6*   Develop an equation for determining the velocity of uniform flow similar to the Manning equation using SI (metric) units.

11-7 A 4-ft flat-bottom channel with 2:1 side slopes is covered with some weeds with heavy brush on its banks. The slope is 0.5 percent and the depth is 5 ft. Compute the velocity of flow by the Manning equation. Check your results by Figure 1⧸-3.

11-8 Determine the depth and velocity of flow in a trapezoidal channel ($n = 0.030$) with 2:1 side slopes and 4-ft bottom width discharging 60 ft³/sec on a slope of 2.0 percent. Determine the critical depth, velocity, and slope for these conditions.

11-9 Determine the depth and velocity of flow in the channel of Problem 10-7 for a value of $n = 0.013$.

11-10 (a) Given a 50-ft-long concrete pipe culvert with a groove-edge entrance to carry 100 ft³/sec on a slope of 0.2 percent, determine the size of pipe required if the limiting headwater depth is 6 ft. (See Figure 1⧸-10.)

(b) What size of concrete pipe would be required to carry 100 cfs through a long culvert (360 ft) on a slope of 0.3 percent if the allowable headwater depth was only 5.5 ft?

(c) How many 36-in. corrugated metal pipes with projecting entrances would be required to carry 100 ft³/sec if the pipes were 50 ft long and laid on a slope of 0.5 percent? Assume that the flow in a multiple-barrel culvert is divided equally among the barrels and the limiting headwater depth is 5.5 ft.

11-11 Given the same conditions as in Problem 10-11a, determine the diameter of culvert required for (a) inlet control (Figure 10-8) and (b) culvert flowing full (Figure 10-9).

REFERENCES

1. *Drainage of Highway Pavements.* Highway Engineering Circular No. 12, Federal Highway Administration, Washington, D.C. (1984).

2. *Hydrology of a Highway Stream Crossing.* Hydraulic Engineering Circular No. 3, Federal Highway Administration, Washington, D.C. (1961).

3. Bock, P., Enger, I., Malhotra, G. P., and Chisholm, D. A. *Estimating Peak Runoff Rates from Ungaged Small Rural Watersheds.* NCHRP Report No. 136, Washington, D.C. (1972).

4. *Rainfall-Frequency Atlas of the United States.* Technical Paper No. 40, U.S. National Weather Service, Washington, D.C. (1961).

5. *Handbook of Steel Drainage and Highway Construction Products.* American Iron and Steel Institute, New York (1967).

6. Kirpich, P. Z. Time of Concentration of Small Agricultural Watersheds. *Civil Engineering*, vol. 10, No. 6 (1940).

7. *Design of Roadside Drainage Channels.* Hydraulic Design Series No. 4, Federal Highway Administration, Washington, D.C. (1965).

8. *Manual on Drainage Design for Highways.* Georgia Department of Transportation, Atlanta (1975).

9. *Design Charts for Open-Channel Flow.* Hydraulic Design Series No. 3, Federal Highway Administration, Washington, D.C. (1973).

10. Normann, Jerome M. *Design of Stable Channels with Flexible Linings.* Hydraulic Engineering Circular No. 15, Federal Highway Administration, Washington, D.C. (1975).

11. *Hydraulic Charts for the Selection of Highway Culverts.* Hydraulic Engineering Circular No. 5, Federal Highway Administration, Washington, D.C. (1965).

12. *Capacity Charts for the Hydraulic Design of Highway Culverts.* Hydraulic Engineering Circular No. 10, Federal Highway Administration, Washington, D.C. (1972).

13. *Handbook of Concrete Culvert Pipe Hydraulics.* Portland Cement Association, Skokie, Ill. (1964).

14. Normann, Jerome M. Improved Design of Highway Culverts, *Civil Engineering* (March 1975).

15. Harrison, L. J., Morris, J. L., Normann, J. M., and Johnson, F. L. *Hydraulic Design of Improved Inlets for Culverts.* Hydraulic Engineering Circular No. 13, Federal Highway Administration, Washington, D.C. (1972).

16. *Hydraulic Design of Energy Dissipators for Culverts and Channels.* Hydraulic Engineering Circular No. 14, Federal Highway Administration, Washington, D.C. (1975).

17. *Concrete Pipe Handboodk.* American Concrete Pipe Association, Vienna, Va. (1981).

18. *Debris Control Structures.* Hydraulic Engineering Circular No. 9, Federal Highway Administration, Washington, D.C. (1971).

19. *Standard Specifications for Highway Bridges.* American Association of State Highway and Transportation Officials, Washington, D.C. (1983).

chapter twelve

TRAFFIC ENGINEERING AND SAFETY

12-1 Introduction

Before a highway is opened to traffic, careful thought must be given to conveying to drivers information concerning its proper use. Such information is given primarily by traffic control devices.

Engineers must also recognize that despite their best efforts to promote traffic safety by well-designed traffic control systems, vehicles will occasionally go out of control and possibly crash. Every effort must be made to prevent traffic accidents from occurring and, failing that, to minimize the injuries and economic losses from those crashes that do occur.

In this chapter, we will examine ways in which traffic engineers promote operational efficiency and safety through the use of traffic control devices and systems. We will also describe other ways of preventing crashes (e.g., by providing clear roadsides) and discuss methods for reducing the losses from crashes by the use of breakaway sign supports and energy attenuation devices.

TRAFFIC CONTROL DEVICES AND SYSTEMS

In order to provide safe and efficient traffic flows, uniform standards have been developed for the use of all public street and highway facilities in the United States. The need for such standards became apparent in the late 1920s, and in 1935 a joint committee of the American Association of State Highway Officials and the National Conference on Street and Highway Safety developed and published a Manual on Uniform Traffic Control Devices (MUTCD). The committee is now called the National Committee on Uniform Traffic Control Devices. The Federal Highway Administration has approved the MUTCD as the national standard for all U.S. highways open to public travel. Much of the material in the following paragraphs has been abstracted from the 1970 and 1978 editions of that manual (1).

Traffic control devices include all signs, markings, and signals placed on or adjacent to a street or highway by public agencies in order to regulate, warn, or guide traffic. If traffic control devices are to be effective, they should

1. Fulfill a need.
2. Command attention.
3. Convey a clear, simple meaning.
4. Command respect of drivers and pedestrians.
5. Be place so as to give users time for proper response.

Traffic control devices should be reasonable and appropriate for the traffic requirements at the location used. The use of a traffic control device at a location where it is not warranted tends to invite motor-vehicle operators to disregard the device and to have less respect for traffic control devices in general.

Traffic control devices should be properly designed. The size of the device, its shape, colors, contrast with the background, and lighting or reflectorization should draw attention. The sign, marking, or signal should simply and forthrightly convey a clear and simple message. The use of uniform devices following the MUTCD (1) simplifies the driver's task of recognizing and understanding the traffic control messages and tends to increase the level of observance.

The National Joint Committee has adopted the following color code for traffic control devices.

Color	Meaning
Red	Stop or prohibition
Green	Indicated movements permitted, direction guidance
Blue	Motorist services guidance
Yellow	General warning
Black	Regulation
White	Regulation
Orange	Construction and maintenance warning
Brown	Public recreation and scenic guidance

Traffic control devices should be placed and operated in a uniform, consistent manner. In this way, motorists can be expected to properly respond to the devices on the basis of previous exposure to similar traffic situations. A control device should be placed within the user's cone of vision and in reasonable proximity to the point, object, or situation to which it applies. Its location and legibility should be such that a driver traveling at normal speed has enough time to respond appropriately.

It is important that devices be maintained to high standards to ensure that legibility and visibility are retained. When no longer needed, traffic control devices should be removed.

12-2 Traffic signs

There are three functional classes of traffic signs: (1) regulatory, (2) warning, and (3) guide signs. Examples of these three classes are shown in Figure 12-1.

Regulatory signs give users notice of traffic laws or regulations. Such signs designate right-of-way (e.g., STOP, YIELD), indicate speed controls (e.g., SPEED LIMIT 50, SPEED ZONE AHEAD), control movements (e.g., NO RIGHT TURN, KEEP RIGHT, ONE WAY), regulate parking (e.g., NO PARKING), control pedestrian movements (e.g., CROSS ONLY AT CROSS-WALKS), and regulate traffic in various other ways.

FIGURE 12-1 Typical highway signs. (Courtesy Federal Highway Administration.)

Warning signs direct attention to conditions on or adjacent to a street or highway that are potentially hazardous to traffic operations. Such signs require motorists to exercise caution, reduce speed, or make some maneuver in the interest of their own safety or that of other motorists or pedestrians.

Typical locations or situations that may warrant the use of warning signs include

1. Intersections, entrances, and crossings.
2. Changes in horizontal alignment.
3. Grades.
4. Railroad crossings.
5. Narrow roadways.
6. Roadway surface conditions.
7. Advance warning of control devices.

Examples of warning signs are curve signs (showing curved arrow), STOP AHEAD signs, PAVEMENT ENDS signs, and advisory speed plates. The latter signs supplement other types of warning signs. Warning signs have a black legend and border placed on a yellow background. With but few exceptions, warning signs have a diamond shape.

Guide signs indicate route designations, directions, distances, points of interest, and other geographic or cultural information. Two types of guide signs (route markers and a destination sign) are illustrated by Figure 12-1. Other examples of guide signs include JUNCTION signs, DETOUR signs, REST AREA signs, and service signs (FOOD, GAS, LODGING, etc.).

The MUTCD (1) gives detailed specifications on color, size, shape, height, and placement of signs as well as warrants for their use. The Federal Highway Administration has published an operating guide (2) that describes methods for fabricating, installing, and maintaining traffic control devices and gives examples of specifications, contract plan sheets, and work orders to aid in the preparation of such documents.

12-3 Traffic markings

Markings consist of paint or some other material placed on the pavement, curb, or object to convey traffic regulations and warnings to drivers. Markings may be used alone or in combination with traffic signs or signals. Although markings are an effective and essential means of traffic control, they tend to be difficult to see in rainy weather and may be obliterated altogether by snow and ice.

There are five general classes of markings: (1) pavement markings, (2) curb markings, (3) object markings, (4) delineators, and (5) colored pavements.

The most common type of markings—pavement markings—includes longitudinal and transverse lines and words and symbols. By choice of color, width, and type of longitudinal markings, the traffic engineer is able to convey a variety of messages to drivers. Briefly stated,

1. Broken lines are permissive in character; solid lines are restrictive; double lines indicate maximum restrictiveness.

2. White lines delineate separation of traffic flows in the same direction; yellow lines delineate separation of flows in opposing directions.

3. The width of the line indicates degree of emphasis.

Transverse markings include pedestrian crosswalks, stop lines, and cross-hatched areas to discourage use of shoulders or to identify object hazards.

Typical application of various classes of pavement markings are given in Table 12-1. Examples of pavement markings are illustrated in Figure 12-2.

Curb markings are used for roadway delineation and for parking regulation. Color-coded markings may be used as a supplement to standard signs to indicate the type of parking regulation.

When potentially hazardous objects such as bridge supports must be located within or adjacent to the roadway, they should be conspicuously marked. The MUTCD *(1)* specifies three alternative types of object marker designs, including reflectors mounted in a prescribed way on a rectangular or diamond-shaped panel and reflectorized all-yellow or yellow and black diagonally striped panels.

Delineators are used to aid night driving and are especially valuable where there is a change in horizontal alignment, where the alignment could be confusing, and where there is a transition in pavement width. Delineators normally consist of reflector elements that have a minimum dimension of about 3 in. and that are mounted on thin posts. White delineators are used along

TABLE 12-1
TYPICAL APPLICATIONS OF PAVEMENT MARKINGS

Type	*Typical Application*
Longitudinal lines	
Broken white line	Lane line of multilane highway
Broken yellow line	Two-lane two-way roadway, overtaking and passing permitted
Solid white line	Pavement edge marking
Double solid white line	Channelizing line in advance of obstructions
Solid yellow line	Used with broken yellow line to indicate no-passing zone for traffic adjacent to solid line
Double broken yellow line	Edges of reversible traffic lanes
Dotted lines	Extension of lines through an intersection or interchange area
Transverse Markings	
Crosshatched shoulder markings	To discourage use of shoulders as a traffic lane
Pairs of solid white lines, 6 in. or more in width, spaced 6 ft or more apart	Pedestrian crosswalks
Solid white lines, 12 to 24 in. in width	Stop lines to indicate where vehicles are required to stop

a—Typical pavement marking with offset lane lines continued through the intersection and optional crosswalk lines and stop limit lines.

b—Typical pavement marking with optional double turn lane lines, pavement messages, crosswalk lines, and stop limit lines.

c—Typical pavement marking with optional turn lane lines, crosswalk lines, and stop limit lines.

FIGURE 12-2 Typical pavement marking applications. (Courtesy Federal Highway Administration.)

through roadways, and yellow delineators are placed alongside interchange ramps, speed-change lanes, and median crossovers.

Colored pavements are sometimes used to supplement other traffic control devices. Red pavements are specified for approaches to a STOP sign that is in continuous use, yellow pavements for medians separating opposing traffic flows, and white for shoulders, crosswalks, and channelizing islands.

12-4 Traffic signals

The most common types of signals are traffic control signals, beacons, and lane-use control signals.

Traffic control signals

Traffic control signals are primarily used to control the movements of vehicular and pedestrian traffic at intersections. The traffic signal device (or head) is an assembly that contains one or more signal *faces*. Each signal face, except in pedestrian signals, is comprised of three to five optical units that display color-coded indications (red, yellow, green, green arrow, etc.) to the road user. Each optical unit consists of a lens, reflector, lamp, and lamp socket assembly.

The optical units in each signal face are usually in a vertical line, arranged red, yellow, and green from top to bottom. The face may be mounted horizontally, with the red at the left, the yellow next, and the green at the viewer's right.

A steady circular red indication is used to prohibit traffic from entering an intersection or other controlled area. A steady circular yellow indication warns traffic of an impending change in the right-of-way assignment. A steady circular green indication permits traffic to proceed in any direction that is lawful and practical *(1)*.

Steady red, yellow, and green arrow indications may be used under certain conditions, for example, where certain movements are prohibited or impossible and where turning movements are protected by the signal from conflicting movements.

The Manual on Uniform Traffic Control Devices *(1)* recommends that at least two signal faces be provided on each approach to a signalized intersection. To ensure adequate visibility, the manual gives specifications for the location and height of signals and minimum visibility distance for various approach speeds.

Traffic signals are operated by electromechanical or electronic *controllers* that switch the signal indications on and off in accordance with a plan selected by the traffic engineer.

Two basic types of control concepts are used for isolated intersections: (1) pretimed control and (2) traffic-actuated control. The following description of these concepts is quoted from Ref. 3 with minor editorial changes.

Pretimed control   Pretimed control is a mechanical operation that gives no recognition to the flow rates of traffic through an intersection as they occur on a minute-by-minute basis. This type of control is characterized by a fixed sequence of right-of-way assignments (phases) that occur in fixed time periods (cycles). The major elements of pretimed control are (1) cycle length, (2) phase lengths, and (3) number and sequence of phases.

The assignment of right-of-way to traffic movements by phase (phase A and phase B) and the common graphical techniques of representing phase movements are illustrated in Figure 12-3.

Traffic-actuated control   Traffic-actuated control of isolated intersections attempts to adjust phase length continuously (and thus adjust cycle length) and, in some cases, the sequence of phasing (through skipping of phases with no traffic demand). This adjustment occurs in accordance with measures of traffic demand obtained from vehicle dectectors* placed on one or more of the approaches to the intersection.

The basic technique in traffic-actuated control utilizes the following two basic timing features:

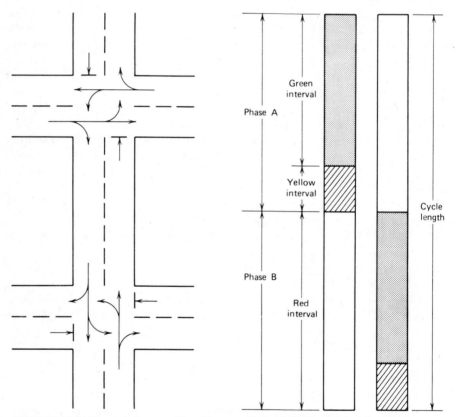

FIGURE 12-3   Signal-timing graphics. (Courtesy Federal Highway Administration.)

---

*A wide variety of devices, called traffic detectors, are available to sense the presence of traffic, including pressure detectors, loop detectors, radar detectors, and sonic detectors. Such detectors are fully described in Ref. 3.

1. Initial vehicle interval.
2. Vehicle extension (passage time).

A typical intersection approach with a vehicle detector is illustrated in Figure 12-4. The distance from the detector to the stop line influences the timing to be allocated to both the initial vehicle interval and the vehicle extension.

The initial vehicle interval should be long enough to permit traffic stored between the stop line and the detector to start and move into the intersection. If no additional vehicles cross the detector, then the phase will be terminated at the end of a minimum green time (initial vehicle interval plus vehicle extension plus a yellow change interval). If a vehicle crosses the detector during the timing of the vehicle extension, then an amount of time indicated by the vehicle extension setting will be added. The length of time provided by the vehicle extension should be sufficient to allow a vehicle to move from the detector and into the intersection (passage time).

Once the vehicle extension interval timing is initiated and an additional vehicle is detected, then the present vehicle extension timing is canceled and a new vehicle extension is initiated. This process is repeated for additional vehicles crossing the detector until (1) no more vehicles are detected or (2) interval timing reaches a preset maximum. In either of these two cases, the timing of a yellow change interval is initiated and the phase is terminated. If the vehicle extension is not completely timed out (because of the maximum override), then a recall situation is set and the timing will return to this phase, at the first opportunity, to prevent the trapping of vehicles.

The basic concept of traffic-actuated control is thus built around the two basic timing features (initial vehicle interval and vehicle extension). There are several forms of traffic-actuated control including semiactuated control and full actuated control.

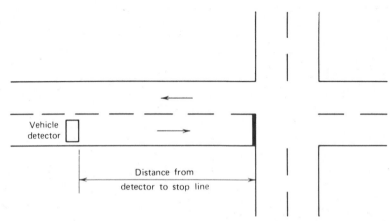

FIGURE 12-4 Traffic detection on intersection approach. (Courtesy Federal Highway Administration.)

In semiactuated control, detectors are placed on one or more, <u>but not all, of the approaches to the intersection.</u> The <u>minor street</u> approaches are usually selected for detector placement. The control assembly for this type of equipment provides for the following:

Minimum green time on major street.

Interruption of major street green only after minimum green interval has been timed following detection of vehicle on minor approach.

Minor street green controlled by timing initial vehicle interval and vehicle extension (up to preset maximum).

Control will dwell in main street green in the absence of minor street actuations.

No provision for automatically varying the length of the vehicle extension.

In <u>full actuated control, detectors are placed on all approaches.</u> The length of each green phase is dependent on timing an initial vehicle interval plus a vehicle extension. If consecutive vehicle detections on a given phase occur at a shorter time interval than the timing of the vehicle extension, then the green will remain on that phase up to a preset maximum. If the detections occur at an interval longer than the vehicle extension—that is, a vehicle gap whose length is greater than a preset amount is detected—then the control will go to the next phase for which there is traffic demand. In the absence of traffic demand (or the setting of a recall switch), the control will dwell in the phase that last received the green (3).

Traffic control signal warrants

A thorough investigation of traffic conditions and physical characteristics of the location should be made to determine the need for a traffic signal and to provide necessary data for the design and operation of the signal where it is warranted. The Manual on Uniform Traffic Control Devices (1) lists eleven sets of conditions that warrant the installation of a traffic signal. Stated in general and abbreviated terms, a traffic signal is warranted:

1. Where traffic volumes on intersecting streets exceed values specified in the MUTCD (1).
2. Where the traffic volume on a major street is so heavy that traffic on a minor intersecting street suffers excessive delay or hazard in entering or crossing the major street. The MUTCD (1) gives traffic volumes on the major and minor street that satisfy this warrant.
3. Where vehicular volumes on a major street and pedestrian volumes crossing that street exceed specified levels.
4. Where inadequate gaps exist for schoolchildren to cross at established school crossings.
5. Where necessary to maintain proper grouping or platooning of vehicles and effectively regulate group speed. This is known at the *progressive movement* warrant, which is subsequently defined.

6. Where the number of reported accidents exceeds a specified value.
7. At a common intersection of two or more major routes within a traffic flow system. The purpose of this warrant is to encourage concentration and organization of traffic flow networks.
8. Where no single warrant is satisfied but two or more warrants, 1, 2, and 3 are satisfied to the extent of 80 percent or more of specified values.
9. Where in each of any four hours of an average day the hourly volumes on the major street (both approaches) and the minor street (one direction) exceed values specified in the Manual (*1*).
10. Where for one hour of the day minor street traffic suffers undue delay in entering or crossing the major street.
11. Where the peak hourly volumes on the major street (both approaches) and the minor street (one direction) exceed values specified in the Manual (*1*).

Reference 1 gives specific and quantitative descriptions of these warrants.

Beacons

Four types of beacons are recommended by the Manual on Uniform Traffic Control Devices: (1) hazard identification beacons, (2) speed limit sign beacons, (3) intersection control beacons, and (4) stop sign beacons. Hazard identification beacons are used in conjunction with appropriate warning or regulatory signs or markers to draw attention to obstructions in or immediately adjacent to the roadway or other potentially hazardous location or condition. Such beacons consist of one or more sections of a standard traffic signal head with a flashing circular yellow indication in each section.

Speed limit sign beacons are used to draw attention to fixed or variable speed limit signs. Such beacons commonly consist of two circular yellow indications that are aligned vertically and that are flashed alternately. The use of a flashing single large circular yellow lens section is also permitted.

Intersection control beacons are used to control traffic at intersections where traffic or physical conditions do not justify conventional traffic signals but high accident rates exist. Such beacons consist of one or more sections of a standard traffic signal head, having circular yellow or circular red indications in each face.

Stop sign beacons consist of one or two sections of a standard traffic signal head with a flashing circular red indication in each section. Stop sign beacons are mounted within 12 to 24 in. above the stop sign.

Lane-use control signals

Lane-use control signals are normally used to permit or prohibit use of a traffic lane. They are most commonly used to control traffic on streets with reversible lanes. Examples of lane-use control signals include a steady downward green arrow, which indicates that a driver may use the lane over which the signal is located, and a steady red X, which means that the driver may not use the lane.

Several other types of signals are used but are not described here, including drawbridge signals, emergency traffic control signals, and train approach signals and gates.

12-5 Signal control for arterial streets

Along arterial streets it is often desirable to operate the signals as a system. By coordination of all of the signals along a street, it is possible to provide progressive movement of traffic through several intersections without stopping. Reference 3 explains:

> The basic approach to arterial street signal control considers that vehicles traveling along the arterial street are released in platoons from a signal and thence travel in platoons to the next signal. Thus, it becomes desirable to establish a time relationship between the beginning of the arterial green at one intersection and the beginning of the arterial green at the next intersection so that traffic platoons may receive a green indication just as they approach a signalized intersection. This permits the continuous flow of traffic along an arterial street and aids in reducing delay.
>
> This traffic flow control concept for arterial streets can be presented graphically by a technique known as a time–space diagram, as shown in Figure 12-5. This figure introduces the need for several definitions, which are presented as follows:

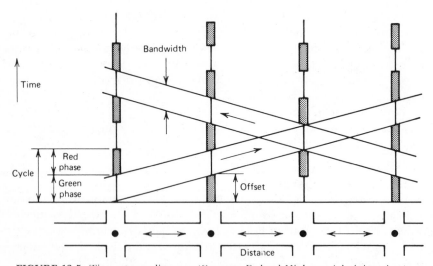

FIGURE 12-5 Time–space diagram. (Courtesy Federal Highway Administration.)

Through-band. The space between a pair of parallel speed lines that delineates a progressive movement on a time–space chart.

Band speed. The slope of the through band representing the progressive speed of traffic moving along the arterial.

Bandwidth. The width of the through band in seconds indicating the period of time available for traffic to flow within the band.

Offset. The time relationship expressed in seconds (or percentage of cycle length) determined by the difference between a defined interval portion of the coordinated phase green and a system reference point.

To establish a control system for an arterial street (or open network), it is necessary to develop a timing plan for all of the signals in the system. Such a timing plan consists of the following elements:

Cycle length. This must be the same for all signals in the system. Cycle length is usually established by examining traffic flow requirements to each intersection and computing the required length for each intersection. The largest cycle length required in the system is the one usually selected.

Cycle splits. A determination of the length of the various signal phases must be made for each individual intersection in the system. Phase lengths (splits) may vary from intersection to intersection.

Offset. A determination of an offset value must be made for each intersection in the system. The offset is usually established with reference to one master intersection in the system. *(3)*.

Several different systems have been employed in providing for progressive movement, including

1. A *simultaneous system,* in which several signals along a given street display the green indication at the same time. This system is usually employed where the block lengths are extremely short or where saturated conditions exist.
2. A *single alternate system,* in which each signal shows an opposite indication to that of the next signal. Half-cycle offsets are used in this system that are alternated at each signal.
3. A *double alternate system,* in which groups of two signals display the same indications simultaneously and adjacent groups display opposite indications.
4. A *limited progressive system,* in which the signal controllers are equipped to operate with a single cycle length, a single cycle split, and a single timing schedule that permits continuous operation of platoons of vehicles along the street at the progression speed.

5. A *flexible progressive system*, in which the common time cycle, splits, and offsets may be adjusted to conform to varying traffic conditions throughout the day or week.

Analyses to determine cycle length, splits, and offsets may be made manually or with the aid of computer programs developed for that purpose. Some work has also been done toward the use of on-line or "real-time" computer procedures that make calculations on the basis of traffic data which are collected by traffic detectors and fed directly to a central computer for storage and analysis *(3)*.

TRAFFIC SAFETY ENGINEERING

In Chapter 5 we described the nature and extent of the highway accident problem. Here we will examine methods and procedures that highway and traffic engineers may use to deal with the problem.

Clearly, engineers can reduce highway accident losses in many ways. One approach is to attempt to prevent crashes from occurring by providing suitable horizontal and vertical alignment, adequate lane widths and shoulders, appropriate signing and signaling, and so forth, as described elsewhere. In the following sections, we describe another approach in which the engineer focuses on the "at-crash" phase of accidents by providing clear roadsides, breakaway sign supports, redirecting barriers, and energy attenuation devices.

12-6 Clear roadside recovery areas

The *Handbook of Highway Safety Design and Operating Practices (4)* describes the need for clear roadside recovery areas as follows:

> To increase safety when vehicles leave the pavement, a
> recovery area free of physical obstacles should be provided. This
> clear roadside area should be as wide, as flat, as rounded, and
> as easy to traverse as it is feasible to construct for the conditions
> along each section of the highway . . . . Corrective programs
> have been and must continue to be undertaken to eliminate
> from the roadside such hazardous fixed objects such as trees,
> drainage structures, sign supports, utility poles, steep slopes, and
> other obstructions that are now exposed to traffic.

The provision of clear roadside recovery areas along all of the nation's 3.8 million miles of streets and highways would be a massive and expensive undertaking. On the basis of research performed at Georgia Institute of Technology *(5)*, it is estimated that there are more than 900 million fixed-object hazards lining U.S. highways. Modification or removal of all of these hazards would require many years and cost tens or even hundreds of billions of dollars. In view of the magnitude of this problem, Wright and Robertson *(5)* undertook a study of 30 fatal roadside object crashes in Georgia and developed priorities for roadside hazard modification. They recommended that top priority be given to roadside hazard modification on or near curves greater than 6°, particularly those accompanied by downhill grades of 2 percent or steeper. Their research indicated that the greatest needs for improvement exist on nonlocal road systems and along the outside of horizontal curves. Figure 12-6 illustrates the greater tendency for drivers to leave the roadway along the outside of horizontal curves.

Research has shown that clear roadside areas about 30 ft wide should permit recovery of about 85 percent of out-of-control vehicles. Where feasible, fixed-object hazards should be removed from a 30-ft-wide zone adjacent to the roadway or else be modified to provide energy attenuation or a breakaway design. Where neither removal nor modification of a fixed-object hazard is feasible, the installation of a traffic barrier may be warranted.

12-7 Breakaway sign supports

Unless properly designed, unshielded traffic signs may pose a hazard to the traveling public. Whenever possible, signs should be placed behind existing or

Excludes 13 cases where curves had more than 1 type of bend or information was missing.

FIGURE 12-6 How out-of-control vehicles leave the roadway. (Courtesy Insurance Institute for Highway Safety.)

planned guardrail or placed on bridges or grade-separation structures. Such is not always possible, and where signs must be located in areas exposed to errant vehicles, breakaway sign supports should be used.

A variety of breakaway sign support designs have been proposed and evaluated on the basis of full-scale crash tests. Large wooden sign supports can be suitably modified by drilling holes through the center of the post parallel to the face of the sign. Such a modification will usually greatly lessen the force of impact without significantly reducing the resistance to forces due to winds.

Two designs that have given satisfactory service for metal sign supports are shown in Figure 12-7. The small post slip-base design shown in Figure 12-7a is used for single post mountings for small signs and route markers. When struck, the connecting bolts shear and the entire assembly rotates over the moving vehicle. The hinged joint and slip-base support shown in Figure 12-7b is used for larger signs. A vehicle striking the post just above the slip base causes it to rotate about the hinged joint and clear the vehicle as it passes (4).

12-8 Guardrail design

In Table 8-6, design features of guardrails that have performed satisfactorily are shown. To be effective, guardrails must be carefully designed and installed. The purpose of guardrails is to redirect errant vehicles away from a hazard rather than to protect or prevent damage to a sign or other highway appurtenance. It should be remembered that guardrails may also constitute a fixed-object hazard. They should be used only where the hazardous feature cannot be removed, relocated, or redesigned to eliminate the hazard. Table 12-2 gives warrants for traffic barriers to shield sign supports, light poles, trees, and other fixed objects.

Guardrails should be placed along roadways with high, steep embankments where the hazard of being redirected by a guardrail is less than that of being permitted access to the slope. As Figure 12-8 illustrates, the height and slope of embankments are the determining factors that will indicate whether guardrails should be installed along embankments (6).

AASHTO (7) recommends that longitudinal barriers be installed with the face of the rail a minimum of 2 to 3 ft beyond the edge of the shoulder in order to provide safe use of the full shoulder width. Generally, the guardrail should be located not more than 2 to 4 ft from the edge of the shoulder (4).

The length of guardrail should be sufficient to prevent vehicle access behind the protective system. To provide adequate shielding along one-way facilities, the minimum required length, exclusive of specially designed terminal or end sections, is

$$L = (1 - A/B)E \qquad (12-1)$$

where

A = lateral distance from pavement edge to guardrail (ft)
B = lateral distance from pavement edge to hazard being shielded (ft)
E = encroachment distance parallel to pavement edge (ft)

FIGURE 12-7 Breakaway sign supports. (Courtesy Federal Highway Administration.)

TABLE 12-2
WARRANTS FOR TRAFFIC BARRIER PLACEMENT AT FIXED OBJECTS

	Traffic Barrier Required	
Fixed Objects Within 30 ft. (9.1 m) of Traveled Way	*Yes[a]*	*No*
1. Sign support (ground mounted):		
(a) Post of breakaway design[b]		X
(b) Wood poles or posts with area greater than 50 in.^{2}	X[c]	
(c) Sign bridge supports	X	
(d) Metal shapes with moment of inertia greater than 3.0 in. for steel, 4.5 in. for aluminum	X	
(e) Concrete base extending 6 in. or more above ground	X	
2. Light poles and supports with breakaway linear impulse:		
(a) Less than 1100 lb-sec[d]		X
(b) Greater than 110 lb-sec	X	
3. Bridge piers and abutments at underpasses	X	
4. Retaining walls and culvert headwalls	X	
5. Trees with diameter greater than 6 in.	X	
6. Wood poles or posts with area greater than 50 in.^{2}	X[e]	

[a]Traffic barrier recommended only if fixed object cannot be removed from 30-ft zone or where breakaway design is not feasible.
[b]Usually breakaway design should be used regardless of distance from traveled way.
[c]Cross-sectional area of large wood members can be reduced to 50 in.^{2} or less by boring holes at about 6 in. above grade. If this is not feasible, traffic barrier is recommended.
[d]Breakaway bases should always be used except where low-speed vehicular traffic or heavy pedestrian traffic is a consideration.

SOURCE: *National Cooperative Highway Research Program Report No. 118.* Highway Research Board, Washington, D.C. (1971).

Special care should be exercised when designing and installing guardrail ends and transitions. Unless properly designed, exposed guardrail ends can "spear" a vehicle and injure or kill its occupants. Generally speaking, it is advisable to flare guardrail ends away from the pavement and anchor them in an existing or false cut. Another method of treating the approach end of guardrails is to design it to collapse and absorb the energy of a crash. A recommended design employing a breakaway cable terminal is shown as Figures 12-9.

Where guardrails transition to bridge parapets or other type of barrier, connections should be carefully designed. Unless special precautions are taken, there will be a tendency for the guardrail to excessively deflect and "pocket" the vehicle, directing it into the end of the barrier. Some concepts of barrier transition design are shown by Figure 12-10.

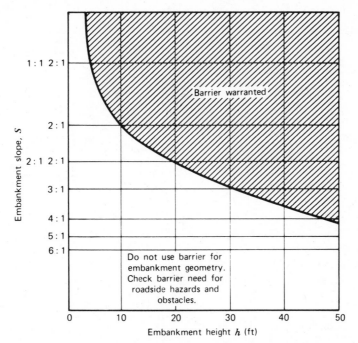

FIGURE 12-8 Barrier requirement for embankment geometry. (Courtesy National Cooperative Highway Research Program, Transportation Research Board.)

12-9 Median barriers

Median barriers are used primarily to prevent vehicles from crossing the median and encroaching into opposing lanes. Although cross-median crashes are rare events, they are usually high-speed crashes that tend to be severe in terms of vehicular damages and personal injuries. It is therefore important that the installation of such barriers be considered, especially where the median width is narrow and traffic volumes and speeds are high. Generally, median barriers are not warranted if the median width exceeds 40 ft, except on the basis of adverse accident experience *(4)*.

Design features of median barrier systems that have performed satisfactorily are given in Table 12-3. Barrier systems perform best when a level surface is provided in front of the barrier. Curbs, dikes, sloped shoulders, and stepped medians can cause a vehicle to vault over the barrier or strike it and overturn *(6)*.

In narrow medians, concrete barriers such as those shown in Figure 12-11 are recommended. Such barriers are often used at retaining walls and along the face of rock cuts. If such a barrier is placed on an existing paved median, 1-in.-diameter dowels 8 in. long should be placed on 4-ft center-to-center spacing as shown in the figure.

TABLE 12-3
DESIGN FEATURES OF MEDIAN BARRIERS THAT HAVE PERFORMED SATISFACTORILY

System	MB1 Cable	MB2 "W" Beam	MB3 Box Beam	MB4W Blocked-Out "W" Beam (Wood Post)
Deflection	11 ft[a]	7 ft	4 ft	2 ft
Post spacing	8 ft, 0 in.	12 ft, 6 in. nominal	6 ft, 0 in.	6 ft, 3 in.
Post	H2-1/4 × 4.1	S3 × 5.7	S3 × 5.7	8×8 in. Douglas fir
Beam	Two 3/4 in. diam. steel cables	Two steel "W" sections	8 × 6 × 1/4 in. steel tube	Two steel "W" sections Two C6 × 8.2 steel sections (rub rails).
Offset brackets				Two 8 × 8 × 14 in. Douglas fir blocks
Mountings	5/8 in. diam. steel "U" bolts	1 5/16 in. diam. bolts	Steel paddles	5/8 in. carriage bolts
Footings	Details vary with application	1/4 in. steel plate welded to post	1/4 in. steel plate welded to post	None
Developed by	California	New York	New York	California

[a]1 ft = 0.3048 m.

Source: *A Handbook of Highway Safety Design and Operating Practice.* Federal Highway Administration, Washington, D.C., revised 1973 and 1978.

Note: Cable to be parallel to guardrail for straight runs of rail. Cable may have angle point at anchor plate if guardrail is curved.

Note: 5'–6" with terminal section. May be less with retu sections if separate rods connect to concrete anchor. For type 3 flare, anchorage is only required for guardrail on side of median where traffic is approaching bridge.

JRE 12-9 End anchorage details for guardrail. (Courtesy Federal Highway Administration.)

12-10 Crash cushions

Crash cushions are used to decelerate errant vehicles to a stop, greatly reducing the severity of head-on impact with a fixed object by spreading the energy of impact over time and space. Such devices are most commonly used in front of retaining walls, bridge piers, and abutments and in off-ramp gore areas. A variety of types of crash cushions have been proposed and tested, including rows of barrels, entrapment nets, and arrays of containers filled with sand or water. One such device is shown in Figure 12-12, and additional ones are described in Ref. 6.

To illustrate the mechanics of crash cushions, consider a 4500-lb vehicle that strikes a barrel crash cushion at a speed of 60 mph (88 ft/sec). The kinetic energy of the vehicle

$$\text{K.E.} = \frac{Wv^2}{2g} = \frac{4500(88)^2}{2(32.2)} = 541,000 \text{ ft-lb}$$

FIGURE 12-10 Barrier transition design. (Courtesy Federal Highway Administration.)

FIGURE 12-11 Concrete median barriers. (*a*) New Jersey barrier. (*b*) General Motors barrier. (Courtesy National Cooperative Highway Research Program, Transportation Research Board.)

Suppose it is desired to bring the vehicle to a stop with an average deceleration of 6g. The deceleration is equal to the rate of change of velocity

$$6g = \frac{v_0 - v_f}{t} = \frac{88 - 0}{t}$$

The time of deceleration $t = 0.45$ sec. Since the average velocity is 44 ft/sec, the distance required to bring the vehicle to a stop

$$D = 44 \times 0.46 = 20 \text{ ft}$$

Elevation

FIGURE 12-12 A barrel-type crash cushion. (Courtesy National Cooperative Highway Research Program.)

This distance is the theoretical minimum and does not take into account crash cushion inefficiency.

Consider another example, abstracted from Ref. 6.

EXAMPLE 12-1. *Crash Cushion Design*   A crash cushion device is to be placed at an elevated expressway gore to safely decelerate a 4500-lb vehicle traveling at a speed of 60 mph. Fifty-five gallon steel barrels with a 7-in.-diameter hole in the center of each end will serve as the basic element. Laboratory studies have indicated that a dynamic force of 9000 lb is required to crush one barrel from its original 2-ft diameter to approximately 0.5 ft. The dynamic energy consumption $e_d = 1.5 \times 9000 = 13,500$ ft-lb.

Determine the number and arrangement of barrels that will stop the car and the average deceleration level.

Solution.   As computed above, the kinetic energy K.E. = 541,000 lb. The minimum number of barrels needed

$$N_b = \frac{\text{K.E.}}{13,500 \text{ ft-lb/barrel}} = \frac{541,000}{13,500} = 40$$

Suppose the barrels are arranged as shown in Figure 12-13. The barrier stopping force is a stepped function corresponding to the number of barrels in a row, as the figure illustrates.

Vehicle penetration into the crash cushion is determined by equating the kinetic energy to the area under the force–penetration curve. The penetration of the single, double, and triple rows of barrels is $1.5 + 3 \times 1.5 + 6 \times 1.5 = 15$ ft. Let X_1 be the penetration of the barrels arranged in four rows and X_2 the total penetration. The kinetic energy

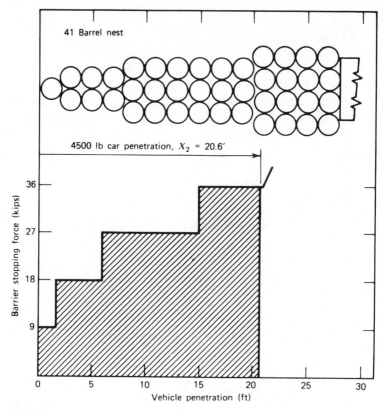

FIGURE 12-13 Analysis of crash cushion system, Example 12-1. (Courtesy National Cooperative Highway Research Program, Transportation Research Board.)

K.E. = 541 ft-kip = $(9 \times 1.5) + (18 \times 4.5) + (27 \times 9.0) + (36X_1)$

$X_1 = 5.6$ ft

$X_2 = 5.6 + 15 = 20.6$ ft

The average deceleration level

$$G_a = \frac{v^2}{2gX_2} = \frac{(88)^2}{2g(20.6)} = 5.84g$$

The selection of an appropriate crash cushion depends not only on its dynamic performance characteristics but also on the space available for the cushion and the installation, maintenance, and damage repair costs.

Additional information on the location, selection, and maintenance of various types of highway traffic barriers is given in Ref. 6.

12-11 Conclusion

In this chapter we have only been able to briefly outline some of the more important aspects of traffic and safety engineering. The reader is referred to the references listed at the end of this chapter for a more complete treatment of these important areas of study.

PROBLEMS

12-1   Compute the minimum length of the guardrail required to shield a hazard located 25 ft from the pavement edge of a one-way facility. The guardrail is to be located 10 ft from the pavement edge. Assume that vehicles will encroach at an angle of 11° to the pavement edge. If the roadway has two-way operation, what minimum length of guardrail will be required?

12-2*   Compute the minimum length of guardrail in meters required to shield a hazard located 8 m from the edge of a one-way facility. The guardrail is to be located 4 m beyond the pavement edge. The encroachment angle is 15°.

12-3   Given the crash cushion described in Example 12-1, how far will a 2000-lb car traveling 55 mph at impact penetrate the crash cushion? Calculate the average deceleration level in g's.

12-4*   Given the crash cushion described in Example 12-1. A 1200-kg car impacts the device at a speed of 75 km/hr. Estimate the penetration in meters and the average deceleration rate in meters per second squared.

12-5   Given the following data, plot a time–space diagram for a one-way street with a progressive signal system. Determine the bandwidth and the band speed. The cycle length is 60 sec.

Signal No.	Split (% green + amber, red)	Distance between Intersections
1	60–40	
2	55–45	300 ft (91 m)
3	62–38	600 ft (183 m)
4	50–50	500 ft (152 m)

12-6*   Given the conditions described in Problem 12-5, determine the bandwidth and band speed in kilometers per hour.

REFERENCES

1. *Manual on Uniform Traffic Control Devices for Streets and Highways, including Revisions 1–4.* U.S. Department of Transportation, Federal Highway Administration, Washington, D.C. (1986).

2. *Traffic Control Devices Handbook.* U.S. Department of Transportation, Federal Highway Administration, Washington, D.C. (1983).

3. *Traffic Control Systems Handbook.* U.S. Department of Transportation, Federal Highway Administration, Washington, D.C. (1976).

4. *Handbook of Highway Safety Design and Operating Practices.* U.S. Department of Transportation, Federal Highway Administration Washington, D.C. (1978).

5. Wright, Paul H., and Robertson, Leon S. *Priorities for Roadside Modification—A Study of 300 Fatal Roadside Object Crashes.* Insurance Institute for Highway Safety, Washington, D.C. (1976).

6. Michie, J.D., and Bronstad, M.E. *Location, Selection, and Maintenance of Highway Traffic Barriers*. National Cooperative Highway Research Program Report No. 118, Washington, D.C. (1971).

7. *Guide on Evaluation and Attenuation of Traffic Noise*. American Association of State Highway and Transportation Officials, Washington, D.C. (1974).

chapter thirteen

SURVEYS AND PLANS

In the relocation or reconstruction of existing highways and the establishment of new ones, surveys are required for the development of project plans and the estimation of costs. The performance of good surveys requires well-trained engineers who have an understanding of the planning, design, and economic aspects of highway location and who are sensitive to the social and environmental impacts of highway development. As Figure 13-1 illustrates, the work of the highway location may include (1) a desk study, (2) a reconnaissance survey, (3) a preliminary survey, and (4) a final location survey. Each phase of the highway location study is discussed in some detail in the following sections.

13-1 Available techniques

Highway surveys may be accomplished either by conventional ground survey methods or by remote sensing techniques. A variety of remote sampling devices have been suggested for highway location studies including side-look radar, infrared photography, and even multispectral devices for recording a scene at several wavelengths simultaneously. Aerial photography with either panchromatic or color film is the most commonly employed remote sensing procedure in highway location studies. References 1 and 2 provide additional information on remote sensing techniques.

Aerial photographs provide the highway location team with both qualitative and quantitative information. By direct study of individual and stereoscopic pairs of photographs and photographic mosaics, trained professionals may glean a great deal of information about landforms, soil and drainage conditions, and the character of land use. With appropriate horizontal and vertical ground controls, accurate photogrammetric maps can be drawn at practically any desired scale, providing quantitative data on ground contours, drainage features, buildings and structures, property boundaries, and existing transportation facilities.

While traditional techniques are still used on projects of medium and small size, the use of photogrammetric surveys on large works is almost universal, and it is becoming more frequent on smaller projects. The adaptability of photogrammetry to computer operations has enabled substantial savings in time and money over conventional methods. Conventional transit and tape surveys, described in Section 13-8, are generally used for final location surveys on projects of all sizes.

FIGURE 13-1 The rural highway location and design process.

13-2 Desk study of area

Before any field or photogrammetric investigation is made, a great deal can be learned from a desk study of the area. The first step in highway location and design is to procure all available pertinent data. These data may be in the form of maps, aerial photographs, charts or graphs, and so on and may require the application of a large variety of engineering, environmental, social, and economic knowledge. The type and amount of data collected during this initial phase will, of course, vary with the functional classification of the road and the nature and size of the project. Sources of available information are listed in Table 13-1.

When all the available data have been assembled, a detailed analysis should reveal much information pertinent to the proposed project. For example, analysis of the available information may allow the engineer to determine the advisability of selecting an entirely new location or improving the existing one. After an exhaustive study of aerial photographs, topographic maps, drainage maps, soil maps, and other data is made in the office, a series of proposed locations may be selected for a field or photogrammetric investigation.

TABLE 13-1
POSSIBLE DATA SOURCES FOR DESK STUDY OF HIGHWAY LOCATIONS

1. Engineering data
   Topographic and geological maps
   Stream and drainage basin maps
   Climatic records
   Preliminary survey maps of previous projects
   Traffic surveys and capacity studies
2. Environmental data
   Agricultural soil surveys indicating soil erodibility
   Water quality studies
   Air pollution studies
   Noise and noise attentuation studies
   Fish and wildlife inventories
3. Social data
   Demographic and land-use information
   Census data
   Zoning plans and trends
   Building permit records
   Motor-vehicle registration records
4. Economic data
   Overall costs of previous projects
   Unit construction cost data
   Agricultural, industrial, commerical activities and trends

13-3 Reconnaissance survey from aerial photographs

For many projects, the data on available maps and photographs will be out of date or incomplete. In some cases, there are no available base maps of suitable scale and accuracy. A reconnaisance survey from aerial photographs will amplify and verify conditions determined from the preliminary study. The function of the survey is to provide sufficient information on the topography and culture of the area to enable the selection of one or more preliminary route locations.

Control points between the two terminal points on the route are determined and the flight lines between these control points established. It is customary for the width of coverage to be about 0.4 to 0.6 the distance between control points. From the aerial photographs a base map is produced, usually to a scale of 1 in. = 200 ft with a 2- or 5-ft contour interval. A stereoscopic examination of the photographs will also permit the determination of ground cover and soil conditions (Figure 13-2). From the base maps and photointerpretation of the surveyed area, likely routes can be designated.

A location team should examine potential routes on the basis of the following requirements:

1. Traffic service for population and industrial areas.

FIGURE 13-2 Stereoplotter used by the Georgia Department of Transportation.

2. Directness of route.
3. Suitability of terrain encountered.
4. Adequacy and economy of crossings at watercourses and at other transportation routes.
5. Extent of adverse social, environmental, and ecological effects.

The reconnaissance is an important portion of the work and should not be treated lightly. During this phase of the location process, the range of alternative routes is narrowed and a small number of routes are selected for further study. The use of photogrammetric base maps often enables an objective study of all possible routes with greater ease than field reconnaissance.

13-4 Selection of preliminary centerline

Possible alternative alignments are plotted on the base map, and from these alignments, with preliminary gradelines, the alternatives are compared for suitability. Such features as design capacity, safety, road user costs, construction costs, and maintenance costs are examined. Possible adverse impacts of the various alignments due to the dislocation of families, farms, and businesses should be carefully considered, as should the effects of the roadways on the environment and the ecology. Opportunities to provide scenic vistas and aesthetic enhancement of the area should not be overlooked. Based on the results of objective analysis, a preliminary alignment is chosen. Figure 13-3 shows a preliminary alignment drawn on a photogrammetric map.

13-5 Mapping for final design

For final design and location, further photogrammetric mapping is necessary. The selected preliminary alignment is used as a guide for the strip area to be mapped. It is becoming customary to use coordinate systems in highway design because of the increasing use of electronic computers for direction and distance calculations. Therefore, at this time, prior to taking low-level aerial photographs, a baseline is established on paper and by monuments on the ground. The low-level aerial photographs are taken with markers on the ground to tie in the baseline. Distances between the baseline monuments are often determined by electronic measuring devices such as the Electrotape or Geodimeter, which have excellent accuracy over distances ranging from a few hundred feet to 20 or 30 miles. Such devices measure distances by precisely measuring the time required for electromagnetic waves to travel the distance being measured. For example, the Electrotape consists of two identical units, each capable of transmitting and receiving microwave signals *(3)*. The Geodimeter employs a modulated beam of light that is transmitted to a passive reflector, which reflects the light pulses back to the instrument *(4)*. The baselines, which run between

FIGURE 13-3 Preliminary alignment drawn on a photogrammetric map.

monuments of calculable coordinates, can be used in the final location survey in locating the alignment in the field.

The low-level photographs are used to produce base maps along the preliminary alignment, usually at a scale of not less than 1 in. = 100 ft. These base maps are more complete than those used for reconnaissance. The preliminary survey taken at a lower level can give virtually complete information on

1. Topography, with respect to changes in elevation and drainage characteristics, and information concerning the soil conditions in the area.
2. Land uses, designating type, intensity, and quality.
3. Transportation facilities, with respect to proximity and effect of proposed location.
4. The effect of the proposed location on property within the area.

For the purposes of design, however, the preliminary survey will produce topographic base maps and information concerning property and utility location that will enable the engineer to proceed with the design of the final alignment.

13-6 Design of the final alignment

The process of the design of the final alignment is one requiring great skill and judgment. It is a trial-and-error process by which the most suitable alignment is obtained. The highway is fitted by hand to the topography and land use until the engineer is satisfied that no better fitting can be achieved. Modern practice makes great use of splines and curve templates. The spline is a flexible plastic guide which can be bent to different positions until the most satisfactory curvilinear alignment is achieved. Curve templates can come in such simple forms as the old-fashioned highway curves, which are boxes of circular segments usually in increments of 15 min (degree of curve). Currently these are supplemented by transparencies, which to a standard scale display circular curves, three-center compound curves, and transition spirals.

To facilitate computation, tangent alignments can be set at even-degree angles. This speeds computations by permitting the use of design tables, which can be generated within the computer. In urban areas, where small changes in alignment may cause property encroachment, the use of whole-degree azimuths may not be feasible.

The method of design usually followed is to adjust the spline until a smooth, consistent alignment is achieved that is satisfactory from the viewpoint of grade, curvature, cross sections, drainage, and stream crossings. The hand-fitted line is then converted into a defined line by fitting to it tangent lengths, transition spirals, and simple and compound curves, using the visual aids of the templates. The standardized templates contain information that can be used directly to compute coordinates of the controlling points on the curvilinear alignment and for the computation of tangents.

13-7 Final location survey

Conventionally, the method of establishing the final location in the field has been by direct chaining along the tangents, first setting the points of intersection of the tangent lines (PI's) and then closing out the horizontal curves between the tangents, using deflection angles and chaining to set the curve stations. This method is still in use, although the electronic distance-measuring devices have to a great degree replaced chaining procedures. Moreover, the use of coordinates in highway location means that the computer can be used to calculate azimuths between control points on the curvilinear alignment and points of known coordinates on the baseline. In this case, the errors along the alignment are not additive. The points of curvature (PC's) and points of tangency (PT's) can be set and checked by the intersection of three azimuths instead of two. Chaining between the control points in the alignment is done to set stakes at the even 100-ft stations, and stakes on the curves can be set from the PC or PT by normal deflection angle methods.

In general, it is unnecessary for ground crews to take cross sections along the alignment. Such cross sections can be obtained directly from the photogrammetric maps with as much accuracy as ground surveys usually achieve.

13-8 Conventional ground techniques

Where photogrammetric surveying methods are unavailable, or where the size or nature of project makes the use of aerial photographs unsuitable, it may be necessary to use the older, conventional ground surveys. Such instances are becoming less common, as the development of photogrammetry has led to the widespread availability of aerial photographic equipment. This can generally be used to economic advantage.

The steps involved in ground surveys are similar to those where photogrammetry is used:

1. Desk study.
2. Reconnaissance survey.
3. Preliminary location survey.
4. Final location survey.

A desk study includes the preliminary steps of evaluation of all available data as covered in Section 13-2. The reconnaissance survey consists of a field investigation that usually provides a means of verification of conditions as determined from the preliminary study. For example, building symbols on maps do not indicate the true value of property under consideration, and this information can usually be secured by field investigation. A study is made of the profiles and grades of all alternative routes and cost estimates made for grading, surfacing, structures, and right-of-way. A comparison of alternative routes in this fashion will aid the final selection of the most likely location.

Next, a preliminary survey is made to gather information about all the physical features that affect the tentatively accepted route. In general, the work is carried out by a regular survey party consisting of party chief, instrument men, chainmen, and rodmen. A primary traverse or baseline is established. This is an open traverse consisting of tangent distances and deflection angles following approximately the line recommended in the reconnaissance report. When the preliminary line has been established, the topographic features are recorded. The extent to the right and left of the traverse line to which the topography should be determined will vary, and it is usually left to the judgment of the party chief in charge of the work. It will certainly never be less than the proposed width of right-of-way.

The amount of level work on a preliminary survey should be kept to a minimum. In flat areas no centerline profiles are needed; in rolling or mountainous terrain a profile of the traverse centerline is often sufficient. A few cross sections are sometimes of value in exceptionally rugged areas where centerline profiles do not give proper information.

Using the preliminary survey as a basis, a preliminary survey map is drawn. The preliminary map should show all tangents with their bearings and distances, all deflection angles, ties to property corners or state control monuments, and all witness points of the alignment. Certain topographic features are shown on the map, such as streams, water courses, lakes, hills, and ravines, and man-made features such as buildings, drainage structures, power lines, and other public facilities.

The design of the final alignment when ground survey methods have been used is essentially the same as the procedure indicated in Section 13-6. Location of the centerline requires considerable skill and judgment. Trial lines are drawn on the map and are related to control points such as road intersections, and building corners. The trial lines should avoid badly drained ground, rough or impassable areas, valuable property, or other obstructions. The use of splines and curve templates is recommended. Rough estimates of excavation and embankment may also influence the placing of the trial lines on the map. Essentially, the only difference encountered between conventional ground methods and aerial surveys is that it is more likely that with the latter the whole project will be tied to a coordinate grid system to facilitate the use of electronic computers in determining the distances and azimuths. On the smaller projects where ground methods are still used, the older system of chaining along the centerline, tangent distances, and deflection angles may be the essential data given for the final location survey.

The final location survey serves the dual purpose of permanently establishing the centerline and collecting the information necessary for the preparation of plans for construction. The line to be established in the field should follow as closely as is practical the line drawn on the preliminary map, conforming to the major and minor control points and the alignment that was previously determined.

The first step in the final location survey requires the establishment of the centerline, which is used as a survey reference line, upon which property descriptions are based for the purpose of purchasing right-of-way. Modifications of the final location may also be made in the field as a result of incomplete or inaccurate preliminary information. Centerline stakes are placed at 100-ft intervals, and each stake is marked with the proper station number. Curves are staked out by deflection angles, and the error of closure is determined by an independent stakeout of the P.T. from the P.I. Particular care must be taken in the chaining, and a closure should be effected where possible on control hubs for the preliminary survey, since errors along the centerline are cumulative.

Level work is of the utmost importance, since the grade line, earthwork, and drainage are designed from the level notes. Elevations may be obtained from U.S. Geological Survey or other bench marks, or elevations may be assumed for a project. Bench marks are permanent reference points, and extreme care should be used when establishing benches along the route of the proposed project. The bench marks should be placed approximately 1000 ft apart in fairly level country and closer in hilly country. Benches should always be placed at least 200 ft from any proposed structure, if possible. Bench mark elevations should be established by the loop system of levels or by any other approved method where a high degree of accuracy can be obtained. The error of closure between benches should never be greater than 0.01 ft.

The importance of keeping good notebooks on alignment and level surveys cannot be overemphasized. Notes taken in the field during the final location survey usually become part of the permanent record of the project.

Finally, cross-section levels are taken at even 100-ft stations and at any intermediate points with abrupt slope changes. This information is necessary for accurate determination of existing grade line, design of drainage structures, computation of earthwork quantities, and, in general, estimation of quantities of work to be done. The manner in which the cross sections are taken and the form of notes used are fairly well established. In order to obtain a fairly representative cross section at a station, an elevation is always obtained at the centerline stake. Then elevations and distances at right angles to the centerline are recorded, first on one side and then on the other. Cross sections should also be taken at least 500 ft beyond the limits of the point of beginning and the point of ending of a project. Centerline profile elevations should also be obtained at some distance beyond these points in order to establish proper grades for connecting projects. The final location survey is complete when all the necessary information is available and ready for the designer to use. This includes all information pertaining to (1) alignment, (2) topography, (3) bench mark levels, (4) cross sections, (5) section corner ties and other land ties, (6) plats and subdivisions, (7) soil surveys, and (8) drainage and utilities.

13-9 Automated procedures

Much of the work involved in route location is repetitive and requires little exercise of the engineer's professional skill. Therefore, in the past few years great emphasis has been placed on increasing the individual engineer's productivity by the use of automated procedures that rely extensively on electronic computers.

The photogrammetric aerial survey methods already described are readily adaptable to automation. Digital terrain models can be stored in computer memory by linking the stereoplotter to a device that automatically records the x, y, and z coordinates of points on the stereo model. Input of a proposed alignment will result in the computation of horizontal centerline geometry, vertical profile geometry, and earthwork quantities. This allows rapid evaluation of the proposed alignment and permits the engineer to make small modifications without the tedious and expensive repetition of manual calculations.

From the stored terrain data and the input of a proposed alignment, it is possible to obtain cross-section data directly in numerical form or, by linking the computer to an automatic plotting device, to obtain cross sections in a graphical form.

By the use of more sophisticated stereoplotters, the amount of ground control work required for the aerial survey can be greatly reduced and time and cost of the survey minimized. The stereoplotter is an example of a machine that uses *instrumental aerotriangulation* in a process called "bridging," which eliminates the need for ground control points for each pair of stereo photographs forming a stereo model. The computer can also be used for analytical solutions of aerotriangulation problems. Bridging is a process whereby the horizontal and, to some degree, the vertical control on one stereo model can be extended through successive models until additional ground control is reached.

Figure 13-4 shows a Wild Aviolyt BC2 analytical compiler with an Aviotab TA10 automatic plotting table. This system provides high accuracy in photogrammetric measurements for applications in graphical compilation, digital mapping, and computer-controlled profiling.

Most highway agencies use computer programs for the adjustment of field survey traverses and for the computation of earthwork volumes from cross-section data. Modern techniques of establishing horizontal alignment rely extensively on the use of transparent templates for curve design and intersection layout. Precomputed data from these templates are compiled into design tables that can be used as input for the computation of the horizontal alignment referenced completely to a coordinate system. The use of templates and design tables relieves the designer of much computation and (possibly more important) ensures more standardization and higher design standards. Figure 13-5 shows the use of a transparent design template in the design of an interchange ramp.

FIGURE 13-4 Wild Aviolyt BC2 analytical compiler with an Aviotab TA10 automatic plotting table. (Courtesy Wild Heerbrugg Instruments, Inc.)

FIGURE 13-5 Use of a transparent template in ramp layout. (Courtesy DeLeuw, Cather & Co.)

Extensive use has also been made of the computer in the layout of interchanges and bridges. The intersection or bridge is related to the coordinate system of the project. From this relation, the azimuths between any two points are readily available by machine calculation. Field layout can be done by measuring along an azimuth from a known point or by determining the point of intersection of azimuths from at least two other points. The latter method has been found successful for setting bridge piers by using three azimuths from three points, where chaining is difficult.

13-10 Problems of highway location in urban areas

In rural areas, the location of a highway is dictated chiefly by the desired end points of travel, topography, and geology. The road should serve a desired travel pattern with an alignment that provides consistent conditions within acceptable limits of curvature and grade over terrain that is capable of supporting the proposed construction without excessive costs.

In urban areas the location process is more involved, reflecting the complexities of urban conditions. Among the conditions that control urban locations are

1. Traffic service.
2. Land use.
3. Off-street parking.
4. Other transportation systems.
5. Topography and geology.
6. Sociological considerations.

The traffic planner, as a result of extensive investigations of the travel desires and existing travel patterns in an urban area, can predict the effect on travel patterns of alternative locations of proposed highways. Traffic assignment techniques can be used to analyze the probable effect of new links in the highway network. This permits the location engineer to determine how well alternative locations will fit with the existing network. Travel desire lines often can show the major elements of a desirable arterial highway system in an urban area.

Land use is a major factor in the pattern of traffic generation in an urban area. Most cities have a central commercial core surrounded by areas of residential, industrial, and commercial development. Travel patterns vary with time of day, day of week, and season of year. Trips from the home to work are heavy to the commercial downtown area and the industrial areas during the morning peak hours, with a reversed pattern during the evening rush hour. During the day, shopping trips from the residential areas to the outlying commercial areas and the center core predominate. The pattern of land use affects location in other ways. Industrial and commercial areas rely heavily on truck transportation and, therefore, need service by arterial routes. Health and safety requirements may indicate that heavily traveled routes should not be located through residential areas. Aesthetic considerations may indicate that an alignment which cuts through a public open space should be avoided if the space can be preserved by a skirting alignment. Most cities have recognized the interaction between highway location and land use and have integrated a major thoroughfare plan with the city master plan.

Parking is of major importance in urban areas. New highway locations must be considered in relation to existing or planned parking facilities for the vehicles attracted to the new facility. Ideally, the location should be close to existing and potential parking areas to minimize the amount of travel on existing streets.

Interference and interaction with other transportation systems must be considered. The aim of any new facility is to increase the overall level of service of transportation in the urban area, and a location should not unnecessarily disrupt existing service. Overall coordination of the various transport modes is essential in the development of the balanced transportation system so urgently needed in complex urban areas.

As in rural areas, topography and geology have a great effect on urban highway locations. Topographic barriers such as ridges, valleys, rivers, and lakes affect construction costs and determine the economy of a route. Similarly, soil and ground water conditions may render an otherwise feasible location uneconomic. The effect of topography and geology is somewhat lessened in an urban area owing to the high cost of right-of-way in certain areas. For example, the cost of a location through a heavy industrial area might greatly exceed costs involved in remedying the swampy conditions of an alternative alignment. The controls of topography remain, however, imposing grade and curvature controls on any designated alignment.

The complex of utilities that serves an urban area is usually almost totally absent from rural areas. Extensive relocation of utilities where depressed highways and underpasses are involved can add greatly to the cost of urban projects and require considerable advance planning. In certain areas within large cities, the underground maze of utility lines and subways can make the cost of a depressed highway prohibitive.

The social effects of a highway project must be of special concern in urban areas. Proposed locations should not sever residential neighborhoods or create barriers between residence and schools and other community services. Open public land should be retained where possible; park and residential land may be used for highways only if no feasible and prudent alternative to the use of such land exists. On the other hand, a highway can serve as a stabilizing force by screening industrial areas from residential communities.

PREPARATION OF PLANS

Road plans are necessary for making estimates and receiving bids for construction.

The steps involved in the preparation of plans are fairly uniform in all the state highway agencies. Questions related to the type of surfacing, width of grade, preliminary right-of-way requirements, and other design elements are decided before the detailed design is started. It is also determined whether the contemplated improvement can be accomplished at one time, or whether provision should be made to complete it by stages. This stage development may, for example, limit the length to be constructed at any particular time or it may provide for additions to the highway cross section by the addition of future lanes or other improvements.

Generally speaking, plans may be prepared in two phases: (1) the preparation of preliminary plans and (2) the preparation of final plans. The first phase consists of a partial investigation and preliminary design. A field inspection is then made before final plans are prepared. Design standards as discussed in Chapter 8 are applied to all steps when designing a given project.

13-11 Plan and profile

With the information obtained from the final location survey, a plan and a profile of the proposed route are drawn on standard plan and profile sheets, as shown in Figure 13-6. The plan of the road is drawn to some suitable scale, 1 in. equaling 100 ft being the most common. When greater detail is desired, as is often the case in urban areas, the scale may be increased to 1 in. equals 50 ft. The plan should show the proposed construction centerline with the limits of right-of-way and all important topographic features such as fences, buildings, streams, railroads, and other structures on the right-of-way. All survey information necessary for the establishment of the survey centerline should also be placed on the plan. This should include compass or computed bearings of all tangents, together with all curve data, including the point of beginning, point of ending, degree of curvature, and so on, and all witness information to the survey points. Bench marks, with their elevations, are also shown on the plan sheet. Where electronic computers have been used to determine azimuths between control points of known coordinates for the purpose of final location, the coordinates of these control points should be shown on the plan.

A notation is made on the plan of the sizes and types of all existing structures and the manner in which they are to be utilized or removed. Other details are added, such as type and depth of ditches, slopes, right-of-way and borrow requirements, and so forth.

The centerline profile is plotted on the lower portion of the plan and profile sheet. The horizontal scale is usually the same as that used for the plan. The vertical scale, however, is distorted, with scales of 1 in. equals 5 ft and 1 in. equals 10 ft being most commonly used. The profile usually shows the profile of the existing ground line and the profile of the proposed construction centerline, the tangents and vertical curves forming one continuous profile. All grade lines show the percentage of grade and elevation points where changes of grade occur, with the lengths of the vertical curves used at these points of intersection.

All existing drainage structures are generally shown on the plan and profile sections. These include culverts, catch basins, inlets, manholes, and so on. Any special information pertaining to the profile and affecting the design may also be added, such as curb grades, gutter grades, and sodded slopes.

13-12 Grades and grade control

All controls that affect the position of the grade line should be given thorough study. Ideal grades are those that have long distances between points of intersection, with long vertical curves between tangents to provide good visibility and smooth riding qualities.

One consideration in grade control is to try to fit the grade line to the natural ground surface as closely as possible. Good drainage is usually accomplished in this manner. Intersections with railroads and existing bridges are points that are generally fixed and the grade must be set to meet them. Intersections with trunk lines, other surfaced roads, unimproved roads, drives, trails, and field entrances may also be grade controls. The effect of soils on grade is also very important. In swamp or peat areas it is necessary to have fills high enough so that waste material can be disposed of at the sides of the roadway. The grade must also be above the water table, where possible.

The placing of the grade line requires much more detailed study in urban areas than in rural areas. Changes in grade effect store entrances, driveways, and sidewalks. It is often necessary to warp or tilt the pavement to meet existing curbs and to provide proper drainage.

Slight distance is of fundamental importance in both rural and urban areas.

Ruling grades for the economical operation of motor vehicles have been fairly well established. These have been indicated in Chapter 8. When existing conditions indicate a new grade, care and judgment should be exercised to obtain the most economical one possible.

When establishing a grade line, an effort should be made to provide an economical balance between the quantities needed for fill sections and those required in cut sections. Changes in grade from a plus grade to a minus grade should be accomplished in cuts, and changes in grade from a minus grade to a plus grade should be placed in fills. This usually gives a good design and often avoids the appearance of building hills and producing depressions contrary to the existing slope of the land. All hauls should be downhill, if possible, and not too long.

A common method of laying the grade is to staple the various profile sheets together; with pins as points of intersections, a string or thread is drawn taut from end to end to produce a continuous profile. Shifting of the pins to various positions will given an opportunity to study the effect of different grades without drawing lines on the paper. When the grade is established, the pin points are used as control points and the elevations are noted from the profile elevations.

Horizontal control is usually well established when the location survey is made. However, certain refinements may make it necessary to change the alignment when adjusting grade controls for vertical alignment. A combination of grades and horizontal alignment may give insufficient sight distance, poor appearance, or both. Flattening a curve may be necessary to eliminate or reduce superelevation or to lengthen the sight distance in a cut. Special efforts should be made to provide long grades and proper sight distances on curves.

When designing the grade line, consideration must be given to the grades of the side ditches. The depth and width of the ditch and its side slopes affect the width of the highway cross section and right-of-way requirements.

13-13 Grade inspection of preliminary plans

When the grade line has been determined, final completion of the plans is delayed until a field inspection is made. In the process of preparing road plans, arrangements are usually made for two inspections in the field. The first inspection is for checking the partially prepared plans with field conditions and for supplementing the survey information with such additional data as are necessary for completing final plans and preparing a correct estimate of quantities. The field investigation can serve as a check for the general design and should reveal whether or not the width of right-of-way is sufficient for the proposed cross section; how drainage structures, ditches, and slopes are affected by field conditions; and other data that may be necessary for completing the plans. Reproductions are made of the partially completed plans for field inspection purposes.

The field investigator makes an inspection by walking the entire length of the project with a set of plans in hand and checking the various items of the plans, giving approval or making notations about recommendations for a change in design. When the inspection is completed, the plans that were used for the field investigation are returned to the designer. Where the surveys have been conducted by photogrammetric means, the amount of field inspection at this stage can be reduced by a restudy of the mosaics constructed from the aerial photographs.

It may be mentioned here that a final inspection of the proposed project is made when the final plans have been completed. This inspection, called the "plans, specifications, and estimates" inspection and commonly referred to as the P. S. & E. inspection, is made in order to verify the final design before the project is approved to be advertised for bids.

PREPARATION OF FINAL PLANS

After the field inspection is made with the partially completed plans, or "preliminary plans" as they are sometimes called, the final design steps are carried out. These include (1) plotting the cross sections of the original ground and placing "templates" on original ground sections; (2) computing earthwork quantities; (3) preparing construction details for bridges, culverts, guardrails, and other items; (4) preparing summaries and estimates of quantities; and (5) preparing specifications for the materials and methods of construction.

13-14 Cross sections and templates

One important phase of design is determining the amount of earthwork necessary on a project. Earthwork includes the excavation of material and any hauling required for completing the embankment. Payment for earthwork is based on excavated quantities only and generally includes the cost of hauling the materials to the embankment. An additional item of payment called "overhaul" is often used to provide for hauling of the excavated material beyond a certain free-haul distance. The excavated material may be obtained from within the area of the highway cross section or from some distance outside the proposed highway. When excavation is obtained outside the proposed highway limits it is called "borrow excavation."

In order to determine earth excavation and embankment requirements by manual means, a section outline of the proposed highway, commonly referred to as a "template section," is placed on the original ground cross section; the areas in cut and the areas in fill are determined; and the volumes between the sections are computed. Figure 13-7 shows various conditions that may be encountered when plotting these template sections. "Cut" and "fill" are the terms that are usually used for the areas of the sections, and the terms "excavation" and "embankment" generally refer to volumes.

Cross sections are plotted on standard cross-section paper to any convenient scale. A scale of 1 in. equals 5 ft vertically and horizontally is common practice. Each cross section should show the location or station of the original ground section and template section, the elevation of the proposed grade at that station, and the areas of cut and fill for each section. The computed volumes of excavation and embankment may also be placed on the cross-section sheet between two successive cross sections.

To facilitate the plotting of these cross sections, a template of the roadway that conforms to the outline of the proposed highway cross section is made of some transparent material. The outline can be quickly drawn on the cross-section paper. When templates are placed on the cross-section paper, consideration must be given to superelevations and transitions due to horizontal curvature.

FILL SECTION

CUT SECTION

CUT AND FILL SECTION

FIGURE 13-7 Original ground line and template sections.

The areas of cut and fill may be measured by use of a planimeter, a computation method using coordinates, or some other suitable method. The areas of cut and fill are recorded in square feet. These are then used to compute the volumes of excavation and embankment between the sections, from which a summation can be made of the total volumes of excavation and embankment for the entire project.

The average-end-area method, which is used by a majority of state highway departments, is based on the formula for the volume of a right prism whose volume is equal to the average end area multiplied by the length, or

$$V = \frac{\frac{1}{2}L(A_1 + A_2)}{27}$$

(13-1)

where

V = volume (cu yd)

A_1 and A_2 = area of end sections (ft^2)

L = distance between end sections (ft)

Volumes computed by the average-end-area method are slightly in excess of those computed by a prismoidal formula when the sections are not right prisms. This error is small when the sections do not change rapidly, which is usually the case when a small degree of curvature is used. On sharp curves, however, a prismoidal correction should be applied.

When a section changes from a cut section to one of fill, a point is reached when a zero cut and zero fill occurs. This usually happens along an irregular line across the roadway. In order to properly compute the volumes at these locations, it is necessary to compute volumes at intermediate points. The usual practice is to determine end areas for three sections. One section would be at the centerline, where the cut and the fill equal zero. The other two sections would be at the right and left of the edge of the grade, where the change from cut to fill or fill to cut occurs.

Manual methods for computing earthwork have largely been supplanted by methods based on the use of electronic computers. In typical cases and for earthwork quantities at the design stage, the computer input includes the cross-section data, centerline grades, and design template data with provision for superelevation, curve widening, and varying side slopes. The computer produces a tabulation of the cut and fill volumes at each station and the difference between them adjusted for shrinkage and swell, the cumulative volumes of cut and of fill, and mass diagram ordinates. Figure 13-8 shows such a tabulation.

EARTHWORK QUANTITIES CALCULATION PROCESS
IDENTIFICATION

EARTHWORK QUANTITIES LIST FOR ROADWAYS A B

BASELINE STATION NUMBER	PRISM SHRINK/ SWELL FACTOR	STATION CUT (SQ-FT)	STATION CUT (CU-YD)	ADJUSTED STATION CUT (CU-YD)	STATION FILL (SQ-FT)	STATION FILL (CU-YD)	ADJUSTED STATION FILL (CU-YD)	MASS ORDINATE (CU-YD)
10050+00.00	0.8000	0.00	0	0	1653.66	5846	5846	-100156
10051+00.00	0.8000	95.48	177	141	1765.60	6332	6332	-106347
10052+00.00	0.8000	483.19	1072	857	1539.15	6120	6120	-111609
10053+00.00	0.8000	128.07	1132	906	1742.23	6077	6077	-116780
10054+00.00	0.8000	4.64	246	197	1607.84	6204	6204	-122787
10055+00.00	0.8000	7.12	22	17	1725.86	6174	6174	-128944
10056+00.00	0.8000	0.22	14	11	1972.13	6848	6848	-135781
10057+00.00	0.8000	0.00	0	0	1898.75	7168	7168	-142949
10058+00.00	0.8000	0.00	0	0	2204.06	7598	7598	-150547
10058+59.00	0.8000	0.00	0	0	2590.40	5238	5238	-155785
10059+00.00	0.8000	0.00	0	0	2598.81	3940	3940	-159725
10060+00.00	0.8000	0.00	0	0	2559.77	9553	9553	-169278
10061+00.00	0.8000	0.00	0	0	2419.85	9222	9222	-178499
10062+00.00	0.8000	0.00	0	0	2645.45	9380	9380	-187880
10063+00.00	0.8000	0.00	0	0	2777.71	10043	10043	-197922
10063+90.00	0.8000	0.00	0	0	2810.29	9313	9313	-207236
10064+00.00	0.8000	0.00	0	0	2873.41	1053	1053	-208288
10065+00.00	0.8000	0.00	0	0	2644.88	10219	10219	-218507
10066+00.00	0.8000	0.00	0	0	2111.44	8808	8808	-227315
10067+00.00	0.8000	0.00	0	0	1904.40	7437	7437	-234752
10068+00.00	0.8000	3.29	6	5	765.89	4945	4945	-239692
10069+00.00	0.8000	1972.60	3659	2927	0.00	1418	1418	-238183

FIGURE 13-8 Computer tabulation of earthwork calculations. (Courtesy Georgia Department of Transportation.)

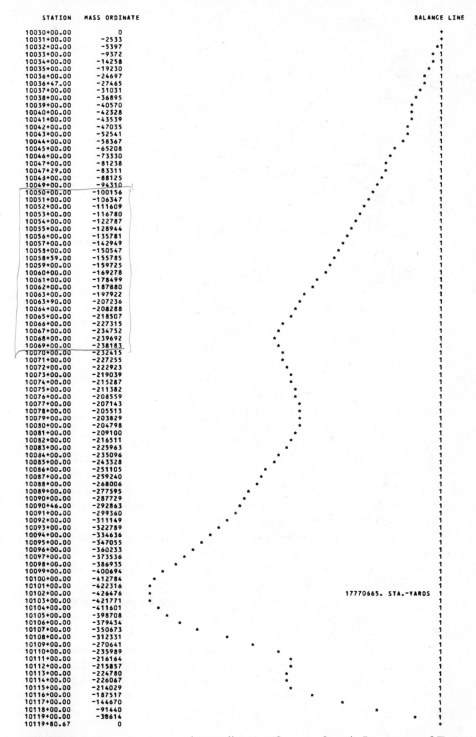

FREE HAUL DISTANCE IS   0.
VERTICAL SCALE =   62000 CU/YDS PER INCH   HORIZONTAL SCALE -- NONE

STATION	MASS ORDINATE		BALANCE LINE
10030+00.00	0		+
10031+00.00	-2533		+
10032+00.00	-5397		+1
10033+00.00	-9372		+1
10034+00.00	-14258		+ 1
10035+00.00	-19230		1
10036+00.00	-24697		1
10036+47.00	-27465		1
10037+00.00	-31031		1
10038+00.00	-36895		1
10039+00.00	-40570		1
10040+00.00	-42328		1
10041+00.00	-43539		1
10042+00.00	-47035		1
10043+00.00	-52541		1
10044+00.00	-58367		1
10045+00.00	-65208		1
10046+00.00	-73330		1
10047+00.00	-81238		1
10047+29.00	-83311		1
10048+00.00	-88125		1
10049+00.00	-94310		1
10050+00.00	-100156		1
10051+00.00	-106347		1
10052+00.00	-111609		1
10053+00.00	-116780		1
10054+00.00	-122787		1
10055+00.00	-128944		1
10056+00.00	-135781		1
10057+00.00	-142949		1
10058+00.00	-150547		1
10058+59.00	-155785		1
10059+00.00	-159725		1
10060+00.00	-169278		1
10061+00.00	-178499		1
10062+00.00	-187880		1
10063+00.00	-197922		1
10063+90.00	-207236		1
10064+00.00	-208288		1
10065+00.00	-218507		1
10066+00.00	-227315		1
10067+00.00	-234752		1
10068+00.00	-239692		1
10069+00.00	-238183		1
10070+00.00	-232415		1
10071+00.00	-227255		1
10072+00.00	-222923		1
10073+00.00	-219039		1
10074+00.00	-215287		1
10075+00.00	-211382		1
10076+00.00	-208559		1
10077+00.00	-207143		1
10078+00.00	-205513		1
10079+00.00	-203829		1
10030+00.00	-204798		1
10081+00.00	-209100		1
10082+00.00	-216511		1
10083+00.00	-225963		1
10084+00.00	-235096		1
10085+00.00	-243328		1
10086+00.00	-251105		1
10087+00.00	-259240		1
10088+00.00	-268006		1
10089+00.00	-277595		1
10090+00.00	-287729		1
10090+46.00	-292863		1
10091+00.00	-299360		1
10092+00.00	-311149		1
10093+00.00	-322789		1
10094+00.00	-334636		1
10095+00.00	-347055		1
10096+00.00	-360233		1
10097+00.00	-373536		1
10098+00.00	-386935		1
10099+00.00	-400694		1
10100+00.00	-412784		1
10101+00.00	-422316		1
10102+00.00	-426476		17770665. STA.-YARDS 1
10103+00.00	-421771		1
10104+00.00	-411601		1
10105+00.00	-398708		1
10106+00.00	-379434		1
10107+00.00	-350673		1
10108+00.00	-312331		1
10109+00.00	-270641		1
10110+00.00	-235989		1
10111+00.00	-216164		1
10112+00.00	-215857		1
10113+00.00	-224780		1
10114+00.00	-226067		1
10115+00.00	-214029		1
10116+00.00	-187517		1
10117+00.00	-144670		1
10118+00.00	-91440		1
10119+00.00	-38614		1
10119+80.67	0		+

FIGURE 13-9 Computer-generated mass diagram. (Courtesy Georgia Department of Transportation.)

379

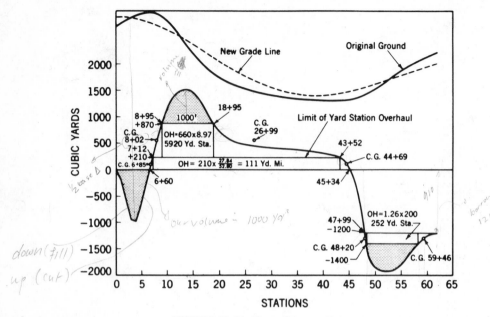

FIGURE 13-10 Example mass diagram.

excavation balances the volume in the embankment. Any line drawn parallel to the baseline and intersecting two points within the same curve will also indicate a balance of cut and fill quantities between these two points.

Further analysis of the curve shows other balance points at stations 45 + 34, 47 + 99, and 62 + 00. The earth excavation from 6 + 60 to 14 + 00 is used to construct the embankment between stations 14 + 00 and 45 + 34, and the direction of haul is to the right. The embankment between stations 45 + 34 and 47 + 99 is constructed from borrow. The embankment between stations 47 + 99 and 52 + 00 is constructed from excavation between stations 52 + 00 and 62 + 00, and the direction of haul is right to left.

Where earth excavation and embankment quantities balance on the complete project, the curve of the mass diagram would end on the baseline at the zero point. This condition does not always occur. In this case the mass ordinate shows a minus 1200 cu yd that must be borrowed from some other source in order to complete the embankment. When the mass ordinate ends in a plus volume and the last point on the curve is above the baseline, there is excess material that must be wasted.

Referring to Figure 13-10 it is seen that the quantities balance between 0 + 00 and 6 + 60, a distance of 660 ft. This is within the free-haul distance and need not be considered for overhaul. It also indicates that 977 cu yd are to be excavated and placed in the embankment and that no payment will be made for hauling this material. Payment is made, however, for 1311 cu yd of excavation within this balance point.

The next balance occurs between stations 6 + 60 and 45 + 34. A scaled horizontal line 1000 ft long is drawn parallel to the baseline and found to intersect the curve at station 8 + 95 and 18 + 95. This represents the free-haul distance, and all excavation and embankment above this line can be eliminated from further consideration as to overhaul. The length of this balance from 6 + 60 to 45 + 34 equals 3874 ft, which is over 3640 ft and indicates that there will be both yard stations and yard miles for overhaul. A horizontal line 3640 ft long is drawn parallel to the baseline and is found to intersect the curve at stations 7 + 12 and 43 + 52. The overhaul above this line will be yard stations and below this line it will be yard miles.

The 1000-ft free-haul line intersects the volume ordinate at +870 at stations 8 + 95 and 18 + 95; the 3640-ft line intersects the volume ordinate line at 210 at stations 7 + 12 and 43 + 52.

Perpendicular lines are drawn from the limits of free haul to the limits of the yard station overhaul line at stations 8 + 95 and 18 + 95; this vertical line represents the volume ordinate +210 to +870, which represents 660 cu yd of excavation from stations 7 + 12 and 8 + 95 that must be placed in the embankment between stations 18 + 95 and 43 + 52. The cubic-yard stations are determined by multiplying the excavation, which in this case is 660 cu yd by the distance from the center of mass of excavation to the center of mass of embankment, less 1000 ft free haul.

In order to determine the number of cubic-yard miles, perpendiculars are drawn from the limit of yard station overhaul at station 7 + 12 and station 43 + 52; this vertical line represents the volume ordinate 0 to +210, which is 210 cu yd from station 6 + 60 to station 7 + 12 that must be placed in the embankment between stations 43 + 52 and 45 + 34. The cubic-yard miles are determined by multiplying 210 cu yd by the distance in feet from the center of mass excavation to the center of mass embankment, less 1000 ft free haul, divided by 5280.

There are three methods in common use for determining the center of mass of excavation and embankment when mass diagram calculations are done manually:

1. The graphical method.
2. The method of moments.
3. The planimeter method.

Various computer programs have been developed to perform these calculations by machine, but using a similar approach.

13-17 The graphical method

The volume ordinate line from +210 to +870 represents 660 cu yd. If a line is drawn bisecting this line and extended horizontally to intersect the curve between stations 7 + 12 to 8 + 95 and 18 + 95 to 43 + 52, the length of this line would represent approximately the average length of haul. This method is fairly accurate when the volume curve has a fairly uniform slope such as exists between stations 7 + 12 and 8 + 95, but appreciable error is evident when finding the center of mass between stations 18 + 95 and 43 + 52. No attempt will be made to complete the analysis by the graphical method because of its limitations. The more accurate method of moments will be explained in detail below.

13-18 Method of moments

To find the center of mass between stations 7 + 12 and 8 + 95, moments may be taken about station 7 + 12. A moment is computed for each volume that makes up the volume curve between 7 + 12 and 8 + 95. Dividing the total moments by the total volume will give the distance to the center of mass from 7 + 12. From the mass diagram of Figure 13-10 it can be determined that the volumes and distances are as indicated in the table below.

Station	Volume (cu yd)	Distance (ft)
7 + 12 to 8 + 00	329	44
8 + 00 to 8 + 95	331	135.5
Total	660	

Taking moments about station 7 + 12, we have the distance to the center of mass equal to

$$\frac{44 \times 329 + 135.5 \times 311}{660} = 90 \text{ ft}$$

The location of the center of mass equals 7 + 12 + 90 ft = 8 + 02.

In the same manner the center of mass of the embankment between stations 18 + 95 and 43 + 52 is determined by taking moments about station 18 + 95 and using the volumes given in volume sheet (not shown) between stations 18 + 95 and 43 + 52. The sum of these moments divided by the total yardage equals 804 ft. The center of mass is therefore 804 ft from 18 + 95 or at station 26 + 99.

With the center of mass of the excavation and that of the embankment being located at stations 8 + 02 and 26 + 99, respectively, the distance between these points equals 1897 ft. Subtracting 1000 ft as free haul, the overhaul distance is 897 ft or 8.97 stations. The overhaul will equal 660 × 8.97 or 5920 cubic-yard stations.

Overhaul on a cubic-yard mile basis is computed in the same manner, except that the overhaul distance is expressed in miles instead of stations. In this case the centers of mass of excavation and embankment are found to be at stations 6 + 85 and 44 + 69, a distance of 3784 ft. The cubic-yard miles will then be 210(3784 - 1000/5280) or 111 cubic-yard miles.

13-19 Planimeter method

To determine the center of mass by use of the planimeter, the following formula is used:

$$\text{Distance to center of mass} = \frac{\text{area} \times \text{horizontal scale} \times \text{vertical scale}}{\text{volume ordinate}}$$

where

area = planimetered area under the volume curve (in.^{2})
horizontal scale = horizontal scale of the mass diagram (stations/in.)
vertical scale = vertical scale of the mass diagram (cu yd/in.)
volume ordinate = vertical ordinate between the baseline and limits of haul (cu yd)

For example, to determine the center of mass of the embankment from station $18 + 95$ to $43 + 52$, the area under the volume curve between these stations to the line-marked limit of yard station overhaul was planimetered and found to be 2.122 in.^{2} Using the above formula, the distance of the center of mass from station $18 + 95$ is as follows:

$$\text{distance to center of mass} = \frac{2.122 \times 5 \times 500}{660} = 8.04 \text{ stations}$$

Center of mass = station $18 + 95$ plus $8 + 04 = 26 + 99$, which is the same as the result obtained by the method of moments.

13-20 Borrow overhaul

The volume ordinate between stations $45 + 34$ and $47 + 99$ indicates 1200 cu yd of fill for which suitable borrow must be provided. When it becomes necessary to obtain excavation from a borrow pit in order to complete the embankment, the overhaul distance is determined from the center of mass of the borrow pit to the center of mass of the embankment as measured along the haul route, less the amount of free haul.

A tabulation of earth excavation, borrow excavation, and overhaul, as determined from the mass diagram of Figure 13-10, is given below.

	Earth Excavation	Borrow Excavation	Overhaul Cu-yd Sta.	Cu-yd Mi
$0 + 00$ to $6 + 60$	1311			
$6 + 60$ to $45 + 34$	4473		5920	111
$45 + 34$ to $47 + 99$	107	1200	[a]	[a]
$47 + 99$ to $62 + 00$	1567	____	252	____
Total	7458	1200	6172	111

[a]Overhaul will have to be paid for the 1200 cu yd of borrow excavation if it is obtained beyond the free-haul limits.

SUMMARY OF QUANTITIES AND BID UNIT PRICES

STATE AID PROJ. NO. SAP-1135-A (2).
COUNTY ATKINS
LOCATED BETWEEN on the Pearson to Ray City Road, S.R. 61. 6.315 miles grading and bituminous surfacing

Item	Approx. Quan.	Unit	Unit Price	Amount
Clearing and Grubbing Roadway	76.550	Acre	LUMP	29,750.00
Clearing and Grubbing	7	Acre	650.00	4,550.00
Unclass. Excav. & Bor. Incl. Ditches & Shldrs.	139,000	Cu. Yd.	1.00	139,000.00
Overhaul on Excavation	250,000	Sta. Yd.	0.01	2,500.00
Special Subgrade Compaction	6.315	Mile	500.00	3,157.50
Excavation for Culverts & Minor Structures	500	Cu. Yd	2.30	1,150.00
Selected Matl. for Clvt. Found. Backfill	360	Cu. Yd.	5.25	1,890.00
Class "A" Concrete Culverts	132	Cu. Yd.	135.00	17,820.00
Bar Reinforcement Steel	10,600	Lb.	0.34	3,604.00
18" Class "B" Concrete Pipe S.D.	980	Lin. Ft.	7.50	7,350.000
24" Class "B" Concrete Pipe S.D.	48	Lin. Ft.	12.20	585.60
18"Std. Str. Reinf. Concrete Pipe C.D.	264	Lin. Ft.	9.70	2,560.80
24" Std. Str. Reinf. Concrete Pipe C.D.	1,344	Lin. Ft.	17.40	23,385.60
30" Std. Str. Reinf. Concrete Pipe C.D.	96	Lin. Ft.	19.25	1,848.00
36" Std. Str. Reinf. Concrete Pipe C.D.	192	Lin. Ft.	26.80	5,145.60
Culvert Pipe Removed S.D. or C.D.	330	Lin. Ft.	3.30	1,089.00
Culvert Pipe Relaid S.D. or C.D.	50	Lin. Ft.	3.30	165.00
Sand-Cement Bag Rip Rap	460	Sq. Yd.	16.00	7,360.00
Concrete Right-of-Way Markers	48	Each	15.00	720.00
Sprigging	74,000	Sq. Yd.	0.06	4,440.00
Loose Sod Rip Rap	450	Sq. Yd.	1.50	675.00
Water for Grassing	75	M. Gal.	8.00	600.00
First Application Fertilizer	10	Ton	150.00	1,500.00
Second Application Fertilizer	2,160	Lb.	0.75	1,620.00
Removal of Existing Wood Bridge	Lump Sum	LUMP	LUMP	500.00
Reflect. Railroad Warning Signs—Type "F"	2	Each	160.00	320.00
Reflect. Railroad Stop Signs—Type "F"	2	Each	160.00	320.00
6" Soil-Bitum. Stabilized Base Course	78,800	Sq. Yd.	0.58	45,704.00
Cut-Back Asphalt—RC-3	325,200	Gal.	0.46	149,592.00
Aggr. for Penetration Mac. Base-Railroad Crossing	25	Ton	15.00	375.00
Bit. Matl. for Penetration Mac. Base-Railroad Crossing	400	Gal.	0.46	184.00
Asphalt Cement Surf. Treat.-Stone Size 4-Type-2	75,000	Sq. Yd.	0.55	41,250.00
Asphalt Cement Bituminous Seal	75,000	Sq. Yd.	0.42	31,500.00
Finishing and Dressing	55,000	Sq. Yd.	0.05	2,750.00
		Total		534,961.10
		Contingency		53,438.90
				588,400.00

FIGURE 13-11 Hypothetical engineer's estimate.

13-21 Preparation of construction details

Earth excavation quantities, overhaul, and borrow requirements are generally the first items that are determined when drawing the plans for a project, and then the other details are prepared.

Many of the structures and miscellaneous items included in the project may be standard items for which designs have previously been approved and adopted for use. When such a design can be used, it is made a part of the plans by reference. If it cannot be used, the structure has to be designed to meet existing conditions.

13-22 Preparation of quantities and estimates

When the plans are completed, a summary is made of all the items of work, estimated quantities, and units of measurement. These are placed on one of the plan sheets. Some states provide regular quantity sheets for this purpose that show all items as per plan, with provision for inserting quantities as constructed. These sheets are useful for computing final quantities on the completed project. The summary of quantities may be arranged in groups such as grading items, drainage structures, surfacing items, and incidentals.

A summary of quantities is tabulated and an estimate is made of the probable cost of the project, with an allowance made for contingencies. An engineer's estimate is shown in Figure 13-11.

13-23 Preparation of specifications

When the plans have been completed, a proposal and specifications are then prepared for the project. When this is done, the project is ready to be advertised for bids and contracts prepared for the construction of the project according to the prepared plans. These matters are discussed in the next chapter.

13-24 Completed final plans

A complete set of plans will contain any number of sheets, depending on the length and complexity of the project. Most state highway departments specify a definite order in which the plan sheets are to be arranged, and this varies considerably with the several states. A complete set of plans should contain the following.

1. *Title sheet.* The first sheet of a set of plans, which is known as the title sheet, shows a map of the area in which the project is located. The map should show the point of beginning and point of ending; the main topographic features; township, county, and state lines; and possible detour routes. The name and number of the project are also indicated. The index, legend, and tabulation of standard plans used are generally included in the title sheet. Provision is also made on this sheet for the affixing of signatures approving the plans that are contained in the completed set.

2. *Typical cross-section sheet.* The typical cross-section sheet shows all earth-work and finished road sections. These should include ditch sections and any treatment needed for the subgrade section.

3. *Note and summary sheet.* Notes or special provisions consist of special requirements, regulations, or directions prepared to cover the work which is not covered by the standard specifications. The summary of quantities is a tabulation of the items of work, quantities, and units of measurement for that particular project.

4. *Plan and profile.* The plan and profile sheets contain most of the details in regard to the work. These have been previously discussed. Their number depends on the length of the project.

5. *Mass diagram.* This sheet gives the contractor and the engineer a compre-hensive graphic representation of the source and disposition of all earth moved on the project.

6. *Drainage structures.* All drainage structures are usually shown in detail. These will include detail drawings of culverts, inlets, manholes, and so on and will show the flow lines, sizes, and locations of the various structures. Bridge plans are generally not included in these drawings, but are in a separate set of plans.

7. *Special details.* Special details may include slab or box culverts, special intersection treatment, railroad crossings, or any special detail not cov-ered by the standard plans or the specifications.

8. *Utility sheet.* On some projects, the number of utilities involved may be considerable, and a special sheet is devoted to their treatment.

9. *Quantity sheets.* These sheets contain all the items of work involved on a project. They have been previously discussed. They differ from the note and summary sheet by having more detail, such as stations for the location of culverts and so forth.

10. *Cross sections.* Prints are usually made of the cross sections used to determine earth excavation quantities and are made part of the plans. On projects of considerable length many of these sheets may be required. Many states do not attach these sheets to each set of plans but keep them at some central or division office where they are available to the contractor or others desiring this information.

When the plans are completed, a final inspection of plans, surveys, and estimates, known as the P.S. & E., is made, and when the necessary approvals are obtained the project is ready for construction.

PROBLEMS

13-1 The 1200 cu yd of borrow needed in Figure 13-10 can be secured from 900 ft right of station 77 + 40. What is the overhaul?

13-2 Assuming that the free-haul distance is 1500 ft and the cost of excavation is $1.60 per cubic yard, what is the limit of economical haul when the unit price of overhaul is $0.17 per cubic-yard station?

13-3 Given the following end areas in cut and fill, complete the earthwork calcula-tions, using a shrinkage factor of 20 percent. Prepare the mass diagram and indicate the direction of haul and whether you will have to borrow or waste any material in order to complete the project.

Station	End Areas Cut	End Areas Fill	Station	End Areas Cut	End Areas Fill
286	60	15	296	23	148
287	54	22	297	77	160
288	48	50	298	109	69
289	24	70	299	156	40
290	12	53	300	112	27
291	0	66	301	83	20
292	6	51	302	68	9
293	7	63	303	56	5
294	13	97	304	54	0
295	28	110	305	42	0

13-4 If you as a highway engineer had to lower the grade through a cut and waste a large amount of earth to obtain a passing sight distance, how would you justify the additional cost?

13-5 Given below are the ordinates of the mass diagram for a highway project. No haulage is possible past station 38 + 50, so this point may be considered the end of the project for this purpose. The free-haul distance is 500 ft. Draw the mass diagram, and determine (a) the volume of excavation and embankment, (b) the quantities of overhaul (station yards), waste, and borrow; and (c) the distribution of haul.

13-6 Consult current literature, and write a report on the use of (a) aerial photogrammetry and associated devices, (b) electronic surveying devices, or (c) electronic computers in highway location studies by your state highway department or some other highway agency.

Station	Mass Ordinate (cu yd)	Station	Mass Ordinate (cu yd)
0	0	11	+4,931
1	+481	12	+3,214
2	+982	13	+865
3	+1,343	14	−437
4	+2,022	15	−2,892
5	+3,589	16	−4,504
6	+5,567	17	−6,075
7	+7,810	18	−7,111
7 + 40	+8270	19	−7,686
8	+7,782	20	−9,122
9	+7,465	21	−10,437
10	+6,632	22	−9,333

23	−8,138	32	−27,397
24	−8,985	33	−29,025
25	−11,214	34	−26,991
26	−12,482	35	−24,018
27	−14,076	36	−18,212
28	−17,451	37	−12,222
29	−20,519	38	−7,537
30	−23,048	38 + 50	−4,238
31	−25,110		

REFERENCES

1. *Remote Sensing and Its Application to Highway Engineering.* Highway Research Board Special Report No. 102 (1969).

2. *Remote Sensing for Highway Engineering. Highway Research Record 421* (1972).

3. Warren, Robert J. Electrotape Use in Establishing Basic and Supplemental Control for Aerial Surveys. *Highway Research Record 65* (1965).

4. Radmanovich, D. Evaluation of the Model 4 Geodimeter. *Highway Research Record 65* (1965).

chapter fourteen

CONTRACTS AND SUPERVISION

Practically all highway construction projects in the United States are public works, which are constructed with public funds. The agency authorizing this construction may be a federal, state, municipal, or county governmental unit. However, the greatest number of highway construction projects today are authorized through the various state highway departments. More than 95 percent of the construction done under state highway supervision is done by contract. The remaining 5 percent is done by the state's own forces organized and equipped to do this work. The contract system has proved beneficial, and it is very unlikely that it will be changed.

Under the contract system, plans and specifications are prepared and are submitted for competitive bids. As is usually the case when public funds are used, the methods and procedures for carrying out public works projects are prescribed by law in each state and must be strictly adhered to. It is the intent to include here the procedure generally followed by most state highway agencies in preparing contractual documents and in supervising construction.

14-1 Unit price system
Nearly all highway contracts are prepared on a unit price basis. The unit price contract includes an estimate of the number of units of each type of work and the price for each unit. The unit price contract has many advantages and is well adapted to highway construction, because usually it is not possible to determine exact quantities of some items of work when plans are prepared. Sometimes, after construction has started, unusual conditions are encountered. In either event, the various quantities can be adjusted; that is, they may be increased or decreased to the quantities actually constructed. Under the unit price system, each unit includes the materials that are to be furnished, the work to be done, the method of measurement of the work completed, and the basis on which payment is to be made. These units are described in detail in the specifications.

14-2 Lump-sum bids and force-account work
It is not always feasible to have a breakdown of all the work on a unit price basis. It may be necessary to have minor items done on a lump-sum basis, especially when some piece of work is too intangible for unit prices to be computed. Some highway organizations now use lump-sum bidding to cover items that traditionally have been bid on a unit price basis. These items include clearing and grubbing, structural excavation, structural concrete, and the like, where estimated and final quanuties can be expected to be essentially the same. In some

cases, it is preferred to have this type of work done on a force-account basis; that is, the work may be done by the contractor, who is reimbursed on a "cost-plus" basis, with the work paid for at actual cost plus a percentage of the actual cost as the contractor's profit. By "actual cost" is meant the cost of labor and materials actually required on that particular operation. The percentage of profit allowed varies. The contractor is compensated for the use of equipment on a rental basis, generally at an agreed-on daily or hourly rate. Small tools are generally excluded in the cost of force-account work.

When large sums are involved, force-account work of this nature requires good supervision. Good practice dictates a close auditing of all work and materials. The audit should include certified copies of payrolls, original re-ceipted bills of materials, and evidence of all other charges, such as freight bills. All force-account work should be checked and approved by both parties daily.

It is possible to have a contract in which the work to be done will include unit price items, a lump-sum bid item, and even provisions for a force-account item.

CONTRACTUAL DOCUMENTS

In order to put a highway construction project under contract, it is necessary to take various steps and to prepare certain documents. The contractual docu-ments generally include (1) the advertisement, (2) instructions to bidders, (3) the proposal, (4) the agreement or contract, (5) plans, (6) the specifications, (7) bonds, and (8) all written or printed agreements and instructions pertaining to the method and manner of furnishing materials and performing the work under the contract.

14-3 The advertisement

The advertisement or notice to contractors is a written notice to inform pro-spective bidders that bids are requested for certain types of work and that a contract may be entered into with the successful bidder. The advertisement is usually inserted in a newspaper or magazine that serves as a medium for supplying such information to persons interested in this particular type of construction. Advertisements or invitations for bids are also mailed to pre-qualified contractors. Highway departments are required by law to provide ample time between publication of the notice to contractors and the taking of bids. This period may vary from two to several weeks, and it should allow contractors sufficient time to prepare their estimates, obtain prices for materi-als, and make essential arrangements necessary for filing a proper bid. On large or complex projects, arrangements often are made for a "prebid confer-ence," a meeting at which details of the work are discussed with prospective bidders.

The advertisement should be brief and still include the information needed by the prospective bidder. The information usually includes the following:

1. A brief description of the type of work and the location of the project.
2. The date, time, and place where bids will be received.
3. Where plans, proposals, and specifications may be seen or obtained and the amount of fee or deposit required if any.
4. Approximate quantities of the work involved.
5. Amount of bid bond, certified check, surety, or security required and provisions for its return to the unsuccessful bidder.
6. Any restrictions related to the qualifications of bidders.
7. Reservations for the rejection of any or all bids or waiver in bidding formalities.
8. A statement regarding minimum or prevailing wage rates and other conditions and regulations pertaining to labor.

14-4 Instructions to bidders

Some state highway departments issue additional "Instructions to bidders," a document in which more information is given about the procedure to be followed in the preparation and submission of bids, details of any formalities required in the acceptance or rejection of bids, and miscellaneous instructions related to the work.

This particular instrument is used to expand the information given in the advertisement, and it is sometimes used instead of an advertisement. When this is done, it is necessary to have all the information given in the advertisement contained in the instructions to bidders. The instructions to bidders may also include requirements for the submission of the bidder's experience record or prequalification data, the time for the starting and completion of the work, and a reference to any legal considerations in regard to the work.

It may be pointed out here that many highway departments omit the instructions to bidders as a separate document but include this information in other contractual documents such as the advertisement, the proposal, and their standard specifications that have been previously adopted and that satisfy any legal requirements included in the instructions to bidders.

14-5 The proposal

The form on which the bidder submits a bid is called the "proposal" and is prepared by the highway or transportation department. The proposal is arranged so that all the items of work and the quantities are tabulated, with spaces for the bidder to insert prices and extensions for each item and a space for the total amount for all items of work. The proposal also contains the statement that the bidder has carefully examined the plans and specifications and is fully informed as to the nature of the work and the conditions related to its performance. The proposal also states that the bidder agrees to do the work in the time proposed and agrees to forfeit his "proposal guarantee" in case his bid is accepted and he fails to enter into a contract. The bidder also agrees to furnish the necessary bonds or security for carrying out the work.

The amount of the proposal guarantee (certified check, bid bond, or other security) required for submission with the proposal and the manner in which it is to be returned are also specified in the proposal. Space may also be provided for the listing of any subcontractors the bidder may require, for the proper signatures of the bidders, and for the address of their place of business.

A proposal generally includes the specifications and plans. As previously mentioned, all state highway agencies have prepared and adopted standard specifications for road and bridge construction. The standard specifications are generally in book form and include general requirements and covenants, items of work, materials, methods of measurement, and payment of quantities. The standard specifications are always referred to in the other contract documents and in this case become a part of the proposal by reference. Due to certain conditions in the design or construction of a particular project, it is often necessary to supplement existing specifications as well as to add certain special provisions necessary for the project. It is quite convenient to add the supplemental specifications and any special provisions to the proposal. This gives the bidder the information needed in order to prepare the bid.

14-6 The agreement or contract

The agreement is the formal part of the contract documents, and from a legal point of view it is the strongest of them all. The advertisement, instructions to bidders, and the proposal are generally classed as contract documents; but they are not, strictly speaking, parts of a contract but are preliminary to it. If any statements of contractual importance are included in any of these documents, they should be made a part of the agreement or contract by reference or they must be repeated in the formal contract.

The contract must be in accord with the form prescribed by law and must contain the following:

1. The declaration of the agreement, names of the parties, their legal residence, and the date of execution of the agreement.
2. The consideration for work to be done, the items of work, quantities, and unit prices with reference to the plans, specifications, and proposal.
3. The time for starting and the date for completion of the work and provisions for damages, if any.
4. Signatures and witnesses.

14-7 Plans

Plans are approved drawings or reproductions of drawings pertaining to the work covered by the contract. They are usually complete in all details before construction is started. The plans, together with the specifications, are an essential part of the contract documents.

14-8 Specifications

The specifications are the written instructions that accompany or supplement the plans and form a guide for standards required in the prosecution of the work. These standard specifications, which are the result of experience and knowledge acquired over a period of years, cover the quality of materials, workmanship, and other technical data. Standard specifications are often revised from time to time in order to keep up with improved processes and the use of different materials. It is sometimes necessary to deviate from the standard specifications, and this is generally done by adding supplemental specifications either to the plans or to the proposal.

14-9 Bonds

All contracts require some form of security as a guaranty of faithful performance of the work. This security may be in the form of collateral deposited with the contracting agency by the contractor, or it may be in the form of surety bonds.

Private individuals may serve as sureties, but as a general rule surety bonds are issued by bonding companies. Bonds issued in connection with highway construction projects are of three common types: (1) bid bonds, (2) performance bonds, and (3) lien bonds.

The bid bond, previously mentioned in connection with the advertisement and proposal, is one form of the proposal guaranty required when the contractor submits a bid. This is the contractor's guaranty that he will enter into a contract if his bid is accepted. In the event that the contractor fails to enter into a contract if his bid is accepted, the surety or bonding company is required to pay the owner damages in the amount specified in the bond. Bid bonds generally range from 5 to 10 percent of the bid amount or 5 to 10 percent of the estimated cost of the project, based on the engineer's estimate prepared for the owner.

Performance bonds, as the name implies, guarantee that the work will be completed as required by the state.

Lien bonds, or "payment bonds," are used by some organizations and guarantee that the contractor will pay all obligations for labor and materials incurred in the performance of the contract.

The amount of bonds required is generally 100 percent of the bid or contract price. Some states require performance bonds only, in the amount of 100 percent of the bid price, with provisions in the bond for the payment of all bills and protection against all liens. Some states require bonds in the amount of 100 percent of the bid or contract price, but make a distribution of 50 percent for performance bonds and 50 percent for lien or payment bonds. Some states require both, each in the amount of 100 percent of the contract price.

Surety bonds are generally furnished before or at the time the contract is executed. When bonds are executed, they require the signatures of the three parties: (1) the owner, (2) the contractor, and (3) the surety.

Surety bonds are contracts between the owner and the surety. Since most highway construction contracts provide for changes and modifications in the contract, it is essential that the surety be informed of any modification of the contract between the owner and the contractor. In case the contractor fails to fulfill the terms of the contract, it becomes the responsibility of the bonding company to do so. This is usually done by letting the contractor complete the work with his own forces under the direction of an engineer supplied by the surety. This method has been found to be of maximum benefit to the owner and to the contractor in that the work is completed with a minimum of delay and expense to the owner and at the same time lessens the loss entailed by the contractor. If this procedure is not followed, the bonding company may let a new contract for the remainder of the work or may elect to pay the penalty and let the owner complete the work, provided that the owner wishes to do so.

14-10 Supplemental agreements

After the contract has been executed and construction work has been started, it is often necessary to make changes in the contract. Conditions commonly arising that require supplemental agreements are the time necessary to complete the work, changes in the quantities of work involved, and the addition of extra work not provided for in the original contract. The standard specifications usually provide for these conditions by granting an extension of time and authorizing any changes or extra work necessary to complete the project. All supplemental agreements must conform to the same basic requirements as the original contract with respect to competent parties, monetary consideration, legality of subject matter, and mutual agreement. All supplemental agreements should be in writing. In addition, unforeseen conditions may force an adjustment in the contract to accommodate these "changed conditions" (see Section 14-14).

14-11 Extension of time

The contract usually designates the time when work is to begin and when it shall be completed. The time to begin work may be designated as a certain number of days after the contract has been executed. The completion date may be a specific calendar date or it may be a specific number of calendar days in which the contractor is to complete the work. In most states it is understood and agreed that time is of the essence of the contract, and failure to complete the contract on time may make the contractor liable for damages. These damages may be a definite amount per day for each day beyond the time required for completion of the work. These are known as "liquidated" damages. In lieu of liquidated damages, the owner may charge actual damages incurred, such as for engineering costs and supervision, maintenance of necessary detours, or any other direct charges caused by the delay. The American Association of State Highway and Transportation Officials recommends a schedule of liquidated damages ranging from $63 to $980 per workday, depending on the size of the contract (1).

From a legal point of view, no work can be done beyond the date of completion, as listed in the contract. It is necessary to make a new agreement or extend the original agreement by requesting an extension of time to complete the contract. This should be done within a reasonable time before the completion date of the contract arrives. If the contractor requests an extension of time due to extra work or other extenuating circumstances beyond his control, such time should be extended without penalty. The agreement extending the time is also a formal document and requires all the necessary elements of a legal contract. When properly executed, it is as binding as the original contract.

14-12 Authorizations for changes in the work

It is impossible to determine exact quantities for some items of work (e.g., cubic yards of excavation) when designing a project. When a contractor bids on the various unit price items, as set up in the proposal, he is reasonably certain that the final quantities may be increased or decreased from the bid amount. To make the necessary adjustment from the quantities as stated in the contract to the quantities as actually constructed, written authorizations are made in accordance with methods of procedure set forth in the specifications.

On major items where large quantities are involved, some specifications provide for an increase of the unit price if the quantities are drastically reduced, and, conversely, if the quantities are materially increased, the unit price may be decreased. Some states require an adjustment of the unit price when the quantities vary more than 20 or 25 percent from estimated quantities and when the value of the work exceeds a certain sum.

14-13 Authorization for extra work

Some states make a distinction between a change authorization and an extra authorization. A "change authorization" may be defined as one in which there is a change in the quantities at the unit price as listed in the contract, while an "extra authorization" is for extra work for which no unit price is listed. The unit price for extra work is usually agreed on at the time the extra work is discussed, and it should always be approved before any extra work is started.

14-14 Changed conditions

Highway contracts may include a "changed conditions clause" to allow for alteration of plans or character of work provided such alteration falls within the general scope of the contract. For example, specifications with such a clause would allow the agency to make changes in alignment or grade of the road or structure. Unless the alterations materially change the character or cost of the work to be performed, the altered work is paid for at the same unit prices as other parts of the work. If the character or unit costs of the work are materially changed, the specifications may permit the engineer to make an allowance in pay for the changed conditions and to extend the contract time.

AASHTO suggests (1) the following optional changed conditions clause:

> Should the contractor encounter or the department discover during the progress of the work subsurface or latent physical conditions at the site differing materially from those indicated in this contract, or unknown physical conditions at the site of an unusual nature, differing materially from those ordinarily encountered and generally recognized as inherent in work of the character provided for in the contract, the engineer shall be promptly notified in writing of such conditions before they are disturbed. The engineer will thereupon promptly investigate the

conditions and if the engineer determines that they do materially differ and cause an increase or decrease in the cost of, or the time required for, the performance of the contract, an equitable adjustment will be made and the contract modified in writing accordingly.

14-15 Subcontract

The use of a subcontract in highway work is often necessary. A subcontract may be defined as a contract for the performance of a part or all the work previously contracted for. When this is done, the subcontract should have all the elements of a formal contract. The scope of the work to be performed, the compensation to be received, and the terms and conditions of the original contract with which the subcontractor must comply should be clearly stated. The owner holds the original contractor directly responsible for the work of the subcontractor. The subcontractor, however, has the legal right to hold the owner responsible for any default of payment due him for labor or supplies by the contractor. The owner usually protects himself by having the contractor furnish the necessary bonds that guarantee payment of all labor and materials.

On state highway contracts the amount of work that a contractor may sublet is usually limited; in some cases, it is up to 50 percent of the value of the work contracted for. State highway and transportation departments usually reserve the right to approve or disapprove a subcontract, and they may require information in regard to the qualifications of the subcontractor and his ability to do the work. Copies of all subcontracts, including prices, are usually required before a subcontract may be approved.

14-16 Prequalification of contractors

On most public works, contracts are awarded to the lowest responsible bidder and it is the problem of the state agency to determine the responsibility of the lowest bidder after the bids are taken. This is necessary because contractors with limited financial resources and limited experience may be competing with reliable contractors for an award of a contract. Such an award to an unqualified contractor often results in unnecessary delay, unsatisfactory work, and sometimes in failure to complete the work, with the corresponding extra expense to the state agency requesting bids. In order to avoid or reduce these dangers, many states have adopted prequalification laws and regulations. The fundamental purpose of the prequalification requirements is to determine beforehand whether or not the contractor is responsible and competent to bid on a project and to permit only those properly qualified to do so. A typical procedure followed in prequalifying contractors is described below.

To comply with prequalification requirements, the prospective bidder is requested to submit a comprehensive experience and financial statement. From the information thus supplied, the prospective bidder is given a numerical rating that indicates the maximum amount of work in dollar volume he is permitted to have at any one time, and is further classified for the various types of work he is permitted to bid on. The major rating factors for determining a prospective bidder's qualifications are his experience record, the amount of equipment he owns, and his financial resources at the time of filing his statement.

The numerical rating and classification required for each project may be given in the advertisement. Prequalification of contractors is highly desirable in that there is more time to investigate the bidders' qualifications beforehand, and in that it saves the time and expense of contractors who are not qualified to do the work. The legality of prequalification has been established by the rulings of the various courts. Opponents of prequalification object to the lengthy questionnaires and the time and expense of preparing statements, and they state that it restricts competition to those with large resources. Additional objection to prequalification based on equipment owned is raised when suitable equipment may often be rented.

14-17 State-and-federal agreements

On state projects in which the federal government is to participate in the cost, certain conditions and agreements are necessary to qualify for these funds. In the apportionment of these funds for highway purposes, each state is required to submit for approval a program of projects to be constructed on a predetermined system of federal-aid highways, the total cost of which must not exceed that state's apportionment. Plans and specifications are then prepared and a detailed estimate based on current prices is submitted to the Federal Highway Administration for approval. Approval of this agency is required before the project is advertised and before the award is made to the low bidder. After the award is made by the state to the contractor, a detailed estimate is made using contractor's unit bid prices, and it is upon this estimate that the "project agreement" is to be based. The contractor's bid, plus a contingency not to exceed 10 percent of the bid, is usually the amount of the agreement. This project agreement is the formal document or contract between the state and the federal government. There is one other formal agreement that may be mentioned here and that is the "maintenance agreement." In this document the state agency agrees to the maintenance and upkeep of the project after it is constructed. The maintenance agreement is usually made after plans are completed for the particular project and before federal approval is given to advertise the project.

14-18 Mechanics of bidding procedure

When plans, specifications, and estimates are completed and the necessary approvals are obtained, the project is ready for advertising. It is always preferable to have several projects ready for advertising at the same time. The advertisements are written with the required information previously discussed, and the date is set in accordance with statute requirements. Plans and proposals are then requested by those desiring to submit bids on the projects advertised. Bids may be mailed or submitted in person at the designated time and place. At large lettings, when there are many projects and bidders, much detailed work is necessary in order to check the many details that arise when bids are received.

The sealed bids are opened, the bid security is checked, and the proposed details are scrutinized for irregularities. The total amount of the bid is then read aloud. Any irregularities found before the reading of the bid are usually a cause for rejection. For example, when bid bonds are requested to accompany the proposal, the bidder may neglect to submit his bid bond or it may not be in the amount requested. The omission of a unit price may also be a cause for rejection. Rejection at this point is desirable in that the bid security can be returned without undue delay.

After all bids are read aloud and the apparent low bidders are determined, all bids are checked for errors. The bids are then tabulated in the order of first, second, third, and so forth. After the low bidder is determined on each project, the detailed unit prices of the successful bidder are then read aloud. Electronic computers are also being used successfully in this phase of the work of the state highway departments, with great savings in time and manpower. The contract is then prepared, bonds are furnished by the successful bidder, and an award is made within a reasonable time. On federal-aid projects concurrence in the award of a contract to a contractor must also be received from the federal agency. When the contractor furnishes the necessary bond, arrangements are made for the return of the other bidders' security. A majority of states return the bid security of all bidders except the lowest two or three immediately after a letting. Up to 30 days is sometimes required after the low bidder has been determined in order to make an award of a contract.

14-19 The unbalanced bid

On unit price contracts it is expected that each bid item will carry its proportionate share of the cost as well as the profit. It is possible for a contractor to raise certain unit prices on some items and lower unit prices on other items without changing the total of his bid. The result is usually an unbalanced bid. Generally speaking, unbalanced bids are undesirable and, when detected, are usually a cause of rejection. An example of an unbalanced bid may be shown in the bidding of the following items:

	Balanced		Unbalanced	
	Unit Price	Amount	Unit Price	Amount
12,000 cu yd earth excavation	$2.00	$24,000	$ 1.00	$12,000
500 cu yd rock excavation	6.00	3,000	30.00	15,000
Total cost		$27,000		$27,000

Assuming that when final quantities were determined the amount of earth excavation was 10,000 cu yd and the rock excavation was 2500 cu yd, the final cost would be as follows:

	Balanced		Unbalanced	
	Unit Price	*Amount*	*Unit Price*	*Amount*
10,000 cu yd earth excavation	$2.00	$20,000	$ 1.00	$10,000
2,500 cu yd rock excavation	6.00	15,000	30.00	75,000
Total cost		$35,000		$85,000

14-20 Right-of-way agreements

Contracts between a state highway agency and a property owner are often necessary in order to obtain the right-of-way for highway construction purposes. These contracts may be for the outright purchase of the land needed or as an easement for necessary grading operations. Most of these agreements are negotiated agreements between the property owner and a representative of the state highway agency.

SUPERVISION OF CONSTRUCTION

After a contract is awarded, arrangements are made by the contractor to move equipment and personnel to the job site in order to start construction operations. The time it takes to do this will depend on the distance involved. However, the contractor must comply with contract requirements, which usually state that construction work must start within so many days of the award of the contract. The manner in which supervision of construction is carried out is described below and is considered representative of the general practice of many highway agencies.

14-21 Project supervision

The work of the project conducted under the supervision of a state highway or transportation department is usually under the administration of the project engineer. It is the engineer's duty to see that all phases of the work are carried out in accordance with plans and specifications. The engineer receives instructions from, and is usually responsible to, a district or division engineer. The project engineer is assisted by a number of trained persons, including a survey crew, inspectors, office engineer, and others. The number of people obviously depends on the length, type, and complexity of the project.

On certain federal-aid projects, FHWA engineers make periodic inspections of the work as it progresses, as well as a final inspection. On projects performed under "certification acceptance" procedures (see Section 2-2), routine inspections and audits are performed by state personnel, but FHWA engineers make a final inspection of the work. Their approval is required for acceptance of all federal-aid highway projects.

14-22 Duties of the project engineer

It is necessary that the project engineer secure a high standard of work, and this can be done only by observing whether the contractor complies with the contract and specifications. This function is usually delegated to the various inspectors on the project who are responsible to the project engineer.

In recent years, congressional investigations and other critical reviews of the highway program have focused attention on the fact that absolute compliance with typical plans and specifications is neither desirable nor necessary in this type of work. Many state highway agencies are including in their specifications the "substantial compliance" clause recommended by AASHTO *(1)*. This clause reads as follows:

> All work performed and all materials furnished shall be in reasonably close conformity with the lines, grades, cross sections, dimensions, and material requirements, including tolerances, shown on the plans or indicated in the specifications.
>
> Plan dimensions and contract specification values are to be considered as the target values to be strived for and complied with as the design values from which any deviations are allowed. It is the intent of the specifications that the materials and workmanship shall be uniform in character and shall conform as nearly as realistically possible to the prescribed target value or to the middle portion of the tolerance range. The purpose of the tolerance range is to accommodate occasional minor variations from the median zone that are unavoidable for practical reasons. When either a maximum and minimum value or both are specified, the production and processing of the material and the performance of the work shall be so controlled that material or work will not be preponderantly of borderline quality or dimension.
>
> In the event the Engineer finds the materials furnished, work performed, or the finished product not within reasonably close conformity with the plans and specifications but that reasonably acceptable work has been produced, the Engineer shall then make a determination as to whether the work shall be accepted and remain in place. In this event, the Engineer will document the basis of acceptance by contract modification which will provide for an appropriate adjustment in the contract price for such work or materials necessary to conform to his determination.
>
> In the event the Engineer finds the materials or the finished product in which the materials are used not within reasonably close conformity with the plans and specifications, but that reasonably acceptable work has been produced, the Engineer will make a determination as to whether the work will be accepted and remain in place. In such event, the Engineer will document the basis of acceptance by a contract modification which will provide for an appropriate adjustment in the contract price for such work or materials as the Engineer determines to be appropriate, based on engineering judgment.

In the event the Engineer finds the materials or the finished product in which the materials are used are not in reasonably close conformity with the plans and specifications and have resulted in an unsatisfactory or unacceptable product, the work or materials shall be removed and replaced or otherwise corrected by and at the expense of the Contractor in a manner satisfactory to the Engineer.

It is the project engineer's duty to see that all engineering work, such as the setting of stakes for alignment and grade, is performed well in advance of the contractor's work so that there will be no delay. It may be necessary to effect a change in design during the progress of the work, and each state defines the procedure to be followed in such matters.

The necessary reports required by the state highway or transportation department and the Federal Highway Administration are the responsibility of the project engineer. These reports include the preparation of estimates for partial payments and also for final payment.

As a representative of the highway department on a project, the project engineer must keep others properly informed as to the progress of the work; for instance, municipalities and utility companies must frequently be contacted on most construction projects. The proper authority should be informed of any additional borrow requirements or on matters involving disputes in regard to the right-of-way. Good administration of a project usually results in a well-constructed project.

14-23 Project engineer – contractor relationship

The highest standards of work are developed when there is complete cooperation and accord between the contractor's employees and those of the state highway agency. It must be remembered that the contractor is entitled to a fair profit and is due every consideration consistent with the specification requirements and quality of work demanded of him.

In case of a claim by the contractor, a careful record should be made of all work in question and the information made available to those responsible for the adjustment of claims.

The contractor should make available at all times any information that may be required, such as transcripts of payrolls and receipted invoices. The project engineer should see that the contractor complies with all regulation regarding labor, equipment, and so on.

14-24 Construction and inspection details

All materials used on project must be inspected for compliance with applicable specifications. These materials may be inspected at the site of the project, or if quantities are sufficient to warrant it, they may be tested and approved before shipment is made to the project. When this is done, the testing is frequently done by an approved commercial testing laboratory. This does not mean that materials may not be rejected if they become damaged during shipment and cannot comply with specifications or if a defect was not detected in the inspection at the source.

Most state highway agencies maintain well-equipped testing laboratories of their own. The contractor is required to submit samples of the materials to be used, for testing purposes, and to name the sources of the materials. For concrete and bituminous paving mixtures, proportioning of the various ingredients is usually under laboratory control and the various details are carried out by inspectors assigned to the project.

The number and kind of inspectors needed for each project will depend on the kind and type of work to be done. The project may require the services of a grade inspector, a culvert inspector, a gravel plant inspector, a plant inspector for the preparation of various types of plant mixtures, and others. The inspector not only inspects the materials but also exercises control over the manner in which the work is done. The importance of the work of the inspector should never be underrated.

14-25 Quality control of materials and construction

Quality control of materials and workmanship by tests on small samples has been the traditional practice in highway construction. However, interpreting the results of such tests has been complicated by the wide variability of the test results. Certain variations in the quality of a highway component obey the laws of probability and occur strictly by chance. Other variations are systematic and can be attributed to assignable causes such as differences in equipment, materials, construction methods, testing conditions, and so forth.

State highway agencies are increasingly turning to the science of statistics to understand the variability of quality control tests. Statistical concepts can be used to establish specifications in terms of maximum tolerances from an established "target" value. Research studies have shown that quality control test data tend to be grouped symmetrically about the mean value, and distribution of the data can generally be satisfactorily described by the familiar bell-shaped normal curve. (See Figure 14-1).

A graphical procedure, employing a *control chart* is recommended for

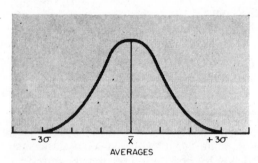

FIGURE 14-1 Distribution of chance variations in a test sample measure of quality. (Source: *Public Roads*, Vol. 35, No. 11, December 1969.)

statistical quality control of highway construction. To construct a control chart, samples of some highway characteristic are randomly taken and tested. Examples of test results include subgrade densities, the percentages of asphalt in pavement mix samples, the fractions of aggregate passing a certain sieve, etc. As Figure 14-2 illustrates, the test values are plotted as ordinates along a horizontal scale showing time or unit of production.

When charts of individual test values are prepared, upper and lower control limits are usually established 3 standard deviations* away from a central or target value $\overline{X}$.

$$\text{upper control limit, UCL} = \overline{X} + 3\sigma \qquad (14\text{-}1)$$

$$\text{low control limit, LCL} = \overline{X} - 3\sigma \qquad (14\text{-}2)$$

where σ is the standard deviation. When averages of n samples per lot are used as the control statistic, the control limits are

$$\text{UCL} = \overline{X} + \frac{3\sigma}{\sqrt{n}} \qquad (14\text{-}3)$$

$$\text{LCL} = \overline{X} - \frac{3\sigma}{\sqrt{n}} \qquad (14\text{-}4)$$

Consider the example illustrated by Figure 14-2. The test characteristic is the percentage of aggregate in a bituminous mix passing the No. 30 sieve, and the mean or target value is 38.0 percent. If the standard deviation $\sigma = 1.93$ and $n = 5$,

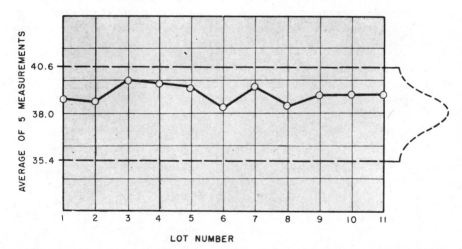

FIGURE 14-2 Typical control chart. (Source: *Public Roads*, Vol. 35, No. 11, December 1969.)

*Some engineers prefer a control limit of 2.33 standard deviations, which corresponds to a 98 percent probability interval.

$$UCL = 38.0 + \frac{3 \times 1.93}{\sqrt{5}} = 40.6$$

$$LCL = 38.0 - \frac{3 \times 1.93}{\sqrt{5}} = 35.4$$

As long as the plotted test values fall within the control limits and are randomly distributed about the average, the process is said to be in control. If the points fall outside the control limits or are consistently above or below the average line, the reasons for the irregularities should be investigated. Granley (2) has published the following warning signals to indicate processes that are out of control:

Individual charts
1 test value greater than 2.33 standard deviations from centerline
3 consecutive test values greater than 1 standard deviation above or below the centerline
11 consecutive test values on same side of centerline job mix target

Average and range charts
1 average or range value outside control limit lines on R and $\bar{X}$ chart
2 consecutive averages outside 2σ limits above or below the centerline
7 consecutive values on either side of centerline for $\bar{X}$ charts or above centerline for R charts

---

Control charts can be constructed from test data in the form of individual measurements, averages, standard deviations, and ranges (R charts) or some combination of these. Some states employ the moving average of the five most recent individual test results as the control statistic.

> Diagnosis of the reason for the process being out of statistical control is an engineering consideration. The chart can only point out that something is wrong . . . . The use of control charts will not alter the fact that the engineer has the final decision in accepting construction. The charts, however, are effective tools to visually forewarn that undesirable trends may be developing and to help the engineer and contractor decide when to take action and when not to take action. (2)

Additional information on the important topic of statistical quality control of highway construction is contained in Ref. 3.

14-26 Construction surveys
The project engineer and his party are usually assigned to a construction project at about the same time that the plans and specifications are completed and when it is reasonably certain that bids will be received within a short time.

This permits the construction party to do necessary engineering work before the contractor arrives.

The first problem in construction surveys is to check the alignment to see if it is in accord with the final plans. In many cases this means reestablishing the centerline of the entire project. It is also necessary to check all witness points of alignment; and if there is a possibility that these witness points may be in the way of construction, they should be reestablished beyond construction limits. Consideration should also be given to the establishment of additional reference points on the road centerline to facilitate the instrument work when setting stakes for alignment and grade.

All bench marks should also be checked for location and elevation. Any benches that have been damaged or destroyed should be reestablished. All benches should be outside the limits of construction.

14-27 Staking requirements

After the road centerline and the bench marks have been checked or reestablished, the major portion of the survey work consists of setting stakes for the guidance and control of construction operations. All stakes that are set by the engineer should be marked in such a manner that they can be easily interpreted.

The first stakes that are set are usually clearing and grubbing stakes. The distances out from the road centerline are generally determined by using the plans and scaling the distances to the limits of construction. The stakes may consist of small saplings cut to 4- or 5-ft lengths, with a blaze mark facing the road centerline. These stakes are usually placed at even 100-ft intervals.

14-28 Cross sections and slope stakes

The construction survey party is responsible for all earthwork calculations. It is therefore important that they take cross sections over the entire project in many cases. Such a procedure may not be necessary in cases where cross sections (and slope stake positions) are determined from aerial photographs and by the use of computers, as discussed in Chapter 13. The original cross sections provide a record of original ground surface elevations that can never be obtained again after construction has started. The cross sections taken on the project that served as a basis for the final design may not be suitable. The period of time between the taking of the cross sections for design purposes and letting a contract for construction may vary from several weeks to two years or more. Taking cross sections just before construction operations begin eliminates one of the many possible disputes over earthwork quantities.

A slope stake is set at the point where the proposed fill slope or cut slope intersects the original ground. The stakes may be set when cross sections of the original ground are taken. Slope stakes for grading are usually set on both sides of the road at 50-ft intervals at the station and half-station points. All information necessary for grading operations should be shown on the stake. It is sometimes necessary to offset slope stakes. When this is done, the amount of offset, which is the distance from the offset location of the stake to its intended location, is noted on the stake. The cut or fill reading is taken from the ground location of the stake, as set. Figure 14-3 shows typical cross sections and positions of slope stakes.

When setting slope stakes, the cut or fill is usually marked on the center-line stakes to serve as an aid to grading operations. Many states set slope stakes in heavy cuts only and control grading operations in other areas by the use of earth-grade stakes.

14-29 Earth-grade stakes

When the rough grading operations have been completed, it is generally necessary to set finishing grade stakes to control the finishing of the earth grade. These stakes are generally set on the shoulder line at every station and half station. Some operators require that the finish grade stakes be set on the road centerline instead of the shoulders. The grade stakes may be set in such a manner that the top of the stake is at grade. The stake may be left high and the amount of cut or fill noted from a keel mark on the stake.

14-30 Final cross sections

When grading operations are complete, it is necessary to take final cross sections of the completed work in order to determine the final quantities of

FIGURE 14-3 Sketch showing slope stakes.

excavation and embankment completed on the project. The final cross sections should be taken at the same points as the original sections. Final cross sections may be taken as soon as grading operations are completed. This permits the calculation of earth quantities as the work progresses and generally shortens the time necessary for the preparation of the final estimate when the project is completed.

14-31 Borrow-pit measurements

When additional earth for embankment sections must be obtained from borrow pits, it is necessary to take cross sections before and after the removal of earth in order to determine the quantities used. The area from which the earth is to be excavated should be staked with one or more baselines, and cross sections should be taken at 25-ft intervals. The base lines should be well referenced so that they may be relocated for the final cross sections. Sketches of borrow-pit layout and reference lines should always be made. After the borrow earth has been removed, cross sections are again taken on the original lines. Both original and final cross sections are plotted, and the volume of earth removed is determined by the average-end area method or some other acceptable method. Figure 14-4 shows a typical method of staking a borrow pit for cross-sectioning. Quantities of borrow material may also be determined by weighing loaded hauling units.

14-32 Side borrow

Most borrow pits are located at some distance from the roadway. Sometimes it may be feasible to obtain additional earth from the side of the roadway in large cuts by extending the side slopes. In order to obtain an accurate measurement

FIGURE 14-4 Method of staking and cross-sectioning a borrow pit.

of the earth, sections should be taken at 25-ft intervals, just as in borrow pits, and the method of computing volumes may be the same.

14-33 Grade and line stakes for pavements

The first stakes required for a surface paving project are generally for trenching or for the establishment of a uniform grade in accordance with the plan centerline profile. These stakes may be placed on the side of the road and may be used later for form stakes when placing concrete pavement. The form stakes are usually placed 2 ft from the edge of the slab unless otherwise requested by the contractor. The stakes should be set 50 ft apart on tangents and 25 ft apart on horizontal and vertical curves. Form stakes are placed on both sides of the road. On many projects, final finishing of the subgrade, base, or subbase is done by a machine with automatic controls operating on a taut wire stretched between stakes or pins; the field survey party must establish and check the position of the wire.

14-34 Culvert stakes

The plans give the size, location, and lengths of all culverts on the project. They also give the elevations of the flow line and the angle of crossing of the road centerline. Field conditions should always be checked to see whether or not the proposed installation will be satisfactory. Figure 14-5 shows two typical culvert installations, one at right angles to the centerline and the other at an angle of less than 90°.

The stakes required for a culvert installation consist of the centerline stakes of the culvert and reference stakes that usually indicate the length of the culvert. The flow-line elevations are generally given on the centerline reference stakes, as indicated in Figure 14-5. All stakes should be offset so they will not be disturbed during construction. On large box culverts the same method of staking is used.

14-35 Miscellaneous engineering work

The engineer may be required to furnish stakes for the location of manholes, catch basins, and inlets, with the necessary elevations for flow lines and tops of covers. Curb-and-gutter stakes and sidewalk stakes showing grade and alignment may also be required.

14-36 The preparation of estimates

As the work progresses, the contractor is entitled to partial payment for the work completed. These payments are generally shown on forms called "estimates" and are prepared by the project engineer. Estimates may be made monthly or bimonthly. On lump-sum projects, the amount of the estimate will be a percentage of the whole. On unit price contracts, the individual items of work completed are listed and the value of the work is computed.

It is sometimes possible to pay for materials delivered to the site of the project or furnished for later use, such as gravel (stockpiled), reinforcing steel,

FIGURE 14-5 Method of staking culverts.

and so forth. However, such a procedure should be specified in the proposal or specifications, or definite means for providing authorizations should be made.

On partial payments the full amount is not always paid to the contractor, but a percentage, usually 10 percent, is retained. This protects the owner in case of an underrun of estimated quantities, as a guaranty for faulty work, or for minor claims for materials. On large contracts, 10 percent sometimes results in a large amount of capital being tied up for reserve purposes. Some highway departments arrange a reduction of this reserve to 5 percent when the work has been completed and accepted and pending the preparation of the final estimate. When the final estimate is complete (and certain delays in office procedure delay its processing), this reserve may be reduced to a small lump sum.

In the preparation of the final estimate, all items of work are measured as constructed in the field. If these quantities differ from those in the contract as authorized and explained under changes in the work, it is necessary to prepare authorizations that decrease or increase the quantities, as the case may be.

14-37 Final acceptance

When all work has been completed, inspection of the whole project is made by the authorized party for the highway agency. If the project is one in which federal funds were used, the final inspection is made by an engineer of the Federal Highway Administration. When all parties indicate their acceptance, the final estimate, which is usually in the process of completion, is also approved, and recommendation for final payment is made.

During construction, it is the responsibility of the project engineer to supervise all phases of the work, approve or disapprove all authorizations for changes or extra work, and see that the work follows the schedule of operations as set up in the contract. As no work can be done beyond the completion date, the project engineer must notify the contractor in due time and prepare to extend the contract by supplemental agreements, such as an extension of time. When all the work is completed, the recommendations of the project engineer as to the contractor's performance, quality of work, and cooperation are necessary for the contractor's rating factor, as mentioned under the prequalification of contractors. The preparation of weekly and monthly reports, a history of the project, and administrative functions are always incidental to the supervision of a construction project.

PROBLEMS

14-1   At a central mixing plant, tests are made on asphaltic mix yielding the following consecutive percentages of asphalt: 6.1, 5.5, 7.1, 6.3, 5.8, 6.6, 6.3, 5.9, 6.9, and 7.3. Each of these values represents an average of three samples per lot. The central or target value is 6.3 percent. If the standard deviation of such tests is 0.5 percent, plot a control chart and determine if the process is in control.

14-2   A nuclear density gauge is used to determine the density of an embankment, yielding the following consecutive individual values: 110, 110, 114, 106, 103, 102, 112, 115, 107, and 118 lb/ft^3. The central or target value is 108 lb/ft^3, and the standard deviation is 2.0. Plot a control chart and determine if the process is in control.

REFERENCES

1. *Guide Specifications for Highway Construction.* American Association of State Highway and Transportation Officials, Washington, D.C. (1984).

2. Granley, Edwin C. Quality Assurance in Highway Construction: Part 6—Control Charts. *Public Roads,* Vol. 35, No. 11 (1969).

3. Willenbrock, Jack H. *Statistical Quality Control of Highway Construction,* Vols. I and II. U.S. Department of Transportation, Federal Highway Administration, Washington, D.C. (1976).

chapter fifteen

HIGHWAY MATERIALS

Each year in the United States, enormous quantities of construction materials are used for improvements to the public roadway system. Such projects require annually over 370 million tons of aggregates, 8 million tons of bituminous materials, and 10 million tons of cement, as well as vast quantities of steel, lumber, explosives, and petroleum products *(1)*.

In this chapter, we briefly describe some of the physical characteristics and quality control tests for soils, aggregates, bituminous materials, and portland cement. Detailed material specifications and tests for these and other highway construction materials have been published by the American Association of State Highway and Transportation Officials *(2)*.

SOILS

Basically, the engineering definition of the word "soil" is a very broad one. Soil might be defined as all the earth material, both organic and inorganic, that blankets the rock crust of the earth. Practically all soils are products of the disintegration of the rocks of the earth's crust. This disintegration, or "weathering," has been brought about by the action of chemical and mechanical forces which have been exerted on the parent rock formations for countless ages. Included among these forces are those of wind, running water, freezing and thawing, chemical decomposition, glacial action, and many others.

Soils may be described in terms of the principal agencies responsible for their formation and position. Thus a *residual soil* is one that, in its present situation, lies directly above the parent material from which it was derived. Soils formed by the action of wind are known as *aeolian soils*; a typical example of a windblown soil deposit is seen in the very considerable deposits of loess in the Mississippi Valley. *Glacial soils* occur in many parts of the United States. An example of such a soil deposit is a glacial till, which is a deposit of tightly bonded materials containing particles ranging in size from boulders down to very finely divided mineral matter.

Soils formed through the action of water are generally termed *sedimentary soils*. Typical sedimentary soils are formed by the settling of soil particles (or groups of particles) from a suspension existing in a river, lake, or ocean. Sedimentary soils may range in type from beach or river sands to highly flocculent clays of marine origin.

Soils may also be described in terms related to the amount of organic material contained in them. Soils in which the mineral portion (soil particles) predominates are properly called *inorganic soils*. Those in which a large amount of organic matter is contained are called *organic soils*. Organic soils are usually readily identified by their dark brown to black color and distinctive odor.

One of the most important facts regarding soils and soil deposits is their normal lack of homogeneity. Due to the more or less random process of their formation, soils vary greatly in their physical and chemical composition at different locations over the surface of the earth. Although, generally speaking, soils derived from the same parent material under similar factors of geographic location, climate, and topography will be similar wherever they are found, several soil types may, and usually do, exist within a comparatively small area. Soil deposits are also characteristically varied with depth. Deposits of sedimentary soils in which layers of varying fineness were created during the process of their formation are said to be stratified, and some degree of stratification must be expected in any sedimentary soil deposit.

GENERAL SOIL TYPES

Certain general soil types are of importance to the highway engineer. No attempt will be made in this section to cover all the soil types that the engineer may encounter. However, certain soil types occur generally throughout the country and will be discussed here, along with the terms which are used in describing them.

15-1 Classification by grain size

It is possible to classify soils into several principal groups and many subgroups on the basis of their grain size. By grain size is meant the average dimension (or diameter, in a general sense) of a soil particle contained within a soil mass. Principal divisions of any classification system based on grain size are usually gravel, sand, silt, and clay. Several different classification systems have been proposed, and one in general usage among highway engineers is shown in Figure 15-1.

It must be emphasized that although in many cases a knowledge of the distribution of grain sizes in a soil is valuable to an understanding of soil action, this knowledge alone is not enough to give an accurate picture of the properties of a soil and its behavior under various conditions. For example, a fine silt or a rock flour may have a large percentage of particles that fall within the range of clay sizes and yet have entirely different properties from those exhibited by an inorganic clay having about the same range of particle sizes. Similarly, two sands composed of particles of about the same size may be radically different in such a property as shearing resistance if one is a beach sand composed of smooth, rounded grains and the other is composed of grains that are sharp and angular. It should therefore be recognized that the terms "clay" and "silt," at least, carry meanings that go beyond mere reference to particle size; and that for any soil, additional information must be given if the description is to prove adequate from an engineering viewpoint.

Obviously, many soil deposits will contain particles that fall into two or

*Materials smaller than 0.001 mm are called colloids.

FIGURE 15-1 Grain-size classification. (AASHTO Specification M146)

more of the groups described above. Thus a soil might contain percentages of sand, silt, and clay, as defined by the chart of Figure 15-1. Terms such as "sandy clay," "silty clay," "silty sand," and many others have been employed to describe such soils.

15-2 Common soil types

Sand and *gravel* are coarse-grained soil types possessing little or no cohesion. They are readily identified by visual inspection and are distinguished, generally speaking, by their relative stability under wheel loads when confined, by their high permeability, and by their failure to shrink or expand in detrimental amounts with change in moisture content. The term *gravel* is usually applied to natural pit, river, or bank gravels consisting largely of rounded particles; *crushed gravel,* or *crushed stone,* is the term applied to the products of crushing larger rocks into gravel sizes. The term *dirty sand* is frequently applied to sands which contain small amounts of fine material.

Silt is the term applied to fine-grained soils of low to medium plasticity, intermediate in size between sand and clay. Silts generally possess little cohesion, undergo considerable shrinkage and expansion with change in moisture content, and possess a variable amount of stability under wheel loads. If they contain large percentages of flat, scalelike particles, such as mica flakes, they are likely to be highly compressible and somewhat elastic in nature. *Organic silts* contain appreciable amounts of decomposed organic matter and are generally highly compressible and unstable.

Clays are distinguished by the occurrence of very fine grains, since many clays are principally composed of particles that are colloidal in size. Clays generally possess medium to high plasticity, have considerable strength when dry, undergo extreme changes in volume with change in moisture content, and are practically impervious to the flow of water. *Lean clay* is the term given to silty clays or clayey silts, while fine, colloidal clays of high plasticity are called *fat clays.* Clays may be further distinguished by the fact that, although they may possess considerable strength in their natural state, this strength is sharply reduced and sometimes completely destroyed when their natural structure is disturbed, that is, when they are remolded.

Loam is an agricultural term used to describe a soil that is generally fairly well graded from coarse to fine, that is easily worked, and that is productive of plant life. This name frequently appears in engineering literature in combination with other terms. Thus a soil may be called a *sandy loam*, a *silty loam*, or a *clay loam*, depending on the size of the predominating soil fraction.

Loess is a fine-grained aeolian soil characterized by its nearly uniform grain size, predominantly silt, and by its low density. Highway cuts through loess deposits usually resemble those made in rock, in that this soil will stand on a nearly vertical slope, while it is readily eroded by rainwater if flatter slopes are used.

Muck is soft silt or clay, very high in organic content, which is usually found in swampy areas and river or lake bottoms.

Peat is a soil composed principally of partially decomposed vegetable matter. Its extremely high water content, woody nature, and high compressibility make it an extremely undesirable foundation material.

BASIC SOIL PROPERTIES

To have an understanding of soil action, an engineer must be familiar with certain basic soil properties. We are all familiar with the basic properties of other engineering materials, such as steel, wood, and concrete. A soil engineer must have similar knowledge relative to soils. The job of the soil engineer is complicated by the fact that many soils are quite complex in nature, both physically and chemically, and that soil deposits are likely to be extremely heterogeneous in character. It must be remembered that the properties of any given soil depend not only on its general type but also on its condition at the time when it is being examined.

15-3 Moisture content

Water is an extremely important constituent of soils. The moisture content is defined as the weight of water contained in a given soil mass compared with the oven-dry weight of the soil and is usually expressed as a percentage. In the laboratory, moisture content is usually determined by selecting a small, representative sample of soil and placing it in a "tared" container, such as a pair of matched watch glasses or a moisture can. The weight of the "wet" soil sample and container is determined, and the sample is then brought to constant weight by drying in an oven at a temperature between 100° and 110°C. After constant weight has been attained, the sample is taken from the oven, cooled in a desiccator, and the oven-dry weight of the sample is determined. All weights are recorded in grams, and the following expression is used to determine the moisture content:

$$w(\%) = \frac{W_1 - W_2}{W_2 - W_c} \times 100$$

$$(15\text{-}1)$$

where

$w(\%)$ = moisture content (percent)
W_c = tare weight of the container (g)
W_1 = weight of "wet" soil and container (g)
W_2 = weight of oven-dry soil and container (g)

If the void spaces in a soil are completely filled with water, the soil is said to be saturated. The moisture content of a soil may then be 100 percent or more, as might be the case in a saturated clay, muck, or peat soil.

Water in soils may be present in its normal liquid form, as when filling or partly filling the voids of a sand mass, or it may be present in the form of adsorbed water existing as films surrounding the separate soil particles or groups of particles, as in the case of the water remaining in a partially dried clay mass. The water films existing in the latter case may have properties sharply differing from those exhibited by water in its normal form. Properties of fine-grained soils are greatly dependent on the properties and behavior of the adsorbed water films.

15-4 Specific gravity

By the term "specific gravity" (G), as applied to soils, is meant the specific gravity of the dry soil particles or "solids." The specific gravity is frequently determined by the pycnometer method, the determination being relatively easy for a coarse-grained soil and more difficult for the finer soils. Values for the specific gravity refer to the ratio of the unit weight of soil particles to the unit weight of water at some known temperature (usually 4°C) and range numerically from 2.60 to 2.80. Within this range, the lower values for the specific gravity are typical of the coarser soils, while the higher values are typical of the fine-grained soil types. Values of the specific gravity outside the range of values given may occasionally be encountered in soils derived from parent materials that contained either unusually light or unusually heavy minerals.

15-5 Unit weight

The unit weight of a soil is the weight of the soil mass per unit of volume and is expressed in pounds per cubic foot. As commonly used in highway engineering, the term "wet unit weight" refers to the unit weight of a soil mass having a moisture content which is anything different from zero, while "dry unit weight" refers to the unit weight of the soil mass in an oven-dry condition. The wet unit weight, dry unit weight, and moisture content are related by the following expression:

$$\text{dry unit weight} = \frac{\text{wet unit weight}}{\dfrac{(100 + w\%)}{100}} \qquad (15\text{-}2)$$

The wet unit weight of a soil may vary from 90 lb/ft^3 or less for saturated, organic soils to 140 lb/ft^3 or more for well-compacted granular materials.

15-6 Shearing resistance

Failures that occur in soil masses as a result of the action of highway loads are principally shear failures. Therefore, the factors that go to make up the shearing resistance of a soil are of importance. Shearing resistance within soil masses is commonly attributed to the existence of "internal friction" and "cohesion."

A simplified explanation of these properties is most easily accomplished by consideration of two extremely different types of soils; first, a cohesionless sand, and second, a highly cohesive clay in which the internal friction is assumed to be negligible. In a cohesionless sand, the force required to overcome shearing resistance on any plane is assumed to be given by the expression $S_r = \sigma \tan \phi$ where σ is the normal force (stress) on the plane being considered and ϕ is the angle of internal friction. The value of the angle of internal friction is assumed to include the factors of resistance to sliding (or rolling) of the soil particles over one another and any interlocking that may have to be overcome before a slip can occur. For a dry sand, ϕ is primarily dependent on density (void ratio); the lower the void ratio the higher the value of ϕ. Grain shape is also important, as is surface texture; ϕ is higher for a rough, angular sand than for a smooth, rounded sand having the same void ratio. Gradation of a sand is also important, with ϕ being generally higher for sands that are well graded from coarse to fine. The angle of internal friction is relatively independent of the moisture content for sands; ϕ for a wet sand will be only slightly, if any, lower than ϕ for a dry sand, other conditions being the same.

In a saturated clay mass it may be assumed, for the purposes of explanation, that the angle of internal friction is equal to zero and that the resistance to sliding on any plane is equal to the cohesion, C (usually expressed in pounds per square foot). In a simple explanation, C is supposed to include both "true" cohesion, that due to intermolecular attraction, and "apparent" cohesion, that due to surface tension effects in the water contained in the clay mass. The shearing strength of most fine-grained soils decreases when their moisture content is increased and is frequently sharply reduced when their natural structure is destroyed. The interpretation of the factors influencing the shearing strength of cohesive soils is probably the most complex problem in soil mechanics, and no comprehensive explanation will be attempted here. Factors of importance include density, water content, loss of strength with remolding, drainage conditions of the clay mass subjected to stress, variation of cohesion with pressure, and variation in the angle of internal friction.

For the large majority of soils as normally encountered in the field, shearing resistance is made up of both cohesion and internal friction. For these soils the shearing resistance on any plane is frequently, although somewhat empirically, given by Coulomb's law

$$S_r = \sigma \tan \phi + C \qquad (15\text{-}3)$$

where the symbols have the meanings previously indicated.

Shearing resistance may be evaluated in the laboratory by use of the unconfined compression test, the direct shear test, or the triaxial compression test. Samples may be tested in an undisturbed condition or under conditions similar to those expected in the field. Direct measurements of the shearing resistance of subgrade soils may be made in the field through the use of loaded circular plates (plate bearing test). In addition, various semiempirical tests (such as the California bearing ratio test) have been developed to measure shearing resistance more or less directly in connection with the design of flexible and rigid pavements. Design methods in which these tests are utilized are fully described in later chapters.

15-7 Other soil properties

Several other soil properties, including those listed and briefly defined below, may influence the behavior of soil masses and affect the performance of highway subgrades and structure foundations. For a more complete description of these properties, the reader is referred to one of several textbooks in geotechnical or materials engineering *(3, 4)*.

Permeability is the property of a soil mass that permits water to flow through it under the action of gravity or some other applied force.

Capillarity is that property which permits water to be drawn from a free water surface through the action of surface tension and independent of the force of gravity.

Shrinkage of a soil mass is a reduction in volume that occurs when the moisture content is reduced from that existing when it is partially saturated or saturated.

Swelling is a term used to describe the expansion in volume of a soil mass that accompanies an increase in the moisture content.

Compressibility is the property of a soil that permits it to consolidate under the action of an applied compressive load.

Elasticity is the property of a soil that permits it to return to its original dimensions (or nearly so) after the removal of an applied load.

SOIL CLASSIFICATION FOR HIGHWAY PURPOSES

The objective behind the use of any soil classification system for highway purposes is to be able to predict the subgrade performance of a given soil on the basis of a few simple tests performed on the soil in a disturbed condition. On the basis of these test results and their correlation with field experience, the soil may be correctly identified and placed into a group of soils, all of which have similar characteristics and properties.

The student should realize, however, that placing a soil into the correct group in some classification system, although desirable in many respects, is not the final objective of the soil engineer. That is, classifying a soil should not be regarded as an end in itself but as a tool to further our knowledge of soil action and behavior.

The principal soil classification system in use in this country by highway engineers is the AASHTO classification system, first proposed in 1931 and later revised. Principal tests utilized by this and other classification systems are mechanical analysis and various routine laboratory tests. The mechanical analysis is performed on the entire sample and has as its objective the determination of the proportion of particles of various sizes in the given soil. The routine tests mentioned above are performed on the "soil binder" or fraction passing a No. 40 sieve. Most important of the routine tests or "soil constants" are the Atterberg limits, that is, liquid limit, plastic limit, and shrinkage limit. The plasticity index is also of significance and is calculated from the results of the liquid limit and plastic limit determinations. Other routine tests that may be used for classification purposes include other shrinkage factors (shrinkage ratio, volumetric shrinkage, and lineal shrinkage), the field moisture equivalent, and the centrifuge moisture equivalent. The routine tests are intended to describe definite physical properties of the soil and are performed with standardized laboratory procedures. Test procedures are briefly described in the following paragraphs. For a full description, the student is referred to the appropriate publications of the American Society for Testing and Materials and the American Association of State Highway and Transportation Officials.

15-8 Mechanical analysis

Separation of the soil into its fractions may be done by shaking the dry, loose material through a nest of sieves of increasing fineness, that is, successively smaller openings. The sieve analysis may be performed directly on soils that contain little or no fines, such as a clean sand or a soil in which the fines may readily be separated from the coarser particles. Soils that have little dry strength and can be easily crushed between the fingers would generally fall in the latter category. If the character of the fines is such that the fine material adheres to the coarser particles and is not removed by drying sieving action, the sample is *prewashed* and the fine material removed. Material that is retained on the No. 200 sieve during the washing process is then dried and subjected to sieving as before.

The practical lower limit for the use of sieves is the No. 200 sieve, which has openings 0.074 mm square. Consequently, the mechanical analysis used in many soil laboratories is a combined sieve and hydrometer analysis. A representative sample of air-dry soil is selected and separated into two parts by passing it over a No. 10 sieve. The coarse material retained on the No. 10 sieve is subjected to sieve analysis and a portion of the remainder of the sample is dispersed and put into suspension in water, usually in the presence of a deflocculating agent. Changes in the specific gravity of the suspension with the passage of time are then noted by means of a hydrometer. Calculations based on Stokes' law can then be employed to determine the percentages of grains of various average diameters remaining in suspension (or which have settled out). In other words, for each of the time intervals specified in the test procedure, the percentage of the dispersed sample finer than (or coarser than) a certain grain size may be determined. Following completion of the hydrometer analysis, which may require 24 hr or more if the percentage of particles of colloidal size is to be determined, the soil is washed over a No. 200 sieve. Material retained on the No. 200 sieve is oven dried and passed over a nest of sieves

FIGURE 15-2 Typical grain-size distribution curves.

(frequently Nos. 20, 40, 60, 140, and 200 are used). The results of the sieve analysis and hydrometer analysis are then combined to give the complete grain-size distribution curve for the soil being tested. One method of presenting this information is shown in Figure 15-2.

Curves of four different soils are shown. Curve A is that of a uniform sand; B is that of a poorly graded gravelly soil; C is a well-graded soil, with a good distribution of particles from coarse to fine; and D is coarse aggregate such as used in concrete.

In many cases the entire grain-size distribution curve is not required. In such circumstances, a "wet sieve analysis" may be employed, in which the soil is washed over a nest of sieves of selected sizes. One such procedure requires washing the sample over the Nos. 10, 40, and 200 sieves and subsequent calculation of the percentage of soil retained on each of these sieves as well as that which passes the No. 200 sieve. In this method the material passing the No. 200 sieve is designated as "combined silt and clay."

15-9 Liquid limit

The liquid limit may be defined as the minimum moisture content at which the soil will flow under the application of a very small shearing force. At this moisture content the soil is assumed to behave practically as a liquid. The liquid limit is usually determined in the laboratory by use of the mechanical device shown in Figure 15-3a. When this device is used, the liquid limit may be further defined as the moisture content at which a groove of standard dimensions will just close upon application of 25 blows or "shocks," applied by raising the cup of the liquid limit device and allowing it to fall a distance of 1 cm. In performing the test it is usually not practicable to establish the exact moisture content corresponding to 25 blows. Consequently, a number of trials are made with the soil at moisture contents both above and below the liquid limit, and the plot of Figure 15-3b is prepared. The moisture content at which the straight line (or "flow curve") crosses the line corresponding to 25 blows is then taken to be the liquid limit. Laboratories which perform a very large number of routine tests have developed various short-cut methods of determining liquid limits within permissible ranges of accuracy.

15-10 Plastic limit

The plastic limit may be defined in general terms as the minimum moisture content at which the soil remains in a plastic condition. This lower limit of plasticity is rather arbitrarily defined, and the plastic limit may be further described as the lowest moisture content at which the soil can be rolled into a thread of $\frac{1}{8}$-in. diameter without crumbling (Figure 15-4). If a cohesive soil is wetter than the plastic limit, threads may be rolled out to diameters less than $\frac{1}{8}$-in. without crumbling; if drier, the soil will crumble before this diameter can be reached. When the soil has a moisture content equal to the plastic limit, a thread can be rolled out by hand pressure to $\frac{1}{8}$-in. diameter and then will crumble or break into pieces. Certain soils, such as clean sands, are nonplastic, that is, no plastic limit can be determined.

FIGURE 15-3 Liquid limit test. (Courtesy Soiltest, Inc.)

FIGURE 15-4 Plastic limit test.

15-11 Plasticity index

The plasticity index (PI) of a soil is defined as the numerical difference between the liquid and plastic limits. It thus indicates the range of moisture content over which the soil is in a plastic condition. Sandy soils and silts, particularly those of the rock-flour type, have characteristically low PIs, while clay soils show high values of the plasticity index. Generally speaking, soils which are highly plastic, as indicated by a high value of the PI, are also highly compressible. It is also evident that the plasticity index is a measure of cohesiveness, with a high value of the PI indicating a high degree of cohesion. Soils that do not have a plastic limit, such as cohesionless sands, are reported as being nonplastic (NP).

15-12 AASHTO classification system

The AASHTO classification system was first presented in 1931 by the Bureau of Public Roads in the research magazine *Public Roads*. In the original system, subgrade soils were grouped into eight principal groups, designated by the symbols A-1 through A-8. Since 1931, the system has been modified and refined and adopted by the American Association of State Highway and Transportation Officials. The following description of the classification system has been quoted from AASHTO Recommended Practice M145 *(2)* with but minor editorial changes.

This recommended practice describes a procedure for classifying soils into seven groups based on laboratory determination of particle-size distribution, liquid limit, and plasticity index. Evaluation of soils within each group is made by means of a "group index," which is a value calculated from an empirical formula. (See Section 15-13.) The group classification including group index should be useful in determining the relative quality of the soil material for use in earthwork structures, particularly embankments, sub-grades, subbases, and bases. However, for the detailed design of important structures, additional data concerning strength or performance characteristics of the soil under field conditions will usually be required.

Test procedures

The classification is based on the results of tests made in accordance with the following standard methods of AASHTO:

Amount of material finer than 0.075 mm (No. 200) sieve in aggregate	T11
Sieve analysis of fine and coarse aggregates	T27
Mechanical analysis of soils	T88
Determining the liquid limit of soils	T89
Determining the plastic limit and plasticity index of soils	T90

Classification

The classification is made by using the test limits and group index values shown in Table 15-1. If a more detailed classification is desired, a further subdivision of the groups shown in Table 15-1 may be made. An example of the classification with such subgroups is shown in Table 15-2. The liquid limit and plasticity index ranges for the A-4, A-5, A-6, and A-7 soil groups are shown graphically in Figure 15-5.

Classification procedure   With required test data available, proceed from left to right in Table 15-1 or Table 15-2 and the correct group will be found by the process of elimination. The first group from the left into which the test data will fit is the correct classification. All limiting test values are shown as whole numbers. If fractional numbers appear on test reports, convert to nearest whole number for the purpose of classification. Group index values should always be shown in parentheses after group symbols as A-2-6(3), A-4(5), A-6(12), A-7-5(17), and so forth. A description of the AASHTO classification groups is given in Table 15-3.

15-13 The group index

The group index is calculated from the following formula:

$$\text{group index} = (F - 35)[0.2 + 0.005(LL - 40)] + 0.01(F - 15)(PI - 10) \tag{15-4}$$

where

F = percentage passing 0.074-mm (No. 200) sieve, expressed as a whole number; this percentage is based only on the material passing the 75-mm (3-in.) sieve

LL = Liquid limit

PI = plasticity index

TABLE 15-1
CLASSIFICATION OF HIGHWAY SUBGRADE MATERIALS

General Classification	Granular Materials (35% or less passing No. 200)			Silt–Clay Materials (more than 35% passing No. 200)			
Group Classification	A-1	A-3[a]	A-2	A-4	A-5	A-6	A-7
Sieve analysis, percent passing							
No. 10 (2.00 mm)	50 max.						
No. 40 (0.425 m)	25 max.	51 min.					
No. 200 (0.075 mm)		10 max.	35 max.	36 min.	36 min.	36 min.	36 min.
Characteristics of fraction passing No. 40							
Liquid limit				40 max.	41 min.	40 max.	41 min.
Plasticity index	6 max.	NP		10 max.	10 max.	11 min.	11 min.
General rating as subgrade	Excellent to good			Fair to poor			

[a]The placing of A-3 before A-2 is necessary in the "left to right elimination process" and does not indicate superiority of A-3 over A-2.

TABLE 15-2
CLASSIFICATION OF HIGHWAY SUBGRADE MATERIALS (WITH SUGGESTED SUBGROUPS)[a]

General Classification	Granular Materials (35% or less passing No. 200)							Silt–Clay Materials (more than 35% passing No. 200)			
	A-1		A-3	A-2				A-4	A-5	A-6	A-7
Group Classification	A-1-a	A-1-b		A-2-4	A-2-5	A-2-6	A-2-7				A-7-5, A-7-6
Sieve analysis, percent passing											
No. 10 (2.0 mm)	50 max.										
No. 40 (0.425 mm)	30 max.	50 max.	51 min.								
No. 200 (0.075 mm)	15 max.	25 max.	10 max.	35 max.	35 max.	35 max.	35 max.	36 min.	36 min.	36 min.	36 min.
Characteristics of fraction passing No. 40											
Liquid limit				40 max.	41 min.	40 max.	41 min.	40 max.	41 min.	40 max.	41 min.
Plasticity index	6 max.		NP	10 max.	10 max.	11 min.	11 min.	10 max.	10 max.	11 min.	11 min.[b]
Usual types of significant constituent materials	Stone fragments, gravel, and sand		fine sand	Silty or clayey gravel and sand				Silty soils		Clayey soils	
General rating as subgrade	Excellent to good							Fair to poor			

[a]Classification procedure: With required test data available, proceed from left to right on above chart and correct group will be found by process of elimination. The first group from the left into which the test data will fit is the correct classification.
[b]Plasticity index of A-7-5 subgroup is equal to or less than LL minus 30. Plasticity index of A-7-6 subgroup is greater than LL minus 30 (see Figure 15-5).
[c]See group index formula and Figure 15-5 for method of calculation. Group index should be shown in parentheses after group symbol as: A-2-6(3), A-4(5), A-6(12), A-7-5(17), and so forth.

FIGURE 15-5 Liquid limit and plasticity index ranges for the A-4, A-5, A-6, and A-7 subgrade groups.

The group index should be reported be reported to the nearest whole number.

When the calculated group index is negative, it should be reported as zero. When calculating the group index of A-2-6 and A-2-7 subgroups, only the PI portion of the formula should be used.

The following are examples of calculations of the group index.

Assume that an A-6 material has 55 percent passing the 0.075-mm sieve, liquid limit of 40, and plasticity index of 25. Then,

$$\text{group index} = (55 - 35)[0.2 + 0.005(40 - 40)] + 0.01(55 - 15)(25 - 10)$$
$$= 4.0 + 6.0 = 10$$

Assume that an A-2-7 material has 30 percent passing the 0.075-mm sieve, liquid limit of 50, and plasticity index of 30. Then,

$$\text{group index} = 0.01(30 - 15)(30 - 10)$$
$$= 3.0 \text{ or } 3$$

(Note that only the PI portion of the formula was used.)

The empirical group index formula devised for approximately within-group evaluation of the "clayey granular materials" and the "silt−clay materials" is based on the following assumptions:

1. Materials falling within groups A-1-a, A-1-b, A-2-4, A-2-5, and A-3 are satisfactory as subgrade when properly drained and compacted under

TABLE 15-3
DESCRIPTION OF CLASSIFICATION GROUPS

Granular Materials—*Containing 35 percent or less passing the 0.075-mm sieve.*

Group A-1.   The typical material of this group is a well-graded mixture of stone fragments or gravel, coarse sand, fine sand, and a nonplastic or feebly plastic soil binder. However, this group also includes stone fragments, gravel, coarse sand, volcanic cinders, and so forth, without soil binder.

Subgroup A-1-a includes those materials consisting predominantly of stone fragments or gravel, either with or without a well-graded binder of fine material.

Subgroup A-1-b includes those materials consisting predominantly of coarse sand either with or without a well-graded soil binder.

Group A-3.   The typical material of this group is fine beach sand or fine desert blown sand without silty or clay fines or with a very small amount of nonplastic silt.

Group A-2.   This group includes a wide variety of "granular" materials that are borderline between the materials falling in groups A-1 and A-3 and the silt−clay materials of groups A-4, A-5, A-6, and A-7. It includes all materials containing 35 percent or less passing the 0.075-mm sieve that cannot be classified as A-1 or A-3 due to fines content or plasticity or both in excess of the limitations for those groups.

Subgroups A-2-4 and A-2-5 include various granular materials containing 35 percent or less passing the 0.075-mm sieve and with a minus 0.425 mm portion having the characteristics of the A-4 and A-5 groups.

Subgroups A-2-6 and A-2-7 include materials similar to those described under subgroups A-2-4 and A-2-5, except that the fine portion contains plastic clay having the characteristics of groups A-6 and A-7.

Silt−Clay Materials—*Containing more than 35 percent passing 0.075-mm sieve.*

Group A-4.   The typical material of this group is a nonplastic or moderately plastic silty soil usually having 75 percent or more passing the 0.075-mm sieve.

Group A-5.   The typical material of this group is similar to that described under group A-4, except that it is usually of diatomaceous or micaceous character and may be highly elastic, as indicated by the high liquid limit.

Group A-6.   The typical material of this group is a plastic clay soil usually having 75 percent or more passing the 0.075-mm sieve. The group also includes mixtures of fine clayey soil and up to 64 percent of sand and gravel retained on the 0.075-mm sieve. Materials of this group usually have a high volume change between wet and dry states.

Group A-7.   The typical material of this group is similar to that described under group A-6, except that it has the high liquid limits characteristic of the A-5 group and may be elastic as well as subject to a high volume change.

Subgroup A-7-5 includes those materials with moderate plasticity indexes in relation to liquid limit and which may be highly elastic as well as subject to a considerable volume change.

Subgroup A-7-6 includes those materials with high plasticity indexes in relation to liquid limit and which are subject to an extremely high volume change.

moderate thickness pavement (base and/or surface course) of a type suitable for the traffic to be carried, or can be made satisfactory by additions of small amounts of natural or artificial binders.

2. Materials falling within the clayey granular groups A-2-6 and A-2-7 and the silt–clay groups A-4, A-5, A-6, and A-7 will range in quality as subgrade from the approximate equivalent of the good A-2-4 and A-2-5 subgrades to fair and poor subgrades, requiring a layer of subbase

TABLE 15-4
UNIFIED SOIL CLASSIFICATION SYSTEM

Major Divisions			Group Symbols	Typical Name	Field Identification Procedures (Excluding Particles Larger Than 3 Inches and Basing Fractions on Estimated Weights)		
1		2	3	4	5		
Coarse-grained Soils More than half of material is larger than No. 200 sieve size.	Gravels More than half of coarse fraction is larger than No. 4 sieve size.	Clean gravels (little or no fines)	GW	Well-graded gravels, gravel-sand mixtures, little or no fines	Wide range in grain sizes and substantial amounts of all intermediate particle sizes		
			GP	Poorly-graded gravels, gravel-sand mixtures, little or no fines	Predominantly one size or a range of sizes with some intermediate sizes missing		
		Gravels with fines (appreciable amount of fines)	GM	Silty gravels, gravel-sand-silt mixtures	Nonplastic fines or fines with low plasticity (for identification procedures see ML below)		
			GC	Clayey gravels, gravel-sand-clay mixtures	Plastic fines (for identification procedures see CL below)		
	Sands More than half of coarse fraction is smaller than No. 4 sieve size. (For visual classification, the ¼-inch size may be used as equivalent to the No. 4 sieve size.)	Clean sands (little or no fines)	SW	Well-graded sands, gravelly sands, little or no fines	Wide range in grain size and substantial amounts of all intermediate particle sizes		
			SP	Poorly-graded sands, gravelly sands, little or no fines	Predominantly one size or a range of sizes with some intermediate sizes missing		
		Sands with fines (appreciable amount of fines)	SM	Silty sands, sand-silt mixtures	Nonplastic fines or fines with low plasticity (for identification procedures see ML below)		
			SC	Clayey sands, sand-clay mixtures	Plastic fines (for identification procedures see CL below)		
Fine-grained Soils More than half of material is smaller than No. 200 sieve size. The No. 200 sieve size is about the smallest particle visible to the naked eye.	Silts and Clays	Liquid limit less than 50			Identification procedures on fraction smaller than No. 40 sieve size		
					Dry strength (crushing characteristics)	Dilatancy (reaction to shaking)	Toughness (consistency near PL)
			ML	Inorganic silts and very fine sands, rock flour, silty or clayey fine sands with slight plasticity	None to slight	Quick to slow	None
			CL	Inorganic clays of low to medium plasticity, gravelly clays, sandy clays, silty clays, lean clays	Medium to high	None to very slow	Medium
			OL	Organic silts and organic silty clays of low plasticity	Slight to medium	Slow	Slight
	Silts and Clays	Liquid limit greater than 50	MH	Inorganic silts, micaceous or diatomaceous fine sandy or silty soils, elastic silts	Slight to medium	Slow to none	Slight to medium
			CH	Inorganic clays of high plasticity, fat clays	High to very high	None	High
			OH	Organic clays of medium to high plasticity, organic silts	Medium to high	None to very slow	Slight to medium
Highly organic soils			Pt	Peat and other highly organic soils	Readily identified by color, odor, spongy feel and frequently by fibrous texture		

(1) *Boundary classifications*: Soils possessing characteristics of two groups are designated by combinations of group symbols. For example

Adopted by Corps of Engineers and Bureau of Reclamation, January, 1952.

material or an increased thickness of base course over that required for other soil classes in order to furnish adequate support for traffic loads.

3. The assumed critical minimum percentage passing the 0.075-mm sieve is 35 neglecting plasticity and 15 as affected by plasticity indexes greater than 10.
4. Liquid limits of 40 and above are assumed to be critical.
5. Plasticity indexes of 10 and above are assumed to be critical.

TABLE 15-4 *(continued)*

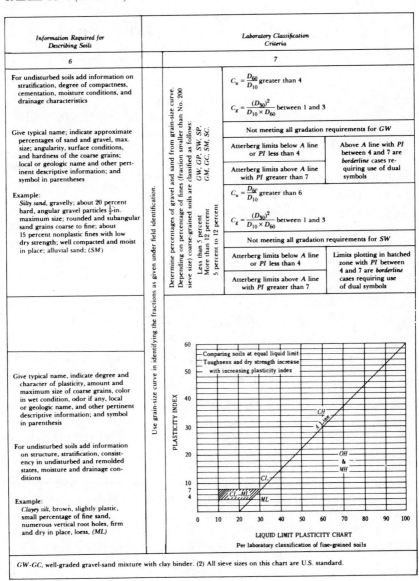

Information Required for Describing Soils	Laboratory Classification Criteria
6	7

For undisturbed soils add information on stratification, degree of compactness, cementation, moisture conditions, and drainage characteristics

$C_u = \dfrac{D_{60}}{D_{10}}$ greater than 4

$C_g = \dfrac{(D_{30})^2}{D_{10} \times D_{60}}$ between 1 and 3

Give typical name; indicate approximate percentages of sand and gravel, max. size; angularity, surface conditions, and hardness of the coarse grains; local or geologic name and other pertinent descriptive information; and symbol in parentheses

Not meeting all gradation requirements for *GW*

Atterberg limits below *A* line or *PI* less than 4 — Above *A* line with *PI* between 4 and 7 are *borderline* cases requiring use of dual symbols

Atterberg limits above *A* line with *PI* greater than 7

Example:
Silty sand, gravelly; about 20 percent hard, angular gravel particles ½-in. maximum size; rounded and subangular sand grains coarse to fine; about 15 percent nonplastic fines with low dry strength; well compacted and moist in place; alluvial sand; (*SM*)

$C_u = \dfrac{D_{60}}{D_{10}}$ greater than 6

$C_g = \dfrac{(D_{30})^2}{D_{10} \times D_{60}}$ between 1 and 3

Not meeting all gradation requirements for *SW*

Atterberg limits below *A* line or *PI* less than 4 — Limits plotting in hatched zone with *PI* between 4 and 7 are *borderline* cases requiring use of dual symbols

Atterberg limits above *A* line with *PI* greater than 7

Determine percentages of gravel and sand from grain-size curve. Depending on percentage of fines (fraction smaller than No. 200 sieve size) coarse-grained soils are classified as follows:

GW, GP, SW, SP
GM, GC, SM, SC

Less than 5 percent
More than 12 percent
5 percent to 12 percent

Use grain-size curve in identifying the fractions as given under field identification.

Give typical name, indicate degree and character of plasticity, amount and maximum size of coarse grains, color in wet condition, odor if any, local or geologic name, and other pertinent descriptive information; and symbol in parenthesis

For undisturbed soils add information on structure, stratification, consistency in undisturbed and remolded states, moisture and drainage conditions

Example:
Clayey silt, brown, slightly plastic, small percentage of fine sand, numerous vertical root holes, firm and dry in place, loess, (*ML*)

LIQUID LIMIT PLASTICITY CHART
Per laboratory classification of fine-grained soils

GW-GC, well-graded gravel-sand mixture with clay binder. (2) All sieve sizes on this chart are U.S. standard.

6. For soils that are nonplastic and when the liquid limit cannot be determined, the group index shall be considered zero (0).

There is no upper limit of group index value obtained by use of the formula. The adopted critical values of percentage passing the 0.075-mm sieve, liquid limit, and plasticity index are based on an evaluation of subgrade, subbase, and base course materials by several highway organizations that use the tests involved in this classification system.

Under average conditions of good drainage and thorough compaction, the supporting value of a material as subgrade may be assumed as an inverse ratio to its group index; that is, a group index of 0 indicates a "good" subgrade material and group index of 20 or greater indicates a "very poor" subgrade material.

15-14 Unified soil classification system

The Unified Soil Classification System is based on the Airfield Classification System developed by Professor A. Casagrande of Harvard University during World War II. The system was modified slightly and became the Department of the Army Uniform Soil Classification System. In turn, this system was slightly modified and adopted by the Corps of Engineers and the Bureau of Reclamation in January 1952.

The Unified System is based primarily on characteristics that determine how a soil will behave when used as a construction material. Table 15-4 is the master chart for the Unified System; it contains procedures that are to be followed in identifying and classifying soils under this system.

Soils may be classified and placed in one of the 15 major soil groups in the Unified System by using either laboratory or field identification procedures. Soils are classified on the basis of (1) percentage of gravel, sand, and fines (fraction passing the No. 200 sieve), (2) shape of the grain-size distribution curve, and (3) plasticity and compressibility characteristics.

The coefficients of uniformity (C_u) and gradation (C_g) are used to judge the shape of the grain-size distribution curve of a coarse-grained soil. The term "D_{10}" which appears in the table means the grain size (diameter) that corresponds to 10 percent on a grain-size distribution curve of the type shown in Figure 15-2. D_{30} and D_{60} have similar meanings.

The influence and relationship of the liquid limit and plasticity index are reflected in the plasticity chart of Table 15-4.

In general terms, clays (C) plot above the "A" line of the plasticity chart and silts (M) plot below the "A" line. The silt (M) and clay (C) groups are further divided on the basis of low (L) or high (H) liquid limit; a high liquid limit is associated with high compressibility.

Further details of the classification procedure will be revealed by close study of Table 15-4. Further information is contained in Ref. 5.

15-15 Classification of typical soils

Table 15-5 gives the results of laboratory tests on three inorganic soils. Each soil may be classified under the AASHTO and Unified Soil Classification systems, as follows:

Soil No. 1   AASHTO system. By Eq. 15-4, group index = 1. Entering Table 15-2, and using a left-to-right elimination process, soil cannot be in one of the *granular materials* groups, since more than 35 percent passes a No. 200 sieve. It meets the requirements of the A-4 group. It is therefore A-4(1).

Unified. Soil is fine-grained since more than 50 percent passes No. 200 sieve. Liquid limit is less than 50, hence must be ML or CL, since it is inorganic. On plasticity chart it falls below the A line, therefore ML.

Soil No. 2   AASHTO system. By Eq. 15-4, group index = 27. This soil falls in the A-7 group. Referring to Figure 15-5, it falls in the A-7-6 subgroup; hence, A-7-6(27).

Unified. Soil is fine-grained since more than 50 percent passes No. 200 sieve. Liquid limit is more than 50, hence must be MH or CH. On plasticity chart, it falls above the A line, therefore CH.

Soil No. 3   AASHTO. This is one of the soils described as *granular material*. It will not meet the requirements of the A-1 group since it contains practically no fines. It does meet the requirements of the A-3 group. Equation 15-4 yields a negative group index for this soil. The classification is A-3(0).

Unified. Soil is coarse-grained, since very little passes No. 200 sieve. All passes a No. 10 sieve, hence sand. The soil contains less than 5 percent passing No. 200; therefore it must be either an SW or SP. Value of $C_u = 2$ will not meet requirements for SW, hence SP.

TABLE 15-5
RESULTS OF LABORATORY TESTS ON THREE INORGANIC SOILS

	Soil Number		
Sieve Size	*1*	*2*	*3*
Mechanical Analysis, percent passing, by weight			
No. 10	100.0	100.0	100.0
No. 40	85.2	97.6	85.0
No. 60	—	—	20.0
No. 200	52.1	69.8	1.2
No. 270	48.2	65.0	—
C_u	—	—	—
C_g	—	—	—
Plasticity Characteristics, percent, by weight			
Liquid limit	29.2	66.7	21.3
Plasticity index	5.0	39.0	NP

SOIL SURVEYS FOR HIGHWAY PURPOSES

Soil surveys are made in connection with highway location, design, and construction. Many sources of information concerning soils generally are available for the area in which a highway project is to be carried out. These sources include geological and topographic maps and reports, agricultural soil maps and reports, aerial photographs, and the results of previous soil surveys in the area. Information from such sources is of importance in two general ways. First, a study of this information will aid in securing a broad understanding of soil conditions and associated engineering problems that may be encountered. Second, such information is of great value in planning, conducting, and interpreting the results of detailed soil surveys that are necessary for design and construction. Modern techniques emphasize the use of all the available information about a given area in order to minimize the amount of detailed field and laboratory work necessary for a given project.

MOISTURE–DENSITY RELATIONSHIPS

Practically all soils exhibit a similar relationship between moisture content and density (dry unit weight) when subjected to dynamic compaction. That is, practically every soil has an optimum moisture content at which the soil attains maximum density under a given compactive effort. This fact, which was first stated by R. R. Proctor in a series of articles published in *Engineering News-Record* in 1933, forms the basis for the modern construction process commonly used in the formation of highway subgrades, bases, and embankments, earthen dams, levees, and similar structures. In the laboratory, dynamic compaction is achieved by use of a freely falling weight impinging on a confined soil mass; in the field, similar compaction is secured through the use of rollers or vibratory compactors applied to relatively thin layers of soil during the construction process. In Figure 15-6 is shown the relationship between moisture and density for a typical soil subjected to dynamic compaction; the meaning of the terms "maximum density" and "optimum moisture" is clearly shown. This curve was obtained in the laboratory. In the field an attempt is usually made to maintain the soil at optimum moisture and bring the soil to maximum density or some specified percentage thereof.

In the laboratory, compaction is usually performed under what has come to be called the "standard Proctor" or "standard AASHTO" methods (T99). In the basic procedure used in this test (method A), the portion of the soil that passes a No. 4 sieve is placed in a mold 4 in. diameter which has a volume of $\frac{1}{30}$ ft^3. The soil is placed in three layers of about equal thickness, and each layer is subjected to 25 blows from a hammer weighing 5.5 lb and having a striking face 2 in. in diameter, falling freely through a distance of 12 in. (12,375 ft-lb/ft^3). In recent years, heavier compaction equipment has come into widespread use, with the result that under certain conditions a greater compactive

FIGURE 15-6 Moisture–density relationship for a typical soil under dynamic compaction.

effort may be required in the laboratory. An increased compactive effort that is frequently used is known as the "modified Proctor" or "modified AASHTO" compaction (T180); the soil is compacted in the Proctor mold, using 25 blows from a 10-lb hammer dropping through a distance of 18 in. on each of five equal layers (56,250 ft-lb/ft^3). These are "impact" methods of compaction; some work has been done with "kneading-type" compactors, which simulate the action of a rubber-tired roller, and vibratory compaction of granular materials.

Regardless of the compactive effort used, the optimum moisture and maximum density are usually found in the laboratory by a series of determinations of wet unit weight and the corresponding moisture content. The series is started with the soil in a damp condition, at a moisture content a few percent below optimum. The soil is compacted, the wet unit weight determined, and a sample selected from the interior of the compacted soil mass for the determination of moisture content. The soil is then broken up (or a new sample of the same soil selected); the moisture content is increased by 1 or 2 percent, and the compaction process is repeated. This process is continued until a decrease is noted in the wet unit weight or until the soil definitely becomes wet and shows an excess of moisture. After the moisture content determinations are completed the dry unit weights may be calculated and plotted as shown in Figure 15-6. The relationship between the dry unit weight, wet unit weight, and moisture content is as given in Eq. 15-2.

The behavior of the soil mass under dynamic (impact) compaction is of interest and may be explained in the following fashion. In the early stages of the compaction run, when the moisture content is considerably less than optimum, the soil does not contain sufficient moisture to flow readily under the blows of the hammer. As the moisture content is increased, the soil flows more readily under the "lubricating" effect of the additional water, and the soil particles move closer together, with a resulting increase in density. This effect is continued until the optimum moisture content is reached, at which point the soil has achieved its greatest density. The condition at optimum may be explained in terms of the degree of saturation (the ratio between the volume of water and the volume of voids); at this moisture content as large a degree of saturation has been attained as can be reached by compaction. This limiting degree of saturation is due to the existence of air entrapped in the void spaces and around the soil particles. Further increase in moisture tends to overfill the voids but does not materially decrease the air content. As a consequence, the soil particles are forced apart and the unit weight decreases. At moisture contents well beyond optimum, the soil mass tends to become somewhat elastic, the hammer "bounces," and excess moisture is generally forced out of the bottom of the mold.

Also shown on the plot of Figure 15-6 is the "zero air voids curve" for the soil in question. Points on this curve represent the theoretical density that this soil would attain at each moisture content if all the void spaces were filled with water, that is, if the soil were completely saturated or $S = 100$ percent. The position of points on the curve may be calculated from the following equation if the specific gravity of the soil is known or may be assumed:

$$\text{dry unit weight} = \frac{62.4G}{1 + \dfrac{wG}{100}} \text{ lb/ft}^3 \tag{15-5}$$

where
G = specific gravity of solids
w = moisture content (percent)

Although the densities shown on the zero air voids curve represent theoretical values that are practically unattainable, since all the air can never be exhausted from any soil system by dynamic compaction alone, the relative position of the moisture–density curve with respect to the zero air voids curve is of consequence, as the distance separating the two curves is indicative of the air voids remaining in the soil at various moisture contents. For example, for the soil shown, the sample at 17.8 percent moisture has the same dry weight as it has at 15.1 percent moisture, although the percentage of air voids at the lower moisture content (on the dry side of optimum) is much greater. Thus, if the soil is compacted at the lower moisture content and then subjected to inundation or "flooding," it might be expected that a larger percentage of air would be displaced by entering moisture and greater expansion would occur than if the soil were compacted at the higher moisture content.

Another item of importance about the compaction process is that maximum density and optimum moisture are a function of the compactive effort. In general terms, an increase in compactive effort results in an increase in maximum density and a decrease in optimum moisture. The effect of variation in compactive effort is shown graphically in Figure 15-7. Examination of this figure shows that if soil is to be compacted to a specified density efficiently, careful control of the moisture content is required. Suppose, for example, that the specifications required a density of 116.5 lb/ft^3. If the soil were too dry (say a moisture content of 7 percent), about 70 percent (6.63/3.98 × 100 percent = 170 percent) more compactive effort would be required to achieve the specified density than would be required at a moisture content of 10.4 percent. On the other hand, as greater and greater compactive effort is used, there is a wider and wider range of moisture contents over which the soil may be compacted to a specified minimum density.

Obviously, the nature of the soil itself has a great effect on the density obtained under a given compactive effort. Different soils react somewhat differently to compaction at moisture contents that are somewhat less than optimum. Moisture content is less critical for the heavy clay than for the slightly plastic sandy and silty soils. Heavy clays may be compacted through a relatively wide range of moisture contents below optimum with comparatively small changes in dry density. On the other hand, the granular soils that have better grading and higher densities under the same compactive effort react sharply to small changes in moisture content, with significant changes in dry density. The relatively clean, poorly graded, nonplastic sands also are relatively insensitive to changes in moisture.

Many other factors influence the results of compaction tests done in the laboratory, including the amount of coarse aggregate contained in the soil. These factors are discussed in detail in Ref. 6.

FROST ACTION IN SOILS

One of the special soil-related problems of interest in many areas of the United States is that of damage caused to roads and streets by the freezing and thawing of subgrades and bases during the winter and spring seasons. This phenomenon is discussed here for convenience and as a background for the discussion of certain design, construction, and maintenance considerations in later chapters.

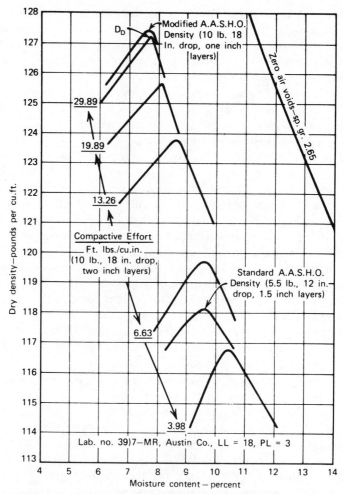

FIGURE 15-7 Effect of compacting effort on density. (Source: Density Control: Its Benefits and Complexities, by Chester McDowell, *Highway Research Record 177* (1967).

Under certain conditions, very severe damage to subgrades, bases, and pavements, both rigid and flexible, may result from frost action. As a simplification, damaging action may be visualized in terms of freezing phenomena, of which an example is "frost heave," and the effects of a sudden spring thaw, which are sometimes categorically called "the spring breakup."

By frost heave is meant the distortion or expansion of the subgrade soil or base material during the period when freezing temperatures prevail and when the ground is frozen (or partially frozen) to a considerable depth. The phenomena of frost action are generally associated with the existence of a shallow water table and a frost-susceptible soil. It would be expected that excess moisture existing in a base or subgrade would freeze during a severe winter season. It might also be expected that the soil mass would expand, since water expands as it freezes. However, the magnitude of the severe frost heaves that occur regularly in some states cannot be explained on the basis of freezing of existing pore water alone. Investigations have shown that, as the water in the upper soil layers freezes, ice crystals are formed, and water may be drawn from a free water surface into the zone where freezing temperatures prevail. This water then freezes, additional water may be drawn to this level, and this process continues until layers of ice or "ice lenses" of considerable thickness may be formed. The volume increase brought about by the formation of these layers of ice is the cause of frost heaving.

It will be noted that a certain combination of circumstances (in addition to freezing temperatures) must usually exist before severe frost heave can occur. As previously mentioned, these include the existence of a shallow water table and a frost-susceptible soil. If the water table is not close to the surface, water is generally not available to the shallow soil layers, and the heave that can occur is limited to that caused by the expansion of freezing water in the voids of the soil. Since the change in volume from water to ice is only about 9 percent, this expansion would generally not be sufficient to cause severe frost damage. However, it might be possible, in somewhat unusual circumstances, that sufficient free water would be available from seepage flows or some other cause for frost heave to occur in a location where the water table is located well beneath the ground surface.

The requirement related to capillarity seems to limit detrimental frost action to soils containing a considerable percentage of fine particles such as silt or clay. Experience has shown that frost action is generally most severe in silt soils, as the upward movement of water in clay soils is usually so slow that the amount of water made available to the freezing layer is quite small. Soils that contain less than 3 percent by weight of material finer than 0.02 mm in diameter are not frost-susceptible. Soils that contain more than 3 percent of material of this size generally are susceptible to frost action, except that uniform fine sands that contain up to 10 percent of material finer than 0.02 mm generally are not susceptible to frost action.

Experience has also shown that frost heave is most severe when temperatures only slightly below freezing prevail for a considerable period of time so that frost penetration comes about somewhat gradually.

During the period of freezing temperatures, under the conditions described above, a considerable amount of water in the form of ice may accumulate in the shallow soil layers beneath a road surface. If a sudden rise in temperature or "thaw" occurs, as it frequently does in the spring in many localities, this accumulated ice may melt rapidly. Generally speaking, it would be expected that the ice in the upper soil layers would melt first, while deeper layers would remain frozen for a longer period of time. As a consequence, excess moisture may be trapped beneath the pavement and a sharp reduction in shearing strength may result. If this happens, the pavement and base may then fail under wheel loads; this type of failure is associated with the term "spring breakup." Damage to pavements and bases incurred during this period results in severe economic losses in many states.

Certain preventive measures are commonly employed to eliminate or minimize frost damage. Probably the most common solution to the problem is to remove soils subject to frost action and replace them with suitable granular backfill to the depth of the frost line. It is not always necessary to treat unsuitable soils to the full depth of frost penetration. For example, experience in Minnesota has shown that treatments to a depth of about 3 ft are sufficient to prevent any breakup in the spring, although some slight heave may still occur. Another solution may be effected by the installation of drainage facilities in areas where frost action is likely to be detrimental, with the objective of lowering the water battle or intercepting seepage flows which may cause excessive amounts of water to be accumulated in certain areas. Insulation or blanket courses may be employed as a construction measure, while some state highway departments regulate truck traffic in the spring months to prevent damage caused by heavy wheel loads on roads known to have been weakened by frost action.

AGGREGATES

The term aggregate refers to granular mineral particles that are widely used for highway bases, subbases, and backfill. Aggregates are also used in combination with a cementing material to form concretes for bases, subbases, wearing surfaces, and drainage structures. Sources of aggregates include natural deposits of sand and gravel, pulverized concrete and asphalt pavements, crushed stone, and blast-furnace slag.

15-16 Properties of Aggregates

The most important properties of aggregates used for highway construction are

1. Particle size and gradation.
2. Hardness or resistance to wear.
3. Durability or resistance to weathering.
4. Specific gravity and absorption.
5. Chemical stability.
6. Particle shape and surface texture.
7. Freedom from deleterious particles or substances.

These properties are discussed in more detail in the following paragraphs.

15-17 Particle size and gradation of aggregates

A key property of aggregates used for highway bases and surfaces is the distribution of particle sizes in the aggregate mix. The gradation of aggregates, that is, the blend of particle sizes in the mix, affects the density, strength, and economy of the pavement structure.

A grain-size analysis is used to determine the relative proportions of various particle sizes in a mineral aggregate mix. To perform this analysis, a weighed sample of dried aggregate is shaken over a nest of sieves having selected sizes of square openings. The sieves are grouped together with the one with the largest openings on top and those with successively smaller openings placed underneath. The aggregate sample is shaken with a mechanical sieve shaker, and the weight of material retained on each sieve is determined and expressed as a percentage of the original sample *(4)*. Detailed procedures for performing a grain-size analysis of coarse and fine aggregates are given in AASHTO Method T27 *(2)*.

The grain size analysis data are usually plotted on an aggregate grading chart, exemplified by Figure 15-8. With the aid of such a chart, engineers determine a preferred aggregate gradation and require that the gradation of aggregates used for highway projects conform to the limits of a specification band, as Figure 15-8 illustrates.

Testing sieves commonly used for highway projects are those with $2\frac{1}{2}$, 2, $1\frac{1}{2}$, 1, $\frac{3}{4}$, $\frac{1}{2}$, and $\frac{3}{8}$ in. square openings for the large fractions and those with 4, 10, 40, 80, 100, and 200 meshes per inch for the smaller fractions. The latter sieves are designated No. 4, No. 10, and so on.

That portion of aggregate material that is retained on a No. 10 sieve (i. e., with particles larger than 2.00 mm) is known as *coarse aggregate*. Material that passes a No. 10 sieve but is retained on a No. 200 sieve (particles larger than 0.075 mm) is known as *fine aggregate*. Material that passes a No. 200 sieve is referred to as *fines*.

15-18 Resistance to wear

Materials used in highway pavements should be hard and resist wear due to the polishing effects of traffic and the internal abrasive effects of repeated load-

FIGURE 15-8 Aggregate gradation specification chart.

ings. The most commonly accepted measure of the hardness of aggregates is the Los Angeles abrasion test. The machine used in the Los Angeles abrasion test consists of a hollow steel cylinder, closed at both ends and mounted on shafts in a horizontal position. See Figure 15-9. A removable steel shelf extending the length of the cylinder is mounted on the interior surface of the cylinder.

To perform the Los Angeles abrasion test, a clean sample of the aggregate to be tested is placed in the cylinder along with a standard weight of steel spheres as an abrasive charge. The drum is then rotated at a speed of 30 to 33 rpm for 500 revolutions, after which the aggregate sample is removed and sieved on a No. 12 (1.70 mm) sieve. The material retained on the sieve is washed, dried to a constant mass, and weighed. The difference between the original mass and the final mass of the sample, expressed as a percentage of the original mass, is reported as the percentage of wear. A detailed procedure for this test is given by AASHTO Method T96 (2).

15-19 Durability or resistance to weathering

The durability of aggregates is commonly measured by a soundness test, as described in AASHTO Method T104 (2). This test measures the resistance of aggregates to disintegration in a saturated solution of sodium or magnesium sulfate. It simulates the weathering of aggregates that occurs in nature.

The test is made by immersing sized fractions of the aggregate to be tested in a saturated solution of sodium or magnesium sulfate. The aggregate is then removed and dried in an oven to a constant mass. This process is repeated for a specified number of cycles, typically five. After the repeated cycles of alternate wetting and drying, the aggregate is divided into fractions by sieving, and the percentage weight loss is determined for each fraction. The percentage loss is expressed as a weighted average. For a given sieve size, the percentage

FIGURE 15-9 Los Angeles abrasion testing machine.
(Courtesy The Asphalt Institute.)

weighted average loss is the product of the percentage passing that sieve and the percentage passing that sieve in the original material. The total of such values is the percent loss test value.

15-20 Specific gravity and absorption

The specific gravity and absorption of aggregates are important properties that are required for the design of concrete and bituminous mixes. The specific gravity of a solid is the ratio of its mass to that of an equal volume of distilled water at a specified temperature. Because aggregates may contain water-permeable voids, two measures of specific gravity of aggregates are used: *apparent* specific gravity and *bulk* specific gravity.

Apparent specific gravity, G_A, is computed on the basis of the net volume of the aggregates, that is, the volume excluding the water-permeable voids. Thus,

$$G_A = \frac{M_D/V_N}{w} \tag{15-6}$$

where

M_D = dry mass of the aggregate
V_N = net volume of the aggregates, excluding the volume of absorbed water
w = density of water

The bulk specific gravity, G_B, is computed on the basis of the total volume of the aggregates including the water-permeable voids:

$$G_B = \frac{M_D/V_B}{w} \tag{15-7}$$

where

V_B = total volume of the aggregates, including the volume of absorbed water

The difference between the apparent and bulk specific gravities accounts for the water-permeable voids of the aggregates. One can measure the volume of such voids by weighing the aggregates dry and in a saturated, surface dry condition, that is, with all permeable voids filled with water. The difference between the two masses is the mass of the absorbed water, M_w. The absorption of water is usually expressed as a percentage of the mass of the dry aggregate,

$$\text{percentage absorption} = \frac{M_w}{M_D} \times 100 \tag{15-8}$$

EXAMPLE 15-1. The dry mass of a sample of aggregates is 1982.0 g. The mass in a saturated, surface dry condition is 2006.7 g. The net volume of the aggregates is 734.4 cm^3. Find the apparent specific gravity, the bulk specific gravity, and the percentage absorption.

Mass of absorbed water = 2006.7 − 1982.0 = 24.7 g

$$\text{Volume of absorbed water} = \frac{24.7 \text{ g}}{1 \text{ g/cm}^3} = 24.7 \text{ cm}^3$$

Bulk volume = 734.3 + 24.7 = 759.1 cm^3

By Eq. 15-6, the apparent specific gravity

$$G_A = \frac{1982.0/734.4}{1 \text{ g/cm}^3} = 2.699$$

By Eq. 15-7, the bulk specific gravity

$$G_B = \frac{1982.0/759.1}{1 \text{ g/cm}^3} = 2.611$$

$$\text{percentage absorption} = \frac{24.7}{1982.0} \times 100 = 1.25\%$$

15-21 Chemical stability of aggregates

Certain aggregates may be unsuitable for a particular highway construction application because of the chemical composition of the aggregate particles. In asphalt mixes, certain aggregates that have an excessive affinity for water may contribute to what is known as film stripping, leading to disintegration of asphalt surfaces.

An aggregate that is "hydrophobic" in nature may be said to be one which exhibits a high degree of resistance to film stripping in the presence of water. The bituminous substance in a bituminous mixture may generally be assumed to be present in the form of thin films surrounding the aggregate particles and filling, or partially filling, the void spaces between adjacent particles. These thin films of bituminous material adhere to the surface of normal aggregates and contribute to the shearing resistance of the mixture, this effect being generally considered as a part of the "cohesion" of the mix. On continued exposure to water, either in the laboratory or in the field, bituminous mixtures containing certain aggregates show a definite tendency to lose shearing resistance or "strength" because of a decrease in cohesion due primarily to replacement of the bituminous films surrounding the aggregate particles with similar films of water. Aggregates that exhibit this tendency to a marked, detrimental degree are termed "hydrophilic" aggregates, hydrophilic meaning "water-loving." Conversely, aggregates that show little or no decrease in strength due to film stripping are called "hydrophobic" or "water-hating."

In judging the relative resistance to film stripping of aggregates, several different laboratory procedures have been used, the most prominent being a stripping test (7) and an immersion–compression test (8). The stripping test involves coating the aggregate with the bituminous material, soaking the coated aggregate in water for 16 to 18 hr, and then observing whether the total area of the aggregate coated with a bituminous film is above or below 95 percent. The immersion—compression test involves comparison of the compressive strengths of cylindrical specimens of a bituminous mixture—prepared, molded, and tested in a standard fashion—with those of duplicate specimens that have been subjected to immersion in water for a definite and standard time.

Aggregates used in portland cement concrete mixes may also cause problems related to chemical stability. In certain areas, considerable difficulty has been experienced with aggregates that contain deleterious substances which react harmfully with the alkalies present in the cement. Detrimental alkali—aggregate reactions generally result in abnormal expansion of the concrete. Methods have been devised (ASTM Methods C227 and C289) for detecting aggregates with these harmful characteristics, and suitable stipulations are included in typical specifications (for example, ASTM C33).

15-22 Other properties of aggregates

Specifications for aggregates used in highway construction commonly have requirements related to the particle shape, surface texture, and cleanliness of the aggregate. Specifications for aggregates used in bituminous mixes usually require that the aggregates be clean, tough, durable in nature, and free of excess amounts of flat or elongated pieces, dust, clay balls, and other objectionable material.

Similarly, aggregates used in portland cement concrete mixes must be clean and free of deleterious substances such as clay lumps, chert, silt, and other organic impurities.

Cleanliness of the aggregate is generally guaranteed by the inclusion in the specifications of requirements relative to the maximum percentages of various deleterious substances which are permitted to be present. Specific requirements in this respect vary somewhat with different agencies.

BITUMINOUS MATERIALS

Bituminous materials, which are widely used in the United State for road purposes, may be generally defined as substances consisting primarily of bitumen or containing a large percentage of bitumen. Bitumen, in turn, may be said to be a complex organic material. This organic material may occur naturally or it may be artificially created during a production process. Bitumen is, by definition, a material that is soluble in carbon disulfide. Bituminous materials are frequently divided on the basis of their consistency into liquid, semisolid, and solid materials.

The definitions given above are somewhat unsatisfactory from the viewpoint of common experience, and these materials may be further defined as "sticky, adhesive substances that are dark brown to black in color, frequently associated with sharp, characteristic odors, and that are usually liquid at the time of their application to mineral aggregates or similar materials in road construction." This latter description is hardly adequate from any scientific point of view, but it may serve to identify these substances from the standpoint of the casual observer who may not associate solubility in carbon disulfide with everyday experience.

Bituminous materials are of value to the highway engineer principally because of their binding or cementing power and their waterproofing properties. They are now used in the construction of a large number of different types of wearing surfaces and bases, ranging in application from low-cost, light-traffic roads to high-type bituminous pavements intended to carry extremely heavy volumes of traffic. A large majority of the surfaced highways and streets existing in the United States today were built by some process involving the use of bituminous materials.

For the purpose of simplification, bituminous materials used in road construction in this country may be divided into

1. Tars.
2. Asphalts: (a) petroleum; (b) native.

15-23 Tars
Although asphalts and tars are somewhat similar in appearance and general composition, they are produced by greatly dissimilar processes and differ sharply in certain properties that are of importance in highway work. Generally speaking, tars are more susceptible to temperature change than asphalts of similar grade, they are toxic in nature, they possess more free carbon than do asphalts, and they harden more rapidly when exposed to the air after being incorporated in a wearing surface or pavement.

Tar may be generally defined as a substance obtained by the condensation of distillates resulting from the destructive distillation of organic substances such as wood, coal, or peat.

Today, tars are not extensively used as binders for highway pavements. Therefore, the subject of tars is only treated lightly here.

PRODUCTION OF PETROLEUM ASPHALTS

As the term implies, petroleum or "oil" asphalts are asphalts produced from petroleum oil. A wide variety of refining processes may be used to produce petroleum asphalts suitable for use in highway construction. However, petroleum asphalts are, in general, obtained by distillation and by blowing with air at high temperatures.

15-24 Distillation of petroleum oils
As previously stated, petroleum asphalts may be produced by distillation. Two basic types of distillation may be briefly explained at this point.

1. *Fractional distillation,* in which removal of the various volatile materials contained in the crude oil is accomplished by more or less direct or "mechani-

cal" separation. That is, the volatile constituents are driven off at successively higher temperatures without substantial chemical change. In the distillation of asphaltic petroleums, the successive fractions obtained would be those which yield gasoline, naphtha, kerosene, diesel oil, and lubricating oil. The residue obtained would be petroleum asphalt. Fractional distillation may be facilitated through the use of steam or a vacuum.

2. *Destructive distillation,* in which the material undergoes chemical change under application of extreme heat and pressure. This process is generally employed when greater yields of the lighter fractions are desired. A refining process known as "cracking" is a form of destructive distillation, and the "cracked residues" or "pressure tars" that are produced may be used in the manufacture of asphaltic paving materials.

15-25 Steam distillation

Both batch and tube (or pipe) stills may be employed in the steam distillation of petroleums to produce petroleum asphalts. Tube or pipe stills are generally preferred in modern refineries because of their greater efficiency and flexibility of operation. However, the process of steam distillation of an asphaltic petroleum in a batch still will be described here because of its greater simplicity (see Figure 15-10).

In this process the crude oil is placed in a large cylindrical still and its temperature raised by application of heat at the bottom of the vessel. Dry steam is introduced into the still to aid in the vaporization of the more volatile constituents of the petroleum and to minimize decomposition of the distillates and residue. The volatile constituents are collected, condensed, and the various fractions stored for further refining if desired.

The distillation process may be continued until the residue attains the desired consistency. The "stopping point" may be determined by sampling the residue, noting the character of the distillate, or measuring the temperature of the residue. The longer the process of distillation is continued, the harder will be the residue that is produced. The time of distillation may be from 12 to 36 hr, and the final temperature of the residue from 600° to 700°F.

The residue produced by this process is, of course, a petroleum asphalt. Other terms that may be used to describe this residual product are "residual oil," which term may be used when the residue is liquid in consistency, and "residual asphalt," when the residue is of semisolid to solid consistency.

A residue of any desired consistency may be obtained in the process by stopping the distillation at the desired point; when this is done the residue is termed a "straight-run asphalt." In common terminology in the highway field, a semisolid to solid residue produced in this way would be called a "steam-refined asphalt cement." If the residue is liquid, a broad term would be "liquid asphalt."

In further explanation, a straight-run product of liquid consistency would most generally fall into the classification commonly known as "slow-curing liquid asphalts." Liquid asphalts (other than emulsions) used in highway

FIGURE 15-10 Simplified flowchart of recovery and refining of petroleum asphalts. (Courtesy The Asphalt Institute.)

work are most frequently obtained by fluxing or "cutting back" a residual asphalt with a distillate. The "cut-back asphalts" thus produced may be applied cold or at slightly elevated temperatures. Generally speaking, when gasoline or some other light solvent is used to temporarily soften or cut-back the residual asphalt, the liquid asphalt produced is called a "rapid-curing (RC) cut-back asphalt." Similarly, when kerosene or light fuel oil is used, "medium-curing (MC) cut-back asphalt" is produced. Likewise, a slow-curing (SC) liquid asphalt may be obtained by fluxing a residual asphalt with a less volatile distillate such as gas—oil. It should also be noted that these slow-curing materials, whether straight-run or fluxed with a relatively nonvolatile material, are known simply as "road oils" in many sections of the country. Figure 15-10 is a simplified flowchart showing the recovery and refining of asphaltic products prepared from petroleum.

15-26 Cracking
Refining processes that have been devised to increase the yield of light distillates, particularly gasoline, from a given crude oil may be referred to as "cracking" methods. Many different and complex processes are now used to "crack" petroleums, and these methods will not be described here. Generally speaking, however, cracking is accomplished by heating the crude oil to a high temperature under pressure. Temperatures as high as 1100°F and pressures in excess of 50 atmospheres are used in some cracking processes. The residual that is produced is called "pressure tar" or "cracked residue" and is of some consequence in highway usage. In general, asphalts prepared from cracked products are regarded as being less weather-resistant and durable than those prepared from uncracked materials.

15-27 Air blowing
Air blowing is a refining process used in the production of asphalts from residual oils. As the term indicates, air-blown ("oxidized") asphalts are produced by blowing air through residual oil at moderately high temperatures. The residual oil is placed in a still and the temperature raised to from 525° to 575°F. Air is introduced into the bottom of the still and blown through the heated mass under pressure or by applying a partial vacuum at the top of the still. This process is continued for several hours or until the residue has attained the consistency desired. Advantages claimed for this type of refining process include greater yields of asphalt, better control of the consistency or "grade" of asphalt that is produced, and the production of a material that is less susceptible to temperature change than comparable residual asphalts produced by steam distillation. It is also claimed that blown asphalts are more weather-resistant than residual asphalts.

15-28 Emulsification
In order that bituminous materials may be used with unheated aggregate, they must be rendered liquid. One method is by fluxing, as previously described for cut-back asphalts. A second method of making asphalts liquid is emulsification. Petroleum asphalts are usually employed in this process, and an emulsified asphalt may be defined as a dispersion of asphalt particles, around 3 μm in size, in water in the presence of an emulsifying agent. The emulsifying agents commonly used are soaps, used alone or in combination with other substances. This type of asphalt emulsion is further generally described as an "oil-in-water" emulsion.

   The principal objective of the liquefaction of asphaltic materials by this method is to allow their application at normal temperatures without the use of heat. The emulsion breaks down when sprayed or mixed with mineral aggregate in a field construction process; the water is removed, and the asphalt remains as a film on the surface of the aggregate. The water simply serves as a temporary transporting medium and may be compared to the volatile constituents of a cut-back asphalt in that both largely disappear during the mixing process, leaving the semisolid asphalt as a binding material. Three principal classes of asphalt emulsions are used for highway purposes. These are called "rapid-setting," "medium-setting," or "slow-setting" emulsions, depending on the speed at which the emulsion breaks down when placed in contact with mineral aggregates.

Emulsified asphalts may be of the anionic or cationic type, depending on the emulsifying agent. The anionic type contains electronegatively charged asphalt globules and the cationic type contains electropositively charged asphalt globules. It is only within the past few years that the latter type has been produced commercially.

The use of anionic and cationic emulsions is important when using various types of aggregates. The anionic emulsions carry a negative charge and are effective in coating electropositive aggregates such as limestones. The positive charge in a cationic emulsion reacts favorably with an electronegative aggregate such as the highly siliceous aggregates.

Asphaltic emulsions of the types described are usually manufactured by use of a high-speed disintegrator or "colloid mill." In this process the asphaltic material, which may be a cold or heated liquid or even a comminuted solid, and the mixture of water and emulsifying agent are fed simultaneously into the disintegrator. Disintegration is accomplished by passage of the material between revolving or fixed blades or between smooth or slightly roughened plate or conical surfaces, one of which rotates while the other is stationary, or which rotate in opposite directions. The contiguous surfaces of the latter type of machine are very closely spaced, the clearance being in the range of 0.002 to 0.003 in., and the surfaces rotate at very high relative velocities. The asphaltic material is reduced to elements in the colloidal size range and is dispersed in the water. The emulsions thus produced are quite stable and may be transported or stored for long periods of time.

The principal disadvantage of asphalt emulsions is that the water occupies space and has weight. The user then pays for a large percentage of water which has no use other than as a transporting medium for the asphalt. This factor may be of special significance in areas where distances are great and freight rates are high.

On the other hand, asphalt emulsions offer certain advantages in construction, particularly when used with moist aggregates or in wet weather. An emulsion does not require a solvent to make it liquid and generally can be used without additional heat. These factors tend to cause emulsions to be more energy-efficient and less costly than cutback asphalts. Asphalt emulsions are steadily replacing cutbacks in may areas because of concerns for energy conservation, environmental pollution, and economy of construction (9).

PRODUCTION OF NATIVE ASPHALTS

Native asphalt may be defined as asphalt that occurs in a pure or nearly pure state in nature. Native asphalts were extensively used in highway construction in this country in the period from about 1880 to the end of World War I. From that time on the use of petroleum asphalts has steadily increased and native asphalts (not including rock asphalts) play a distinctly minor role in highway construction at the present time. Deposits of native asphalt occur in various parts of the world, although few of the deposits are of commercial importance

LABORATORY TESTS

A large number of different laboratory tests are performed on bituminous materials for the purpose of checking compliance with the specifications that are being used. Most of the tests that are performed in the laboratories of the various highway agencies have come to be more or less routine in nature and are conducted in accordance with methods of test established by the American Association of State Highway and Transportation Officials and the American Society for Testing and Materials. Some of the tests are intended to measure specific properties of the material, while others are used primarily as identification tests or in checking uniformity of the material.

In order to permit the student to obtain a better concept of the number of tests in common use, Table 15-6 has been prepared. This lists the names of the most commonly used tests, the corresponding AASHTO and ASTM designations for each method of test, and indicates whether the test is normally performed on asphaltic materials only, on tar products only, or on both. Some of these test methods are briefly outlined in the following pages. The reader is referred to the designated methods of test for more complete descriptions. In the following discussion an attempt is made to give some idea of the significance of the various tests. Tests performed on emulsified asphalts are mentioned separately, and no attempt has been made to discuss the testing of various miscellaneous substances derived from bituminous materials and of use in a minor way in highway work.

GENERAL TESTS

15-29 Specific gravity

The specific gravity of a bituminous material is defined as the ratio of the weight of a given volume of the material at 25°C to that of an equal volume of water at the same temperature. The specific gravity may be determined by use of a hydrometer, by displacement, or by use of a pycnometer. The hydrometer method has been used in the evaluation of thin fluid materials. In this method, the fluid is brought to the desired temperature (25°C), and the specific gravity is measured directly by immersing a suitably calibrated and marked hydrometer in the fluid mass. The displacement method may be used to determine the specific gravity of bituminous materials that are sufficiently hard and solid to be

TABLE 15-6
LABORATORY TESTS OF BITUMINOUS MATERIALS
USED IN ROAD CONSTRUCTION

Name of Test	AASHTO Designation	ASTM Designation	Appli-cability[a]
General			
Specific gravity	T228	D70-D76	3
Flash point (open cup)	T48	D92	1
Flash point ("Tag" open cup)	T79	—	1
Water in petroleum products, etc.	T55	D95	3
Solubility			
Solubility of bituminous materials	T44	D2042	3
Spot test	T102	—	1
Consistency			
Specific viscosity (Engler)	T54	D1665	2
Kinematic viscosity	T201	D2170	3
Absolute viscosity	T202	D2171	1
Float test	T50	D139	3
Penetration	T49	D5	1
Softening point (ring-and-ball method)	T53		3
Ductility			
Ductility	T51	D113	1
Volatility			
Distillation			
Cut-back asphaltic products	T78	D402	1
Tar products	T52	D20	2
Loss on heating	T47	D6	1
Thin-film oven test	T179	D1754	1
Tests for emulsions	T59	D244	1

[a]1, Asphaltic materials only; 2, tar products only; 3, both asphalts and tars.

handled in fragments. The method involves determination of the weight in air of an irregular piece or a ½-in. cube of the material, its complete immersion in distilled water maintained at the standard temperature, and determination of the corresponding "weight in water." A simple calculation may then be employed to determine the specific gravity. The hydrometer and displacement methods are seldom used, the majority of laboratory determinations being made by the pycnometer method.

Suitable pycnometers used in this test procedure are made of glass; two types are illustrated in Figure 15-11. The clean and dry pycnometer, with stopper, is placed on an analytical balance and the weight W_0 determined. The pycnometer is then completely filled with distilled water at the standard tem-

FIGURE 15-11 Pycnometers used for specific gravity determinations.

perature, the stopper inserted, and the weight W_w determined. In the case of liquid materials that are sufficiently fluid to flow readily through the hole in the glass stopper, the water is removed and the pycnometer completely filled with the liquid. This weight may be called W_1. The specific gravity is then

$$G_b = \frac{W_1 - W_0}{W_w - W_0}$$ (15-9)

where all weights are in grams.

In the case of semisolid or very viscous materials, a small portion of the substance is heated gently, poured into the pycnometer, and then allowed to cool. This weight of pycnometer and contents is called W_2. The pycnometer is then completely filled with water, the stopper inserted as before, and the pycnometer weighed again. This weight may be called W_3. The specific gravity is then given by the expression

$$G_b = \frac{W_2 - W_0}{(W_w - W_0) - (W_3 - W_2)}$$ (15-10)

all weights again being in grams.

The specific gravity of asphaltic products prepared from petroleum varies from about 0.92 for thin, watery liquids to about 1.06 for semisolid materials. Refined tars (road tars) used in highway construction vary in specific gravity from about 1.08 to 1.24. The specific gravity of a bituminous material is of some value for identification purposes and for checking uniformity. Its principal use, however, is in converting from volume to weight measurements and vice versa.

15-30 Flash point

Two methods are in common use for the determination of the flash point of native and petroleum asphalts used in highway work. The flash point of cutback asphalts (RC and MC) is generally determined by use of a Tagliabue Open Cup apparatus, while the Cleveland Open Cup is used for flash-point determinations on other asphaltic materials. The two methods are essentially the same. However, in the Cleveland Open Cup method the asphalt is heated in a metal container suspended in an air bath, while in the "Tag" Open Cup method heating takes place in a glass cup held in a water bath.

In both methods, the temperature of the asphaltic material is gradually increased at a uniform rate. As the temperature rises, a small, open flame is passed at specified intervals across the surface of the heated material. Volatile constituents of the asphalt are driven off as the temperature is increased. As the test proceeds, a point will be reached where sufficient volatiles will collect at the surface of the heated asphalt to briefly ignite and cause a distinct flicker or "flash" across the surface of the material. The minimum temperature at which this flash occurs is called the flash point.

The flash point is an indirect measurement of the quantity and kind of volatiles present in the asphalt that is being tested. Generally speaking, rapid-cure cutback asphalts will have very low flash points (100°F or less); medium-cure cutbacks will have slightly higher flash points; slow-curing liquid asphalts higher; and semisolid materials, such as asphalt cements, will have flash points of 450°F or more. In addition to giving some information as to the amount of volatile constituents that are present in the material, the flash point is indicative of the temperature to which the asphalt can be safely heated under working conditions, especially where heating is being accomplished in an open vessel over an open flame.

15-31 Determination of water

Specifications for bituminous materials to be used in highway construction usually state that the material "shall be free from water," or the amount of water that may be present is limited to a small percentage of the volume of the material. This requirement tends to aid in ensuring careful manufacture, storage, and shipment and has the purpose of preventing "boiling" or "foaming" of the substance when it is heated. Excessive foaming may readily occur when bituminous materials containing any sizable amount of water are heated to normal working temperatures.

The amount of water present in a bituminous substance may be determined by heating a sample of the material that has been combined with an equal amount of suitable petroleum (or coal tar) distillate in a metal still or glass flask. As the temperature is increased, the volatiles, including the water, are driven off and pass through a condenser-and-trap arrangement. The process is continued until all the water has been removed from the sample. The volume of water may be measured in the trap and the percentage, by volume, of water in the sample calculated.

SOLUBILITY TESTS

15-32 Solubility in carbon disulfide and other organic solvents

As previously noted, bitumen is, by definition, a substance that is soluble in carbon disulfide. Petroleum asphalts are practically 100 percent soluble in carbon disulfide, while most native asphalts and tars contain considerable percentages of nonbituminous materials, or "impurities," which may be either organic or inorganic in nature. Obviously, then, this test has value in identifying the nature and source of a given bituminous material.

In the laboratory this test is conducted by dissolving a known weight of the material in a given volume of carbon disulfide and filtering the solution through an asbestos mat formed in the bottom of a Gooch crucible. Insoluble material retained in the crucible is dried and cooled, and the increase in weight of the crucible determined. If only a small portion of the material is retained, the test may be stopped at this point. Further steps that may be taken include ignition of the residue in the crucible and evaporation and ignition of the filtrate. Results are reported simply as "percent bitumen" or "percent soluble in carbon disulfide."

Because carbon disulfide is extremely flammable and highly toxic, other solvents such as carbon tetrachloride, benzene, and trichloroethylene are sometimes used for this test. These substitute solvents are not recommended for tars.

CONSISTENCY TESTS

Several consistency tests are commonly used in connection with the laboratory examination of bituminous materials. The preferred method for measuring the viscosity of liquids is by use of kinematic viscometers, as described in the following section. Consistency tests for bituminous materials in the solid or semisolid state include the float test, the penetration test, and the ring-and-ball test.

15-33 Kinematic viscosity and absolute viscosity

Kinematic viscosity is the quotient of the absolute or dynamic viscosity divided by the density, both at the same temperature (2). It is measured in dimensions of centimeters squared per second, or "stokes." For liquid asphalts, the kinematic viscosity may be measured with a gravity-flow viscometer (AASHTO Designation T201). According to the AASHTO procedure, the time for the fluid to flow between two points in a capillary tube under carefully controlled conditions of temperature and head is measured. Using the measured time in seconds and the viscometer calibration constant, it is possible to compute the viscosity of the material in fundamental units, stokes or centistokes. A kinematic capillary test apparatus is shown as Figure 15-12.

Some agencies measure absolute viscosity of asphalt cements, in poises, using standard test procedure AASHTO Designation T202. This test is usually performed at a temperature of 60°C (140°F). The test involves measuring the time required for a fixed volume of the liquid to be drawn through one of several specially designed capillary tubes by means of a vacuum. The absolute viscosity in poises is determined by multiplying the flow time in seconds by the viscometer calibration factor.

15-34 Float test

As previously indicated, the float test is used for determining the consistency of semisolid to solid road tars. Specified grades of heavier road tars are based on this test, performed at a temperature of either 32° or 50°C. The test is also useful in evaluating the consistency of the residues obtained in the distillation of slow-curing liquid asphalts.

FIGURE 15-12 Apparatus for kinematic capillary viscosity test. (Courtesy The Asphalt Institute.)

This test, which is quite arbitrary in nature, involves use of the apparatus shown in Figure 15-13. In conducting the test, the brass collar B, shown in Figure 15-13, is filled with the material to be tested and brought to a temperature of 5°C by immersing it in ice water. A water bath is prepared, and the temperature of the water in the bath is brought to the desired level (32° or 50°C). The brass collar containing the chilled bituminous material is then screwed into the bottom of the float and the assembly is floated in the water bath maintained at the desired temperature. As time passes, the material in the plug softens, and eventually the water breaks through the plug of bituminous material into the upper part of the float. The time, in seconds, which elapses between the instant when the assembly is floated in the water and that when the water breaks through is the reported value. This time is held to be a measure of consistency.

15-35 Penetration

The penetration of a bituminous substance may be defined as the distance (in hundredths of a centimeter) to which a standard needle penetrates the material under known conditions of time, loading, and temperature. This test is used for evaluating the consistency of asphaltic materials; it is not regarded as suitable for use in connection with the testing of road tars because of the high surface tension exhibited by these materials and the fact that they contain relatively large amounts of free carbon.

The standard penetration test procedure involves use of the standard needle under a load of 100 g for 5 sec at a temperature of 25°C. In Figure 15-14 is shown a so-called precision-type penetrometer that may be used for laboratory measurements of penetration. Simpler and more rugged types of pene-

FIGURE 15-13 Apparatus for float test.

FIGURE 15-14 Penetrometer.

trometers may also be used. The grade of semisolid and solid asphaltic materials is usually designated by the penetration. For example, penetration ranges such as 30–40, 40–50, 50–60, 60–70, 70–85, and 85–100 may be used in specifying the desired grades of asphalt cements prepared from petroleum.

At the beginning of this chapter it was mentioned that bituminous materials could be classed as liquid, semisolid, or solid on the basis of their consistency. This classification may be arbitrarily based on the penetration test, as shown in Table 15-7.

15-36 Softening point (ring-and-ball method)
Since the softening of a bituminous material does not take place at any definite temperature, but rather involves a gradual change in consistency with increasing temperature, any procedure that is adopted for determining the softening point must be of a somewhat arbitrary nature. The procedure in common use in highway materials laboratories is known as the "ring-and-ball method" and may be applied to semisolid and solid materials.

In this test procedure, the sample is melted and poured into a brass ring

TABLE 15-7
PENETRATION LIMITS OF CONSISTENCY

Classification	Penetration	Load (g)	Time (sec)	Temperature (°C)
Liquid	351 or more	50	1	25
Semisolid	⎰ 350 or less	50	1	25
	⎱ 11 or more	100	5	25
Solid	10 or less	100	5	25

which is of $\frac{5}{8}$-in. inside diameter and is $\frac{1}{4}$ in. high. After cooling, the ring is suspended in a water bath maintained at a temperature of 5°C, in such a fashion that the bottom of the filled ring is exactly 1 in. above the bottom of the bath. A steel ball $\frac{3}{8}$ in. in diameter is placed on the surface of the bituminous material contained in the ring, and the temperature of the water is elevated at a standard rate. As the temperature is increased, the bituminous material softens and the ball sinks through the ring, carrying a portion of the material with it. The behavior of the ball is observed, and the softening point is taken to be the temperature at which the bituminous material touches the bottom of the container. The ring-and-ball method is used to determine the penetration index and, in conjunction with penetration and loading time, can be used to estimate viscoelastic properties of both the asphalt and a paving mixture. The method can also be used to advantage to predict high-temperature shear resistance of the paving mixture (3).

DUCTILITY

By ductility is generally meant that property of a material which permits it to undergo great deformation (elongation) without breaking. In regard to bituminous materials, ductility may be further defined as the distance, in centimeters, to which a standard sample or "briquette" of the material may be elongated without breaking. The test is applicable only to semisolid (or solid) asphaltic material that is melted by gentle application of heat and poured into a standard mold. Dimensions of the mold are such that the minimum cross section of the briquette thus formed is exactly 1 cm². The mold is then immersed in a water bath maintained at the desired temperature of the test, which is usually 77°F.

After the material has attained the desired temperature, the sample is placed in a ductility machine. The machine is so arranged that one end of the mold is held in a fixed position while the other end is pulled horizontally at a standard rate. The behavior of the "thread" of material is noted, and elongation is continued until the thread breaks. The distance (in centimeters) that the machine has traveled is the "ductility" of the material. Many asphalt cements have a ductility of 100 or more.

In interpreting the results of this test, it is generally assumed that ductility is a measure of the cementing power of the asphaltic material. Since high cementing qualities are desirable in most applications, it is held that an asphalt that is to be used as a binder should be ductile. However, it should be noted that the exact value of the ductility is not as important as the mere presence or lack of ductility.

VOLATILITY TESTS

15-37 Distillation

The objective of a distillation test performed on a bituminous material is simply to separate the volatile from the nonvolatile substances. By fractional distillation the relative proportions of volatile materials driven off at various specified temperatures may be determined, and the residue may be tested to ascertain the properties of the residual substance contained in the original fluid material.

In testing cut-back asphaltic products, including slow-curing liquid asphalts, the apparatus shown in Figure 15-15 is used. As the distillation proceeds, the volatiles are condensed and collected in the graduated cylinder of Figure 15-15. The original volume of the material contained in the flask is 200 ml, and the temperature of the material is increased at a gradual rate. It may be desirable to know the proportion of distillate driven off at various selected temperatures. The percentage of volatiles collected to any given temperature may be determined on a volumetric basis. The residue is cooled and set aside for additional testing.

The distillation test supplies a very considerable amount of information,

FIGURE 15-15 Distillation apparatus assembly.

as it permits the testing engineer to ascertain the kind and amount of volatile material that has been used in fluxing or "cutting back" a semisolid to produce a liquid substance. Furthermore, by removal of the volatile substances, the engineer creates in the laboratory a material resembling that which will exist in the field; that is, the volatiles are removed, leaving the residual that will go to make up the binder in a mixture of the bituminous material and mineral aggregate. The residue may then be tested to determine whether or not it has the qualities the engineer deems desirable and necessary. Asphalt emulsions are also tested by volatility tests to determine the water content and the amount of the residue from a sample.

15-38 Loss on heating

In determining the percentage of volatile material that will be removed from the asphaltic substance being examined under the imposed conditions of the loss on heating or "volatilization" test, 50 g of the selected material is carefully weighed into a standard flat cylindrical container. The sample is then placed in a specially constructed oven maintained at 163°C and allowed to remain there for a period of 5 hr. At the end of this period the sample is removed from the oven, cooled, and weighed. The loss in weight may then be determined, and the percentage loss on heating, based on the weight of the original sample, may then be calculated. Detailed procedures for this test are given by AASHTO Designation T47.

15-39 Thin-film oven test

Relatively high temperatures are used in the plant mixing of asphalt cements and aggregates. Excessively high temperatures, however, are detrimental, hardening the mixture and reducing pavement life.

In measuring the amount of hardening of an asphalt cement that may be expected to occur during plant mixing, the standard penetration test is made before and after the thin-film oven test (AASHTO Designation T179). Hardening is recorded as the percentage of penetration of the sample before the test. A representative sample, 50 cm^3 in volume, is poured into a pan of specified dimensions such that a film approximately ⅛ in. thick is tested. This sample is then placed on an aluminum rotating shelf in a ventilated oven at 325°F and rotated at a specified rate for 5 hr. It is then poured into a standard container and its penetration determined. The percentage of the penetration before the thin-film oven test is calculated and reported. Specifications usually prescribe minimum values for the percentage of retained penetration for the various grades of asphalt cement.

TESTING OF EMULSIFIED ASPHALTS

Emulsified asphalts are specified on the basis of the results of a complete series of tests, some of which are of a specialized nature. Standard methods of testing emulsified asphalts are designated as AASHTO T59 and ASTM D244. Tests of the following categories are included: composition, consistency, and stability. Composition tests include water content, in which a 50- or 100-g sample is mixed with an equal volume of water-immiscible solvent, distilled, and the quantity of water determined by volume; residue by distillation, in which the percentage by weight of residue is obtained by distillation of a 200-g sample in an iron still to 500°F; and residue by evaporation, which is less frequently used. Consistency is measured by use of the standard Saybolt viscometer with the sample at 77° or 122TF using the Furol tip (AASHTO Designation T72).

Stability tests include demulsibility, in which a 100-g sample is mixed with a dilute solution of calcium chloride and the percentage by weight of asphalt that fails to pass a No. 14 wire cloth determined; settlement, in which the difference in asphalt content of top and bottom samples is determined on a 500-ml sample after allowing it to stand for 5 days; cement mixing, in which the sample is diluted to 55 percent residue with water, mixed with cement, and the percentage by weight of coagulated material which fails to pass a No. 14 sieve determined; a sieve test, in which the percentage by weight that fails to pass the No. 20 sieve is determined; a coating test, in which the ability of a sample to coat a specified stone is judged visually after mixing; and miscibility with water, in which approximately 50 ml of sample is mixed with 150 ml of water and the visible coagulation after 2 hr observed. Tests are also performed on the residue, including specific gravity, ash content, solubility in CS_2, penetration and ductility.

A particle charge test is made to identify cationic emulsions and a pH test is performed to determine their acidity.

CLASSIFICATION OF BITUMINOUS MATERIALS

Bituminous materials are commonly grouped in various classes or grades based on consistency. The classification of these materials facilitates the development of specifications and provides the framework for the establishment of product testing and quality controls.

15-40 Grades of asphalt cements

Until about 1970, the consistencies of asphalt cements were based on the penetration test (see Section 15-35), and specifications were commonly written in those terms. The penetration grades of asphalt cement are

Asphalt Cement Grade	Penetration Range (100 g, 5 sec)
AC 40–50	40–50
AC 60–70	60–70
AC 85–100	85–100
AC 120–150	120–150
AC 200–300	200–300

The grades of asphalt cements are now more commonly based on a standard capillary viscometer test, AASHTO Designation T202 (2). The grades and corresponding viscosity ranges are

Asphalt Cement Grade	Viscosity, 60°C (140°F) (poise)
AC 2.5	250 ± 50
AC 5	500 ± 100
AC 10	1000 ± 200
AC 20	2000 ± 400
AC 30	3000 ± 600
AC 40	4000 ± 800

Another grading procedure for asphalt cements has been adopted by several Western states. It uses the results of vicosity tests made on the residue from a rolling thin-film oven test (AASHTO Designation T240). The grading system is

Asphalt Cement Grade	Viscosity, 60°C (140°F) (poise)
AR 10	1,000 ± 250
AR 20	2,000 ± 500
AR 40	4,000 ± 1,000
AR 80	8,000 ± 2,000
AR 160	16,000 ± 4,000

15-41 Grades of cut-back asphalts

It will be recalled from Section 15-25 that cutback asphalts are classified as rapid-curing (RC), medium-curing (MC), and slow-curing (SC), depending on the type of distillate used to flux or "cut back" the asphalt cement. The amount of distillate determines the viscosity and thus the grade of the liquid asphalt. The more viscous products contain as little as 15 percent diluent, while the less viscous cutbacks may contain up to 40 percent diluent. The grades of cutback asphalts are based on kinematic viscosity (AASHTO Designation T201) and are designated by the lower measure of the viscosity range. The classification system is

Cutback Grade	Kinematic Viscosity (60°C, centistokes)
MC30	30–60
RC, MC, and SC 70	70–140
RC, MC, and SC 250	250–500
RC, MC, and SC 800	800–1600
RC, MC, and SC 3000	3000–6000

15-42 Grades of asphalt emulsions

It was mentioned previously that the consistency of emulsified asphalts is measured by the Saybolt viscosity test, AASHTO Designation T72 *(2)*. This test measures the time required for a 60-ml sample to flow through a calibrated orifice, and the viscosity is reported in seconds. The ranges of such viscosities for various grades of emulsions are not given here but can be found in AASHTO Designation M140 *(2)*.

Emulsions are generally classified as rapid-setting (RS), medium-setting (MS), and slow-setting (SS). Cationic emulsions (defined in Section 15-28) are designated CRS, CMS, and CSS for rapid-, medium-, and slow-setting emulsions, respectively. Emulsions are further categorized according to viscosity grade, for example, RS-1, RS-2, MS-1, and MS-2. Certain emulsions contain the letter "h" (e.g., MS-2h) following certain grades, which means that a harder base asphalt was used.

A new standardized type of medium-setting emulsion is known as "high-float" emulsion and is designated as "HF." The major difference between high-float emulsions and conventional medium-setting emulsions is the high-float characteristic measured on the asphalt residue by the high-float test, AASHTO Designation T50 *(2)*. High-float emulsions are less susceptible to temperature changes. They tend to soften less in the summer and do not harden as much in the winter. These emulsions have a quality that permits a thicker asphalt film on the aggregate particles with a minimum probability of drainage *(9)*.

SPECIFICATIONS FOR BITUMINOUS MATERIALS

A larger number of different specifications for bituminous materials are in use by the various highway agencies in the United States. Some organizations have developed their own specifications on the basis of local experience and to permit the use of local materials. Others have adopted the specifications of AASHTO, ASTM, or some similar body. In any case, the objective of the specifications is the same—to ensure that materials will be secured that will be satisfactory for the use for which they are intended. Recommended specifications for bituminous materials are given by Ref. 2.

USES OF BITUMINOUS MATERIALS

The bituminous materials that have been described in the preceding portions of this chapter find many and varied uses in highway construction and maintenance. Typical uses of various asphalt products are given in Table 15-8. Recommended spraying the mixing temperatures for these substances are given in Table 15-9. Additional information on the nature, uses, and testing of asphalt is given by Ref. *10* and other publications of the Asphalt Institute.

TABLE 15-8
TYPICAL USES OF ASPHALT

Type of Construction	Viscosity graded-original					Viscosity graded-residue					Penetration graded					Rapid curing (RC)				Medium curing (MC)					Slow curing (SC)				Emulsified — Anionic							Emulsified — Cationic					
Grade	AC-40	AC-20	AC-10	AC-5	AC-2.5	AR-16000	AR-8000	AR-4000	AR-2000	AR-1000	40-50	60-70	85-100	120-150	200-300	70	250	800	3000	30	70	250	800	3000	70	250	800	3000	RS-1	RS-2	MS-1	MS-2	MS-2h	SS-1	SS-1h	CRS-1	CRS-2	CMS-2	CMS-2h	CSS-1	CSS-1h
Asphalt-Aggregate Mixtures																																									
Asphalt concrete and Hot laid plant mix																																									
Pavement base and surfaces		x	x			x	x	x	x[b]	x[d]		x	x	x	x[d]																										
Highways		x	x	x		x	x	x				x	x	x																											
Airports	x	x	x	x		x	x	x				x	x																												
Parking areas		x	x	x	x	x	x	x				x	x																												
Driveways			x	x	x	x	x				x																														
Curbs					x	x					x																														
Asphalt-aggregate applications					x						x																														
Surface treatments																																									
Single surface treatment			x	x	x				x					x	x		x	x	x				x	x					x	x						x	x				
Multiple surface treatment			x	x	x								x	x	x	x	x	x	x			x	x	x					x	x						x	x				
Aggregate seal			x	x	x												x	x	x			x							x	x						x	x				
Sand seal				x					x								x												x							x					
Slurry seal																																		x	x[d]					x	x[d]
Penetration macadam																																									
Pavement bases			x											x					x				x	x																	
Large voids																		x	x				x	x					x	x						x	x				
Small voids																x	x				x	x							x	x						x	x				
Asphalt applications																																									
Surface treatment																																									
Fog seal																																		x[c]	x[c]					x[c]	x[c]
Prime coat, open surfaces																					x	x												x		x					
Prime coat, tight surfaces																				x	x																				
Tack coat																x													x					x[c]	x[c]					x[c]	x[c]
Dust laying																x				x	x													x[c]						x[c]	
Crack filling																																									
Asphalt pavements											x[e]																							x[d]	x[d]					x[d]	x[d]
Portland cement concrete pavements	x[e]																																								

[a] For use in cold climates
[b] For use in bases only in cold climates
[c] Diluted with water
[d] Slurry mix
[e] Rubber asphalt compounds

SOURCE: Asphalt Institute.

TABLE 15-9
TYPICAL TEMPERATURES FOR USES OF ASPHALT[a] (°F)

Type and Grade of Asphalts	Pugmill Mixture Temperature[b]		Spraying Temperatures[c]	
	Dense-Graded Mixes	Open-Graded Mixes	Road Mixes	Surface Treatments
Asphalt Cements				
AC-2.5	235-280	180-250	—	270+
AC-5	250-295	180-250	—	280+
AC-10	250-315	180-250	—	280+
AC-20	265-330	180-250	—	295+
AC-40	270-340	180-250	—	300+
AR-1000	225-275	180-250	—	275+
AR-2000	275-325	180-250	—	285+
AR-4000	275-325	180-250	—	290+
AR-8000	275-325	180-250	—	295+
AR-16000	300-350	180-250	—	—
200-300 pen.	235-305	180-250	—	265+
120-150 pen.	245-310	180-250	—	270+
85-100 pen.	250-325	180-250	—	280+
60-70 pen.	265-335	180-250	—	295+
40-50 pen.	270-350	180-250	—	300+
Cutback Asphalts (RC, MC, SC)[d]				
30 (MC only)	—	—	—	85+
70	—	—	65+	120+
250	135-175[e]	—	105+	165+
800	165-210[e]	—	135+	200+
3000	180-240[e]	—	—	230+
Emulsified Asphalts	50-160[f]		70-160	70-185

[a]Temperatures for asphalt cements and cutback asphalts are guides only.
[b]Temperature of mixture immediately after discharge from the pugmill rather than temperature of asphalt cement or cutback asphalt.
[c]The maximum temperature (asphalt cement and cutback asphalt) shall be below that at which fogging occurs.
[d]Application temperatures may, in some cases, be above the flash point of the material. Caution must therefore be exercised to prevent fire or an explosion.
[e]Rapid curing (RC) grades are not recommended for hot pugmill mixing.
[f]Temperature of the emulsified asphalt in the pugmill mixture.

PORTLAND CEMENT

Portland cement is a material that reacts chemically with water by a process called hydration to form a stonelike mass. It was patented by an English mason named Joseph Aspdin in 1824. He named his product portland cement because it produced a concrete that resembled the natural limestone on the Isle of Portland.

Portland cement is composed of four principal compounds: tricalcium silicate, dicalcium silicate, tricalcium aluminate, and tetracalcium aluminoferrite. Raw materials used in the manufacture of portland cement include lime, iron, silica, alumina, gypsum, and magnesia. A wide variety of natural and industrial materials may be used as sources of raw materials in the manufacture of portland cement.

When mixed with water, sand, and gravel, portland cement produces concrete for pavements, bridges, drainage structures, buildings, and other structures. The process by which portland cement concretes are made is described in Chapter 20.

In the manufacture of portland cement, raw materials are crushed and blended to produce the desired proportions of minerals and then burned in a large rotary kiln. The high temperatures (1400° to 1650°C) in the kiln produce rocklike particles called clinker. The clinker is then cooled, blended with gypsum, and ground to a fine powder. The resulting pulverized material is ready for use as portland cement.

The blend of the constituent compounds in portland cement determines how rapidly it sets or hardens in a concrete mix, how much heat is generated by the hardening process, and how it reacts with other chemicals in soils and ground waters.

The American Society for Testing and Materials provides specifications for eight types of portland cement in ASTM Designation C150. Three of these cements are frequently used in highway construction: types I, II, and III.

Type I is what might be termed standard or "normal" portland cement and is intended for use in general concrete construction where the cement is not required to have special properties. Type I cement is normally supplied by the manufacturer, unless one of the other types is specified.

Type II cement is also regarded as a standard type of portland cement and is used for general concrete construction. It is specifically recommended for use in situations in which the concrete will be exposed to moderate sulfate action or where a moderate heat of hydration is required. Types I and II are both widely used by highway agencies at the present time in paving and structural work.

Type III cement is "high early strength" cement. It differs from the standard types described in that concrete made from it attains, in a much shorter period of time, compressive and flexural strengths that are comparable to those attained by concrete in which the same amount of one of the standard types is used. It is useful, for example, in paving operations in which the surface must be opened to traffic very soon after construction is complete.

Specifications for air-entraining cements cover three types that are comparable to types I, II, and III above, except that they contain air-entraining agents which are interground with the cement at the mill. The properties of air-entrained concretes made by using cements of these types will be discussed more fully in Chapter 20, but the use of such a cement may be said to result in the trapping of a small percentage of air which is present in the form of small, disconnected bubbles, distributed uniformly throughout the mix. Air-entrained concretes are, among other things, more durable than comparable concrete mixtures which do not contain air. A number of state highway agencies permit (or require) the use of air-entraining cement, while others secure air-entrainment by the use of an admixture that is incorporated into the concrete during the mixing operation.

Common units of measurement of cement quantity are the sack, which weighs 94 lb and has an approximate loose volume of 1 ft^3, and the ton. The barrel, which is equivalent to four sacks or 376 lb of cement, is seldom used today as a unit of measurement. Shipment of cement is sometimes made in sacks or bags but more commonly in bulk quantities.

PROBLEMS

15-1   Given the following information calculate the group index, and classify each of these soils into the proper subdivision of the AASHTO classification system.

	Sieve Analysis			Characteristics of Soil Binder	
		Percent Passing			
Sample No.	10 (2.00 mm)	40 (0.425 mm)	200 (0.075 mm)	Liquid Limit	Plasticity Index
1	100.0	97.5	56.1	62.3	12.7
2	100.0	73.1	5.4	20.1	NP
3	100.0	46.2	15.9	16.9	NP
4	77.2	37.1	28.2	33.1	6.8
5	100.0	100.0	87.6	60.2	21.7
6	100.0	100.0	83.7	54.2	33.6
7	100.0	56.3	19.1	24.8	10.6
8	100.0	96.3	75.6	33.7	8.9
9	100.0	100.0	83.1	50.6	8.6
10	100.0	100.0	95.0	30.3	11.3
11	100.0	73.7	29.0	42.9	12.7
12	37.1	21.1	8.6	12.3	NP

15-2   Classify each of the soils of Problem 15-1 under the Unified Soil Classification System. The following additional data are available for certain of the samples.

Sample No. 2	$C_u = 3.5$	
Sample No. 4	Passing No. 4 sieve	86.2%
	Passing $\frac{3}{8}$-in. sieve	100.0%
Sample No. 12	Passing No. 4 sieve	49.9%
	Passing $\frac{1}{2}$-in. sieve	80.6%
	Passing 1-in. sieve	100.0%
	$C_u = 5$	
	$C_g = 2$	

15-3 Given the following information from a compaction test performed in the laboratory by the standard Proctor compaction procedure:

$$\text{weight of mold} = 2456 \text{ g}$$

$$\text{volume of mold} = \tfrac{1}{30} \text{ ft}^3$$

Trial No.	Weight of Compacted Soil Plus Mold (g)	Moisture Content (percent)
1	4136	9.4
2	4205	11.2
3	4308	13.1
4	4408	13.9
5	4398	15.8
6	4354	17.9

Draw the moisture–density curve and determine optimum moisture and maximum density for this soil.

15-4 Assuming that the soil of Problem 15-3 has a specific gravity of 2.74, make the necessary computations and plot the "zero air voids curve" on the drawing prepared in Problem 15-3.

15-5 The dry mass of a sample of aggregates is 1206.0 g. The mass in a saturated, surface dry condition is 1226.8 g. The volume of the aggregates, excluding the volume of absorbed water, is 440.6 cm^3. Calculate the bulk specific gravity, the apparent specific gravity, and the percentage absorption.

15-6 The following weights are recorded during the determination of the specific gravity of a bituminous material by the pycnometer method:

Weight of pycnometer, empty	34.316 g
Weight of pycnometer, filled with water	60.000 g
Weight of pycnometer, filled with bituminous material	58.202 g
Temperature (all determinations)	25°C

Calculate the specific gravity of this substance.

15-7 The following are the results of laboratory tests performed on a sample of medium-curing liquid asphalt (MC800) for the purpose of checking compliance with the specifications:

Flash point	120°F
Kinematic viscosity at 140°F	805 centistokes
Distillation	
To 437°F	15.0%
To 500°F	34.5%
To 600°	45.0%
Residue from distillation to 680°F, by volume	75%
Tests on residue from distillation	
Penetration (standard conditions)	110
Ductility, 77°F	100 cm
Solubility in CCL_4	99.1%

Consult Ref. 2 and determine if this material complies with AASHTO specifications.

15-8 Prepare a report on the significance of the following tests for bituminous materials.

(a) Penetration test.
(b) Kinematic viscosity test.
(c) Ductility test.
(d) Solubility in carbon disulfide test.
(e) Thin-film oven test.

REFERENCES

1. *Highway Construction Usage Factors for Cement, Bitumens, Concrete Pipe and Clay Pipe* and *Highway Construction Usage Factors for Aggregates 1981−82−83.* Federal Highway Administration, Washington, D.C. (1984).

2. AASHTO Materials, Part I, Specifications, Part II, Tests, 13th ed. American Association of State Highway and Transportation Officials, Washington, D.C. (1982).

3. Sowers, G. F. *Introductory Soil Mechanics and Foundations*, 4th ed. Macmillan, New York (1979).

4. Atkins, Harold N. *Highway Materials, Soils, and Concretes*, 2nd ed. Reston Publishing Co., Reston, Va. (1983).

5. *Materials Testing*. Joint publication of the U.S. Army, Navy, and Air Force, TM5-530/ NAVFAC MO-330/AFM 89-3 (1971).

6. *Factors Influencing Compaction Test Results*. Highway Research Bulletin No. 319, Highway Research Board, Washington, D.C. (1962).

7. *AASHTO Materials*, Part II, Tests. Designation T 182-82, American Association of State Highway and Transportation Officials, Washington, D.C. (1982).

8. *AASHTO Materials*, Part II, Tests. Designation T 84-81, American Association of State Highway and Transportation Officials, Washington, D.C. (1982).

9. *A Basic Asphalt Emulsion Manual*. Manual Series No. 19, The Asphalt Institute, College Park, Md. (1979).

10. *A Brief Introduction to Asphalt and Some of Its Uses*. Manual Series No. 5, The Asphalt Institute, College Park, Md. (1984).

chapter sixteen

FLEXIBLE PAVEMENT DESIGN

Generally speaking, pavements (and bases) may be divided into two broad classifications or types: rigid and flexible. As commonly used in the United States, the term "rigid pavement" is applied to wearing surfaces constructed of portland-cement concrete. A pavement constructed of concrete is assumed to possess considerable flexural strength that will permit it to act as a beam and allow it to bridge over minor irregularities which may occur in the base or subgrade on which it rests; hence the term "rigid." Similarly, a concrete base that supports a brick or block layer might be described as "rigid." The structural design of this type of pavement deserves special consideration and is discussed in detail in Chapter 20.

16-1 General discussion of flexible pavements

All other types of pavements have traditionally been classed as "flexible." A commonly used definition is that "a flexible pavement is a structure that maintains intimate contact with and distributes loads to the subgrade and depends on aggregate interlock, particle friction, and cohesion for stability" (1). Thus, the classical flexible pavement includes primarily those pavements that are composed of a series of granular layers topped by a relatively thin high-quality bituminous wearing surface. Typically, the highest-quality materials are at or near the surface. It should be pointed out that certain pavements that have an asphalt surface may behave more like the classical "rigid" pavements, for example, pavements that have very thick asphalt surfaces or that have base courses composed of aggregate treated with asphalt, cement, or lime-fly ash. However, for convenience of presentation, these pavements will be considered to be in the flexible class and their design will be considered in this chapter. In Figure 16-1 is shown a typical flexible pavement cross section. The principal elements of this structure are shown to be the pavement that is composed of a "wearing surface," base, subbase (not always used), and subgrade. The wearing surface and the base often comprise two or more layers that are somewhat different in composition and that are put down in separate construction operations. On many heavy-duty pavements, a subbase of select material is often placed between the base and subgrade. The wearing surface may range in thickness from less than 1 in. in the case of a bituminous surface treatment used for low-cost, light-traffic roads to 6 in. or more of asphalt concrete used for heavily traveled routes. The wearing surface must be capable of withstanding the wear and abrasive effects of moving vehicles and must

Handwritten annotations in figure:
- to usually. 2:1 thickness. base → surface course.
- used in a very heavy density (freeway) or soft soil or frozen soil (i.e small voids) acceptable

FIGURE 16-1 Typical flexible pavement cross-section (Courtesy Transportation Research Board.)

possess sufficient stability to prevent it from shoving and rutting under traffic loads. In addition, it serves a useful purpose in preventing the entrance of excessive quantities of surface water into the base and subgrade from directly above.

The base is a layer (or layers) of very high stability and density. Its principal purpose is to distribute or "spread" the stresses created by wheel loads acting on the wearing surface so that the stresses transmitted to the subgrade will not be sufficiently great to result in excessive deformation or displacement of that foundation layer. The base must also be of such character that it is not damaged by capillary water and/or frost action. Locally available materials are extensively used for base construction, and materials preferred for this type of construction vary widely in different sections of the country. For example, the base may be composed of gravel or crushed rock or it may be a granular material treated with asphalt, cement, or lime-fly ash stabilizing agents.

A subbase of granular material or stabilized material may be used in areas where frost action is severe, in locations where the subgrade soil is extremely weak, or where a construction working table is needed. It may also be used, in the interests of economy, in locations where suitable subbase materials are cheaper than base materials of higher quality.

The subgrade is the foundation layer, the structure that must eventually support all the loads which come onto the pavement. In some cases this layer will simply be the natural earth surface. In other and more usual instances it will be compacted soil existing in a cut section or the upper layer of an embankment section. In the fundamental concept of the action of flexible pavements, the combined thickness of subbase (if used), base, and wearing surface must be great enough to reduce the stresses occurring in the subgrade to values that are not sufficiently great to cause excessive distortion or displacement of the subgrade soil layer.

16-2 Status of thickness design

Before World War II, the determination of the combined thickness of flexible pavement and base required in any set of circumstances was largely a matter of judgment and experience. This is attributed (2) to the fact that (1) this method appeared to be giving satisfactory results, (2) much of the basic knowledge required for a more scientific approach to the problem was not available, and (3) the methods of stage construction then in wide use did not seem to require the evolution of more scientific design methods. However, in the period immediately prior to the war and during its early stages, the necessity for economically designing and constructing a truly enormous mileage of airport runways and access roads to military installations focused attention on this design problem. Efforts directed toward the solution of the problem of economical design of flexible pavements and bases were further emphasized by the increased cost of road building during and after the war years.

As a consequence, many governmental agencies entered on comprehensive investigations that led to development of a large number of different design methods. There was (and is) no universal agreement among practicing engineers as to which of these is the best.

The start of construction of the Interstate System in 1956 greatly accelerated efforts to improve pavement design methods. The AASHO road test (Section 16-9) was the most extensive field test in road-building history. One major phase of the road test, which was completed in 1961, was concerned with flexible pavements. Since that time, the results of the road test have been distributed widely.

The AASHTO Interim Guide for Design of Flexible Pavement Structures was originally published in October 1961. Based on a review of this guide and research and experience since that time, a new interim guide was published in 1972 (3). In 1977, it was reported (4) that 32 state highway agencies used the interim guide directly for pavement design.

For many years, some engineers have wanted to design flexible pavements on an analytical basis, that is, by applying the theory of elasticity or similar concepts to the behavior of the layered system that is a flexible pavement. They have been hampered by the complexity of the problem and by the lack of certain parameters to fit into the design equations. At present, aided by the accumulation of vast amounts of experimental data and extensive research activity, implementable flexible pavement designs based on analytical methods are available (5). The Asphalt Institute method, discussed in Section 16-13, is based on an analytical approach.

16-3 Elements of thickness design

From a somewhat simplified viewpoint, the principal factors entering into the problem of the thickness design of flexible pavements are

1. Traffic loading.
2. Climate or environment.
3. Material characteristics.

A number of other elements must also be considered in order to arrive at a final thickness design. These include cost, construction, maintenance, and design period. Thus, the student should realize that the design process is complex, and it is highly unlikely that any extremely simple method of approach will prove entirely successful under all conditions.

16-4 Traffic loading

Protection of the subgrade from the loading imposed by traffic is one of the primary functions of a pavement structure. The designer must provide a pavement that can withstand a large number of repeated applications of variable-magnitude loadings. The primary loading factors that are important in flexible pavement design are

1. Magnitude of axle (and wheel) loads.
2. Volume and composition of axle loads.
3. Tire pressure and contact area.

The magnitude of maximum loading is commonly controlled by legal load limits. Traffic surveys and loadometer studies are often used to establish the relative magnitude and occurrence of the various loadings to which a pavement is subjected. Prediction or estimation of the total traffic that will use a pavement during its design life is a very difficult but obviously important task. Most design procedures provide for an increase in traffic volume on the basis of experience by using some estimated growth rate.

16-5 Climate or environment

The climate or environment in which a flexible pavement is to be established has an important influence on the behavior and performance of the various materials in the pavement and subgrade. Probably the two climatic factors of major significance are temperature and moisture.

The magnitude of temperature and its fluctuations affect the properties of certain materials. For example, high temperatures cause asphaltic concrete to lose stability whereas at low temperatures asphaltic concrete becomes very hard and stiff. Low temperature and temperature fluctuations are also associated with frost heave and freeze–thaw damage. Granular materials, if not properly graded, can experience frost heave. Likewise, the subgrade can exhibit extensive loss in strength if it becomes frozen. Certain stabilized materials (lime, cement, and lime-fly ash treated) can suffer substantial damage if a large number of freeze–thaw cycles occur in the material.

Moisture also has an important influence on the behavior and performance of many materials. Moisture is an important ingredient in frost-related damage. Subgrade soils and other paving materials weaken appreciably when saturated, and certain clayey soils exhibit substantial moisture-induced volume change.

16-6 Material characteristics

Proper design of flexible pavement systems requires a thorough understanding of important characteristics of the materials of which the pavement is to be composed and on which it is to be founded. Depending on the nature of the design procedure, the required material characteristics may vary, but in general the following are desirable:

1. **Asphalt surface.** Strength or stability (possibly repeated load properties).
2. **Granular base and subbase.** Gradation, strength, or stability (shear strength and/or possibly repeated load properties).
3. **Treated or stabilized layers.** Strength (flexural, compressive) and repeated load properties such as fatigue.
4. **Subgrade.** Strength or stability, soil classification, and possibly repeated load properties.

Various standard test methods are available for determining the desired properties. Many of the testing procedures are covered in ASTM and AASHTO Testing Standards (6, 7) while methods for conducting repeated load-type tests on paving and subgrade materials are discussed in Ref. 8. Some of the traditional testing methods are covered in this book. For example, the Marshall stability test for asphalt mixtures is discussed in Section 19-10. The California bearing ratio (CBR) and plate bearing tests are presented in Sections 16-7 and 16-8. Shear strength is discussed briefly in Section 15-6.

16-7 California bearing ratio test : *measure the quality of soil to take a load.*

The basic testing procedure employed in the determination of the California bearing ratio was developed by the California Division of Highways before World War II and was used by that agency in the design of flexible pavements. The basic procedures of this test were adopted by the Corps of Engineers of the U.S. Army during the early stages of the war and served as a basis for the development of design curves that were used for determining the required thickness of flexible pavements for airport runways and taxiways. Certain modifications were made in the test procedure that had been used in California. The modified method adopted by the Corps of Engineers has come to be regarded as the standard method of determining the California bearing ratio. It will be briefly described. The description given here will be further limited to that pertaining to the testing of disturbed samples in the laboratory.

The selected sample of subgrade soil is compacted in a mold that is 6 in. in diameter and 6 to 7 in. high. The moisture content, density, and compactive effort used in molding the sample are selected to correspond to expected field conditions. After the sample has been compacted, a surcharge weight equivalent to the estimated weight of pavement and base is placed on the sample, and the entire assembly is immersed in water for 4 days. At the completion of this soaking period the sample is removed from the water and allowed to drain for a period of 15 min. The sample, with the same surcharge imposed on it, is immediately subjected to penetration by a piston 1.95 in. in diameter moving at a speed of 0.05 in. per minute. The total loads corresponding to penetrations of 0.1, 0.2, 0.3, 0.4, and 0.5 in. are recorded.

A load–penetration curve is then drawn, any necessary corrections made, and the corrected value of the unit load corresponding to 0.1-in. penetration determined. This value is then compared with a value of 1000 lb/in^2 required to effect the same penetration in standard crushed rock. The California bearing ratio (CBR) may then be calculated by the expression

$$\text{CBR (\%)} = \frac{\text{unit load at 0.1-in. penetration}}{1000} (100)$$

(16-1)

Figure 16-2 shows the apparatus required for conducting the California bearing ratio test. It should be noted that this penetration test may also be

CBR: 2% — 33%
quick sand / hard rock

FIGURE 16-2 Apparatus for California bearing ratio test. (Courtesy Soiltest, Inc.)

performed in the field or on "undisturbed" samples. In certain cases the bearing ratio is computed at 0.2-in. penetration rather than at the standard 0.1 in. The value for standard crushed rock at 0.2-in. penetration is 1500 lb/in^2. Various modifications of the basic test procedure have been adopted by highway agencies which use a CBR-based method of design.

16-8 Plate bearing test method

During 1945 and 1946 the Department of Transport of Canada undertook a very comprehensive investigation of runways at a number of principal airports in Canada. This work was reported by Norman W. McLeod (9). Later, McLeod expanded the results of this study and related it to pavement design methods based on the theory of elasticity (10, 11). The plate bearing test is used to evaluate the support capability of subgrades, bases and, in some cases, complete pavements.

The schematic arrangement of equipment used in performing the plate load test is shown in Figure 16-3. Although various sizes of steel plates are used, most tests are conducted with a circular steel plate 1 in. thick and 30 in. in diameter. A thin bed of sand is commonly used to ensure an even bearing between the plate and the underlying material. Load is applied to the plate by means of a large-capacity hydraulic jack equipped with pressures gages graduated in increments of 1000 or 2000 lb. Reaction to the load applied by the jack can be supplied in each test by a loaded tractor-trailer unit, scraper, or other heavy equipment.

Deflection of the bearing plate is measured to the nearest 0.001 in. by means of three Ames dials set at third points around the circumference of the plate.

The results of plate bearing tests are often expressed in terms of a modulus of subgrade reaction as determined by

$$k = p/\delta \qquad (16\text{-}2)$$

where

k = modulus of subgrade reaction (lb/in^3)
p = plate pressure (lb/in.^{2})
δ = plate deflection (in.)

FIGURE 16-3 Equipment used in performing plate bearing tests.

16-9 The AASHO road test

The AASHO road test* was a $27 million project undertaken cooperatively by 49 of the states, the District of Columbia, Puerto Rico, the Bureau of Public Roads, and various industry groups; the test was administered by the Highway Research Board.

Construction began in April 1956, and test traffic began in October 1958. Except for some special tests, traffic ceased on the test road in November 1960.

The test road was located near Ottawa, Illinois, about 80 miles southwest of Chicago. Major portions of the test dealt with flexible pavements, rigid pavements, and short-span bridges. Only one subgrade soil, an A-6, was used in the pavement tests.

The test sections comprised four major loops and two smaller ones. Each loop was a segment of a four-lane divided highway with the parallel roadways connected by a turnaround at each end of the loop; tangent sections of the large loops were 6800 ft long.

Each tangent was built as a succession of pavement sections arranged in such a way that pavement design could be varied from one section to another. The minimum length of any section in the main loops was 100 ft.

In the principal flexible pavement test sections, the surface course was bituminous concrete; the base course, a well-graded crushed limestone; and the subbase, a uniformly graded sand–gravel mixture.

Major design factors in the principal experiments were surface, base, and subbase thicknesses. In the main factorial experiments, three levels of surface thickness existed in combination with three levels of base thickness; each of these nine combinations existed in combination with three levels of subbase thickness. Over the test road as a whole, surface thickness ranged from 1 to 6 in., base thickness from 0 to 9 in., and subbase thickness from 0 to 16 in.

Special experiments on the flexible pavements included three other design variables—base type, bituminous surface treatment, and shoulder paving. Besides, the crushed limestone, bases included a well-graded uncrushed gravel, a bituminous plant mixture, and a cement-treated aggregate.

Test traffic included both single- and tandem-axle vehicles with 10 different axle arrangement–axle load combinations. Single-axle loads ranged from 2,000 to 30,000 lb, tandem-axle loads from 24,000 to 48,000 lb. Each pavement section was tested with one of the 10 combinations; each section was subjected to thousands of load repetitions before it was taken out of the test.

Hundreds of thousands of "bits" of data were gathered by investigators during the road test. These included observations and measurements of pavement condition (cracking and patching done to keep the section in service), longitudinal profile, and transverse profile (to determine rutting and other transverse distortions). Other measurements included surface deflection under loaded vehicles moving at creep speed, deflections at different levels in the pavement structure under vehicles operated at various speeds, curvature of the

---

*At the time of the test, AASHTO was called the American Association of State Highway Officials (AASHO).

pavement under vehicles operated at various speeds, pressures transmitted to the surface of the subgrade soil, and temperature distribution in the pavement layers.

Data gathered in the test were subjected to exhaustive analysis by the road test staff and by other interested organizations.

16-10 The pavement serviceability concept

One of the products of the AASHO road test was the "pavement serviceability concept." In essence, this involves the measurement, in numerical terms, of the behavior of the pavement under traffic, its ability to serve traffic at some instant during its life (12).

Such an evaluation can be made on the basis of a systematic, but subjective rating of the riding surface by individuals who travel over it. Or pavement serviceability can be evaluated by means of certain measurements made on the surface, as was done at the test road. The serviceability rating was on a scale of 0 to 5 with 0 corresponding to a Very Poor rating and 5 representing a Very Good rating.

For flexible pavements, researchers at the test road established that the serviceability index (p) at any time is a function of roughness or slope variance in the two wheel paths, the extent and type of cracking (and patching) of the pavement, and the pavement rutting displayed at the surface. The slope variance is an expression of variations in the longitudinal profile, or longitudinal roughness.

At the start of the test, researchers at the test road determined the average initial serviceability index to be 4.2 for the flexible pavements. In general, the value of the index declined gradually under traffic. When the serviceability index (as measured every 2 weeks) dropped to 1.5 for any section, that section was taken out of test. A value of p of, say, 2.5 is an intermediate one between initial construction and failure to render adequate traffic service.

Analysis of relationships among the flexible pavement variables considered in the test resulted in a series of equations, none of which is given here. Figure 16-4 is one way of showing the relationships among the thickness index, axle load, and number of load applications. The thickness index expresses the total effectiveness of the layered pavement structure (see Figure 16-4).

As an example of the use of this chart, assume that it is desired to determine a pavement structure that would survive 1 million applications of a 22,400-lb single-axle load before its serviceability index dropped to 2.5. From Figure 16-4, the thickness index is about 4.5. Many combinations of asphaltic concrete surface, base, and subbase will meet the conditions of the thickness index equation. One such combination is 4 in. of asphaltic concrete, 10 in. of crushed stone base, and 12 in. of sand–gravel subbase. Figure 16-4 is not intended for use in design, but it can serve as a basis for the development of design procedures such as the one explained in Section 16-11.

Other important conclusions about flexible pavements from the road test were

1. One inch of asphaltic concrete surfacing was about three times as effective

FIGURE 16-4 AASHO road test relationship between thickness index and axel loads at $p = 2.5$. (Courtesy Transportation Research Board.)

as an inch of granular base and about four times as effective as an inch of subbase in improving pavement performance in the range studied.

2. Four types of granular base were studied—crushed stone, gravel, cement-treated gravel, and bituminous-treated gravel. Researchers found the performance of the treated gravel base to be "definitively superior" to that of the untreated crushed stone. Most of the untreated gravel bases failed early in the test.

3. Asphaltic concrete overlays were very effective in extending the service life of flexible pavement sections.

16-11 AASHTO design chart   *skip.*

Coincident with the release of results of the road test, the AASHO Operating Committee on Design released recommendations for design methods for both flexible and rigid pavements (*13*). In 1972 a revised Interim Design Guide was published (*3*).

The AASHO committee accepted the pavement serviceability concept (Section 16-10) and the basic equations developed at the test road without modification. To make the procedure practically useful, however, it had to extend the equations to cover mixed traffic and locations where soils, materials, and climate are different from those at the test site.

One result of the committee's work was the design chart of Figure 16-5, which applies to primary highways with an acceptable value of the pavement serviceability index (*p*) of 2.5 after 20 years of service. The chart incorporates

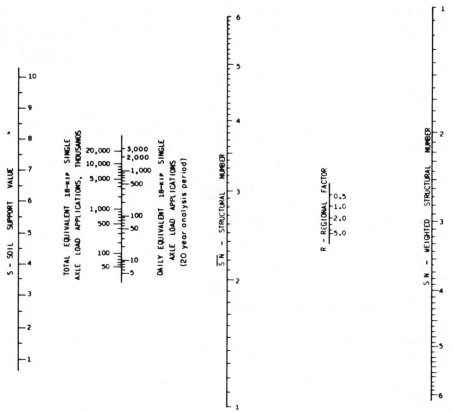

FIGURE 16-5 AASHTO flexible pavement design nomograph. (Courtesy American Association of State Highway and Transportation Officials.)

scales for regional factor (R), equivalent daily or total (for design period) 18-kip (kilopound) single-axle load applications, soil support value (S), structural number ($\overline{SN}$), and weighted structural number (SN). Important facts relative to each of these quantities are summarized below.

1. Regional factor

The regional factor was included in this design method to make it applicable for design of pavements in areas with climatic and environmental conditions different from those at the road test site. The regional factor recognizes that the load-carrying ability of a pavement is a function of its environment; for example, greater damage may be inflicted by traffic during the spring breakup, when the subgrade soil is saturated, than at other times. The committee suggested factors for weighting traffic applications at different seasons of the year to take into account general variations of this type in the strength of subgrade soils.

In the plot of Figure 16-5, a regional factor of 1.0 would have no effect on the structural number. A value above 1.0 would increase the structural number, that is, call for a thicker pavement for the same values of soil support and traffic applications. A regional factor of less than 1.0 is a favorable condition, permitting a thinner pavement structure. The committee suggested that values could range from 0.5 to 3.0 for the continental United States. Figure 16-6 illustrates regional factor contours recommended from results of a recent study (4).

The regional factor can also take into account such unfavorable conditions as a high water table, steep grades, and areas of concentrated turning or stopping movements. Adjustment of the factor for these conditions is a matter of judgment.

2. Load applications

The committee handled the problem of mixed traffic by first adopting an 18,000-lb single-axle load as a standard and then developing a series of "equivalence factors" for each axle weight group. If the estimated traffic to be used in design can be broken down into axle load groupings, the number of load applications in each group can be multiplied by the equivalence factor to determine the number of 18,000-lb axle loads that would have an equivalent effect on the pavement structure (if desired, some axle load group other than 18,000 lb could be chosen as a standard).

The traffic scale in Figure 16-5 is expressed as daily or total (for design period) 18-kip single-axle load applications. The committee tabulated equivalence factors for a range of axle loads, structural numbers, and p values. The factors can be found in Ref. 4.

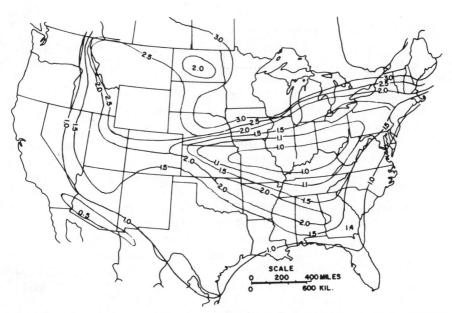

FIGURE 16-6 Generalized regional map of the United States. (Courtesy National Cooperative Highway Research Program, Transportation Research Board.)

3. Soil support value

The soil support value scale shown in Figure 16-5 was established by experience and two numerical values established at the test road—3.0 for the subgrade there and 10.0 for the crushed stone base course.

If one knows the soil support value, the equivalent traffic, and the regional factor, one can find a structural number from Figure 16-5. The procedure recommended by the committee for determining the thicknesses of the various layers in a flexible pavement from the structural number is the same as that explained in Section 16-10. The following equation can be used to determine the thickness of the various layers that have specified layer coefficients:

$$SN = a_1 D_1 + a_2 D_2 + a_3 D_3 \qquad (16\text{-}3)$$

where

SN = structural number (same as thickness index; Figure 16-4)

a_1, a_2, a_3 = layer coefficients representative of surface, base, and subbase courses, respectively

D_1, D_2, D_3 = actual thickness of surface, base, and subbase courses, respectively (in.)

Note that at the AASHO Road Test, the layer coefficients were 0.44, 0.14, and 0.11, respectively, for the asphaltic concrete, crushed limestone, and sand–gravel. Typical layer coefficients for various paving materials are presented in Table 16-1.

16-12 California method

Still another method of thickness design is now being used by the California Department of Transportation (*14, 15*). Original development of this design method was based on the following considerations.

Rather than assuming that the wheel load is distributed or "spread" over some area that varies with depth of the pavement, base, or subbase, this method assumes that the surface and base layers must be designed to resist the potential upward thrust of the subgrade at points adjacent to, but outside, the area which is actually under load. This approach is based on the observed tendency for particles in any layer to be displaced along a curved path and thus exert an upward thrust against the bottom of the base or surface. The principle involved in this concept is shown in Figure 16-7 for a failure involving displacement of the subgrade soil. The concept is that the effect of applying a load over a limited area on a bed of granular material is to force a cone-shaped mass of the material downward, displacing the adjacent material laterally and upward. In this analysis, the pressure on some plane beneath the surface may be greater than at the surface. Present California practice stems from empirical relationships developed from test track and other experiments, plus observation of roads in service.

**TABLE 16-1
STRUCTURAL LAYER COEFFICIENTS
PROPOSED BY AASHO COMMITTEE ON DESIGN,
OCTOBER 12, 1961**

Pavement Component	Coefficient[a]
Surface course	
Road mix (low stability)	0.20
Plant mix (high stability)	0.44[b]
Sand−asphalt	0.40
Base course	
Sandy gravel	0.07[c]
Crushed stone	0.14[b]
Cement-treated (no soil-cement)	
Compressive strength at 7 days	
650 lb/in^2 or more	0.23[c]
400 lb/in^2 to 650	0.20
400 lb/in^2 or less	0.15
Bituminous-treated	
Coarse-graded	0.34[c]
Sand asphalt	0.30
Lime-treated	0.15−0.30
Subbase course	
Sandy gravel	0.11[b]
Sand or sandy clay	0.05−0.10

[a]It is expected that each highway agency will study these coefficients and make such changes as experience dictates.
[b]Established from AASHO Road Test data.
[c]This value has been estimated from AASHO Road Test data, but not to the accuracy of those values marked with the letter b.

SOURCES: *AASHO Interim guide for Design of Flexible Pavement Structures* (1961) and *National Cooperative Highway Research Report 128* (1972).

The general equation used in the California method is

$$GE = 0.0032(TI)(100 - R) \tag{16-4}$$

where

TI = traffic index
R = resistance value of the underlying soil or layer
GE = thickness of the pavement in terms of the gravel equivalent (ft)

The resistance value R of the soil is determined by the stabilometer test methods (*15*). The stabilometer, in simple explanation, is a cell that is capable of measuring the lateral pressure produced by application of a vertical load to a

FIGURE 16-7 Basement soil failure, California method (Hveem).

confined soil mass. The test specimen is 2.5 in. high and 4 in. in diameter. Materials are rated on a scale of 0 to 100 in proportion to their ability to sustain load without transmitting pressure to the sidewalls of the instrument. A liquid, therefore, would have an R value of 0 in the range of loads applied.

Pavements are normally designed to serve the estimated one-way truck traffic for a design period of 20 years following construction. Truck traffic estimates are based on actual counts that have been classified by axle groups. In order to compute the truck volumes for the 20-year design period, the applicable truck counts are expanded to the average year of the design period. The average year would be 10 years after the project is constructed and opened for traffic.

The traffic for each axle group (two axles, three axles, etc.) occurring over the design period is expressed in terms of equivalent 18-kip axle loads (*EAL*). Calculation of *EAL* values involves multiplication of the expanded average daily truck traffic by the following equivalent axle load constants:

Number of Axles	EAL 20-Year Constant
2	1,380
3	3,680
4	5,880
5 or more	13,780

The *EAL* values are then summed and expressed in terms of a traffic index (*TI*), using a conversion table or the following equation:

$$TI = 9.0(EAL/10^6)^{0.119} \qquad (16\text{-}5)$$

The design procedure described here results in a theoretical total thickness expressed in terms of gravel equivalence (*GE*). The theoretical thickness of individual layers must be converted to actual thickness by the use of gravel equivalence factors such as those given in Table 16-2.

TABLE 16-2
GRAVEL EQUIVALENTS OF STRUCTURAL LAYERS IN FEET

Actual Thickness of Layer (ft)	Asphalt Concrete — Traffic Index (TI)											Class B CTB, ATB, CS	Class A CTB	Aggregate base	Aggregate subbase	LCB
	5 and below	5.5 6.0	6.5 7.0	7.5 8.0	8.5 9.0	9.5 10.0	10.5 11.0	11.5 12.0	12.5 13.0	13.5 14.0	14.5 15.0 & up					
(Gf)	2.50	2.32	2.14	2.01	1.89	1.79	1.71	1.64	1.57	1.52	1.50	1.2	1.7	1.1	1.0	1.9
	Gravel Equivalent Factor (Gf)															
0.10	0.25	0.23	0.21	0.20	0.19	0.18	0.17	0.16	0.16	0.15	0.15	.....	.....	.....	.....	.....
0.15	0.38	0.35	0.32	0.30	0.28	0.27	0.26	0.25	0.24	0.23	0.22	.....	.....	.....	.....	.....
0.20	0.50	0.46	0.43	0.40	0.38	0.36	0.34	0.33	0.31	0.30	0.30	.....	.....	.....	.....	.....
0.25	0.63	0.58	0.54	0.50	0.47	0.45	0.43	0.41	0.39	0.38	0.37	.....	.....	.....	.....	.....
0.30	0.75	0.70	0.64	0.60	0.57	0.54	0.51	0.49	0.47	0.46	0.45	.....	.....	.....	.....	.....
0.35	0.85	0.81	0.75	0.70	0.66	0.63	0.60	0.57	0.55	0.53	0.52	0.42	0.60	0.39	0.35	0.67
0.40	1.00	0.93	0.86	0.80	0.76	0.72	0.68	0.66	0.63	0.61	0.60	0.48	0.68	0.44	0.40	0.76
0.45	.....	1.04	0.96	0.90	0.85	0.81	0.77	0.74	0.71	0.68	0.67	0.54	0.77	0.50	0.45	0.86
0.50	.....	1.16	1.07	1.01	0.95	0.90	0.86	0.82	0.79	0.76	0.75	0.60	0.85	0.55	0.50	0.95
0.55	.....	.....	1.18	1.11	1.04	0.98	0.94	0.90	0.86	0.84	0.82	0.66	0.94	0.61	0.55	1.05
0.60	.....	.....	.....	1.21	1.13	1.07	1.03	0.98	0.94	0.91	0.90	0.72	1.02	0.66	0.60	1.14
0.65	.....	.....	.....	1.31	1.23	1.16	1.11	1.07	1.02	0.99	0.97	0.78	1.11	0.72	0.65	1.24
0.70	.....	.....	.....	.....	1.32	1.25	1.20	1.15	1.10	1.06	1.05	0.84	1.19	0.77	0.70	1.33
0.75	.....	.....	.....	.....	.....	1.34	1.28	1.23	1.18	1.14	1.12	0.90	1.28	0.83	0.75	1.43
0.80	.....	.....	.....	.....	.....	1.43	1.37	1.31	1.26	1.22	1.20	0.96	1.36	0.88	0.80	1.52
0.85	.....	.....	.....	.....	.....	1.52	1.45	1.39	1.33	1.29	1.27	1.02	1.45	0.94	0.85	1.62
0.90	.....	.....	.....	.....	.....	.....	1.54	1.48	1.41	1.37	1.35	1.08	1.53	0.99	0.90	1.71
0.95	.....	.....	.....	.....	.....	.....	.....	1.56	1.49	1.44	1.42	1.14	1.62	1.05	0.95	1.81
1.00	.....	.....	.....	.....	.....	.....	.....	1.64	1.57	1.52	1.50	1.20	1.70	1.10	1.00	1.90
1.05	.....	.....	.....	.....	.....	.....	.....	.....	1.65	1.60	1.57	1.26	1.79	1.16	1.05	2.00

Notes: CTB, cement-treated base; ATB, asphalt-treated base; CS, soil cement; LCB, lean concrete base. R values: Class B cement-treated base, 80; aggregate base, 78; aggregate subbase, class 1, 60; aggregate subbase, class 2, 50; aggregate subbase, class 3, 40.

SOURCE: *Highway Design Manual*, Part 7-651. California Department of Transportation (December 1, 1981).

The design formula for flexible pavement design does not allow for deficient layer thicknesses that could result from allowable tolerances in the state specifications. It is therefore necessary to add thickness to the designs, and this is done by adding a safety factor stated in terms of gravel equivalent. It is not necessary to apply a safety factor to the total section thickness; rather, the gravel equivalent of the surface layer or base is increased, and the gravel equivalent of the subbase is decreased. The safety factors recommended by the California Department of Transportation are shown in Table 16-3.

Two cases are encountered in California design practice. The first case is that of an expansive subgrade material. The design procedure involves the determination of various R values for the expansive material, which will vary with moisture content and assumed structural sections. These designs are worked out in the laboratory by a procedure that is too detailed to cover here.

The second, and more usual, case is one in which R values of base and subbase are specified and the minimum R values of the basement soil can be determined, because under the weight of any structural section based on traffic load the basement soil will not be expansive. In such a case, it is not necessary to assume a structural section to perform the laboratory calculations for the R value. Thus, various designs can be determined readily and the most economical section can be determined.

Consider the following illustrative example.

EXAMPLE 16-1. *Design of a Pavement by the California Method* A flexible pavement is to be designed for an eight-lane divided highway, given the following conditions:

R value for the aggregate base $= 78$
R value for aggregate subbase $= 50$
R value for basement soil $\quad = 10$

TABLE 16-3
SAFETY FACTORS RECOMMENDED FOR THE CALIFORNIA METHOD

Base Type	Gravel Equivalent Increase (ft)	Layer Applied to
Class A cement-treated base	0.24	Cement-treated base
Class B cement-treated base	0.18	Asphalt concrete
Asphalt-treated base	0.18	Asphalt concrete
Lime-treated base	0.18	Asphalt concrete
Soil cement	0.18	Asphalt concrete
Aggregate base	0.16	Asphalt concrete

SOURCE: *Highway Design Manual*, Part 7-651. California Department of Transportation (December 1, 1981).

Expanded average daily trucks for the outside lanes* are

Vehicle Type	Average Daily Trucks
2-axle trucks	935
3-axle trucks	550
4-axle trucks	225
5-axle or more trucks	1025

Solution. The equivalent 18-kip axle loads (*EAL*) are computed as follows:

		Outside Lanes	
(1) Vehicle Type	*(2)* EAL 20 Year Constants	*(3)* Expanded Average Daily Trucks	*(4)* Total 20 Year EAL (Col. 2 × Col. 3)
Two-axle trucks	1,380	935	1,290,300
Three-axle trucks	3,680	550	2,024,000
Four-axle trucks	5,880	225	1,323,000
Five-axle or more	13,780	1,025	14,124,500
Totals	—	—	18,761,800

By Eq. 16-5,

$$TI = 9.0(18.7618)^{0.119} = 12.75$$

The gravel equivalent, *GE*, required for the asphalt concrete surface layer is computed by Eq. 16-4:

$$GE = 0.032(12.75)(100 - 78) = 0.89 \text{ ft}$$

From Table 16-3, a safety factor of 0.16 is added, resulting in an adjusted *GE* value of 1.05 ft. From Table 16-2, under the column heading (*TI*) 12.5 and 13.0, the nearest value to 1.05 is 1.02. Reading horizontally to the left in the first column, the actual thickness of the layer is 0.65 ft.

Using Eq. 16-4 again, the required gravel equivalent of the combined surface and base is 2.04 ft. Adding the 0.16-ft safety factor required for the surfacing, the total gravel equivalent required over the subbase is 2.20 ft. Subtract 1.02 (the *GE* of the actual 0.65 ft of asphalt concrete selected) from 2.20 for a required gravel equivalent of 1.18 ft for the aggregate base.

From Table 16-2, under the column heading "Aggregate Base," the

---

*For projects with three or more lanes in one direction, it is recommended that separate designs be made for inside and outside lanes, resulting in steps at the bottoms of the pavement and base.

nearest value to 1.18 ft is 1.16, the last figure in the column. This corresponds to an actual thickness of 1.05 ft of aggregate base.

Using Eq. 16-4 and the R value of 10 for the basement soil, the total gravel equivalent is 3.67 ft. Subtracting the sum of the gravel equivalents for the pavement and base, the required gravel equivalent is 1.49. Since the gravel equivalent factor (G_f) is 1.0 for subbase and all layers are determined to the nearest 0.05 ft, the actual thickness of subbase is 1.50 ft.

The resulting pavement section is:

Surface	0.65 ft
Base	1.05 ft
Subbase	1.50 ft

Alternative designs can be prepared with other available materials, costs can be compared, and a final design can then be selected.

16-13 The Asphalt Institute method

The Asphalt Institute has published a pavement thickness design manual (*17*) which characterizes asphalt pavement as a multilayered elastic system. Using established theory, experience, and test data, the institute's engineers developed a structural thickness design method suitable for a variety of asphaltic pavements.

The method is based on two assumed stress–strain conditions:

1. The wheel load, W, is transmitted to the pavement surface through the tire as a uniform vertical pressure P_0. The stresses are then spread through the pavement structure to produce a reduced maximum vertical stress, P_1 at the subgrade surface. (See Figure 16-8.)
2. The wheel load, W, causes the pavement structure to deflect, creating both compressive and tensile stresses in the pavement structure. (See Figure 16-9.) In developing the design procedure, Asphalt Institute engi-

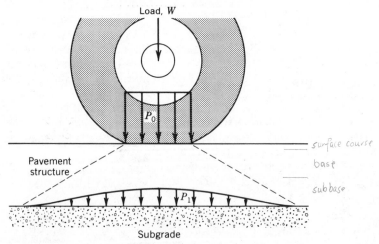

FIGURE 16-8 Spread of wheel load pressure through pavement structure. (Courtesy The Asphalt Institute.)

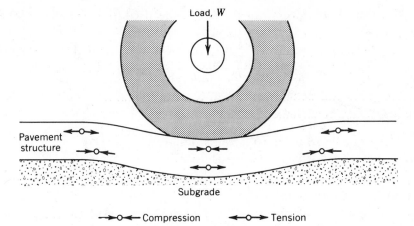

FIGURE 16-9 Pavement deflection results in tensile and compressive stresses in pavement structure. (Courtesy The Asphalt Institute.)

neers calculated induced horizontal tensile strains, ϵ_t, at the bottom of the asphalt layer and vertical compressive strains, ϵ_c, at the top of the subgrade. (See Figure 16-10.)

Thus, the method addresses two of the most commonly occurring traffic-related modes of flexible pavement distress: fracture or cracking of the asphalt-treated layer, especially the surface layer, and distortion or rutting of the subgrade and the other layers of the pavement system.

FIGURE 16-10 Strains in a flexible pavement.

A computer program (called DAMA) was used to determine the thicknesses for the two strain criteria. The materials in the pavement layers were characterized by a modulus of elasticity and a Poisson ratio. Specific values of these characteristics were selected on the basis of experience and extensive laboratory tests. Two thicknesses, one for each critical strain value, were calculated for various combinations of subgrade and loading conditions. The larger of the two values was used to prepare pavement design charts such as that shown as Figure 16-11.

The Asphalt Institute's thickness design manual (*17*) includes design charts for 10 types of pavement structures:

1. Full-depth asphalt concrete utilizing an asphaltic concrete for both the surface and the base.
2. Emulsified asphalt mix (type 1) mixes constructed with processed, dense graded aggregate and an asphaltic emulsion.
3. Emulsified asphalt mix (type 2) made with semiprocessed, crusher-run, pit-run, or bank-run aggregates and an asphaltic emulsion.
4. Emulsified asphalt mix (type 3) made with sands or silty sands and an asphaltic emulsion.
5. Asphalt concrete over an untreated aggregate base with thicknesses of 4, 6, 8, 10, 12, and 18 in.

Traffic analysis

The Asphalt Institute recommends that the effects of traffic on pavement structural design be expressed in terms of the number of equivalent 18,000-lb single-axle loads (EAL). This approach, which is similar to that recommended by AASHTO (*3*) involves the following steps:

1. Estimate the number of vehicles of different types (such as passenger cars, single-unit trucks, and multiple-unit trucks of various sizes and configurations) expected to use the proposed pavement over the design period. During the design period, the pavement is assumed to support the cumulative effects of traffic and provide a satisfactory riding quality without major rehabilitation. The designer should make allowance for traffic growth, basing the growth rate on historic records for the given road or comparable facility. Assuming a compounded rate of growth, the total volume of traffic expected during the design period is

$$T = \left[\frac{(1 + r)^n - 1}{r} \right] T_1 \tag{16-6}$$

where

T_1 = traffic volume during the first year
r = rate of growth expressed as a fraction
n = design period (years)

2. Estimate the percentage of total truck traffic expected to use the design lane. (The design lane is the lane expected to receive the most severe service.)

FIGURE 16-11 The Asphalt Institute design chart for pavement with an 8-in. untreated aggregate base. (Courtesy The Asphalt Institute.)

This estimate can be made from traffic observations for the specific facility or from Table 16-4.

(3.) For each weight class, determine the *truck factor*, the number of equivalent 18,000-lb single-axle load applications contributed by the passage of a vehicle. Truck factors can be computed or estimated from Table 16-5. The truck factors are computed by multiplying the number of axles in each weight class by an appropriate *load equivalency factor*, summing the products for the various weight classes, and dividing the sum by the total number of vehicles involved. The load equivalency factor is the number of equivalent 18,000-lb single-axle load applications contributed by the passage of an axle. Typical load equivalency factors are given in Table 16-6.

EXAMPLE 16-2. *Computation of Equivalent 18,000-lb Load Applications* During the first year of service, a pavement is expected to accommodate the following numbers of vehicles in the classes shown. Estimate the equivalent 18,000-lb single-axle load applications (*EAL*).

Vehicle Type	No. of Vehicles		Truck Factors		Product
Single-unit trucks					
Two-axle, four-tire	87,600	×	0.02	=	1,750
Two-axle, six-tire	23,600	×	0.19	=	4,480
Three-axle or more	4,400	×	0.56	=	2,460
Tractor semis and combinations					
Three-axle	2,100	×	0.51	=	1,180
Four-axle	7,300	×	0.62	=	4,530
Five-axle or more	50,200	×	0.94	=	47,190
			EAL = Sum	=	61,590

TABLE 16-4
PERCENTAGE OF TOTAL TRUCK TRAFFIC IN DESIGN LANE

Number of Traffic Lanes (Two Directions)	Percentage of Trucks in Design Lane
2	50
4	45 (35−48)[a]
6 or more	40 (25−48)[a]

[a]Probable range.

TABLE 16-5
DISTRIBUTION OF TRUCK FACTORS FOR DIFFERENT CLASSES OF HIGHWAYS AND VEHICLES IN THE UNITED STATES

Vehicle Type	Highway System Type		
	Interstate Rural	Other Rural	Urban
Single-unit trucks			
Two-axle, four-tire	0.02	0.02	0.03
Two-axle, six-tire	0.19	0.21	0.26
Three-axle or more	0.56	0.73	1.03
Tractor-semitrailers			
Three-axle	0.51	0.47	0.47
Four-axle	0.62	0.83	0.89
Five-axle or more	0.94	0.98	1.02

SOURCE: *Thickness Design—Asphalt Pavements for Highways and Streets.* Manual Series No. 1, The Asphalt Institute, College Park, Md. (1981).

TABLE 16-6
TYPICAL LOAD EQUIVALENCY FACTORS

Gross Axle Load (lb)[a]	Single Axles	Tandem Axles
5,000	0.00500	—
10,000	0.0877	0.00688
15,000	0.478	0.0360
20,000	1.51	0.1206
25,000	3.53	0.308
30,000	6.97	0.658
35,000	12.50	1.23
40,000	21.08	2.08
45,000	34.00	3.27
50,000	52.88	4.86
55,000		6.93
60,000		9.59
65,000		12.96
70,000		17.19
75,000		22.47
80,000		28.99

[a] 1 lbf = 4.4482 N.

SOURCE: *Thickness Design—Asphalt Pavements for Highways and Streets.* Manual Series No. 1, The Asphalt Institute, College Park, Md. (1981).

EXAMPLE 16-3. *Design EAL for 20-Year Design Period*   If the traffic using the pavement grows at an annual rate of 4 percent, determine the design *EAL* for a 20-year design period.

By eq. 16-6

$$\text{design } EAL = \left[\frac{(1 + 0.04)^{20} - 1}{0.04} \right] 61{,}590 = 1{,}834{,}000$$

Note that if the traffic is expected to grow nonuniformly among weight classes, Eq. 16-6 should be applied to each weight class using appropriate rates of growth.

Materials evaluation

In the Asphalt Institute pavement design procedure, the subgrade strength is characterized by its modulus of elasticity or *resilient modulus*. In the standard method of test for the resilient modulus of subgrade soils (7), a specially prepared and conditioned specimen is subjected to repeated application of axial deviator stress of fixed magnitude, duration, and frequency. During the test, the specimen is subjected to a static all-around stress in a triaxial pressure chamber. The test is intended to simulate the conditions that exist in pavements subjected to moving wheel loads. Detailed procedures for the test are given by AASHTO Designation T274-82 (7).

The Asphalt Institute has also published equations by which it is possible to estimate the resilient modulus from the standard CBR or *R* value test results (17). The equations are

$$M_r \text{ (psi)} = 1500 \text{CBR}$$

$$M_r \text{ (psi)} = 115 + 555(R \text{ value})$$

The Asphalt Institute method recommends that the design subgrade resilient modulus M_r be based on the expected level of traffic, expressed in *EAL*s. To ensure a more conservative design, lower values of M_r are used for higher volumes of anticipated traffic. Specifically, it is recommended that the resilient modulus be obtained for six to eight samples of the subgrade. The results are then arranged in ascending order and plotted as a cumulative distribution as exemplified by Figure 16-12. The design subgrade resilient modulus is then chosen from the curve as the value that 60, 75, or 87.5 percent of the values are equal to or greater than. The applicable percentile depends on the traffic level (*EAL*) as shown in Table 16-7.

16-14 National Crushed Stone Association design method   ‌$5 \, Kip$

The National Crushed Stone Association (NCSA) (18) has adopted a flexible pavement design method patterned very closely after the CBR method developed by the U.S. Army Corps of Engineers. The basis of this method is to provide adequate thickness and quality of material to prevent repetitive shear deformations within any layer. In addition, the climatic factor of frost, if pertinent to the particular design, can be taken into account.

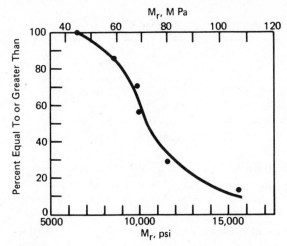

FIGURE 16-12 Typical distribution of subgrade resilient modulus test results. (Courtesy The Asphalt Institute.)

TABLE 16-7
SUBGRADE DESIGN LIMITS

Traffic Level EAL	Design Subgrade Percentile Value
10^4 or less	60
Between 10^4 and 10^6	75
10^6 or more	87.5

SOURCE: *Thickness Design—Asphalt Pavements for Highways and Streets.* Manual Series No. 1, The Asphalt Institute, College Park, Md. (1981).

The factor of traffic is incorporated in the design by use of the design index. Based on the number of equivalent 18-kip single-axle loads (EALs) per lane per day, various categories of design index are assigned. Either the AASHTO load equivalencies or an estimate of the EAL can be made. The appropriate design index is then obtained from Table 16-8.

The basic design thickness of the flexible pavement for a given class of service is dictated by the degree of soil support available. The CBR test is used to evaluate subgrade strength. In the NCSA procedure, the design strength is that CBR equaled or exceeded by 75 percent of the subgrade. NCSA has established four categories of subgrade soil support; these are shown in Table 16-9.

When the soil support and traffic conditions are properly evaluated, an appropriate basic design thickness for normal or temperate climatic conditions can be ascertained. Even for cold regions, this step should be completed prior to any thickness adjustment. Figure 16-13 is entered with the design subgrade

TABLE 16-8 (*18*)
DESIGN INDEX CATEGORIES FOR TRAFFIC

Design Index	General Character	Daily EAL[a]
DI-1	Light traffic (few vehicles heavier than passenger cars, no regular use by group 2[c] or 3 vehicles)	5 or less
DI-2	Medium-light traffic (similar to DI-1, maximum 1000 VPD,[b] including not over 5% group 2, no regular use by group 3 vehicles)	6−20
DI-3	Medium traffic (maximum 3000 VPD, including not over 10% group 2 and 3, 1% group 3 vehicles)	21−75
DI-4	Medium-heavy traffic (maximum 6000 VPD, including not over 15% group 2 and 3, 1% group 3 vehicles)	76-250
DI-5	Heavy traffic (maximum 6000 VPD, may include 25% group 2 and 3, 10% group 3 vehicles)	251-900
DI-6	Very heavy traffic (over 6000 VPD, may include over 25% group 2 or 3 vehicles	901-3000

[a]EAL equivalent 18-kip axle loads in design lane, average daily use over life expectancy of 20 years with normal maintenance.
[b]VPD = vehicles per day, all types, using design lane.
[c]See text for description of Groups 1, 2, and 3 vehicles.

TABLE 16-9 (*18*)
SOIL SUPPORT CATEGORIES

General Soil Description	Strength-CBR
Excellent	15 plus
Containing a uniformly high percentage of granular materials	
Unified Soil Classes: GW, GM, GC, GP; some SM, SP, and SC	
AASHTO Soil Groups: A-1, A-2, some A-3's	
Good	10−14
Containing some granular materials intermixed with silt and/or light clay	
Unified Soil Classes; SM, SP, SC; some ML, CL, CH	
AASHTO Soil Groups: A-2, A-3; some A-4's, a few A-6's or A-7's	
Fair	6−9
Sand clays, sandy silts or light silty clays if low in mica content; may have some plasticity	
Unified Soil Classes: ML, CL; some MH, CH	
AASHTO Soil Groups: Ranging from A-4 to A-7 (low group indices)	
Poor	5 or less
Plastic clays, fine silts, very fine or micaceous silty clays	
Unified Soil Classes: MH, CH, OL, OH; (PT unsuitable)	
AASHTO Soil Groups: Ranging from A-4 to A-7 (higher group indices)	

CBR

Design Thickness in Inches

FIGURE 16-13 Thickness design chart. (From U.S. Army Engineers Manual TM 5-822-5; figure courtesy National Crushed Stone Association.)

CBR and the appropriate design index. The heavy dashed line and arrows represent the method to be followed with Fig. 16-13 to obtain a basic design thickness in inches. Alternatively, Table 16-10 can be used to estimate the basic design thickness.

The basic design should now be checked for adequacy if the pavement is to be in an area of severe climate. A severe climate is one in which detrimental frost action can occur and where the depth of frost penetration is normally greater than the basic design thickness. For such detrimental frost to occur, three conditions must prevail: moisture is available, soils or layers of the pavement itself are frost-susceptible, and freezing conditions penetrate into these materials. The Army has classified the frost susceptibility of soils of various types as shown in Table 16-11.

TABLE 16-10 (*18*)
BASIC DESIGN THICKNESS TABLE (NORMAL CLIMATIC CONDITIONS)

Subgrade Soil		Design Thickness (in.) for Indicated Traffic Intensity Categories					
Class	CBR	DI-1	DI-2	DI-3	DI-4	DI-5	DI-6
Excellent	15+	5	6	7	8	9	10
Good	10−14	7	8	9	10	11	12
Fair	6−9	9	11	12	14	15	17
Poor[a]	5 or less	Subgrade improvement recommended					

[a]Poor subgrade soil should be improved to "fair" or better by protecting them with available "select" materials or by stabilization. The depth or improvement required should be adequate to provide protection to unimproved soil beneath (determined from Figure 16-13).

TABLE 16-11 (18)
FROST GROUP CLASSIFICATION

Frost Group	Percentage Finer than 0.02 mm	Unified Soil Classifications[a]	Frost Susceptibility
F-1			
(a) Gravelly soils	3−10	GW, GP, GW-GM, or GP-GM	Low
F-2			
(a) Gravelly soils	10−20	GM, GW-GM or GP-GM	Low to
(b) Sands, sand clays	3−15	SW, SP, SM, SW-SM, or SP-SM	medium
F-3			
(a) Gravelly soils	Over 20	GM or GC	
(b) Sands, coarse to medium	Over 15	SM or SC	High
(c) Clays, PI>12	—	CL or CH	
F-4			
All silts, very fine silty sands, clays w/PI<12, etc.	Over 15	ML, MH, SM, CL, CL-ML, CH and alternately banded deposits	Very high

[a]Reference ASTM Standard D2487.

Appropriate design thicknesses for pavements to be placed on frost-susceptible soils in areas of deep frost penetration and readily accessible moisture are given in Table 16-12. These design thicknesses do not provide complete protection from any frost action but do take into account reduced soil support or CBR during certain seasons of the year.

If the thickness required for frost conditions is greater than the basic thickness, then it should be used; otherwise the basic thickness should be used.

Having determined the proper total design thickness for the specific conditions, the type and thickness of component layers must be ascertained. Basically the total design thickness is divided into surface, base, and subbase (the latter not always used). The surface course should be high-quality asphaltic concrete and the base and subbase (if used) should be high-quality granular material. Table 16-13 contains surface thicknesses recommended by NCSA. The total base and subbase (if used) thicknesses are determined by subtracting the surface thickness from the total design thickness.

TABLE 16-12 (*18*)
DESIGN THICKNESS, FROST GROUP BASIS[a,c]

Subgrade Soil Frost Group	Design Thickness (in.)[b] for Indicated Traffic Intensity Categories					
	DI-1	DI-2	DI-3	DI-4	DI-5	DI-6
F-1	9	10	12	13	15	17
F-2	10	12	14	16	18	20
F-3	15	18	22	25	28	30
F-4	Subgrade improvement recommended					

[a]Table devised from Figure 19, U.S. Army TM 5-818-2.
[b]1 in. = 2.54 cm.
[c]Design thicknesses may be conservative except where both adverse moisture conditions and deep freezes are common. F-4 soils should be upgraded to F-3 or better prior to construction. This operation should be extended to the full depth of frost penetration.

TABLE 16-13 (*18*)
SURFACING THICKNESS RECOMMENDATIONS

Traffic Intensity Category	Minimum Surfacing Required
DI-1	1 in.[a] (use surface treatments)
DI-2	2 in.
DI-3	2.5 in.
DI-4	3 in.
DI-5	3.5 in.
DI-6	4 in.

[a]1 in. = 2.54 cm.

16-15 Conclusion

In recent years, a great deal of progress has been made in developing more reliable pavement structural design methods. Formerly, design procedures relied heavily on subjective engineering judgments based on evaluations of pavement performance under service conditions. As more is learned about the mechanical properties of paving and subgrade materials and the mechanics of pavement failure, engineers are turning to more rational approaches to pavement design. The efforts of researchers to refine and improve pavement design methodology are certain to result in pavements that are more economical and long-lasting.

REFERENCES

1. *Standard Nomenclature and Definitions for Pavement Components and Deficiencies.* Special Report No. 113, Highway Research Board, Washington, D.C. (1970).

2. *Highway Practice in the United States of America.* Public Roads Administration, Washington, D.C. (1949).

3. *AASHTO Interim Guide for Design of Pavement Structures—1972.* American Association of State Highway and Transportation Officials, Washington, D.C. (1972).

4. Van Til, C. J., et al. *Evaluation of AASHO Interim Guides for Design of Pavement Structures.* National Cooperative Highway Research Program Report No. 128, Washington, D.C. (1972).

5. Monismith, C. L., and Finn, F. W. Flexible Pavement Design: State of the Art 1975. *Transportation Engineering Journal,* Vol. 103, No. TEI (1977).

6. *Annual Book of Standards: Bituminous Materials, Soils, Skid Resistance,* Part 11. American Society for Testing Materials, Philadelphia, Pa.

7. *Standard Specification for Transportation Materials and Methods of Sampling and Testing.* American Association of State Highway Officials, Washington, D.C. (1982).

8. *Test Procedures for Characterizing Dynamic Stress—Strain Properties of Pavement Materials.* Special Report No. 162, Transportation Research Board, Washington, D.C. (1975).

9. McLeod, N. W. *Airport Runway Evaluation in Canada.* Highway Research Board Research Report No. 4B (1947) and Supplement (1948).

10. *Flexible Pavement Design.* Highway Research Record 13 (1963).

11. *Flexible Pavement Design—1963 and 1964.* Highway Research Record 71 (1965).

12. *The AASHO Road Test: Pavement Research.* Special Report No. 61E, Highway Research Board, Washington, D.C. (1962).

13. *International Conference on the Structural Design of Asphalt Pavements.* Preprint volume (supplement), University of Michigan, Ann Arbor (1962).

14. Hveem, F. N., and Carmany, R. M. The Factors Underlying the Rational Design of Pavements. Proceedings, Vol. 28, Highway Research Board, Washington, D.C. (1948).

15. Resistance R-Value and Expansion Pressure of Compacted Soils, AASHTO Designation T190−78. *Standard Specifications for Transportation Materials and Methods of Sampling and Testing,* American Association of State Highway and Transportation Officials, Washington, D.C. (1982).

16. *Highway Design Manual 7-651-*State of California, Department of Transportation, Sacramento (1981).

17. *Thickness Design—Asphalt Pavements for Highways and Streets.* Manual Series No. 1, The Asphalt Institute, College Park, Md. (1981).

18. *Flexible Pavement Design Guide for Highways.* National Crushed Stone Association, Washington, D.C. (1972).

chapter seventeen

EARTHWORK OPERATIONS AND EQUIPMENT

Nearly all highway construction jobs, especially those in new locations, involve a considerable amount of earthwork. In general terms, earthwork operations are those construction processes that involve the soil or earth in its natural form and that precede the building of the pavement structure itself.

Basic earthwork operations may be classified as follows: clearing and grubbing; roadway, borrow, structural, and several special forms of excavation; the formation of embankments; and finishing operations. Any or all of these construction processes may be performed on a given highway project, and they may overlap to a certain extent. This chapter includes a description of the various procedures used in this phase of highway construction. It should be remembered that the discussion which follows will deal largely with average practice and that the specific procedures followed by any one organization may differ considerably from those given here. It should also be noted that the construction specifications which are discussed are intended to be typical of those used by highway agencies in the direction and control of contract work.

Before describing typical earthwork procedures, let us briefly consider the principal tools that are used in earthwork operations.

17-1 Equipment used in earthwork operations

In modern highway practice in the United States, most earthwork operations are accomplished by the use of a large number of highly efficient and versatile machines. The development and application of such machines in the past five decades has been an outstanding phase of highway progress. Machines have been developed that are capable of efficiently and economically performing every form of earthwork. Some machines are tailored to perform a specific operation or series of operations, while others have a wide variety of uses. The proper selection and efficient use of these machines are an important part of the highway scene.

To the beginner in this subject, the large number and almost infinite variety of machines used in earthwork operations are undoubtedly bewildering. Not only are there many different basic types of machines and modifications of them, but there are many manufacturers engaged in the production of machines of similar characteristics and uses. Space does not permit an extensive description of equipment used in highway earthwork operations. The most common types of equipment are listed in Table 17-1 along with an indication of

TABLE 17-1
TYPICAL APPLICATIONS OF HIGHWAY CONSTRUCTION EQUIPMENT

Equipment Type	Typical Uses
Crawler tractor	Various pushing/pulling operations such as push-loading scrapers
Bulldozer units	Ripping and loosening of rocky earth in preparation for excavation; earthmoving operations for short distances, especially in rough, rocky ground and on steep slopes; tree removal
Rubber-tired tractor scraper units	Earthmoving operations—digging, loading, transporting, dumping, and spreading earth material
Farm tractor with front-end loader	Miscellaneous operations associated with earthwork, such as filling trenches, excavation in limited areas, and cleaning up along aggregate stockpiles
Shovel	Excavation at or above existing ground level in firm or hard material, e.g., excavating material from cut slope
Backhoe	Excavation below existing ground in firm or hard material, as in digging trenches
Crane with clamshell	Placing and handling aggregates at a stockpile; dredging operations
Crane with dragline	Excavation in soft materials at or below existing ground level
Crane with hook block	Lifting various heavy objects
Trucks and wagons	Various hauling operations
Motor grader	Shaping of subgrades, shoulders, ditches, and backslopes; maintenance of construction roads
Steel wheeled roller, vibratory compactor	Compaction of gravels and gravel–sand mixtures
Pneumatic-tired roller	Compaction of wide range of soils of low to moderate plasticity including silty gravels, sands, silty sands, and clayey sands
Sheep's-foot roller	Compaction of soils of moderate to high plasticity including silts, clays, clayey sands, silt–clay mixtures, clayey gravels, and silty gravels
Supercompactors (pneumatic-tired rollers weighing 20–50 tons)	All types of soil

typical uses. Various classes of equipment commonly used in earthwork operations are illustrated in Figures 17-1 and 17-2.

For additional information on equipment used in earthmoving operations, the reader should consult with equipment manufacturers or dealers and refer to the various construction magazines.

FIGURE 17-1 Common types of earthwork equipment. (*a*) Bulldozer. (*b*) Rubber-tired front-end loader. (*c*) Rubber-tired tractor scraper unit. (*d*) Truck being loaded by shovel. (Courtesy Terex Division, General Motors Corporation.)

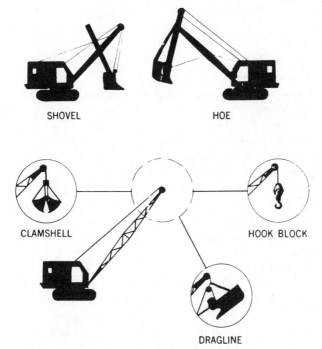

FIGURE 17-2 Shovel-crane units.

CLEARING AND GRUBBING

Clearing and grubbing is generally the first operation to be undertaken on any project involving earthwork, and thus it precedes any excavation.

Clearing and grubbing may be defined as the removal of trees, stumps, roots, snags, down timber, rotten wood, rubbish, and other objectionable material from an area marked on the plans or otherwise designated by the engineer. The area designated generally includes the limits of proposed excavation and embankment as well as all other areas where the obstructions that are to be removed would interfere with the conduct of the proposed work.

"Clearing" refers to the removal of materials above existing ground surface, and "grubbing" means the removal of roots, stumps, and similar objects to a nominal depth below the surface. Frequently, clearing and grubbing comprise a single contract item and include the removal of topsoil to a shallow depth.

In cut areas, in addition to the removal of vegetation and other obstructions from the surface of the ground, specifications generally require that all stumps and roots be removed to a depth of not less than 1 ft below the proposed grade. This requirement also usually holds in embankment areas where the height of the fill is less than 4 or 5 ft. In embankment sections, when the embankment height is to be more than 5 ft, trees and stumps may be left in place and cut off at ground level or at a height of 3 to 6 in. above the existing ground surface. Some agencies require that all large trees (18 in. or more in diameter) be removed, regardless of the height of the fill. It is usually provided that excavations made for the purpose of removing stumps and similar obstructions be filled to a level consistent with the surrounding ground before construction proceeds.

Large trees usually receive special consideration and may be left in place at the discretion of the engineer. In such cases trees which overhang the roadway are usually required to be properly trimmed to furnish a clearance of 14 to 16 ft. Frequently, areas within the limits of the right-of-way but beyond the limits of excavation and embankment are subjected to "partial" or "selective" clearing in which certain trees and shrubs are left standing. Provisions are usually made for the storing and salvage of marketable timber cleared from the rights-of-way. Debris removed from the designated area is generally disposed of by burning or, if burning is prohibited, by burying it in waste disposal pits or other approved areas. It should also be noted that certain organizations include the removal and disposal of buildings, fences, and so on in clearing and grubbing, while others include a special contract item when obstacles of this nature are to be removed.

Measurements of this item for the purpose of payment are generally made in acres; that is, the bid unit is 1 acre. Where only clearing operations are involved, special payment may be made for large trees, that is, so much per tree. Clearing and grubbing operations are generally performed largely by machines, with the bulldozer and its various attachments being the basic devices, or by companion tools such as rock rakes, rippers, and tree-'dozers. A considerable amount of hand labor may also be necessary, and liberal use is made of power saws and similar tools.

EXCAVATION

In the interests of simplicity the following discussion will deal with three types of excavation: roadway and drainage excavation, excavation for structures, and borrow excavation. Before beginning the discussion of each of these types, it is desirable to discuss briefly the subject of "classification of excavation."

17-2 Classification of excavation

It seems obvious that the costs of excavation of materials that require blasting for removal, such as solid rock, will generally be greater than those encountered in the removal of earth materials. Similarly, the removal and disposal of muck and other unsuitable materials may require special treatment and cost more than ordinary earth excavation. Some highway agencies find it advantageous to divide excavation into three separate classes. These classes are generally termed "rock excavation," "common or earth excavation," and "muck excavation," defined by AASHTO (1) as follows:

> *Rock excavation.* Material that cannot be excavated without blasting or the use of rippers, and all boulders or other detached stones each having a volume of 2 cu yd or more.
>
> *Common excavation.* All excavation not included as rock excavation or excavation which is otherwise classified and paid for.
>
> *Muck excavation.* The removal and disposal of deposits of saturated or unsaturated mixtures of soil and organic matter not suitable for foundation material regardless of moisture content.

The method of classification presented above has certain disadvantages, as borderline cases frequently arise in which there may be some dispute as to the proper classification of a portion of the work. Many organizations obviate this difficulty by using the term "unclassified excavation" to describe the excavation of all materials, regardless of their nature. The latter procedure is now preferred by a majority of highway agencies.

17-3 Roadway and drainage excavation

Roadway and drainage excavation, as defined here, means the excavating and grading of the roadway and ditches, including the removal and disposal of all excavated material and all work needed for the construction and completion of the cuts, embankments, slopes, ditches, approaches, intersections, and similar portions of the work.

All excavated materials that are suitable are used in constructing the elements of the roadway structure, including embankments, shoulders, subgrade, slopes, backfill when required for structures, and so on. Provision is made for the disposal of unsuitable or surplus excavated material. It must again be emphasized that, under the definitions that we have given here and in accordance with normal practice in this country, payment made for excavation items is not just payment for excavating the earth. On the contrary, it frequently includes nearly all the work needed in many jobs for forming the earthwork portions of the highway structure. This, again, would normally include such items as the formation of embankments and the finishing operations that are to be discussed later. However, practice is not uniform in this respect, and some agencies pay separately for compaction, addition of water during compaction, fine grading, and other items.

Many miscellaneous provisions are generally included in specifications for roadway and drainage excavation. Provision is usually made for the removal of vegetation if the area has not previously been cleared and grubbed and also for the salvage of portions of the excavated materials that are to be used in miscellaneous sections of the roadway. The latter item, for example, might provide for the use of suitable subgrade materials, saving of topsoil for use on side slopes, and work of a similar nature. When unsuitable materials are encountered in cut or embankment sections, provision is generally made for their removal and replacement with satisfactory materials.

The specifications may also include instructions for grading the roadbed, intersections, and so forth in accordance with the required alignment, grade, and cross section. In connection with this portion of the work, the roadway is usually required to be well drained at all times during the course of construction operations. Provision is also made for the grading of shoulders, slopes, ditches, and so forth in accordance with the plans, sometimes including requirements relative to the flattening, rounding, and warping of side and ditch slopes where applicable. When rock is encountered in cut sections, it is generally required that it be excavated to a depth of at least 6 in. below the grade line, and that suitable backfill material then be used in bringing the cut section up to the proposed grade. Another provision frequently included is that of maintenance of the graded roadbed until final acceptance of the work or until construction of the base or surface course begins. Provisions under this item might also include the scarification and compaction of cut sections, although they would not normally include subgrade soil stabilization.

Measurement of excavated material is usually based on a determination of the volume, in cubic yards in original position, of material actually excavated and disposed of as directed. Volumes are usually determined by the average-end area method, and payment is made for the number of cubic yards of roadway and drainage excavation and for overhaul.

Construction methods used in performing the operations required in this type of excavation may involve nearly every one of the types of equipment previously mentioned. For example, tractor-scraper units are normally used in the basic operation of excavating, hauling, and spreading oridinary soils; if hard, compacted materials are encountered, rippers may be required for breaking up the material so that the scrapers may load. Scrapers with crawler tractors are used for relatively short hauls or in rough terrain; rubber-tired units are used for longer hauls where suitable haul roads can be established. In the case of rock excavation, drilling and blasting will be necessary, and a shovel may be needed to load the blasted rock into trucks or rock wagons for transportation to its disposal point. A dragline is frequently used for excavation in swampy land, and a backhoe for trench excavation. In excavation of this type 'dozer units may be extremely useful in performing excavations involving very short hauls, for spreading dumped materials, and for similar operations. A motor grader may be used for shaping ditches, trimming slopes, and maintaining haul roads, as well as for a host of other functions.

17-4 Excavation for structures

Structural excavation refers to the excavating of material in order to permit the construction of pipe culverts, concrete box culverts, foundations for bridges, retaining walls, and practically all other structures that may be required in a particular job. Payment is usually made separately for this item, except in the case of structures like manholes and catch basins, where the payment made for completion of the structure includes payment for excavation. Some organizations also include in this item the satisfactory placing of backfill around the completed structure.

Specifications of the various state highway agencies covering this item provide for the satisfactory removal of all material, regardless of its nature, which is encountered in structural excavation. In certain cases, particularly where large excavations are required, as in the case of the construction of piers and abutments for large bridges, it may be necessary to employ caissons or cofferdams to facilitate the excavation process. The construction and subsequent removal of these items are also generally provided for in the specifications. Structural excavation is measured and paid for by the cubic yard; the volume on which payment is based is usually the actual volume of material excavated in order to place the given foundation structure. The pay quantity is generally limited, however, to a volume computed by the imagined placing of planes parallel to the lines of the foundation but located 1 to 2 ft outside these lines, as shown on the plans. Provision is also frequently made for an increase in depth of the foundation, with consequent increased payment to the contractor.

Suitable materials taken from excavations for structures are used either in backfilling around the completed structure or in other portions of the roadway structure. Unsatisfactory or surplus excavated materials are, of course, wasted. Requirements related to the placing of suitable backfill are usually quite rigid, particularly with regard to the production of a satisfactory density in the backfill. This density may be obtained by rolling or vibrating the material.

Both machines and hand methods are used in the performance of structural excavation, with more hand labor being required in this operation than in the other types of excavation described in this chapter. When working in average soils, the clamshell is particularly adapted to this type of work because of its ability to work to vertical lines and in close proximity to forms and protective devices such as the bracing and sheathing required in deep excavations. Smaller units of excavating machinery, such as the tractor shovel, also find application in this work. The bulldozer, which is present on nearly every earthwork operation, can also be used to good advantage in excavating and backfilling for structures. The vibratory and tamping tools previously discussed are very useful for compacting backfills, especially in restricted spaces.

17-5 Borrow excavation

When sufficient material for the formation of embankments and other elements of the roadway structure is not available from excavation performed within the limits of the right-of-way, additional suitable material is generally taken from borrow pits. The contracting agency may indicate and furnish the contractor with suitable borrow pits, or the contractor may obtain specified and suitable materials from locations of his own choosing. Specifications related to borrow excavation are generally about the same as those related to roadway excavation; that is, they provide for satisfactory excavation of material from an indicated area and for incorporation of that material into the roadway structure. Additional requirements usually pertain to the condition of the borrow pit during and after construction. For example, it may be stipulated that when construction operations are complete, borrow pits are to be left in such a condition that water will not collect or stand in them. Payment for borrow excavation is generally by volume in cubic yards of borrowed material, measured in place (original volume) and actually incorporated into the designated sections of the work.

The opening and working of borrow pits is frequently a large-scale operation that requires the best in mass-production methods of earth moving. If the bulk of the material is to be incorporated into embankment, and if it is soil as distinguished from rock or rocklike materials, there is little doubt but that borrow-pit excavation in most cases may be most efficiently effected by the use of scraper units. This is, of course, because of the inherent nature of the scraper unit, which, with one operator and one power unit, will dig, load, haul, and spread material. Either crawler or rubber-tired tractor units may be used; the latter units are generally preferred where long hauls (above about 1200 ft for large units) are to be made. Where rock or cemented soils are to be excavated, the shovel is usually the best tool unless the material can be ripped economically; the shovel loads into trucks or wagons that haul the material to the fill, where the load is dumped and spread by a 'dozer or a motor grader. Shovel–truck groups are capable of very large production in suitable circumstances, but the interdependence of the units involved makes their comparison with scraper units generally unfavorable for this operation, except when the excavation is being made in materials or under conditions unsuited to scraper operations.

EMBANKMENTS

Embankments are used in highway construction when it is required that the grade line of the roadway be raised some distance above the level of the existing ground surface in order to maintain design standards or prevent damage to the highway through the action of surface or ground water. Both rolled-earth and hydraulic-fill embankments are built, with the vast majority of highway embankments being of the former type. From the standpoint of average practice it is also likely that the large majority of fills used in highway construction are 15 ft or less in height. Many fills are 2, 3 or 4 ft in height, and fills as high as 383 ft have been reported (2).

Before discussing construction procedures related to embankments, it seems desirable to discuss briefly some of the basic factors involved in the design of such structures. These design elements may be classed as height, stablity of slopes, stability of foundations, and the selection of embankment materials.

17-6 Height

As has been indicated, the height of an embankment is generally fixed by considerations related to the general location of the highway in the area. Thus an embankment may be needed to maintain the grade to some fixed point, as in the approaches to a bridge or a large culvert structure, or to maintain desirable grades in rolling or mountainous country. Similar considerations related to alignment may also force the use of embankments rather than the possible use of side-hill sections. In rolling country it is frequently possible to reduce embankment construction to a minimum on secondary roads by the simple expedient of following closely the shape of the existing ground. Such a procedure is usually not desirable on primary routes because of sight distance requirements.

In low-lying areas where the water table is at or close to the surface of the ground, the minimum height of embankment is frequently established by the desirability of preventing the intrusion of ground water into the subgrade and base. In such cases the elevation of the top of the subgrade is generally required to be at least 2 ft above the water table, and it may be considerably more when soils subject to capillarity are used in the construction of the embankment or in areas subject to frost. If free water is expected in the area crossed by the embankment, the minimum distance above the expected water level may be further increased in the interests of protecting the embankment and pavement structure.

17-7 Embankment slopes

Essentially, the cross section of a highway embankment consists of a flat, horizontal top section with generally symmetrical slopes on either side that begin at the top and continue until they intersect the natural ground surface. The width of the top section depends on the required pavement and shoulder dimensions and is typically a minimum of 40 ft for a two-lane roadway.

Generally, flat side slopes are preferred to satisfy standards and to facilitate maintenance. Typical standard slopes established on the basis of such criteria are shown in Table 17-2. At certain locations, it may be necessary to depart from the standard cross section because of restrictions in rights-of-way or unstable soil conditions.

TABLE 17-2
TYPICAL STANDARD DESIGN SIDE SLOPES
FOR MAJOR CLASS TWO—LANE HIGHWAY
WITH 10 ft^a STABILIZED SHOULDER^b

Height of fill (ft)^b	Distance from edge of shoulder	
	0-20 ft	Beyond 20 ft
≤10	6:1	6:1
>10 ≤30	6:1	4:1
>30	2:1	2:1

^{a}1 ft = 0.3048 m.
^{b}Assumed design hourly volume is less than 850.

SOURCE: *Construction of Embankments.* NCHRP Synthesis of Highway Practice No. 8, Washington, D.C. (1971).

Generally speaking, embankments constructed of high-quality materials (e.g., A-1 and A-3 classifications) can be constructed with slopes as steep as 1.5:1 without concern for soil stability. For other soil classifications, a maximum slope of 2:1 has been recommended, and where the embankment will be subjected to flooding, a 3:1 slope is preferred *(3)*.

All major fills should be subject to stablity analyses and designed in accordance with established principles of soil mechanics.

17-8 Embankment foundations

As has been indicated, a portion of the analysis required for the design of a highway embankment consists of checking the stability of the side slopes or the body of the embankment itself. The design should also include an examination of the soil beneath the embankment proper, or the embankment "foundation." An embankment may fail because the stresses imposed on the underlying soil layer due to the weight of the fill are greater than the shearing resistance of the foundation soils. In such a case, the underlying soil layer generally would flow laterally with resulting subsidence of the fill. The embankment on which a pavement surface has been placed may also fail in its supporting function by continued settlement due to consolidation of the underlying soil layers. Either of these situations is likely to occur when a high fill is founded on cohesive soil or where the foundation material is a soft, compressible, fine-grained soil, such as an organic silt or clay, peat, or muck. Where any doubt exists as to the stability of the embankment foundation, the shearing stresses that will be created in the foundation should be compared with the available shearing resistance of the soils involved, the consolidation characteristics of the layer determined, and a settlement analysis made. The measurement of the shearing resistance and the consolidation characteristics must both be based on the laboratory testing of undisturbed samples taken from the foundation soil.

Construction procedures required for improvement of the structural properties of the foundation are generally involved and costly. Relocation of the route should be considered in every possible case where extremely weak soils are encountered to considerable depth. In the event that relocation is not feasible, several methods have been evolved for the treatment or displacement of the underlying soil layer. These methods are discussed in detail in later portions of this chapter.

17-9 Selection of embankment materials

Many different soils may be satisfactorily used in the construction of rolled-earth highway embankments. In terms of the AASHTO classification system outlined in Chapter 15 and in general statement, materials of the A-1, A-2-4, A-2-5, and A-3 groups should be used when available. Soils of the A-2-6, A-2-7, A-4, A-5, A-6, and A-7 groups are generally regarded as being less desirable for embankments, and when such materials must be used, special attention should be given to the design and construction of the embankment (4).

The A-1 soils are highly desirable for inclusion in embankments since they may be compacted to a high degree of stability and density by the use of normal compaction equipment. A-2 soils are also generally suitable for this purpose, although they require a little more careful control of the compaction process during construction. The A-3 group consists of cohesionless sands that can generally be satisfactorily used in embankment construction, although they cannot be satisfactorily compacted by the use of sheep's-foot rollers. Adequate compaction may be obtained by the use of pneumatic-tired rollers, construction equipment, or a vibratory compactor.

With regard to groups A-4 and A-6, the sandy and silty soils included in these groups, as well as some of the inorganic clays, may be satisfactory for inclusion in embankments under certain conditions. Such conditions include low height of fill, careful control of the compaction process, and utilization in areas where the moisture content of the soil might be expected to remain the same or less during service than that which was utilized in the construction process. These soils are more difficult to compact than those of the first three groups because of their high moisture-retaining characteristics. The moisture content of these soils during construction must generally be maintained within relatively narrow limits in order to secure adequate density and stability. The elastic soils included in the A-5 and A-7 groups are generally regarded as unsatisfactory for embankment construction. In addition, the clay soils of the A-7 group are subject to very high volume change with change in moisture content. Despite some of the faults exhibited by several of the soils included in the A-4, A-5, A-6, and A-7 groups, conditions are sometimes such that these are the only soils available within a reasonable haul distance and their use is therefore dictated by considerations of economy. These soils must all be carefully handled during the construction process.

CONSTRUCTION OF ROLLED-EARTH EMBANKMENTS

17-10 Formation

Rolled-earth embankments are constructed in relatively thin layers of loose soil. Each layer is rolled to a satisfactory degree of density before the next layer is placed, and the fill is thus built up to the desired height by the formation of successive layers or "lifts." The majority of state highway agencies at the present time require layers to be 6 to 12 in. thick before compaction begins, when normal soils are involved. Specifications may permit an increase in layer thickness where large rocks are being used in the lower portion of a fill, up to a maximum thickness of 24 in.

The layers are required to be formed by spreading the soil to approximately uniform thickness over the entire width and length of the embankment section at the level concerned. "End-dumping" from trucks without spreading is usually specifically prohibited. An exception to this latter requirement may be permitted when the embankment foundation is of such a character that it cannot support the weight of spreading and compacting equipment. In such case the lower portion of the fill may be placed by end-dumping until sufficient thickness is developed to permit the passage of equipment. Where unstable soils compromise the foundation, the special construction measures described later in the chapter may be required.

Two general construction methods are used in embankment formation. The first of these is the direct dumping and spreading of the soil by scraper units in one operation. This, of course, is a function for which these units are ideally suited. The second method involves the dumping of the material in the proper location by trucks or wagons. The dumped material must then be spread to the required uniform thickness by 'dozer units or occasionally by motor graders. In addition, on very short fills, 'dozer units might conceivably form the embankment, working alone by moving material short distances from cut into embankment sections.

17-11 Compaction

Requirements of the various highway agencies related to the compaction of soils in embankments (and subgrades) have undergone considerable change since World War II. So also have various elements of the understanding of the behavior of soils undergoing compaction and the end results of such compaction. Considerable variation in requirements for compaction exists among the various agencies.

However, the general nature and purpose of compaction are well understood. The general notion is that soils should be compacted by rolling, at or near optimum moisture content, to some percentage of the maximum density established through the use of a known laboratory compactive effort. It has been established that certain advantages may be expected to accrue from the use of this compaction process. Included among the more important of these advantages are the increased shearing resistance or "stability" of the soil, its decreased permeability, and the minimizing of future settlement of the embankment itself (but not necessarily settlement due to consolidation or shearing failure of the embankment foundation).

Present specifications of the various state highway departments related to compaction are not uniform. A large majority of the states require compaction to a certain minimum percentage of the maximum density, as determined by the standard AASHTO laboratory compaction procedure (Method T99). (See Chapter 15.) The most common requirement, employed by about 50 percent of state highway agencies, is 95 percent of the maximum dry density obtained in the AASHTO T99 test (5). In some states, maximum compaction requirements are stated in terms of the maximum laboratory density; for example, for a soil with a maximum density of 99.9 lb/ft^3 or less, the required density in the field must be 100 percent of the laboratory value; for 100 to 119.9 lb/ft^3, 95 percent; and so on. A few agencies still specify the number of passes to be made by specified equipment. Many specifications contain requirements related to the size and nature of compaction equipment. In recent years, the tendency has been toward the incorporation of "end-result" specifications for compaction, with desired end results specified but leaving the choice of equipment up to the contractor. Moisture content during compaction is controlled, sometimes by general specification statements, sometimes by numerical values (e.g., ±2 percent of optimum moisture).

The establishment of compaction requirements by the design engineer is a matter that requires careful study of all the factors involved. A detailed discussion of this and other matters related to the construction of embankments is given in Ref. 5.

State highway department specifications represent a compromise among the many variables and reflect requirements that are reasonable for the conditions which exist in each state. On large or unusual projects, special provisions may be written and used to control the work. General recommendations for density requirements for highway embankments are contained in Table 17-3.

17-12 Control of compaction

Generally speaking, the field compaction process is controlled by making relatively frequent checks of the density and moisture content of the soil that is undergoing compaction. The measured density is the wet unit weight; the dry unit weight is calculated on the basis of this figure and the measured moisture content. The dry unit weight may then be compared with the compaction curve for the soil and compactive effort involved to see if the density being obtained

TABLE 17-3
RECOMMENDED MINIMUM REQUIREMENTS FOR COMPACTION OF EMBANKMENTS AND SUBGRADES

Class of soil (AASHTO M145)	Density, Percent of AASHTO Standard T99	
	Embankments	Subgrade
A-1, A-3, A-2-4, A-2-5	100	95
A-2-6, A-2-7, A-4, A-5, A-6, A-7	95[a]	95[a]

[a]AASHTO recommends compaction within 2 percent of the optimum moisture content. Use of these materials requires special attention to design and construction of embankments.

SOURCE: *Standard Specifications for Transportation Materials*. Designation M57-80, American Association of State Highway and Transportation Officials, Washington, D.C. (1982).

in the field meets the requirement established in the laboratory. If different soils are being placed in different sections of the fill, it is understood that the curve (or figure) that is being used to check the rolling process is the proper one for the soil concerned.

Three methods have been used widely by highway agencies for determining the density of the soil. These may be designated as the sand, balloon, and heavy oil methods. In each of these the density determination is begun by carefully excavating a cylindrical hole in the soil layer; the hole is usually about 4 in. in diameter and the full depth of the layer. All the material taken from the hole is carefully saved, placed in a sealed container, and weighed as quickly as possible. The weight determination may be made immediately on a field balance if desired. The volume of the hole may then be determined by the use of sand, a balloon apparatus, or heavy oil. Some use has also been made of small undisturbed samples taken with a cylindrical drive sampler; in such cases, wet unit weights may be used for the comparison.

The apparatus required for the sand density determination is shown in Figure 17-3. In using this apparatus, the area around the hole is carefully leveled, the jar—with funnel attached and filled with "standard" sand—inverted over the hole, and the valve opened. The sand flows into and fills the hole. When the hole is filled the valve is closed, and the jar and the remaining sand weighed. The weight of sand required to fill the hole may then be determined. Since the same procedure is carefully followed in each determination and the same dry sand is used, a previously established relationship between a given weight of sand and the volume occupied by this amount of sand may be used to ascertain the volume of the hole. Since the weight of soil

FIGURE 17-3 Sand–density equipment. (Courtesy Soil-test, Inc.)

taken from the hole and its volume are then known, the wet unit weight may be calculated. A balloon filled with water may also be used to determine the volume of the hole, and the volume of the hole measured directly. This volume may also be obtained by filling the hole with a heavy oil of known specific gravity. The latter procedure is more widely used for checking the densities of bases than embankments.

If the soil is known to be at or near optimum moisture content, the density check may be made by comparing the measured wet unit weight with the wet unit weight established by laboratory procedure. Generally, however, the procedure used is to make an additional moisture determination so that this factor may be checked and the dry unit weight determined. If a moisture content determination is deemed necessary, then speed is of the essence. The determination may be made by rapid drying over a field stove. Drying is also frequently accomplished by adding alcohol, or some other volatile solvent, to the soil and then removing the water by igniting the solvent. Time is seldom available for the use of more accurate methods because the decision as to whether or not a satisfactory density has been obtained must be made quickly so that construction operations will not be delayed.

Many highway agencies use nuclear devices to measure in-place density of soils. Reference 6 gives the following description of the principle of nuclear density gauges:

> When gamma rays are emitted from a radioisotope source in proximity to a surface they interact with the material and are scattered or absorbed. The count of the gamma rays emerging from the surface at some point is influenced by the density and the composition of the material. A typical gauge consists of a gamma-ray source, a detector with associated counting electronics, and shielding between the two to prevent direct transmission of the gamma rays from the source of the detector. A wide variety of gauge configurations is possible, involving source energy and intensity, type and efficiency of detector, and source–detector separation. The most universally employed method of determining density with a gamma-ray gauge is by use of a calibration curve prepared from the empirically determined relationships between density and response for each individual instrument. The calibration curve for a particular instrument is originally obtained by plotting the response measured by the gauge for a set of calibration standards of known density.

As Figure 17-4 illustrates, there are two basic types of nuclear gauges: (1) the surface or backscatter type and (2) the transmission type. The latter type has a probe, the transmission source, that is lowered into the ground through an access tube driven into the soil. One type of nuclear gauge is shown as Figure 17-5.

Nuclear equipment may also be used to measure moisture content. A neutron moisture gauge consists of a fast-neutron source and a slow-neutron

Backscatter type   Transmission type

FIGURE 17-4 Representative gamma-ray paths for density gauges. (After R. P. Gardner and K. F. Roberts, Density and Moisture Content Measurements by Nuclear Methods, NCHRP Report No. 43, 1967).

FIGURE 17-5 A nuclear moisture density meter. (Courtesy Soiltest, Inc.)

detector. Fast neutrons emitted from the device are slowed principally by collisions with hydrogen atoms. Assuming that most of the hydrogen present is in the form of water, a calibrated gauge can be used to measure the moisture content.

The principal advantage of nuclear gauges is speed. Such devices require as little as one-fifth the time required for conventional tests. They also do not disturb the soil, can be used on a wide range of materials, and require little operator training. There is little danger from the radioactivity, but the devices must be calibrated carefully.

If the density determined as outlined above is equal to or greater than that required, the rolling may generally be judged to be satisfactory and construction of another lift may proceed. If the density is lower than that specified, then additional rolling may be required or the moisture content may have to be adjusted. If these measures fail, then it is reasonable to assume that the rolling equipment being used is not capable of producing the required density or that the moisture—density relationship that is being used is not the one for the soil which is actually being compacted.

The procedure that has been outlined for the control of compaction may seem unduly cumbersome and complicated. Actually, after compaction of a given section has begun and the initial adjustments have been made, the situation is usually considerably simplified. The inspector will, for example, quickly learn to judge the moisture content of the soil by appearance and feel. The inspector may also quickly decide that, with the proper moisture content being maintained, a certain number of passes of the roller—say six or eight—will accomplish the required densification. The job may then be simplified into a count of the roller trips, with occasional density checks being made as judgment dictates.

The number and frequency of density and moisture checks that are required for adequate control are, of course, functions of job conditions. On an average basis, perhaps one check will be required every hour on a fill of moderate length. Some other standard may be applied, such as a minimum of one test per 1000 cu yd of compacted material. Some soils are more sensitive to moisture change than are others. For example, sands usually have a moisture—density curve that shows a sharp peak, with the optimum moisture content being clearly defined, while clay soils usually have a flatter curve so that a small variation in moisture content does not greatly affect the density. It is impossible, generally speaking, to maintain the moisture content exactly at optimum for any substantial length of time, so that some tolerance must be permitted in the moisture content.

17-13 Compaction equipment

Principal factors involved in the selection of compaction equipment for use on a given project include the specifications under which the work is to be carried out and the choice of equipment available to the contractor. More than one type of equipment and compaction procedure can often be used successfully for a given set of circumstances; however, there usually is one combination that will give the most economical result.

A detailed discussion of the most efficient and economical use of a given type of equipment for a given project is beyond the scope of this text. A more complete treatment of this topic is given by Ref. 5. Pertinent factors that must be considered include thickness of lift, total weights and unit pressures exerted on the soil, required number of passes, operating speeds, soil type and moisture content, and others. On large projects, relationships among the variables involved may often be resolved by the construction of test fills prior to the start of actual construction or by trial during early phases of construction. For economical operation, there must be a proper balance between rolling equipment and hauling equipment on the job.

Additional water required for compaction is generally added by ordinary pressure distributors. Rotary speed mixers may be used to mix the water and soil in some instances. Aeration of soils that are too wet for compaction may be accomplished by the use of agricultural tools, such as harrows and disks, by turning the soil with blade graders, or by rotary speed mixers with the hood up. The problem also may be handled by "sandwiching" alternate layers of wet and dry materials. In many cases aeration is accomplished by simply allowing the soil to dry of its own accord.

REINFORCED EARTH EMBANKMENTS

Reinforced earth has been defined as a construction material composed of soil fill strengthened by the inclusion of rods, bars, fibers, or nets which interact with the soil by means of frictional resistance (7). The development of the concept of reinforced earth is credited to Henri Vidal, a French engineer (8, 9).

Figure 17-6 illustrates a common type of reinforced structure. Its components are (1) the soil backfill, typically a cohesionless free-draining material such as sand; (2) reinforcing members consisting of long thin strips called ties, which are arranged in a regular pattern; and (3) a covering called "the skin," which maintains the integrity of the sand at the outface wall.

The earliest reinforced earth system utilized galvanized steel strips for the ties and galvanized steel semicylindrical elements approximately 3 mm in wall thickness for the skin (9). Such systems depend on the frictional resistance between the grains of the fill mass and the reinforcing ties.

Since the introduction of the patented Vidal reinforced earth system in the mid-1960s, other soil reinforcement systems have been developed using bearing rather than friction for the stress transfer or a combination of bearing and frictional forces. One such system employs a welded wire mesh to resist tensile forces within the soil mass (10). Developers claim that on a unit surface area basis the mesh system exhibits five to seven times the pullout capacity of flat strips.

The California Department of Transportation (Caltrans) has experimented with a reinforced earth system that incorporates used automobile tire sidewalls in the soil mass (11). (See Figure 17-7). The tire sidewalls are attached to the tieback elements by a cross arm that extends through the sidewall and is welded to the tiebar. For the wall facing, the Caltrans system specifies used railroad ties that are laid horizontally and supported with the tire–anchor bar assemblies fastened to vertical timbers.

Reinforced earth systems facilitate the support of sections of highway

Thickness = t

FIGURE 17-6 Components and key dimensions of a typical reinforced earth wall. (Courtesy American Society of Civil Engineers.)

(a) 　　　　　　　　　　　　　　　　　　*(b)*

FIGURE 17-7 The California reinforced earth system. (*a*) Detailed view. (*b*) Overall view. (Courtesy Transportation Research Board.)

along steep slopes (Figure 17-8). Such systems are especially advantageous in urban areas where rights-of-way are limited and land is expensive. Reinforced earth structures may cost as little as one-half as much as fills constructed behind reinforced concrete walls.

Reinforced earth fills can be constructed on relatively weak foundations

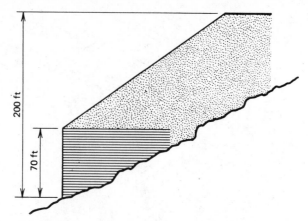

FIGURE 17-8 Typical application of a reinforced earth wall.

since the loads are spread over a large area and can be built to great heights. Reinforced earth structures as high as 70 ft have been constructed.

For more information on the design of reinforced earth embankments the reader is advised to consult Ref. 7 to 12.

SPECIAL TREATMENT OF EMBANKMENT FOUNDATIONS

The foundations of embankments that are constructed in swampy areas—particularly where peat or other highly organic soils are encountered—frequently require special treatment if failures of the embankment or pavement structure due to consolidation or displacement of the underlying soil layer are to be avoided. The methods in general use may be classified as follows:

1. Gravity subsidence.
2. Partial or total excavation.
3. Blasting.
4. Jetting.
5. Vertical sand drains.
6. Reinforcement with engineering fabrics.

Various combinations of these methods may also be used in specific locations.

17-14 Gravity subsidence

In some instances, a fill may simply be placed on the surface of an unsatisfactory foundation soil and allowed to settle as it will with no special treatment of the underlying soil. Where soils of high compressibility and low shearing resistance are present in the embankment foundation, this method is obviously unsuitable for any except secondary roads which carry extremely low volumes of traffic. The fill may settle rapidly during or shortly after the construction period, due to shearing failure and displacement of the foundation soil. Settlement due to consolidation may continue for many years. In either case the satisfactory placing of any wearing surface is virtually impossible and a long period of continuous maintenance must be anticipated, as more and more soil must be added to the fill as time passes until the embankment eventually comes to a stable condition.

In some cases a surcharge of extra height of fill has been used to speed up settlement; the surcharge is later removed. A variation of this method was used by the Louisiana Department of Highways to build a section of Interstate Highway near New Orleans. They used a "rolling" surcharge over a deposit of peat and muck 10 to 20 ft thick. In this process 4 ft of sand surcharge (of a total of $5\frac{1}{2}$ ft) was placed at one end of the fill and allowed to remain in place for a specified time; the sand then was removed and placed at the other end of the fill. Thus, the fill "rolled" slowly ahead in a series of steps.

17-15 Partial or total excavation
Where the undesirable soil is of shallow depth (up to 8 to 12 ft) and is underlain by a soil of satisfactory character or by rock, many highway agencies excavate the entire depth of unsatisfactory soil. In New York State, total excavation has been carried out to a depth of 35 ft. The excavated area is then backfilled with suitable granular material. This method is expensive, of course, but has the advantage of permitting the immediate placing of a pavement on the completed fill. When such treatment is used, the backfill is composed of sand and gravel of suitable gradation.

A procedure sometimes used on secondary roads where deep swamp deposits are encountered consists of partial excavation of the foundation soil. That is, the soil is excavated to a depth of several feet and backfill is placed as before. In this operation it is essential that the backfill be placed as excavation proceeds. This method is open to the same criticisms that are applicable to the method which we have called "gravity subsidence." Displacement of the deep underlying soil may still occur under the weight of the backfill and embankment, while settlements due to consolidation are also to be expected, although they should be somewhat smaller than if the compressible soil had not been partially excavated. It is again usually impossible to place any permanent type of wearing surface without the expectation of very high maintenance costs.

17-16 Blasting
Explosives have been widely used to aid in the placing of embankments in swampy areas. When this method is used, the embankment is generally constructed to a level considerably above the final grade and the displacement of the underlying soil is accelerated by the detonation of deep charges of explosives. The blasts serve a dual purpose in that some of the material may be displaced by the force of the explosions, and the remainder of the unstable soil is liquefied to a greater or lesser extent and can thus be displaced more readily by the weight of the overlying fill. As the underlying soil is displaced, the fill subsides; and it may then be finished to the required elevation and cross

section. This method requires careful control but should result in the formation of an embankment that will be stable in a relatively short period of time.

17-17 Jetting

The process of "jetting" involves the pumping of water into the underlying soil layer in order to liquefy it and thus aid in the displacement of the layer by the weight of the embankment. Jetting is carried out in conjunction with the placing of the embankment material. In typical installations, a single line of jets is placed along one side of the embankment. The jets are spaced 10 to 25 ft apart, and they extend about two-thirds of the way into the underlying layer. The fill settles or "subsides," as before, and must be brought to the desired cross section after a stable condition has been attained. Trenches may be excavated along one or both sides of the embankment to assist in the displacement of the weak soil layer. In some cases the embankment material, rather than the underlying soil, may be jetted; this results in an increase in the weight of the fill.

17-18 Vertical sand drains

Another method used to treat troublesome embankment foundations utilizes vertical sand drains. Vertical sand drains generally consist of circular holes or "shafts" from 18 to 24 in. in diameter, which are spaced from 6 to 20 ft apart on centers beneath the embankment section and are carried completely through the layer of compressible soil. The holes are fashioned by rotary drilling or by the use of a hollow or plugged-end mandrel. The holes are backfilled with suitable granular material. A "sand blanket" 3 to 5 ft thick is generally placed at the top of the drains (and the existing soil) extending across the entire width of the embankment section. The embankment is then constructed by normal methods on top of the sand blanket.

The basic purpose served by the installation of vertical sand drains is the acceleration of consolidation of the compressible soil layer. Consolidation is, of course, due to the forcing of water from the voids of the soil, and the rate of consolidation is dependent, among other things, on the distance which the water has to travel to escape. Without sand drains the water may escape only at the top of the soil layer if there is an underlying layer of impervious material, or at the top and bottom of the layer. If the layer is of considerable thickness, settlements due to consolidation may occur over a period of many years. Vertical sand drains speed up the consolidation process by providing many shortened paths through which the water may escape. Vertical drainage at the top (or top and bottom) continues, and in addition radial horizontal drainage takes place. Water that is thus forced into the sand drains moves up the drains and out through the sand blanket. In addition to the fact that radial drainage is provided, this method takes advantage of the fact that many fine-grained sedimentary soils have a permeability that is several times greater in a horizontal direction than in a vertical direction. This fact also contributes to the accelerated rate of consolidation. Still another advantage lies in the fact that a rapid increase in shearing strength usually accompanies the accelerated consolidation process so that the probability of lateral flow is diminished and the settlement of the fill confined to that caused by consolidation.

Phenomenal results have been reported from the use of this method on many jobs in different sections of the country. In many cases, even where high fills and deep deposits of compressible soil were involved, 90 percent or more of the ultimate settlement has occurred during the construction period or very shortly thereafter. This behavior makes possible the placing of a permanent-type surface within a few months after the fill is completed. Costs of this method have been generally reported as somewhat less than those of most of the methods previously described. In Figure 17-9 is shown a profile of an installation of vertical sand drains used in connection with the construction of an embankment across a swampy area encountered in the construction by the New York Department of Public Works of a highway near Syracuse, N.Y. *(13)*. Fills up to 45 ft high were built on surface organic soils underlain by soft clays to a depth of 35 ft.

17-19 Fill foundation reinforcement with engineering fabrics

In recent years, engineers have found that engineering fabrics can be used to advantage in the construction of low fills over swampy or marshy areas. Other names for engineering fabrics include geotextiles or geotechnical fabric. These products are permeable textile cloths or mats made from a variety of artificial fibers. Engineering fabrics are generaly lightweight, possess moderate to high tensile strength, and withstand large amounts of elongation without rupture *(14)*.

Typically, the engineering fabric is placed on the weak foundation and overlain with the embankment fill. This increases the bearing capacity of the foundation and allows a higher fill to be constructed on it. The use of a fabric in such circumstances has several advantages: (1) it permits construction of embankments on otherwise unsuitable foundations; (2) it eliminates lost construction time due to failures and delays inherent in soft ground construction; and (3) it results in considerable savings of fill material compared to displacement construction methods *(14)*.

FINISHING OPERATIONS

Finishing operations refer to the final series of operations that are required to complete the earthwork operations involved in a typical highway project. These operations are taken here to include such things as trimming of shoulders, side, and ditch slopes; the fine-grading operations required to bring the earthwork sections to their final grade and cross section; the compaction of cut sections; and similar items. Actually, most of these are not separate operations that are performed after all other earthwork operations are completed; rather they are generally carried along and performed as the job approaches completion.

The tool most widely used in the performance of the majority of finishing operations is the motor grader. Its long wheelbase and ability to work to close tolerances make it ideal for this application. 'Dozers, scraper units, and gradalls are also sometimes used in finishing operations. In cut sections, excavation and shaping of the cross section to the dimensions shown on the plans may be all

FIGURE 17-9 Typical section of sand drain installation. (After Lyndon H. Moore; courtesy New York State Department of Public Works.)

that is required. However, many specifications call for scarifying and compaction of the soil in cut sections. The specifications for compaction, generally speaking, are the same as those for the compaction of embankments. There is some tendency to require greater densities in cut sections and in the top layer of embankment sections than are required in normal embankment construction. The required scarification of cut sections may be accomplished by rippers, discs, and harrows of various types or by scarifier attachments on motor graders. The required depth of scarification and compaction in these sections is generally from 8 to 12 in. The compaction operation, as before, may be accomplished through the use of various types of rollers, with the sheep's-foot roller the predominating tool. The section would generally be shaped to final dimensions through the use of a motor grader.

It should be noted that the finishing operations and other earthwork items that have been described are in many cases practically all the work involved in construction prior to the placing of the base and surface. In other cases, additional finishing operations may be required before this phase of construction can begin. For example, the subgrade soil may require stabilization. The placing of high-type pavements and bases, such as portland-cement concrete, usually involves additional subgrade preparation beyond that which has been described in this section. These items and others of a similar nature are discussed in later chapters.

PLANNING AND EXECUTION OF CONSTRUCTION PROJECTS

Since the majority of highway construction is performed by contract, the principles stated in the following sections are primarily intended to apply to the planning and execution of contract work. Most of the items that are discussed are also applicable to work which is performed by force-account. Many contractors prepare tentative job plans as a part of their bid computations, and on large projects it may be required that the job plan be submitted with the bid. In any event, a comprehensive job plan should be prepared by the successful bidder immediately after the contract is awarded and before construction operations begin. Space does not permit lengthy exposition of the many details that enter into the successful planning and conduct of the earthwork operations involved in a typical highway project, but some of the steps that are necessary are presented here.

The first step in the preparation of a schedule of construction operations is a detailed analysis of the amount of work of different types involved in the job. By this is meant such items as the number of acres to be cleared and grubbed, the number of cubic yards of roadway excavation, the haul distances, and so on. The quantities used are generally those of the engineer's estimate. The plans are carefully examined and the location, extent, and conditions affecting the execution of each step of the work determined. A detailed step-by-step analysis is then generally made on the equipment, labor, and material requirements for each phase of the project.

In establishing the equipment estimate sheet, for example, it would be necessary to tabulate each operation and the quantity of work involved in that operation, to select the units of equipment to be assigned to this portion of the job, to estimate the hourly production rate for each of the machines to be used, and to calculate the total number of equipment hours required for completion of the operation. If the time to be consumed in this operation was limited and if sufficient equipment was available, a sufficient number of machines would simply be assigned to the operation to complete it in the given time. The labor required for each operation would probably be estimated at the same time, while material estimates would probably be made separately.

17-20 Job scheduling

On the basis of the information tabulated in the work estimates previously referred to, a detailed construction schedule may be prepared. This schedule may be presented in several different forms, and one method of presentation is shown in Figure 17-10. As will be noted, each of the general operations involved in the project is tabulated, and the time assigned for completion of each operation is shown by the plain bar. The sequence of these operations is also shown. As indicated by the chart, the progress that has been made on the project at any particular time is also shown by the crosshatched bars. Using this sheet, the contractor may quickly determine the status of the work.

It will be noted that certain operations overlap, as would be expected. As a matter of fact, the separate operations must be carefully coordinated and balanced so that they will be completed in the proper sequence and so that equipment on the job will be utilized to the fullest possible extent. Although the

FIGURE 17-10 Schedule of construction operations.

schedule that is illustrated here is for earthwork operations only, a similar schedule would be prepared for every phase of the complete project.

In addition to the construction schedule, contractors on many jobs prepare detailed sheets that show equipment, labor, and material requirements for every day of the project. Preparation of these sheets enables the contractor to further plan operations so as to make use of the available equipment and labor in the best possible way, and they help to maintain a smooth flow of materials, fuel, lubricants, and so on to the job.

17-21 Critical path method

Highway contractors also generally use the critical path method, developed in the mid 1950s by engineers from DuPont and the Univac Division of Sperry Rand, for construction project scheduling and control. State highway agencies also use the method for planning and programming, scheduling of design activities, and other activities.

CPM is an analysis and management tool that is useful in planning and scheduling any project made up of various subunits of work (as are most highway projects), some of which can be done sequentially from start to finish of the project (15). Many engineers feel that the CPM diagram has advantages over the more conventional bar chart described in the preceding section.

The basic concept of CPM is that, among the many operations that form a complete project, there must be one series of sequential operations that takes longer than any other possible series. The shortest possible time in which the project can be completed is the time needed for the longest series. The critical series is the "critical path." Determination of the critical path is in itself an exercise in logic that promotes good management of the project.

Determination of the critical path is aided by the construction of an "arrow diagram," such as the one shown in Figure 17-11. Detailed instructions for the arrow diagram and its application to highway problems are contained in Ref. 16.

In the planning phase of CPM, the first step in drawing the arrow diagram is to break the project into its subunits (activities). On paper, an activity is represented by an arrow drawn from left to right. At each end of the arrow is placed a circle or node to represent an "event." The head end (right end) of the arrow merely indicates the flow of work; the arrow is not to scale.

In CPM terminology, the event at the tail of the arrow is the "i" event; it represents the instant of time at which this event may start and the instant at which the preceding event must end. The event at the head of the arrow is the "j" event; it represents the instant at which the activity indicated by the arrow ends and subsequent activities may begin.

A logical analysis of a project may produce the general form of the diagram of Figure 17-11. In this diagram, activities A, B, and F may proceed simultaneously with C and E. However, note that E cannot begin until C is complete, and H and I, although they can be done concurrently, cannot begin until A, B, E, and F are complete.

Note also that activities D, G, and J must follow one another in that order.

FIGURE 17-11 Typical critical path diagram for a simple highway project. (After Douglas L. Jonas.)

The dashed line between nodes 4 and 7 is a "dummy" activity put there only to show a logic restraint. In this case, it means that K cannot proceed until J is complete (as well as H). It is made necessary on the diagram because of the fact that the arrow for H and I could not be drawn to nodes 4 and 7 at the same time. Activities are commonly designated by connected nodes, for example, 1–3; in that case, additional dummy activities must be used to keep the tabulations straight.)

In Figure 17-11, the remaining steps, K and L, must follow one another, while M can be done concurrently.

In the usual approach, the drawing of a generalized diagram and a check of its logic, forward and backward, complete the "planning" phase of the analysis. The "scheduling" phase may then begin, working with the basic diagram.

The first step in the scheduling phase is to estimate the time required for each activity. The usual procedure is for persons experienced in the work to estimate the normal time required for each operation. At this stage, no attention is paid to other than normal resources of humans, time, machinery, and money. Later, adjustments can be made if desired to obtain the most efficient utilization of one or more of these resources. The normal time, in days, for each of the operations of Figure 17-11 is shown in parentheses along each arrow, for example, "(20)" days for activity F.

Study of the diagram will reveal that, since activity C takes 18 days, activity E cannot begin before that time has elapsed. It follows that, since activity E takes 6 days, activity H cannot begin until 24 days after the start of activity C. In CPM terminology, the earliest start time for activity E would be the end of the 18th day; this is called the "earliest event time" and is shown in a square on the diagram. The "earliest finish time" for E would be 24 days.

Note also that each of the activities concurrent with C and E (i.e., A, B, and F) requires less than 24 days to complete; respectively, they take 5, 4, and 20 days to do. Putting it another way, free time, called "float," is available for each of the activities.

The same sort of analysis can be followed from start to finish of the project to find the earliest finish time for the entire project (in this case, 50 days). Examination also will show that one sequence of activities has no float time; this sequence is, of course, the critical path. In Figure 17-11, the critical path is the horizontal line connecting nodes 1, 2, 3, 4, 8, and 9 (activities C, E, H, K and L).

One further step can be taken at this stage. That is, one can start at the right end of the diagram, with the earliest finish time, and work backward through the diagram to determine the latest time that each activity could start or finish without affecting the project completion date.

Basic construction of the CPM diagram for a simple project thus is a relatively simple matter; calculations may be done manually. On a complex project with hundreds of subitems, the CPM diagram can be highly complex; an electronic computer often is used to speed calculations.

If the time of 50 days is satisfactory, the CPM diagram of Figure 17-11 may be used to control the project. Detailed equipment, labor, and material schedules may be prepared on a day-to-day basis. If job conditions change, the CPM diagram may have to be adjusted to fit the changed conditions. If the time of completion is to be advanced ("crashed"), it is obvious that it is most fruitful to crash the activities on the critical path; crashing noncritical activities will not advance the completion date.

As mentioned previously, one of the greatest values of the CPM approach is in resources planning. Detailed resources planning is beyond the scope of this text. However, its importance cannot be overemphasized. Obvious applications are to make the most efficient use of equipment, to utilize skilled labor in the most productive way, and to determine the optimum job schedule from the standpoint of overall cost. The contractor or owner can, if desired, determine the cost of crashing a project schedule and compare the cost with anticipated benefits. In all these approaches, a computer is an essential tool for a complex project.

REFERENCES

1. *Guide Specifications for Highway Construction.* American Association of State Highway and Transportation Officials, Washington, D.C. (1985).

2. Smith, Travis, and Kleiman, William F. *Behavior of High Earth Embankment on U.S. 101.* Highway Research Record No. 345 (1971).

3. *Compaction of Embankments, Subgrades, and Bases.* Highway Research Board Bulletin No. 58 (1952).

4. *Standard Specifications for Transportation Materials.* Part I. American Association of State Highway and Transportation Officials, Washington, D.C. (1982).

5. *Construction of Embankments.* National Cooperative Highway Research Program Synthesis of Highway Practice 8, Highway Research Board, Washington, D.C. (1971).

6. Gardner, R.P., Dunn, W. L., McDougall, F. H., and Lippold, W. J. *Optimization of Density and Moisture Content by Nuclear Methods.* National Cooperative Highway Research Program Report No. 125, Washington, D.C. (1971).

7. Lee, Kenneth L., Adams, Bobby Dean, and Vagneron, Jean-Marie J. Reinforced Earth Retaining Walls. *Journal of the Soil Mechanics Division,* Vol. 99, No. SM 10, American Society of Civil Engineers, New York (October 1973).

8. Vidal, H. The Principal of Reinforced Earth. *Highway Research Record 282,* Transportation Research Board, Washington, D.C. (1969).

9. Vidal, H. Reinforced Earth Retaining Wall. *Civil Engineering,* Vol. 40, No. 2, American Society of Civil Engineering, New York (February 1970).

10. *Rock and Soil Stabilization Systems.* VSL Corporation, Springfield, Va. (1985).

11. Jackura, K. A. Tire-Anchored Timber Walls. *TR News,* No. 117, Transportation Research Board, Washington, D.C. (March–April 1985).

12. Ingold, T.S. *Reinforced Earth.* American Society of Civil Engineers, New York (1982).

13. Moore, Lyndon H. Summary of Treatments for Highway Embankments on Soft Foundations. *Highway Research Record 133* (1966).

14. Haliburton, T. Allan, Lawmaster, Jack D., and McGuffey, Verne C. *Use of Engineering Fabrics in Transportation-Related Applications.* Prepared for Federal Highway Administration, Washington, D.C. (October 1981).

15. Jonas, Douglas L. Reducing Equipment Needs With CPM. *Western Construction.* (November 1964).

16. Reid, Douglas M. *Project Management System (PMS) Through the Use of the Critical Path Method (CPM).* Federal Highway Administration, Washington, D.C. (1974).

chapter eighteen

BASES, SUBBASES, AND LOW-COST SURFACES

In this chapter and in the following two chapters, we turn our attention to the design and construction of the subbases, bases, and wearing surfaces that comprise the pavement structure. The sequence of the material on pavements generally follows a line of increasing costs. This chapter focuses on relatively low-cost pavements, ranging from soil stabilized roads to bituminous surface treatments and low-cost bituminous plant mixtures.

SOIL STABILIZED ROADS

18-1 Introduction

In a broad sense, soil stabilization may be defined as the combination and manipulation of soils, with or without admixtures, to produce a firm mass that is capable of supporting traffic in all weather conditions.

If a stabilized soil is to be truly "stable," then it must have sufficient shearing strength to withstand the stresses imposed on it by traffic loads in all kinds of weather without excessive deformation. In addition, if the stabilized soil mixture is to be used as a wearing surface, then it must be capable of withstanding the abrasive effects of traffic.

In this branch of highway design and construction the emphasis is placed on the effective utilization of local materials, with a view toward decreased construction costs. In some areas the natural soils are of an unfavorable character and require modification through the use of suitable mineral constituents such as gravel or crushed stone or clay binder. In still other areas, admixtures such as bituminous materials, portland cement, salt, or lime are used for effective stabilization. The type and degree of stabilization required in any given instance are largely a function of the availability and cost of the required materials as well as the use that is to be made of the stabilized soil mixture.

The primary use of stabilized soil mixtures at the present time is in base and subbase construction. A stabilized soil base or subbase may provide the support for a relatively thin wearing surface that will be subjected to light or moderate amounts of traffic, or it may function as a base for a high-type pavement that will be subjected to very heavy volumes of traffic. Certain stabilized soil mixtures, including granular soil stabilized mixtures, serve as wearing surfaces on light-traffic roads.

Stabilized soil mixtures also lend themselves readily to the process of "stage construction," which involves the gradual improvement of the individual units of a highway system as the demands of traffic increase. Thus a properly designed stabilized soil mixture might function briefly as a wearing surface, receive a thin bituminous surface treatment as traffic increases, and eventually serve as a support for a high-type bituminous pavement that will serve a very heavy volume of traffic. Certain naturally occurring soils may require only compaction and drainage for stabilization. Other soils require treatment in various fashions and with various materials in order to satisfactorily perform their intended function. Four general classes of stabilization are briefly described below: (1) soil–aggregate roads and "granular stabilization," (2) bituminous soil stabilization, (3) soil–cement roads, and (4) stabilization by the addition of salt, lime, and various other chemicals.

18-2 Soil–aggregate roads and granular stabilization

Soil–aggregate roads consist of a substantial layer of a properly proportioned and blended mixture of soil and aggregate compacted to form a road that is capable of supporting traffic in all weather conditions. The AASHTO definition of soil–aggregate (dense-graded aggregate) is "natural or prepared mixtures consisting predominantly of stone, gravel or sand and containing silt–clay" (*1*). In addition to serving as wearing courses, generally for light traffic, soil–aggregate mixtures are very widely used as bases and subbases.

In this discussion, the term "aggregate" refers to the porton of a granular mixture or natural soil that is retained on a No. 200 sieve, including, by the AASHTO definitions, "stone," "gravel," and "sand." The function of aggregate in a soil–aggregate mixture is primarily to contribute internal friction. The most suitable aggregates, generally speaking, are those that are well graded from coarse to fine. One approach to the problem of gradation is to establish limits for the various sizes of particles that will result in a combination which gives maximum density; high density, in turn, is associated with high stability. It is not wise to lay too much stress on gradation, however, as many aggregates have been used that are not well graded in the usual sense. Specifications concerning gradation for this purpose are usually quite tolerant, with the idea, again, of making the best possible use of locally available materials.

The "fine-soil fraction" is that portion of the mixture that passes a No. 200 sieve, that is, silts and clays. The function of this portion of the soil–aggregate mixture is to act as a filler for the remainder of the mixture and to supply cohesion that will aid in the retention of stability during dry weather, while the swelling of the clay contained in this mixture serves to retard the penetration of water in wet weather. The nature and amount of this fine material must be carefully controlled if the mixture is to function properly, since an excessive amount of clay and silt may result in excessive volume change with change in moisture content. The properties of the fine-soil fraction are frequently determined by measurements of the plasticity index, or PI. As will be recalled, the PI is the difference between the liquid limit and the plastic limit. Both of these latter tests are customarily performed on the portion of the soil that passes a No. 40 sieve; this is frequently called the "soil binder."

It must again be noted that, for satisfactory results, the soil–aggregate mixture must be properly proportioned. In some cases natural soils may have satisfactory properties and thus may require only compaction and drainage for stabilization. Far more frequently, however, the existing soil will be missing some of the important soil elements. Thus a granular soil or "aggregate" may require the addition of fine material for binder, or a fine-grained soil may require the addition of aggregate for stability and to prevent excessive volume change. In certain cases the soil existing in the roadbed may only require the addition of a small amount of material that may be brought in, distributed along the roadway in the proper amounts, and blended with the existing soil to produce a satisfactory soil–aggregate mixture. In other cases the desired soil–aggregate mixture may be compounded in the proper proportions at a central mixing plant and then deposited on top of the existing roadway soil. In any case the properties of the final mixture are generally judged and controlled by the mechanical analysis, the liquid limit, and the plasticity index. In addition, since the mixture is generally required to be compacted to a high degree of density, the moisture–density relationships for the mixture must be determined for a given compactive effort so that the field compaction process may be properly controlled.

Before presenting detailed recommendations as to the composition of satisfactory soil–aggregate combinations, it seems desirable to discuss briefly the difference in requirements for the composition of mixtures intended for use as bases and those for use as wearing surfaces. If the mixture is to withstand the abrasive effects of traffic, it should have more binder than is required in a base course so that it will be more impervious to precipitation that falls on the surface of the road and can, to a certain extent, replace by capillary moisture what is lost by evaporation. Bases do not require as high a proportion of binder soil, since they may generally be required to have high stability and low capillarity so that they do not tend to soften because of the accumulation of capillary moisture. This consideration becomes of special importance when stage construction is being employed, where the mixture may serve as a wearing surface for a time and then as a base, so that the mixture must be capable of performing both functions satisfactorily during its service life. Where frost action is a factor, the percentage of material passing the No. 200 sieve may be reduced to prevent damage to a base or subbase.

Specifications for granular soil stabilized mixtures

Many different specifications are used for these soil–aggregate mixtures by different highway agencies. Indicative of the requirements are those given in Table 18-1, which is contained in Ref. 1.

It is recommended that the fraction passing the No. 200 sieve not be greater than two-thirds of the fraction passing the No. 40 sieve (*1*). For bases, subbases, and temporary wearing surfaces, the plasticity index should not exceed 6 and the liquid limit should not exceed 25. Maximum size of aggregate for wearing surfaces is recommended as 1 in. In other words, gradings A and B of Table 18-1 are not recommended for wearing surfaces. The quality of the coarse aggregate is controlled by specifying that the percentage of wear by the Los Angeles abrasion test shall not exceed 50. When a mixture of this type is to

TABLE 18-1
GRADATION LIMITS FOR SOIL–AGGREGATE SUBBASE,
BASE, and SURFACE COURSES (1)

Sieve Designation Grading	A	B	C	D	E	F
2 in. (50 mm)	100	100	—	—	—	—
1 in. (25 mm)	—	75–95	100	100	100	100
⅜ in. (9.5 mm)	30–65	40–75	50–85	60–100	—	—
No. 4 (4.75 mm)	25–55	30–60	35–65	50–85	55–100	70–100
No. 10 (2.00 mm)	15–40	20–45	25–50	40–70	40–100	55–100
No. 40 (0.475 mm)	8–20	15–30	15–30	25–45	20–50	30–70
No. 200 (0.075 mm)	2–8	5–20	5–15	5–20	6–20	8–25

be used as a wearing surface for several years, the minimum amount of material passing the No. 200 sieve should be 8 percent, the maximum liquid limit is 35, and the PI should be from 4 to 9.

Use of calcium chloride

Calcium chloride is sometimes used in connection with the construction of soil–aggregate mixes in order to expedite the compaction process by slowing the rate of evaporation of moisture from the mixture during compaction. The incorporation of small amounts of calcium chloride will in many instances produce increased density for a given compactive effort. In wearing surfaces, it aids in maintaining the moisture content at that desired for maximum stability, prevents raveling of the surface, and reduces dust.

Calcium chloride is a hygroscopic material; that is, it has the ability to attract moisture from the air and other sources. Two types of calcium chloride are used for road building and maintenance: (1) regular flake calcium chloride (type 1), which contains a minimum of 77 percent of $CaCl_2$, and (2) concentrated flake, pellet, or other granular calcium chloride (type 2), which has a minimum of 94 percent $CaCl_2$ (1). In new construction, calcium chloride should be added at the rate of 7 to 10 lb of type 1 or 5.6 to 8 lb of type 2 per ton of aggregate (2).

Selection and proportioning of soil elements

It is, of course, frequently desired to use the existing roadbed soil in the stabilized mixture in the interests of economy. Information regarding this material, in both cut and embankment sections, will generally be available from the detailed soil survey for the project under consideration. The results of the mechanical analysis and routine soil tests will then be available for each of the soils encountered in the section concerned before this phase of construction begins. The availability and location of granular and binder materials suitable for use in stabilization will also generally be known.

Assuming that the nature and location of suitable materials are known, the problem is to determine the proportion of "borrow" material that must be combined with the existing soil to produce a granular soil stabilized mixture which will meet the specifications of the organization concerned. It should also be noted that in many cases the entire soil–aggregate mixture may be imported and placed on top of the existing subgrade or previously constructed subbase; the material may come from a single source or may be a blend of materials from two or more sources. Trial combinations of materials are usually made in the laboratory on the basis of the mechanical analyses of the materials concerned. In other words, calculations are made in order to determine the sieve analysis of the combined materials, and the proportions of the various components are adjusted to fall within the limits of the specified gradation.

Numerical proportioning is a simple process for two materials. For example, assume that the materials of Table 18-2 are available and are to be blended to meet the requirements of Table 18-1 (grading D).

By simple trial, it can be established that any blend between the ranges of one part A and one part B to one part A and three parts B will be satisfactory, that is, will meet the grading requirements of Table 18-1. Using a greater amount of material A will abrogate the requirement related to the maximum passing the No. 10 sieve. At the other end of the scale, use of an excessive amount of borrow material will abrogate the maximum passing the No. 40 sieve. Presumably, the borrow material is more expensive than the existing soil; hence, it would be desirable to use as little of this material as possible. In other words, use the 1:1 mix.

The plasticity index of the proposed mixture is then determined in the laboratory and compared with the specified limits. If this value is satisfactory, the blend may then be assumed to be satisfactory. If the plasticity characteristics of the first mixture are not within the specified limits, then additional trials must be made until a satisfactory blend is obtained. The proportions thus secured may then be used in the field-construction process.

TABLE 18-2
EXAMPLE OF NUMERICAL PROPORTIONING PERCENT PASSING (BY WEIGHT)

	Subgrade Soil (A)	Borrow Material (B)	1A, 1B	1A, 3B
Mechanical Analysis				
1 in. (25 mm)	—	100	100	100
$\frac{3}{4}$ in. (19.0 mm)	—	90	95	93
$\frac{3}{8}$ in. (9.5 mm)	100	80	90	85
No. 4 (4.75 mm)	98	62	80	71
No. 10 (2.00 mm)	96	45	70	58
No. 40 (0.475 mm)	63	13	38	25
No. 200 (0.075 mm)	38	1	20	10
Plasticity Characteristics				
Liquid limit, percent	35	20	31	25
Plasticity index, percent	10	2	9	7

In certain cases, where the agency is familiar with the soils concerned, materials of known PI's may be combined in trial proportions established by a design "formula" in order to produce a mixture of given plasticity characteristics. An example of such an equation is as follows:

PI (mixture)

$$= \frac{\{(\%A)(\% \text{ passing No. 40, A})(\text{PI, A}) + (\% \text{ B})(\% \text{ passing No. 40, B})(\text{PI, B})\}}{(\%A)(\% \text{ passing No. 40, A}) + (\%B)(\% \text{ passing No. 40, B})}$$

$$(18\text{-}1)$$

Such proportioning will furnish an approximation of the PI that will serve until an actual test is made. For the 1:1 mixture of Table 18-2, the approximate value is

$$\text{PI} = \frac{50(63)10 + 50(13)2}{50(63) + 50(13)} = 8.6$$

A similar equation may be used to determine an approximate value of the liquid limit; in this case, the result is 31. In any event, a mixture is generally sought that will be well graded from coarse to fine and that will have the desired plasticity. The laboratory examination is generally completed by the determination of the compaction characteristics of the stabilized mixture so that the compaction process may be satisfactorily controlled in the field.

Graphical methods are also available and are particularly useful in the proportioning of more than two materials.

Construction procedures

Three basic procedures are used in the construction of soil–aggregate roads. These methods, which are equally applicable to the production of bases and wearing surfaces as well as to the less frequent purposes of subbase construction and subgrade improvement, are as follows:

1. Road mix construction.
2. Traveling plant construction.
3. Central plant construction.

In all these methods the basic processes are the same. That is, the soil elements must be properly selected and proportioned; the soils pulverized and uniformly blended; water and chemicals, if desired, added and uniformly mixed with the soil; and the blended materials spread in a thin layer of uniform thickness, properly compacted and finished. The several steps may be performed separately, or certain of the steps may be combined in one operation.

Road mix construction

In the basic process of road mixing, the proper proportions of soil elements that are to form the base or wearing surface are mixed directly on the surface of the subgrade or subbase. This may be done by alternately blading the material from one side of the roadway to the other. Special high-speed rotary mixing machines may also be used and will generally provide a more uniform mix.

After the soil elements have been properly blended, the mix must be spread in a uniform layer for compaction. If the moisture content of the mixture is less than the optimum moisture content, water must be added and blended with the soil to form a uniform mixture. After the water has been added, the soil is spread in a uniform layer and compacted. Thickness of layer is quite variable. For example, in the Midwest common practice is to spread the material in thin lifts (4 in. maximum compacted thickness) by a motor grader and begin rolling immediately; each layer is compacted before the next layer is spread. Where large rollers or vibratory compactors are used, material may be spread in much thicker layers, up to about 8 in. loose thickness. Initial compaction of a single layer may be accomplished by the use of sheep's-foot rollers. More commonly, pneumatic-tired rollers are used. Vibratory compactors are very effective on the more granular soil–aggregate mixtures. Following initial compaction, the surface is generally shaped to the proper cross section by a blade grader and final rolling done by pneumatic-tired or smooth-wheel rollers.

Traveling plant construction

When a traveling plant is used in this type of construction, the construction process is not greatly different from the road mix method previously described. However, the operations of mixing the various soil elements, calcium chloride, and water are accomplished by the traveling plant, usually in a single pass of the machine. A photograph of a traveling plant is shown as Figure 18-1.

Central plant method

Many organizations that build a large mileage of granular soil stabilized roads have found it desirable to use central mixing plants for the production of stabilized soil mixtures. Usually mixtures produced in this fashion cost slightly more than those formed by road-mixing methods, but certain advantages accrue from the use of this scheme. Advantages claimed for the central plant method include greater uniformity of the mix, greater ease of control of the proportions of the mixture, greater ease in supplying water to the mix, and fewer delays due to bad weather.

Plant setups vary with the type of aggregates and mixture and with the equipment available. In most plants, the aggregates are carried by conveyor belt into one or more storage bins, whether they are pit or crusher run or a combination of coarse and fine. From the storage bin, the material is conveyed to a pugmill or other mixing equipment. Where clay is to be added to coarser aggregates, a separate setup may be necessary for pulverizing and feeding this material to the mixer.

Usually, calcium chloride is spread on the aggregate in the conveyor through a small hopper. Water is added at the top of the conveyor through spray bars. Most authorities in this field recommend the use of a pugmill mixer.

FIGURE 18-1 A pair of single-pass soil stabilizers process materials for a granular base. (Courtesy The Koehring Co.)

Suitable pugmills are available from various manufacturers. When mixing is complete, the material is discharged into trucks for transportation to the job site. At the job site, steps of spreading, compaction, finishing, and curing are generally the same as those described for in-place construction. However, spreading often is done with a self-propelled aggregate spreader or with a bituminous paver.

18-3 Bituminous soil stabilization

Bituminous materials are used in conjunction with soils (and soil–aggregate mixtures) for two general purposes. In one application the bituminous material may supply cohesion to the stabilized soil mixture, as in the case of the stabilization of sands or very sandy soils. On the other hand, bituminous material may be incorporated into a natural or artificial soil mixture for the purpose of "waterproofing" it. That is, enough bituminous material is added to reduce the detrimental effects of water that may enter the soil during its service life. In some cases the bituminous material may perform the dual functions of supplying cohesion and necessary waterproofing. Bituminous soil stabilized mixtures are most frequently used for base construction.

Material requirements

A wide variety of soils may be stabilized with bituminous materials, including well-graded soils, sands, and even clays. However, generally speaking, the more fine material included in the mixture, the greater will be the amount of bitumen required for satisfactory waterproofing. Very fine, plastic soils generally cannot be stabilized economically with bituminous materials because of inherent difficulties of pulverization and mixing.

Soils that are stabilized with a bituminous material may contain a relatively large percentage of fines, with 25% or greater passing the No. 200 sieve. The fines may possess some plasticity, but for best results it is recommended that the plasticity index be not greater than 10 for subgrades and not more than 6 for bases (3).

In addition to the soil itself, the components of a soil−bitumen mixture usually are water and bituminous material. Water is used to facilitate compaction of the mixture and to aid in the dispersion of the bituminous material uniformly throughout the mix. Typically, the amount of bituminous material used for bituminous stabilized soils varies from 4 to 7 percent. The sum of the percentages of water and bituminous material used should not exceed that which will fill the voids of the compacted mixture. Many different types of bituminous materials have been used in this type of stabilization, including rapid-curing, medium-curing, and slow-curing liquid asphalts. Medium-setting and slow-setting grades of asphalt emulsions increasingly have been chosen in recent years.

The choice of a bituminous material for use in a given project is primarily dependent on local experience and comparative costs.

Of the cutbacks, rapid-curing grades are recommended for use in extremely sandy soils or those containing a minimum of silt and clay particles. As the percentage of silt and clay increases, a medium-curing cutback is recommended, since it will mix better and give a more homogeneous mixture. Slow-curing liquid asphalts have been used successfully with soils that contain 30 to 40 percent silt and clay, particularly in arid regions. As to grade of cutback asphalt, the general rule is "use the heaviest asphalt that can be readily worked into the soil." Asphalt emulsion should be of mixing grade. General recommendations for asphalt emulsion grades are shown in Table 18-3.

Design of mixture   The proportions of soil, water, and bituminous material are generally determined in the laboratory. Soil from the project is usually trial-mixed with various asphalt types at different percentages and tested by some type of strength test to find the best mixture (4). Reference 5 gives additional information on mix design methods for liquid asphalt mixtures.

Construction methods   Methods of mixing soil−bitumen materials are generally the same as those previously described for granular soil stabilized mixtures. An additional step, of course, involves the application of the necessary bituminous material and its uniform distribution through the mixture. In the case of roadmixing, the bituminous material, which sometimes has been heated moderately to the specified temperature for application, is applied in measured amounts by a pressure distributor, along with water applied separately, and the entire mix thoroughly blended by motor graders disc harrows, rotary speed mixers, and so on. Where a traveling or central mixing plant is used, the measured amount of bituminous material is incorporated in the plant.

When aeration is complete, the material is spread in a layer of uniform thickness ready for compaction. The thickness of the layer may vary from 2 to 6 in., depending on job conditions. Compaction usually is done with a sheep's-

TABLE 18-3
RECOMMENDED GRADES OF EMULSIFIED ASPHALTS FOR BITUMINOUS–AGGREGATE MIXTURES

Type of Construction	ASTM Specification D 977					ASTM SPECIFICATION D 2397 (Cationic)			
	MS-1, HFMS-1	MS-2, HFMS-2	MS-2h, HFMS-2h	SS-1	SS-1h	CMS-2	CMS-2h	CSS-1	CSS-1h
Bituminous-aggregate mixtures:									
For pavement bases and surfaces:									
Plant mix (hot) (D 3515)			X[a]						
Plant mix (cold)									
Open-graded aggregate,		X				X			
Dense-graded aggregate			X	X	X		X	X	X
Sand				X	X			X	X
Mixed-in-place:									
Open-graded aggregate	X	X	X			X	X		
Dense-graded aggregate				X	X			X	X
Sand				X	X			X	X
Sandy soil				X	X			X	X
Slurry seal				X	X			X	X

[a]Specification D 3515 permits the use of other emulsion grades by note: "Grades of emulsion other than MS-2h may be used where experience has shown that they give satisfactory performance."

SOURCE: *A Basic Asphalt Emulsion Manual.* Manual Series No. 19, The Asphalt Institute (1979).

foot roller or with pneumatic-tired rollers. Compaction is continued until a specified density (expressed in terms of laboratory density) is achieved. After compaction is complete, it may be required that some "curing" time elapse before the next layer is placed so that the water content after construction may be maintained at the desired figure. It should also be noted, at this stage, that the mixture may be compacted without the addition of water, as such, especially when asphalt emulsions are being used. Rolling of the final layer may be accomplished by the use of steel-wheel or pneumatic-tired rollers, and the base must be finished to the final section shown on the plans.

It should be emphasized that this type of soil stabilized mixture is generally only satisfactory for base construction. Placing of a wearing surface would then generally follow shortly after completion of the base and before any considerable amount of traffic would be permitted to use the road.

Oiled earth surfaces   This type of treatment is quite different from the other types of bituminous soil stabilization that have been discussed. No mechanical mixing is involved in this construction process, as the liquid bituminous material is simply applied to a properly prepared natural soil surface. The material penetrates a short distance into the soil layer, thus preventing dust and forming a thin "stabilized" surface that will be capable of supporting a limited number of light vehicles. Over a period of time, with proper annual retreatment, a stabilized surface may be built up to a thickness of several inches. Its principal use has been in the construction of very lightly traveled local or "service" roads. This type of construction also has been widely used in areas where cracked or uncracked residual oils are cheaply available.

Material requirements

The soil, of course, will be that naturally existing in the section concerned, and the method is principally applicable to silt and clay soils. The bituminous material that is used should generally be a product of low viscosity which contains a wide range of volatile materials or "light fractions." The material that has probably been most widely used is slow-curing liquid asphalt, generally grade SC-70 or SC-250. Medium-curing cutbacks, MC-70 and MC-250, have also been successfully used. It should be noted here that a large number of cracked residuals have also been used, as in the construction of access roads in oil fields. Slow-setting asphalt emulsions are also suitable for oiled earth surfaces. The amount of bituminous material or "road oil" which has been used in initial treatment during the first year is generally about 1 gallon per square yard. Applications during successive years are somewhat less.

Construction methods

The natural soil surface must be adequately prepared by crowning for drainage and by compaction. The surface to which the oil is applied should be free from dust and should be slightly moist in order to facilitate penetration of the oil. Generally speaking, treatment should follow immediately after the final shaping (grading) operation. The oil is usually applied in two or three applications, the total amount applied being as previously indicated. Some time should be allowed to elapse between applications, and traffic should not be allowed on the road until the oil has completely penetrated the surface. After each application, the surface should be compacted under traffic or by the use of pneumatic-tired rollers.

18-4 Soil—cement bases

A type of soil stabilization which has enjoyed increasing popularity in recent years involves the incorporation of portland cement, in amounts generally varying from 7 to 14 percent by volume of the compacted mixture, with naturally occurring or artificially created soils or soil—aggregate mixtures. This type of construction is generally employed in the formation of base courses, usually with thicknesses varying from 4 to 6 in. A soil—cement mixture may serve as a base for a thin wearing surface that will be subjected to light or medium traffic or as a support for a high-type flexible or rigid pavement. Bases of this type have been successfully used in the construction of city streets, and soil—cement is also suitable for use in driveways, shoulders, parking areas, some airport runways, and so on. In some areas, particularly California, this general type of construction is termed "cement-treated base."

Modern use of soil—cement began in South Carolina in 1933 with the construction of experimental sections, and the quantity of soil—cement used in highway construction has steadily increased since that time. Many millions of square yards of soil—cement have been constructed in the United States.

Material requirements

Nearly all subgrade soils, with the exception of those that contain high percentages of organic material, may be stabilized through the use of portland cement. Soils that contain high amounts of fine material, such as silts and clays, generally require large percentages of cement for successful stabilization. Gradation requirements for soils to be used in soil—cement mixtures are nearly nonexistent, since practically any soil may be utilized. Sandy and gravelly soils with 10 to 35 percent silt and clay combined have the most favorable characteristics. Glacial and water-deposited sands and gravels, crusher-run limestone, *caliche*, limerock, and almost all granular materials are good if they contain 55 percent or more material passing the No. 4 sieve. Exceptionally well-graded materials may contain up to 65 percent material retained on the No. 4 sieve and still have sufficient fine material for adequate binding. The maximum size of aggregate in soil—cement mixtures should not be more than 3 in. Materials in old gravel or crushed stone roads or bases make excellent soil—cement. An important factor in determining the suitability of a given soil for soil—cement construction is the ease with which the soil may be pulverized. Soils that contain large amounts of clay will be more difficult to pulverize, will require more cement, and will demand more careful control during the construction process.

The cement used is generally standard portland cement, types 1 and 1A. The third ingredient of soil—cement mixtures is water, which is necessary to aid in the compaction of the loose mixture and for the hydration of the cement in the mix. In this respect, soil—cement mixtures are somewhat similar to concrete in that water is a necessary ingredient, and the loss of excessive amounts of water during the curing period must be prevented if the mixture is to harden properly. Practically any normal source of water may be used, although it should be quite clean and free from excessive amounts of organic matter, acids, or alkalies. Fresh water is, of course, normally used, although seawater has been used in a few instances. The proper amounts of cement and water to be used on a given project are determined by a series of laboratory tests that are briefly described in the following paragraphs. Information secured in the laboratory will also be used in the control of construction, particularly in the control of the compaction of the mixture.

Laboratory tests

Detailed laboratory testing procedures for soil–cement have been developed by the Portland Cement Association (6) and adopted by AASHTO and ASTM. Three general procedures are in existence, depending on the time available for testing and the size of the project.

On major projects, when time is available for testing, and with all soils, a detailed procedure to establish the satisfactory minimum cement content is recommended. As a first step in the laboratory procedure, the subgrade soil encountered in the project would be subjected to routine classification and placed in one of the groups of the AASHTO classification system. With this information as a background, cement contents to be used in the preparation of trial mixtures would then be established on the basis of past experience. For example, it might be decided that mixes would be prepared using 8, 10, and 12 percent of cement by volume. It should be emphasized that the percentage of cement used here is expressed in terms of the volume of the compacted mixture rather than by weight that is, the percentage of a sack of cement per cubic foot of compacted soil cement. Thus, 10 percent by volume means that there is 10 percent of 94 lb, or 9.4 lb, of cement in 1 ft^3 of finished, compacted roadway.

Moisture–density relationships for soil–cement mixtures containing the selected amounts of cement are then determined, using a procedure entitled, "Methods of Test for Moisture–Density Relations of Soil–Cement Mixtures," ASTM Designation D558. The procedure used is very similar to that employed in the standard AASHTO compaction procedure. The percentage of cement required for satisfactory stabilization is based on two series of durability tests, and the moisture–density tests supply information necessary for the formation of the durability test specimens containing the desired amounts of cement. The durability tests used involve the subjecting of the various soil–cement mixtures to cycles of freezing and thawing and of wetting and drying. The test methods used are "Methods of Wetting-and-Drying Test of Compacted Soil–Cement Mixtures," ASTM Designation D559, and "Methods of Freezing-and-Thawing Test of Compacted Soil–Cement Mixtures," ASTM Designation D560.

The detailed laboratory procedures involved in the durability tests are much too lengthy to be presented here. However, they may be described generally. For each of the durability tests and for each of the selected cement contents, two specimens are prepared by compaction at optimum moisture to maximum density. Both samples are cured for 7 days before testing is begun. After curing, both samples are subjected to 12 cycles of alternate wetting and drying. At the end of each cycle, one specimen is vigorously brushed, using a standard procedure. Moisture and density determinations are made on the other, unbrushed specimen during each test cycle. At the end of the test, the specimens are oven-dried to constant weight and the weight losses determined. A similar procedure of alternate freezing and thawing is used to test the two specimens prepared for that test. Although the cement requirements are principally established on the basis of the results of the described durability tests, in some cases the compressive strengths of the test mixtures also are determined. The following are the recommendations of the Portland Cement Association relative to the behavior of the test speciments (6).

1. Losses during 12 cycles of either the wet−dry or freeze−thaw test shall conform to the following standards:
   Soil groups A-1, A-3, A-2-4, and A-2-5, not over 14 percent.
   Soil groups A-2-6, A-2-7, A-4, and A-5, not over 10 percent.
   Soil groups A-6, and A-7, not over 7 percent.
2. Compressive strengths of soil−cement test specimens should increase with age and with increases in cement content in the ranges of cement content producing results meeting the requirements given above.

A short-cut test method has been developed (6) by the Portland Cement Association for determining cement factors for sandy soils. The procedure is designed for use on smaller projects, particularly those where testing facilities and manpower are limited; it results in safe, but not necessarily minimum, cement contents for these soils. The only tests required are grain size analysis, moisture−density tests, and compressive strength tests. Results of the tests are used in conjunction with charts based on past experience involving thousands of test specimens to establish safe cement content.

Construction methods

The basic steps in the construction of a soil−cement base, assuming that the subgrade beneath the base requires no special treatment or has previously been brought to the desired condition, may be listed as follows:

1. Pulverizing of the soil that is to be processed.
2. Spreading of the required amount of cement and mixing with the soil.
3. Addition of the required amount of water and its incorporation with the soil−cement mixture.
4. Thorough compaction, including final rolling and finishing.
5. Curing of the completed soil−cement base.

Soil−cement mixtures may be "processed" by roadmixing equipment, by traveling plants, or by central mixing plants. In the large majority of cases traveling plants are used, including those of the flat type, the windrow type, and rotary speed mixers. The operation of traveling and central mixing plants in other types of soil stabilization has been discussed earlier in this chapter. Their use in this type of construction is not greatly different from that which has been previously described.

Pulverization

With soils that are difficult to process, the soil that is to be used must be thoroughly pulverized before cement is added. If the existing soil is to be used it sometimes must be scarified to the desired depth by use of a ripper or a scarifier attachment on a motor grader. If imported material is being used, it must be hauled in and spread on the existing soil to the required depth. The soil may then be pulverized by the use of offset disk harrows, gang plows, or rotary speed mixers. It is generally required that the soil, with the exception of gravel or stone, be pulverized until at the time of compaction 100 percent of the soil—cement mixture will pass a 1-in. sieve and at least 80 percent will pass No. 4 sieve. It is especially important that the moisture content of fine-grained soils be within rather narrow limits during pulverization. For some soils the correct moisture content for pulverization will be at or near the optimum moisture content. Aeration of a wet soil or addition of water to a dry soil may be necessary, and in some cases steps may be taken to protect the pulverized soil from moisture change before the next step in the construction process is begun.

Incorporation of cement

The proper amount of cement may be spread on the surface of the soil either by hand or by mechanical means. On small jobs, bags of cement are "spotted" by hand along the surface in rows of predetermined spacing. The bags are then opened and the cement spread in uniform transverse rows, also by hand. Spreading is completed by the use of a spike-toothed harrow. On most projects, bulk cement is used and a mechanical spreader is utilized. When spreading of the cement has been properly completed, dry mixing is ready to begin.

Mixing

In the road mix process, blending of the soil and cement may be done by the use of such tools as gang plows, disc harrows, spring-tooth cultivators, and rotary tillers, but usually is done with a traveling plant. When a traveling plant is used, mixing, of course, is carried out within the plant; exact processing procedure will vary with the type of equipment used. With a central plant, mixing is carried out in the plant and the mixture hauled to the job site in trucks; such a procedure is used on projects involving borrow soils. Careful control is necessary in a road mix operation in order that the mixing be carried to the proper depth and that a uniform mix be secured. The equipment available for this purpose will generally be operated in a train, with additional units being added to the train as construction proceeds, to supply water and complete the wet mixing operation. In many cases dry mixing and wet mixing are not essentially separate operations but are carried along together in an integrated series of operations. The proper amount of water is added, and the soil, cement, and water are thoroughly blended to a uniform mixture ready for compaction. The amount of water added is generally enough to increase the moisture content to 1 or 2 percent more than the optimum desired for compaction. This is done because of the loss of water that may occur during road mixing operations as compared with laboratory mixing. In other words, if a slight excess of water is incorporated during mixing, the soil may then generally be expected to be very close to optimum by the time compaction is ready to begin.

Compaction and finishing

Initial compaction of a soil—cement base is generally achieved by the use of sheep's-foot rollers. After the base has been compacted to within 2 or 3 in. of the surface, the layer is brought to shape with a motor grader. Rolling then generally continues until satisfactory density has been secured to within about 1 in. of the surface. In very sandy soils, of course, compaction may not be possible with sheep's-foot rollers, and densification may then be accomplished by the use of pneumatic-tired rollers. Steel-wheel rollers have been used with granular soils, and newer types of compaction equipment, including vibratory compactors, grid, and segmented rollers are finding increasing use. After the soil—cement mixture has been compacted as described, the rollers are taken off and the surface brought to final shape. The final shaping is done in a variety of ways, including steps to remove the compaction planes left in the surface by the rollers. Final compaction is secured by rolling with pneumatic-tired rollers alone or in combination with steel rollers weighing from 3 to 12 tons, depending on the soil involved. Rolling is continued until a tightly packed surface is secured. During all the rolling operations that have been described, it is essential that the moisture content be maintained at optimum. Frequent checks on the density and moisture content are, of course, necessary.

Curing

The water contained in the soil—cement mixture is necessary to the hardening of the cement, and steps must be taken to prevent loss of moisture from the completed base by evaporation. Curing commonly is done by the application of a light coat of bituminous material in most cases; commonly used bituminous materials are RC-250, MC-250, RT-5, and emulsified asphalt (RS-2). Rate of application varies from 0.15 to 0.30 gallon per square yard. A bituminous wearing surface should be placed on the completed soil—cement base as soon as practical.

A sketch showing typical operations for soil—cement road construction is shown as Figure 18-2. Additional information on the construction of soil—cement roads is given in the Portland Cement Association publication *Soil—Cement Construction Handbook*.

18-5 Miscellaneous materials used in soil stabilization

Primarily because of the dictates of economy, many materials other than those that have been discussed here have been used for soil stabilization at various

FIGURE 18-2   Sketch of soil–cement processing operations with multiple-shaft traveling mxiing machine. (Courtesy Portland Cement Association.)

times and in various places. A few of the more important of these materials are discussed in the following paragraphs. The discussion is not intended to be complete but will serve to indicate the wide range of materials that may be used in construction of this sort.

Lime

The incorporation of small amounts of hydrated lime has proved to be effective in the improvement of certain plastic clay soils. Lime has been used—

principally in states along the Gulf Coast—to reduce the plasticity, shrinkage, and swell of clay soils, at the same time somewhat increasing their bearing capacity. In essence, its use has permitted the upgrading of certain marginal and submarginal soils into satisfactory base and subbase materials.

The use of lime also has certain advantages from the standpoint of the construction process by improving the workability of plastic soils (making them easier to pulverize). It tends to waterproof the soil to some extent and allow it to dry out more quickly when saturated, thus speeding construction.

The quantity of lime used in subgrade treatment is generally from 3 to 6 percent. Quantitites as low as 1 percent have been used, while amounts much above 6 percent are not economical. Depth of treatment is usually 6 in. The subgrade soil is scarified and pulverized, followed by spreading of lime, usually with a mechanical spreader or a bulk hauler. Enough water is added to bring the moisture content to 5 percent or more above optimum, and the water is distributed by a rotary speed mixer. The lime−soil mixture is allowed to cure for periods of 1 to 7 days. Mixing and pulverizing then continue until all the material will pass a 1-in. sieve and at least 60 percent will pass a No. 4 sieve. Compaction is done by pneumatic rollers or vibratory compactors, and the compacted layer is allowed to cure for 3 to 7 days before the next layer is placed. On some projects, the lime is added to the soil in the form of a slurry.

Construction procedure is quite similar to that used for soil−cement.

More detailed information on lime stabilization is contained in *Transportation Research Circular No. 180,* published in September 1976. Research on the use of a combination of lime and flyash for soil stablization is described in the NCHRP publication *Synthesis of Highway Practice 57* (1976).

Salt

Rock salt has been successfully used in the stabilization of soils for bases and wearing surfaces subjected to very light traffic. The procedure is very similar to that described for mixtures in which calcium chloride is used. Recommended quantity of salt is generally from 1 to 2 lb per square yard per inch of loose thickness.

Other materials

The use of lignin binder, which is produced as a by-product in the manufacture of paper from wood by the sulfite process, has been reported experimentally, and a limited mileage of roads has been constructed with this material. Similarly, various silicates of soda have been used experimentally to improve certain clay soils.

A number of resinous waterproofing materials, including Stabinol and Vinsol, have been used experimentally; none is in widespread use. A considerable amount of study has been devoted to a variety of resinous bonding materials for soil stabilization. Most successful of these have been a combination of aniline and furfural, and calcium acrylate; both offer considerable promise for the future but are currently too expensive for ordinary civilian use.

MACADAM ROADS

Crushed-stone roads first constructed in Europe during the latter part of the eighteenth and the early part of the nineteenth centuries were the forerunners of modern macadam roads. Construction of this type was introduced in France by Pierre-Marie Tresaguet about 1765 and in England by Thomas Telford shortly after 1800. Both of these early engineers used large pieces of broken stone in the lower course and placed a layer of finer crushed stone as a wearing surface. Telford advocated the use of large, flat pieces of stone, which were hand placed to form the base course. John McAdam, a famous Scottish road builder and engineer, insisted on the use of smaller stone—about 1.5 in. maximum size—for the entire thickness of the "pavement." The first roads of this type were built by him in England early in the nineteenth century. Both Tresaguet and McAdam laid emphasis on the grading and drainage of the subgrade and the crowning of the crushed-rock surface to aid in removal of rainfall from the surface of the road. The technique of macadam construction improved rapidly over the years, and by the end of the nineteenth century many miles of this type of surface had been built and were in service as rural roads and city streets n the United States. With the advent of the motor vehicle, the basic macadam or broken-stone surface became less suited to the demands of traffic, and this type of pavement assumed less importance in highway construction. Roads of the original macadam type were (and are) basically unsuited for use as wearing surfaces carrying large numbers of rapidly moving pneumatic-tired vehicles as they become quite rough and dusty under this type of traffic.

The term "macadam" originally referred to a road surface or base in which crushed or "broken" stone was mechanically "keyed" or locked by rolling and cemented together by the application of stone screenings and water. Later, when "bituminous macadam" roads were built by using a bituminous material as the binder, the original type of road came to be known as a "water-bound macadam." These two types of roads, the most prominent classes of macadam roads, are defined as follows.

Water-bound macadam is a layer composed of broken-stone (or crushed gravel or slag) fragments that are bound together by stone dust and water applied during construction, in connection with consolidation of the layer by a heavy roller or a vibratory compactor. This type of macadam road closely resembles those so widely used in the early days of road building. Now water-bound macadam roads are seldom constructed. Detailed material requirements and construction procedures for such roads are therefore not given here.

Bituminous macadam is crushed-stone or crushed slag base or wearing surface in which the fragments are bound together by bituminous material; the aggregate layer is compacted, and bituminous material is applied to the surface of the layer. The bituminous material then penetrates into the voids of the compacted layer and serves to bind the fragments together. This type is frequently termed simply "penetration macadam."

18-8 Bituminous macadam

As indicated, the term "bituminous macadam" refers to a type of macadam "pavement" in which the aggregates are bonded together by bituminous material. Both wearing surfaces and bases are constructed by this method and, at least as far as primary highway systems in the United States are concerned, this is the most important and extensively used of the macadams. Bituminous macadams have been most widely developed in areas in which suitable aggregates are readily available. A penentration macadam road, described below, is the basic type of bituminous macadam in which "hard" stone is used and the bituminous binder is asphalt cement or one of the heavier grades of road tars.

Material requirements

Detailed requirements for materials used in bituminous macadams vary considerably from state to state, and the following sections are intended only to indicate requirements which are typical of those in current use.

Aggregates   The aggregate that is used is crushed stone or crushed slag. Very hard stone, such as traprock, is used in many areas, while crushed limestone and sandstone are also used. The aggregate is required to be quite uniform in character and free from deleterious amounts of dust, flat or elongated pieces, and other objectionable elements. It may be subject to requirements for soundness. Another typical requirement is that the maximum loss in the Los Angeles abrasion test be 40 percent for wearing surfaces and 50 percent for base courses. In states where unusually hard stone is abundant, the maximum loss permitted may be as low as 25 percent. Some few states still base this requirement on the results of the Deval abrasion test. Crushed slag is generally required to weigh not less than 70 lb/ft.^{3}

The gradation of the aggregate for penetration macadam courses best suited for the work depends on the size of surface voids in the course and fluidity of the bituminous binder. Table 18-4 gives an example of aggregate gradations that have been recommended for such courses (7).

Bituminous materials   The bituminous materials used in this type of macadam are heated to make them liquid at the time of application to the compacted aggregates. Asphalt cements of penetration grades 85 to 100 and 120 to 150 are used, and they are heated to temperatures that generally do not exceed 350°F and applied at temperatures between 275° and 350°F. Some organizations allow use of heavier grades of cutback asphalts in this type of pavement when finer aggregates are used. Road tars have also been used, with grades RT-10,-11, and -12 being preferred. These latter materials are heated to about 250°F and applied at temperatures from about 175° to 250°F.

Construction methods

The construction of a penetration macadam may be generally broken down into the following basic steps:

1. Spreading and rolling of coarse aggregate.
2. Initial application of bituminous material.

TABLE 18-4
GRADING REQUIREMENTS
(PENETRATION MACADAM)

Sieve Designation	Percentage Passing (by Weight)	
Coarse aggregate		
$2\frac{1}{2}$ in. (63 mm)	100	
2 in. (50 mm)	90–100	
$1\frac{1}{2}$ in. (37.5 mm)	35–70	
1 in. (25.0 mm)	0–15	
$\frac{1}{2}$ in. (12.5 mm)	0–5	
	Viscuous Binders	Fluid Binders
Choke aggregate		
1 in. (25.0 mm)	100	
$\frac{3}{4}$ in. (19.0 mm)	90–100	100
$\frac{1}{2}$ in. (12.5 mm)	20–55	90–100
$\frac{3}{8}$ in. (9.5 mm)	0–15	40–75
No. 4 (4.75 mm)	0–5	5–25
No. 8 (2.36 mm)		0–10
No. 16 (1.18 mm)		0–5

SOURCE: *1985 Annual Book of Standards*, Sect. 4, Vol. 04.03. American Society for Testing and Materials.

3. Spreading and rolling of key aggregate.
4. Application of seal coat.

Spreading and rolling of coarse aggregate Coarse aggregate generally is placed between well-compacted shoulders with vertical faces, using mechanical spreaders. The material must be placed in a uniform loose layer of the required thickness. Precautions are taken to avoid segregation of the aggregate and to prevent the material from becoming mixed with dirt or other foreign material during and after the spreading operation. Before rolling begins, the layer is subjected to careful visual inspection and excessively large and/or flat or elongated pieces are removed. Areas that show an excess of fine material are also subject to correction, and the layer is generally brought to a uniform condition.

When spreading of the coarse aggregate has been satisfactorily completed, the layer is rolled with a three-wheel roller weighing not less than 10 tons. Rolling is continued until the surface is thoroughly compacted and the stone keyed together. Material that crushes under the roller in such a way as to interfere with the proper penetration of the bituminous material is removed, and weak spots are corrected and irregularities removed as rolling continues. The finished surface of the coarse layer must be true to the cross section shown on the plans and at the desired elevation. An excessive amount of rolling is not desirable, since enough surface voids must remain to permit ready penetration of the bituminous binder. In Figure 18-3 is shown a close-up view of a properly compacted layer of coarse slag.

FIGURE 18-3 Air-cooled blast-furnace slag shown compacted on the left and bonded with an application of hot asphalt by the penetration method on the right. (Courtesy Macadam Pavements, Inc.)

Initial application of bituminous material The initial application of bituminous material is then made by means of a pressure distributor, with the material at the correct temperature and in the amount prescribed by the engineer. The quantity recommended by the Ashphalt Institute is 1.75 to 2.25 gallons per square yard for a compacted thickness of $3\frac{1}{2}$ to 4 in.; this would correspond to a first application of 350 to 400 lb of aggregate (specific gravity, 2.55 to 2.75) and a second application of 35 to 50 lb. It is important that the bituminous material be spread uniformly over the layer in the correct amounts. Precautions are usually taken to prevent application of excessive amounts of bitumen at the points where successive applications overlap. This may be accomplished by spreading tar paper over the end of the preceding spread of bituminous material so that the sprays may be opened over the paper and uniform distribution over the untreated section assured.

The pressure distributor is the key machine on many types of bituminous paving work, including macadam, surface treatment, and road mix construction. The machine must be capable of applying an accurately measured amount of bituminous material over the distance required to empty the load, regardless of changes that may occur in grade or direction. Distributors are available in several sizes, with capacities from 800 to 5500 gallons.

A typical distributor of this type consists of a rubber-tired truck, truck-drawn trailer, or semitrailer on which is mounted an insulated tank or one that is equipped with heating facilities. Quite frequently an oil burner is used to supply heat, with the burner being located at the bottom and rear of the tank. Heat from the oil burner passes through the interior of the tank, close to the bottom, and out at the top of the tank through a number of heating flues. The distributor is also supplied with an engine-driven pump, so designed that it may handle products ranging from light "oils" that are fluid at air temperature to semisolid materials which are fluid at temperatures of 350°F or more.

The bituminous material is applied to the surface of any desired area through a system of nozzles mounted on a spray bar at the rear of the tank. The fluid material is pumped under pressure through this spray bar, which may be of varying width, with the minimum width being about 6 ft and the maximum being 24 ft. Most modern distributors have a circulating system that keeps the material flowing when it is not being sprayed. The distributor is further equipped with a thermometer installed in the tank and with a hose connection to which a single nozzle may be attached so that the material may be readily applied to a limited area or at a specific point.

When application is to be made through the spray bar and nozzles, rather simple calculations may be made to relate the variables involved in the problem of applying the fluid material at a uniform, specified rate. Assuming that the width of the spray bar, the quantity (gallons) handled by each revolution of the pump, and the speed of the pump are known, the speed at which the truck must be driven to secure a given rate of application may be readily obtained. The rate of application is controlled by a "bitumeter," which comprises a small wheel beneath the truck, connected by a cable to a dial in the cab; the dial usually shows the speed in feet per minute but may show gallons per square yard directly. A schematic drawing of a typical pressure distributor is shown in Figure 18-4. A bituminous distributor in action is shown in Figure 18-5.

The computation of the application rate of a bituminous distributor assumes that all the material pumped goes through the nozzles. Many distributors have circulating spray bars so that part of the material pumped returns to the distributor. For a given pump speed or pressure, a given bituminous material, and a given temperature, a given flow through the nozzles will result. Charts or tables showing these data assist the inspector or driver in the selection of vehicle speed. The application rate is then verified by the distance traveled.

Spreading and rolling of key aggregate   Immediately after the initial application of bituminous material is made, the surface is lightly covered with key aggregate in sufficient quantity to prevent the roller wheels from sticking. The layer is then rolled until thoroughly compacted. The key aggregate is then spread over the surface in the desired quantity so that the surface voids of the layer will be completely filled but there is no excess aggregate remaining on the surface. The surface is then rolled with the same equipment as before. Simultaneously with the rolling, the key aggregate may be lightly broomed to effect an even distribution of the material, and additional small amounts may be added as needed. Vibratory compactors are used in some states in conjunction with rollers. This process is continued until the key aggregate is firmly embedded in the bituminous material and the surface is firmly compacted. In Figure 18-6 is shown a properly bonded bituminous macadam after the application and rolling of the key aggregate. The layer is then ready for the application of the seal coat if it is to function as a wearing surface; if it is to be a base, the seal coat usually is omitted.

Application of seal coat   Before placing of the seal coat is begun, the surface is swept, usually by means of a broom drag, for the purpose of either removing all

FIGURE 18-4 Schematic drawing of a bituminous pressure distributor. (Courtesy The Asphalt Institute.)

FIGURE 18-5 Bituminous pressure distributor in action. (Courtesy E. D. Etnyre & Co.)

loose material or uniformly distributing any loose key aggregate that may remain after the previous rolling. Bituminous material is then applied to the surface in the amount determined by the engineer and is immediately covered with the fine aggregate (stone chips). Rolling and brooming are then begun again and continued until the cover stone is firmly bonded and completely fills the voids in the finished surface. The required amount of rolling cannot be definitely stated, and this process is continued until the surface is completed to the satisfaction of the supervising engineer. Pneumatic rollers may be used on this final layer in some states. The completed surface must, of course, be correct to grade and cross section. The finished surface is tested by templates and straightedges, with deviations up to $\frac{1}{4}$ in. in 10 ft being permitted in the majority of states performing this work. The surface may be opened to traffic soon after the final rolling is complete.

FIGURE 18-6 Close-up of bituminous penetration macadam showing the texture after the first and second penetration applications of bituminous material and choke stone have been applied. (Courtesy Macadam Pavements, Inc.)

PRIME COATS AND TACK COATS

When a bituminous wearing surface is to be constructed, the foundation layer is first sprayed with a liquid bituminous material. Depending on the nature of the foundation layer, the treatment is called either a *prime coat* or a *tack coat*.

18-10 Prime coats

A prime coat is an application of liquid bituminous material to a previously untreated base or wearing surface. The bituminous material penet .es the surface and is generally completely absorbed. A prime coat serves several definite purposes, as follows. First, the prime coat serves to promote adhesion or "bond" between the base and wearing surface; this is probably its most important function. Second, it serves to consolidate the surface on which the new treatment is to be placed, and this may aid in maintaining the integrity of this underlying layer during the construction of a surface treatment or some other type of wearing surface. Third, it may function to a certain extent as a deterrent to the rise of capillary moisture into the wearing surface.

Materials

The material used in the prime coat should be liquid asphalt or road tar of low viscosity. It is important that the material used have high penetrating qualities, and it should be of such a nature that it will leave a high-viscosity residue in the void spaces of the upper part of the surface that is being primed. Of the asphaltic materials, grades MC-30 and MC-70 are widely used, respectively, in the priming of dense tightly bonded surfaces and surfaces having a more open texture. Slow-setting emulsions, grades SS-1, SS-1h, CSS-1, and CSS-1h, are recommended for penetrable surfaces (8).

The amount of bituminous material that is used varies over a considerable range, depending on the material selected and job conditions. Generally, the amount of bituminous material required for priming is from 0.20 to 0.50 gallon per square yard of surface. The lower limit applies to tight, fine-grained surfaces; the upper to loose, open-textured surfaces.

Construction methods

After the foundation layer has been swept to remove loose material, dust, and other foreign matter, the prime coat is applied with a pressure distributor. Best results are obtained when the surface is dry or only slightly damp. The material should be uniformly applied at the recommended spraying temperature. (See Table 15-9.) Priming generally should be done only when the air temperature in the shade is above 55°F and rising or above 60°F if falling.

18-11 Tack coats

A tack coat is defined as a single application of bituminous material to an existing bituminous, portland-cement concrete, brick, or block surface or base. The purpose of a tack coat is to provide adhesion between the existing surface and the new bituminous wearing surface. Liquid asphalts of grades RC-70 through RC-250 are frequently used as tack coats, while medium-curing liquid asphalt of grade MC-250 has also been used. Road tars are used to some extent for this purpose, with grade RT-2 being used by some organizations and the heavier grades, such as RTCB-5 and RTCB-6, by others. Emulsified asphalts diluted with water are specified by some agencies. Recommended spraying temperatures are given in Table 15-9.

The quantity of bituminous material which is normally required for a tack coat is quite small, with the stipulated amount generally ranging from 0.05 to 0.10 gallon per square yard.

Construction methods

Tack coats, as has been indicated, are very frequently applied to old surfaces of various types as a first step in the construction of a new bituminous wearing surface. In many cases the old surface will require extensive correction and remedial treatment before construction of the new surface may begin because of the rough or inadequate condition of the old surface. As in the case of a prime coat, it is absolutely essential that the surface to which the material is applied be clean and dry. In most cases the surface is swept clean of all loose material just before the application is made. If asphalt emulsion is used, the surface may be slightly damp without harmful effect. The bituminous material is then applied at the specified rate with a pressure distributor, with the material being maintained at the proper temperature for application. The treated surface is then protected from traffic and allowed to dry until it reaches the proper stage of stickiness or "tackiness" for the application of the bituminous wearing surface. Any breaks that may occur in the tack coat may be corrected by the application of additional bituminous material at these points.

SINGLE- AND MULTIPLE-LIFT SURFACE TREATMENTS

Single- and multiple-lift surface treatments have been widely used in many sections of the country for the surfacing of primary and secondary roads that carry light to moderate amounts of traffic. The thickness of a single surface treatment is usually from $\frac{1}{2}$ to $\frac{3}{4}$ in. and may be as much as 2.5 in. when multiple-lift treatments are used. A surface constructed by this method is frequently quite "open" in character and has excellent nonskid and visibility characteristics. This type of surface is generally so thin that it has little load-supporting value in itself and must depend on an entirely adequate base and subgrade to function satisfactorily.

Material requirements

Materials that are used in this type of surfacing are extremely varied, with a large number of different aggregates and bituminous materials being successfully used in different areas. In the interests of somewhat simplifying the coverage of this subject, it has been decided to present the detailed material requirements of only one organization, the Federal Highway Administration (9), in the belief that these will be typical of those in current use in this country.

Aggregates   The aggregates that are used in this type of surface are crushed stone, crushed gravel, and crushed slag. The aggregates are required to be clean, hard, tough, and durable in nature, free from an excess amount of flat or elongated pieces, dust, clay films, clay balls, and other objectionable material. They are also required to be hydrophobic in nature, as measured by a stripping test (AASHTO T182).

Six aggregate gradations are included in the specification (9). They are designated gradings A through F and are given in Table 18-5.

Bituminous materials   A wide variety of bituminous materials may be used for surface treatments including medium- and rapid-curing cutback asphalt, rapid-setting emulsified asphalt, penetration grade asphalt cement, and road tar.

TABLE 18-5
AGGREGATE GRADATIONS FOR SURFACE TREATMENTS

Sieve Designation	Percentage Passing (by weight)					
	A	*B*	*C*	*D*	*E*	*F*
$1\frac{1}{2}$ in. (37.5 mm)	100	—	—	—	—	—
1 in. (25.0 mm)	90–100	100	—	—	—	—
$\frac{3}{4}$ in. (19.0 mm)	20–55	90–100	100	—	—	—
$\frac{1}{2}$ in. (12.5 mm)	0–10	20–55	90–100	100	—	—
$\frac{3}{8}$ in. (9.5 mm)	0–5	0–15	40–70	85–100	100	100
No. 4 (4.75 mm)	—	0–5	0–15	10–30	85–100	85–100
No. 8 (2.36 mm)	—	—	0–5	0–10	10–40	60–100
No. 100 (0.150 mm)	—	—	—	—	—	0–10

Composition of surface treatments

The FHWA specification (*9*) contains quantities of materials and indicated sequences of operations for five different wearing surfaces, utilizing six different aggregate gradings and a variety of bituminous materials. See Table 18-6. The surfaces range from a double surface treatment (Designation AT-35) to a four-lift treatment (Designation A-110). The designation refers to the quantity of aggregate per square yard. In the table, the indicated quantities of aggregate are based on a bulk specific gravity of 2.65 and must be corrected if the specific gravity of the aggregate used is more than 2.75 or less than 2.55. The total quantity of aggregate used in each of the surfaces given in the table, corrected for specific gravity and except for some stockpiling if desired, is fixed, while the amount of bituminous material actually used may be varied somewhat to meet job conditions.

Composition of surface treatment (Texas practice)

Fred J. Benson (*10*) has given an authoritative discussion of surface treatment as used in Texas and other southern states. He suggests that the quantity of a given aggregate for single surface treatments and seal coats is best determined by direct measurement. To carry out this measurement, an area 1 yard square is laid off on any reasonably smooth surface, for example, a piece of plywood. A representative sample of the aggregate to be used is then spread over the area so that it is completely covered with a layer one stone thick. The aggregate is then recovered and weighed. The quantity determined in the test is increased by 10 percent in order to compensate for spreading inaccuracies and the failure of all particles to adhere to the bituminous material. Values obtained can be converted to a volume basis by determining the loose unit weight.

The average thickness of the aggregate in the layer can be computed from the spread quantity and the loose unit weight. The calculation is based on the assumption that the particles have the same arrangement in the one-stone spread as they do in the container when the loose unit weight is being determined. For example, assume that the unit loose weight of aggregate is 96 lb/ft^3 and the amount required to cover 1 square yard is 18.5 lb. For a layer 1 in. thick, the required weight of aggregate would be $3(3)96(1/12) = 0.75(96) = 72$ lb. Average mat thickness then is $(18.5/72) = 0.26$ in.

J. P. Kearby of the Texas Highway Department developed a procedure for determining the quantity of bituminous material on the basis of proper embedment of the aggregate; his work was later verified by research workers at Texas A & M. Figure 18-7 shows the required percentage of embedment for aggregate of various sizes on the basis of average mat thickness. In order to compute the quantity of residual bituminous material required to fill the area not occupied by the stone it is necessary to know the average mat thickness, loose unit weight, and the specific gravity of the stone. For example, the aggregate previously considered has an average mat thickness of 0.26 in., unit weight of 96 lb/ft^3, and specific gravity of 2.65. From Figure 18-7, the required percentage of embedment is 32. Thus, the depth of residual bituminous

TABLE 18-6
QUANTITIES AND SEQUENCES OF OPERATIONS FOR BITUMINOUS SURFACE TREATMENTS USING CUTBACK ASPHALT OR ASPHALT CEMENT

Operations (as indicated in sequence)	Quantities of Materials per Square Yard (Bituminous material, gallons; aggregate, 1b) Surface Designation				
	AT-35	AT-50	AT-60	AT-70	AT-110
First layer					
Apply bituminous material	0.22	0.25	0.15	0.30	0.20
Spread aggregate					
Gradation D	25				
Gradation C		35			
Gradation B			40	50	
Gradation A					70
Second layer					
Apply bituminous material	0.13	0.25	0.30	0.35	0.40
Spread aggregate					
Gradation F	10	15			
Gradation D			12	20	
Gradation C					20
Third layer					
Apply bituminous material	—	—	0.15	—	0.20
Spread aggregate					
Gradation E			8	12	
Fourth layer					
Apply bituminous material	—	—	—	—	0.20
Spread aggregate					
Gradation F					8
Totals					
Bituminous material	0.35	0.50	0.60	0.65	1.00
Aggregate	35	50	60	70	110

material required is 0.26 (0.32) = 0.083 in. The volume of bituminous material per square yard is

$$3(3)(1/12)(0.083)\left[1 - \frac{96}{2.65(62.5)}\right] = 0.026\,\text{ft}^3$$

The term 96/62.5(2.65) is the absolute volume of the aggregate, and the term in brackets is the proportion of total volume to be filled with bituminous material. In common terms, the required amount of residual bituminous material would

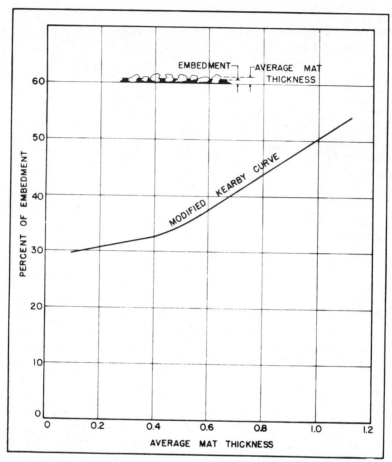

FIGURE 18-7 Relationship between percent embedment and average mat thickness, bituminous surface treatments. (AFter J. P. Kearby.)

be 0.026(7.5) = 0.195 gallons per square yard. Since this is residual material, allowance must be made for the volatile material in a cutback asphalt or the water in an asphalt emulsion when computing the amount to be applied by the distributor in the field.

Later research by John L. Saner and Moreland Herrin, of the University of Illinois, appeared to establish that the volume of voids does not vary linearly with depth within the aggregate layer (*11*). They found that the relationship is curvilinear and varies with different aggregates, and they suggested further research, particularly to relate the volume of voids to the shape of the aggregate.

Specifications for surface treatments have also been developed and published by the Asphalt Institute (*12*).

Construction methods—single- and multiple-lift surface treatments

It will be assumed in this discussion that the surface is to be applied to a granular-type base that has been previously primed, so the application of the prime coat is not included in the procedure presented here. The surface should be free from dust and dry, unless emulsified asphalt is used, when the surface may be slightly damp. No depressions should exist in the base course into which the fluid bituminous material may flow when applied to the surface and collect at these points in excess quantities.

The construction of surface treatments of this type is quite simple in basic procedure, with the surface being built up in lifts through the successive application of bituminous material and progressively smaller aggregate spread in the correct amounts. The layer formed by each application of bitumen and aggregate is rolled, and this basic procedure is continued until the specified number of lifts has been completed.

The initial application of bituminous material to the prepared base is made by a pressure distributor in the specified amount and at the proper temperature of application. It is essential that this application be uniform over the desired width and length of treatment. The desired amount of aggregate is then immediately spread over the bituminous material by means of approved mechanical spreading devices, including self-propelled spreaders. Normally, trucks are used to transport the aggregate to the point of application, and specifications covering this type of work usually require that the trucks be brought into position by traveling over previously spread aggregate. They are not allowed to pass over the portion of the surface on which no cover aggregate has been placed. A special spreading device is normally used to ensure that the aggregate is spread uniformly and in the prescribed quantity.

Rolling begins immediately after the aggregate has been spread. Both pneumatic-tired and steel-wheel rollers are used in this type of work, either alone or in combination with one another. The Asphalt Institute recommends pneumatic rollers because of the more uniform pressure they exert. Where pneumatic-tired rollers are used, they are operated at comparatively low tire pressures in order to avoid displacement and crushing of the aggregate.

The weight of steel-wheel rollers that is most generally used in this type of work is 5 tons, and the weight may vary from 3 to 8 tons under different specifications. Rolling of the aggregate is continued until the aggregate particles are firmly embedded in the bituminous material and sufficiently well bonded that they will not pick up under traffic, but it is discontinued after the asphalt has taken its initial set. Excessive rolling, which results in crushing of aggregate particles or breaking of the asphalt bond, is, of course, undesirable.

If a single-surface tratment is being constructed, this is the end of the construction process and the surface may be opened to traffic. If the wearing surface is to be a multiple-surface treatment, construction continues by the application of another "shot" of bituminous material, which is immediately covered by another spread of aggregate and the layer shaped and rolled as before. This process is simply continued until the desired number of lifts is completed. Surface treatment work is most successfully done in warm weather and should not be attempted when air temperatures are low (say 55°F or less).

18-12 Seal coats

In line with the concepts that have been advanced in this chapter, a seal coat is a very thin single-surface treatment that is usually less than $\frac{1}{2}$ inch thick. Seal coats are applied as a final step in the construction of many types of bituminous wearing surfaces, their primary purpose in this application being to waterproof or "seal" the surface. They are also used to rejuvenate or revitalize old bituminous wearing surfaces, to "nonskid" slippery surfaces, and to improve night visibility. A seal coat usually consists of a single application of bituminous material that is covered by a light spreading of fine aggregate or sand. The selection of the bituminous material and cover aggregate to be used in a given case is usually dependent on the type of wearing surface that is being sealed as well as on the nature of locally available materials. Specifications for seal coats are usually not given separately but are made part of the specification for each particular type of wearing surface. A large number of different bituminous materials are used in this work, and aggregate gradings vary widely in different sections of the country.

18-13 Slurry seal coats

The slurry seal coat, pioneered in California and now used extensively there and in the Southwest, is used principally for the resealing of old bituminous wearing surfaces. The seal coat is quite thin, up to a maximum of $\frac{1}{4}$ in.

The mixture for the slurry consists of a fine sand or crusher dust or a blend of the two, a slow-setting asphalt emulsion, and water. California specifications (13) call for all of the aggregate to pass a No. 4 sieve, 90 to 100 percent of it to pass a No. 8 sieve, and 10 to 20 percent to pass a No. 200 sieve. At least half the material must be a crushed product that meets requirements of the sand equivalent test and a film stripping test. Typical proportions are 1550 lb of aggregate (dry weight), 36 gallons of asphalt emulsion, and 20 gallons of water, including moisture in the aggregates.

The surface to be treated is swept and cleaned prior to application of the seal coat. The work may be done only in warm, dry weather. Before application of the slurry, the surface is sprinkled with water. In some cases, a dilute asphalt emulsion prime coat has been used.

Standard batching equipment can be used to proportion the materials. Batched materials are loaded directly into concrete transit mixers. Mixing takes place enroute to the job site.

Methods of applying the slurry vary somewhat. One method used successfully is as follows. The slurry is poured from the mixer into a rectangular spreader box, 10 to 12 ft wide and approximately 8 ft long. Near the rear of the spreader box there is a neoprene screed, which can be adjusted to control the thickness of the application. The spreader box is pulled forward by the mixer truck and the screed squeegees the slurry into the cracks in the old pavement. Thickness of treatment in one pass varies from $\frac{1}{8}$ to $\frac{1}{2}$ in. A thicker coat should be placed in two passes, with ample curing time between applications.

In recent years, self-contained slurry seal machines have been developed that incorporate a pugmill mixer and do all the mixing and placing operations in one pass.

Traffic should be kept off the new surface for a period of 2 to 4 hr, depending on the weather, thickness of the seal coat, and other factors.

ROCK ASPHALT SURFACES

"Rock asphalt," as the term is applied in highway work in the United States, refers to sandstone or limestone which is naturally impregnated with bitumen. This material should not be confused with native asphalts like Trinidad Lake Asphalt, which are primarily naturally occurring bitumens containing a large percentage of mineral matter. Rock asphalt is primarily mineral matter containing a small amount of natural bituminous binder. The amount of bitumen contained in rock asphalts that have been used in highway construction in this country has generally varied from about 3 to 12 percent.

Rock asphalt occurs in a number of states and has been produced on a commercial basis and in sizable quantity in Kentucky, Texas, Oklahoma, and Alabama. Material of this nature produced in these states generally requires some modification for use in highway work. Both hot and cold paving mixtures have been produced through the use of these rock asphalts. These paving mixtures are used as wearing surfaces on new or old bases, in the reconstruction of old pavements, and for maintenance operations such as patching. Their use at the present time is largely confined to the states in which they are found or in neighboring areas.

ROAD MIX SURFACES

The term "road mix" refers to a type of construction in which the mineral aggregates and bituminous material are intimately blended by mixing on top of the existing surface or base. The upper portion of the existing layer may or may not be included as part of the road mix surface, as circumstances warrant. The surfaces produced by this type of construction are also sometimes designated as "mixed-in place" surfaces.

Formerly, road mix bituminous wearing surfaces were widely used in this country for the surfacing of both primary and secondary roads and to a lesser extent for the surfacing of city streets. The initial development of this type of surface was probably primarily due to the need for the construction of a large mileage of surfaced roads at low cost, using local materials in areas of relatively sparse population, as in certain of the western states. This type of construction is seldom used today in the United States for wearing surfaces.

BITUMINOUS PLANT MIX SURFACES USING LIQUID BINDERS

In many "low-cost" bituminous plant mixtures, liquid bituminous materials—including liquid asphaltic products, asphalt emulsions, and road tars—are used as the binder.

The development of these intermediate surfaces was a natural outgrowth of certain difficulties encountered in the road mix process. Preparation of the paving mixture at a central plant offers such advantages as more careful proportioning of the ingredients, more uniform and thorough mixing with consequent production of more uniform mixtures, less dependence on favorable weather conditions, and the use of more viscous bituminous materials. Mixtures that are being included here under the classification of intermediate-type plant mixes range in nature from those that differ only very slightly from the road mix surfaces to those that very closely resemble the high-type mixtures discussed in the next chapter. They also cover a fairly wide range in cost, being generally slightly more expensive than comparable road mix surfaces and somewhat less expensive than high-type mixtures. The more expensive of the intermediate mixes described here differ from high-type mixtures in the following respects:

1. Softer bituminous materials that require only a moderate amount of heating (or no heating, as in the case of asphalt emulsions) are used in the intermediate types, while the majority of the high-type mixtures utilize semisolid binders and are mixed and laid at elevated temperatures.
2. The nature and grading of the aggregate components of a high-type mixture and the composition of the final mixture are more carefully controlled than for the intermediate type.
3. Requirements related to the density and stability of the compacted mixture are generally much more rigid for the high-type pavement than for the low-cost plant mix surface.

Specifications related to the composition and control of the low-cost bituminous plant mixtures show extreme variation, with all sorts of mixtures that would fall into this classification being utilized by different highway organizations in line with their particular experience and to fit their particular needs.

Consider now the equipment that is used in the construction of plant mix surfaces. The same (or similar) equipment is employed in the preparation of the high-type surfaces described in the next chapter, and this fact will be emphasized at various points in the following discussion.

18-14 Central mixing plants

The term "central mixing plant" refers to the plant or "factory" at which the bituminous paving mixture is produced, in a process beginning with the aggregates and bituminous materials and ending with the discharge of the mixture into hauling units for transportation to the job site.

Central mixing plants for the production of bituminous paving mixtures are frequently described as being portable, semiportable, or stationary in nature. The term "portable" is applied to relatively small units that are self-contained and wheel-mounted, and it is also applied to larger mixing plants in which the separate units are themselves easily moved from one place to another. The term "semiportable" is reserved for those plants in which the separate units must be taken down, transported on trailers, trucks, or railroad cars to a new location, and then reassembled, which process may require only a few hours or several days, depending on the plant involved. "Stationary plants" are those that are permanently constructed in one location and are not designed to be moved from place to place. Portable and semiportable plants are much more numerous than stationary plants and are widely used in the construction of rural highways. The capacities of these two types of plants range up to about 400 tons of mixture per hour.

Two general types of central mixing plants are in common use: (1) batch plants and (2) drum mix plants. In the batch type of plant, the correct amounts of aggregate and bituminous material, determined by weight, are fed into the mixing unit of the plant; the "batch" is then mixed and discharged from the mixer into trucks before additional materials are introduced. In the drum mix plants, the aggregates are proportioned prior to entry into the mixing drum by means of precision belt feeders which control the amount of each class of aggregate entering the drum. The drum mixer not only dries the aggregate but also blends the aggregate and asphalt, discharging the mixture in a continuous flow into a surge silo, where it is temporarily stored and later loaded into trucks. Consider now a more complete description of these two types of mixing plants.

Batch plants

One of the basic units contained in a batch bituminous paving plant is the drier. This unit is a necessary part of all "hot-mix" plants and is also generally used in the preparation of low-cost plant mixtures. The drier may not be needed for some mixes of the latter type, as, for example, those in which asphalt emulsion is used as the binder. The drier consists primarily of a large (from 3 to 10 ft in diameter and 20 to 40 ft long) rotating cylinder that is equipped with a heating unit at one end, usually a low-pressure air atomization system utilizing fuel oil. The drier is mounted at an angle to the horizontal, with the heating element being located at the lower end. Hot gases from the burner pass from the lower end up the cylinder and out at the upper end. The so-called cold aggregate is fed into the upper end of the drier, picked up by steel angles or blades set on the inside face of the cylinder, and dropped in "veils" through the burner flame and hot gases, and it moves down the cylinder because of the rotating action and the force of gravity. The hot aggregate then discharges from the lower end of the drier, generally onto an open conveyor or enclosed "hot elevator" that transports it to the screens and storage bins mounted in conjunction with the mixing unit. A drier of this conventional type is shown in Figure 18-8.

In the preparation of the hot mixtures described in the next chapter, the temperature of the aggregates may be raised to 325°F or more, and practically all the moisture in them removed. In the preparation of "cold" mixtures, such

FIGURE 18-8 Drier used in central mixing plant. (Courtesy Iowa Manufacturing Co.)

complete drying may be neither necessary nor particularly desirable. The moisture content may be reduced to about 1 percent in many cases, and excessively high temperatures are not permissible in the dried aggregates when they are to be combined with, for example, cutback asphalt. The dried aggregates may be cooled in various ways, including movement on an open conveyor to the next step in the process, the use of another drier (no heat) as a cooling unit, longer retention time in the storage bins, and so on.

As indicated above, the dried aggregates are generally transported from the drier into a unit that contains screens, storage bins, and aggregate proportioning devices. (See Figure 18-9.) In this unit the dried aggregates pass over vibrating screens that separate them into the desired number of fractions, usually from two to four. Each fraction is then passed into a storage bin or hopper ready to be used in the mixture. The capacity of the storage bins varies with the size and type of plant being used. Where the aggregates are to be recombined by weight, each storage bin is equipped with a discharge gate which is located directly over the so-called weigh box. In proportioning the batch in this type of plant, the required weight of each size of aggregate is placed into the weigh box. When the desired aggregate batch has been obtained, the contents of the weigh box are then discharged into the mixer.

Facilities must also be provided for the storage, heating, and proportioning of the bituminous material that is to go into the mixture. When heating is necessary, the bituminous material is heated in a storage tank, generally by steam, hot oil, or electrical heaters. Devices are provided for the measurement and control of the temperature of the bituminous material and for agitating the heated mass. The heated material is fed into an "asphalt bucket," with facilities being provided for securing the correct weight of material for each batch and for emptying the bucket into the mixing unit.

FIGURE 18-9 Flow of material through a batch-type hot-mix plant. (Courtesy Barber–Greene Company.)

The final basic unit that will be described here is the mixer itself. The mixer that is most commonly used is a twin pugmill of the type shown in Figure 18-10. In a batch-type plant this unit is mounted directly beneath the weigh box and asphalt bucket and high enough so that it may discharge the mixture into a truck or other hauling unit. The aggregates and bituminous material are dumped into the mixer through the top and the unit is then closed. The mixing is accomplished by the mixing blades mounted on the two shafts, which rotate in opposite directions. The space between blades is generally about $\frac{3}{4}$ in. Mixing is continued for a fixed length of time and the mixture is then discharged into the waiting truck.

Drum mix plants

Although batch plants are widely used, few of these plants are being sold today. They are gradually being replaced by the less expensive drum mix plants

FIGURE 18-10 Twin pugmill mixer. (Courtesy Barber–Greene Company.)

with special control systems that permit production of a variety of mixes. A sketch showing the major components of a drum mix plant is shown as Figure 18-11, while a flow diagram is indicated in Figure 18-12.

Simply stated, dryer drum mixing is a process in which asphalt mixtures are produced in a plant without aggregate gradation screens, hot bins, and a pugmill mixer. The basic plant consists of

1. Aggregate cold-feed bins.
2. Conveyor and aggregate weight system.
3. Drum mixer.

FIGURE 18-11 Major components of a recycle drum mix plant. Reclaimed asphalt pavement is fed from the two rear compartments of the cold-feed unit and mixed with virgin aggregate in the drum mixing unit. (Courtesy Cedarapids, Inc.)

4. Liquid asphalt storage tank and pump.
5. Hot mix conveyor.
6. Mix surge silo.
7. Control van.
8. Dust collection system.

Gradation is controlled in drum mix plants prior to entry of the aggregate into the mixing drum. This is normally accomplished with a multiple-bin cold-feed system equipped with precision belt feeders that control the quantities of each aggregate (*14*). On the basis of a sieve analysis for the aggregate in each bin, the operator is able to calibrate the feed system to determine the proper belt speed and gate openings that will produce the desired aggregate blend.

Drum mix plants normally have a continuous weighing system on the cold-feed conveyor belts. Since the aggregate being fed to the drum is not dry, it is necessary to measure accurately the moisture content and to determine the weight of the dry aggregate to ensure that the amount of asphalt delivered to the drum is appropriate for the amount of dry aggregate.

From the cold-feed conveyor the aggregate is fed into the drum mixer, where it is dried and heated. Specially designed shelves or flights attached to the interior of the rotating drum pick up the aggregates and drop them in a uniform veil through the burner flame. In the lower portion of the drum mixer, liquid asphalt is added and thoroughly blended with the aggregates.

The drum mix plant produces a continuous flow of asphalt mix, which is temporarily stored in a surge silo, from which it is loaded into trucks.

The operation of drum mix plants is monitored and controlled by instruments in a control van.

FIGURE 18-12 Asphalt drier drum mix plant. (Courtesy The Asphalt Institute.)

18-15 Spreaders and finishers

Some low-cost plant mixtures may be spread by the use of long-base blade graders or multiple-blade drags. This procedure may be desirable when some aeration of the mix is necessary to remove some of the volatiles contained in the bituminous binder, as might be the case when a dense-graded aggregate is used with slow-curing liquid asphalt. In the majority of cases, however, plant mixes are placed by use of a "bituminous paver" that spreads the mixture in a uniform layer of the desired thickness and shapes or "finishes" the layer to the desired elevation and cross section, ready for compaction. The use of a machine of this type is desirable in the case of open-graded plant mixtures and other mixtures which utilize highly viscous liquid binders. They are also widely used in the placing of the hot mixes described in the following chapter, which must be placed and finished rapidly so that they may be compacted while still hot. Bituminous pavers have not been previously described and will be briefly discussed in the following paragraphs.

The typical modern bituminous paver combines the functions of spreading and finishing in one machine. A machine of this type operates without the use of side forms, being supported on crawler treads or wheels. The machine is of sufficiently long wheelbase to eliminate the necessity for forms and to minimize irregularities occurring in the existing base or surface. These machines will process pavements from a fraction of 1 in. up to 10 in. in thickness over a width of 6 to 14 ft, with working speeds generally ranging from 10 to 70 ft/min; one model has working speeds up to 102 ft/min, although most construction specifications do not permit this high a laydown speed.

The mixture that is to be placed by the paver is dumped into a receiving hopper from trucks or other hauling units. Material is fed from the hopper toward the finishing section of the machine and is spread and agitated by means of screws or other agitator–distributors that serve the purposes of ensuring uniformity of the spread mixture, spread it uniformly the full width of processing, and loosen or "fluff up" the material. The spread material is then struck off at the desired elevation and cross section by one or more screeds. In some cases a single cutter bar may be used to strike off and compact the mixture to some extent. In other machines, oscillating or vibratory screeds are used to strike off and initially compact the mixture. Several machines of this type employ a tamping mechanism in conjunction with the screed to provide additional compaction. Cutter bars and screeds are usually provided with heating units to prevent pickup of the material during the spreading and finishing operations. All machines of the type that have been described are fully adjustable to ensure a uniform flow of material through the machine and to produce a smooth, even layer of the desired thickness and cross section.

Most bituminous pavers are now fitted with electronic screed control systems, which were introduced in 1961. The various systems differ in detail but perform the same general functions. A sensor operates on a reference profile, senses changes in the position of the floating screed element of the paver or of the reference profile, and then applies any necessary correction to the angle of the screed so that the surface being laid off will be parallel to the reference profile. Usually, the reference profile controls the longitudinal profile of the surface at one side of the machine and a slope sensor controls transverse slope, hence the pavement cross section. However, both sides of the paver can be controlled by reference profiles if desired.

Any of several reference profiles can be used, depending on the smoothness of the base or other surface on which the pavement is being laid, surface smoothness and thickness requirements, and so on. Reference profiles can be fixed or mobile. A typical fixed reference profile is a taut wire or string line carefully stretched between stakes or pins set at close spacings, usually 25 ft or less. The taut wire or string line reference is rarely used today because of the high cost of erecting it. A moving reference profile is attached to the side of the paver; it may be a long (up to 40 ft) rigid or articulated ski that incorporates a taut wire on which the sensor rides, a rolling string line (see Figure 18-13), or a short shoe that rides, for example, on the new surface of an adjacent lane. Figure 18-14 shows a paver using a "Leveler" device that averages longitudinal grade errors.

Some pavers also have automatic control of material feed to the screed; sensing elements at the ends of the augers interlock with the material conveyors to control the flow.

PROBLEMS

18-1 It is desired to combine material A, which is the soil existing in the roadbed, with material B, which may be obtained from nearby borrow sources at low cost, to form a stabilized soil–aggregate surface course conforming to the specification recommended by the AASHTO, Table 18-1, grading B. The sieve analyses of materials A and B are shown in the following tabulation. Determine approximately the limiting proportions of A and B that should be used to produce a stabilized mixture of the desired gradation. That is, determine the greatest proportion of A that can be used in a mixture and still meet the specifications, then the smallest proportions of A that will meet the same objective. Plot the gradation curves of the two blends and select the mixture you think best fits the specifications. What other requirements would normally be included in the specification for this material?

	Percentage Passing (by weight)	
Sieve Designation	Material A	Material B
2 in. (50 mm)	100	—
1 in. (25 mm)	89	100
$\frac{3}{8}$ in. (9.5 mm)	68	45
No. 4 (4.75 mm)	42	28
No. 10 (2.0 mm)	30	17
No. 40 (0.425 mm)	28	7
No. 200 (0.075 mm)	11	3

FIGURE 18-13 Sensor operates from 20-ft-long ski to guide bituminous paver. (Courtesy Iowa Manufacturing Company.)

FIGURE 18-14 A paver using a "Leveler" device that averages longitudinal grade errors. (Courtesy Barber–Greene Company.)

18-2 For the materials of Problem 18-1, the following are the plasticity characteristics. Material A: liquid limit, 30, plasticity index, 9. Material B: liquid limit, 13, plasticity index, 3. Estimate the liquid limit and plasticity index of the aggregate blend selected in Problem 18-1.

18-3 Assuming that the compacted mixture of Problem 18-1 has a dry unit weight of 118 lb/ft3, determine how many tons of material B will be required per mile of road, assuming that the surface is to be 24 ft wide and will have compacted thickness of 8 in. Estimate the number of cubic yards (loose measure) of this material that will be required per mile, assuming that the loose material weighs 105 lb/ft3.

18-4* If the compacted mixture of Problem 18-1 has a dry unit mass density of 1700 kg/m3, estimate the mass in kilograms of material B required per kilometer of road, given road width = 7.0 m and surface thickness = 0.25 m. If the loose material has a mass density of 1600 kg/m3, how many cubic meters of loose material will be needed?

18-5 Assume that a pressure distributor has a spray bar that is 12 ft long. If the pump discharge is at the rate of 260 gallons per minute, with the pump operating at the proper speed for the material that is being used, determine the rate of application of the bituminous material in gallons per square yard when the distributor is traveling at the rate of 850 ft/min. Also determine the reading that must be maintained on the truck tachometer to maintain this rate of application if the tachometer is calibrated to read 100 times the truck speed in miles per hour. Assume that all the pumped material passes out through the spray bars and is applied to the road surface.

18-6* A pressure distributor which has a spray bar 3.5 m wide travels 16 km/hr while spraying at a rate of 15 liters/sec. Determine the rate of application in liters per square meter.

18-7 A new granular-type base is to be covered with a double-surface treatment, meeting the requirements of FHWA Type AT-50, as outlined in the text. Assuming that the base is to be 24 ft wide, compute the quantities of bituminous material and crushed stone which are required for 1 mile of pavement if the specific gravity of the stone is 2.70. Estimate the cost of a double-surface treatment of this type in your area.

18-8* Determine the quantities of bituminous material and aggregate required for 1 km of double bituminous surface treatment meeting the requirements of FHWA Type AT-35, Table 18-6. The pavement is to be 7.0 m in width, and the specific gravity of the aggregate is 2.68.

18-9 Refer to the discussion of the Benson method of determining the amount of asphalt required for a single surface treatment. Applicable data are loose unit weight of aggregate, 98 lb; specific gravity, 2.68. The weight of stone required to cover 1 square yard (one stone thick) is 22 lb. Estimate the residual amount of asphalt required for proper embedment of the aggregate.

18.10 Visit a bituminous paving plant in your area and write a report detailing the layout and equipment being used. Where do the aggregates come from? How are they fed to the plant? How is the mixture transported to the place where it is to be used?

18-11 Consult a highway construction agency in your area and deterine the smoothness requirements being applied to new asphalt pavements. Are contractors using electronic screed controls to meet these requirements? How do the controls work?

REFERENCES

1. *AASHTO Materials, Part I, Specifications*, 13th ed. American Association of State Highway and Transportation Officials, Washington, D.C. (1982).

2. Shearer, W.L. Calcium Chloride Stabilization. In *Materials for Stabilization*. Education and Information Guide HC-100A, American Road Builders Association, Washington, D.C. (1976).

3. *FHWA Implementation Package 80-2*. Federal Highway Administration, Washington, D.C.

4. Johnson, James C. Asphalt Stabilization. In *Materials for Stabilization*. Education and Information Guide HC-100A, American Road Builders Association, Washington, D.C. (1976).

5. *Mix Design Methods for Liquid Asphalt Mixtures*. Supplement to MS-2 (Misc. 74-2), The Asphalt Institute, College Park, Md. (1974).

6. *Soil—Cement Laboratory Handbook*. Portland Cement Association, Skokie, Ill. (1971).

7. *1985 Book of ASTM Standards, Section 4, Construction*. Designation D693, American Society for Testing and Materials, Philadelphia, Pa. (1985).

8. *A Basic Asphalt Emulsion Manual*. Manual Series MS-19, The Asphalt Institute, College Park, Md. (1979).

9. *Standard Specifications for Construction of Roads and Bridges on Federal Highway Projects*. Federal Highway Administration, Washington, D.C. (1985).

10. Benson, Fred J. Surface Treatments for Highways. *Public Works* (May 1957).

11. Saner, John J., and Herrin, Moreland. Voids in One-Size Surface Treatment Aggregates. *Highway Research Record 104* (1965).

12. *Asphalt Surface Treatments—Specifications*. Educational Series No. 11, The Asphalt Institute, College Park, Md. (1982).

13. *Standard Specifications*. State of California, Department of Transportation, Sacramento.

14. *Asphalt Plant Manual*. Manual Series No. 3, The Asphalt Institute, College Park, Md. (1983).

chapter nineteen

HIGH-TYPE BITUMINOUS PAVEMENTS

The bituminous wearing surfaces discussed in this chapter represent the highest types of bituminous pavements that are in current use in this country. They are widely used both on rural highways and on city streets that are subjected to large volumes of traffic and severe service conditions. Properly designed and constructed surfaces of this general type are capable of carrying almost unlimited volumes of passenger, mixed, or truck traffic, provided only that they are supported by adequate foundation structures. The majority of these surfaces might be expected to have an economic life of 20 years or more.

All of the high-type bituminous mixtures discussed in this chapter are so-called "hot mixes" that are prepared in central mixing plants. The thickness of these surfaces may vary from slightly less than 1 in. to several inches, depending on the type of surface and its purpose. These mixtures are, generally speaking, marked by the use of carefully selected and graded aggregates in conjunction with semisolid bituminous binders. The composition of the mixture is more rigidly specified and controlled than for other types of bituminous surfaces. Similarly, the preparation, placing, and finishing of the mixture are performed under more rigid specifications and control, with emphasis being generally placed on the securing of very densely compacted stable mixtures. Surfaces of this type are generally more expensive than those that have been described in previous chapters, and less emphasis is placed on the use of local materials in the preparation of most of the mixes that are described here.

DESIGN OF HIGH-TYPE BITUMINOUS PAVING MIXTURES

It is the purpose of this section to set forth some of the underlying principles on which the design of a high-type bituminous paving mixture is based. Statements made in this section are intended to apply specifically to the design of the hot-mixed, hot-laid bituminous concrete mixtures described in this chapter, although they may also generally apply to those mixtures which do not contain coarse aggregate, such as sheet asphalt and sand asphalt. The subject of the design of the high-type mixtures being considered here is not a simple one, and authorities are not entirely in accord with one another either as to the fundamental principles involved or the detailed procedure to be used in the design of a particular mixture. The subject has been deliberately simplified in some respects in order to reduce the number and complexity of the factors that enter into the problem.

The two fundamental properties of a bituminous paving mixture that are held to be of the utmost importance are stability and durability. By "stability" is meant that property of the compacted mixture that enables it to withstand the stresses imposed on it by moving wheel loads without sustaining substantial permanent deformation. Lack of stability in a compacted mixture may be evidenced by displacement of the completed pavement, as by rutting or shoving. By "durability" is meant that property of the compacted mixture that permits it to withstand the detrimental effects of air, water, and temperature change. For successful results the pavement must be both durable and stable during its entire service life.

Both stability and durability are intimately related to the density of the mixture. The density of a compacted mixture may be expressed in several different ways, as will be shown in later sections, but the fundamental importance of this factor cannot be overemphasized. The density of a given mixture is frequently expressed in this work in terms of the "voids" in the mixture, meaning, in general, the amount of space in the compacted mixture that is not filled with aggregate or bituminous material, that is, is filled with air. Thus a dense mixture would have a low percentage of voids, while a loose mixture would have a high percentage of voids, all other things being the same. Many modern designers stress what might be called the "stability−voids" theory in proportioning bituminous paving mixtures. The fundamental concept underlying this approach to the problem is that mixtures which are high in density (low in voids) will be high in stability and that, in turn, properly designed mixtures which have high stability will also be durable. There are certain limitations to this concept, as will be pointed out in succeeding paragraphs, but this basic principle is generally subscribed to in the discussions that are presented in this chapter.

In addition to stability and durability, a well-designed mixture must also be skid-resistant. Lack of skid resistance is most often associated with an excess of bituminous material, although it may also be caused by aggregates that become excessively smooth or polished under the action of traffic.

The desirable qualities of a bituminous paving mixture are dependent to a considerable degree on the nature of the aggregates used. *Coarse aggregate* may be defined as that portion of the aggregate material contained in a bituminous paving mixture which is retained on a No. 10 sieve. *Fine aggregate* is that portion of the aggregate material which passes a No. 10 and is retained on a No. 200 sieve. It should be noted that the Asphalt Institute (*1*) uses the No. 8 sieve as the dividing line between coarse aggregate and fine aggregate. *Filler* is that mineral material that passes a No. 200 sieve. The function of the coarse aggregate in contributing to the stability of a bituminous paving mixture is largely due to interlocking and the frictional resistance of adjacent aggregate particles to displacement. Thus crushed aggregates that contain a preponderance of rough, angular particles are most generally used, including crushed stone, crushed gravel, and crushed slag. Similarly, the fine aggregate or "sand" contributes to stability through interlocking and internal friction, so that particles which are sharp and angular in nature are preferred. Sands are also held to contribute to stability by their function in filling the voids in coarse aggregate.

The filler itself is largely visualized as a void-filling agent, with limestone dust being most widely used for this purpose. Portland cement, slag dust, local sands, and other minerals have also been successfully used as fillers. Most commercial materials used as fillers contain a small proportion that is retained on the No. 200 sieve.

Aggregates that are well graded from coarse to fine are generally sought in high-type bituminous paving mixtures. The basic concept behind this statement is that well-graded materials produce the most dense mixtures and therefore the most durable, requiring minimum bitumen content for satisfactory results. This statement is regarded as being generally true, even though some very dense, durable, and stable mixtures have been produced by the use of aggregates which were not well graded in the usual sense. In most specifications, each of the separate aggregates—coarse, fine, and filler—is required to be well graded from coarse to fine as is the combination of these materials. The concept simply is that in a well-graded aggregate each smaller size or "fraction" of aggregate serves to fill the voids in the next larger one, with the result that a very dense aggregate combination may be secured.

Mixtures that contain more than about 70 percent of coarse aggregate are likely to be harsh and difficult to place and finish, being subject to segregation. The amount of filler used in bituminous concrete generally varies from about 3 to 10 percent, and up to about 16 percent for sheet asphalt. Excessive amounts of filler may result in a gummy mixture that is difficult to finish and that may become brittle and subject to cracking in cold weather. The filler itself should be well graded to a certain extent. It should be kept in mind that the filler is all the material that passes the No. 200 sieve and comes from all coarse and fine aggregate sizes and the commercial filler added to make the total.

The kind and nature of the bituminous material used obviously have some effect on the properties of the bituminous paving mixture. The mixtures described in this chapter use semisolid asphalt cement as the binder material. In mixtures in which this material is used as the binder, it is rendered temporarily liquid during mixing by the application of heat. The nature and quality of the asphalt cement used are controlled by standard specifications. Any of the viscosity grades of asphalt cement, from AC-2.5 to AC-40, can be used for hot-laid plant mixes. The choice of asphalt grade depends on conditons of climate and other factors.

Another basic factor which influences the behavior of a completed bituminous pavement is the degree of compaction of the mixture. A high degree of compaction is regarded as desirable, being associated with high density and stability. Mixtures of the kind being considered are generally compacted with heavy rollers, with rolling being continued until a high degree of density is achieved. However, rolling should not be continued until the aggregate breaks down or "degrades" under the action of the roller. Field rolling is controlled by requiring the attainment of a certain percentage of the density established in the laboratory by some standard compaction procedure.

At this stage in our outline of the factors that are of importance in the design of a bituminous paving mixture, it will be assumed that the aggregates that are to be used have been selected and their gradation adjusted to comply with the grading requirements for the mixture; the kind of bituminous material to be used has been established; and the desired compactive effort determined. The remaining design element to be determined, then, is the optimum bitumen content (usually asphalt). This is probably the most important single factor in the satisfactory design of a bituminous paving mixture. Detailed procedures to be followed in the determination of the optimum asphalt content for a given case will be given later, but certain general principles can be stated here. In order to explain these principles, consider the following line of reasoning. Consider first the mineral aggregates alone, without any bituminous material, assuming that the mineral aggregates have been compacted to a density equivalent to that which will be attained in the field process. This dense aggregate alone will possess a certain density and a relatively low stability. As small quantities of bituminous material are added to the mixture, both the density and stability will increase. At this stage the bituminous material would generally be present in the form of thin films surrounding, or partially surrounding, the aggregate particles. If more bituminous material is added, a point will be reached at which the aggregate particles are completely covered with bituminous material; additional bituminous material would serve to fill the void spaces between the aggregate particles. For many mixtures, both stability and density would increase until the voids in the mixture had been completely filled. The addition of any more bituminous material would result in a decrease in both density and stability, as the aggregates would generally be forced apart by the excess bituminous material.

It thus appears that the most desirable bitumen content for many mixtures would be that which would just fill the voids in the compacted mixture. This concept is, however, modified by certain practical considerations. In the first place, any bituminous material expands with an increase in temperature. Thus if the voids in the mixture were completely filled at the time of placing, any increase in temperature subsequent to that time would result in overfilling of the voids in the mixture with consequent "bleeding" of the pavement and loss of stability. Similarly, the density of the compacted mixture may increase under the action of traffic until it exceeds that used in the design. If this happened, an excess of bituminous material might again be present. As a result of these conditions and others, a compromise is frequently made in the selection of the optimum bitumen content. Assuming that all mixtures must possess adequate stability, the amount of bituminous material used would be such as to produce a mixture that would have, on the one hand, more than a certain *minimum* void content and, on the other, less than a certain *maximum* void content. The selection of the maximum void content that could be permitted would generally be based on durabilty considerations; air and water permeability may also be considered. For example, the percentage of bituminous material chosen for a given mix might be such that the compacted mixture would have a percentage of voids (calculated as explained later) that was not less than two or more than five. A final fundamental requirement for the design of a mixture is that it be economical.

To summarize, the objectives of asphalt paving mix design are to select and proportion materials to produce a mix that has

1. Sufficient asphalt to ensure a durable pavement.
2. Sufficient mix stability to satisfy the demands of traffic without distortion or displacement.
3. Sufficient voids in the total compacted mix to allow for a slight amount of additonal compaction under traffic without bleeding or loss of stability, yet low enough to prevent the entrance of harmful air and moisture.
4. Sufficient workability to permit efficient placement of the mix without segregation (*1*).

19-1 Classificaton of hot-mix paving

According to The Asphalt Institute (*1*),

> Asphalt paving mixes may be designed and produced from a wide range of aggregate blends, each suited to specific uses. The aggregate composition may vary from coarse to fine particles. Many different compositions are specified throughout the world—the ones in any given locality generally being those that have proven adequate though long usage and, in most cases, these should be used. For a general classification of mix compositions, however, The Asphalt Institute recommends consideration of mix designations and nominal maximum size of aggregate: 37.5 mm ($1\frac{1}{2}$ in.), 25.0 mm (1 in.), 19.0 mm ($\frac{3}{4}$ in.), 12.5 mm ($\frac{1}{2}$ in.), 9.5 mm ($\frac{3}{8}$ in.), 4.75 mm (No. 4), and 1.18 mm (No. 16), as specified in ASTM* Standard Specification D 3515 for Hot-Mixed, Hot-Laid Bituminous Mixtures. The grading and asphalt content limits of these dense mixes agree, in general, with overall practice but may vary from the practice of a given area.

A broad but widely used class of asphalt paving mixes is referred to as asphaltic concrete. Asphaltic concrete may be generally defined as an intimate mixture of coarse aggregate, fine aggregate, mineral filler, and asphalt cement. These mixtures may be called *open graded, coarse graded, dense graded,* or *fine graded,* depending on the gradation of the aggregates. Other classes of hot asphalt paving mixtures include sheet asphalt and sand asphalt, which contain no coarse aggregate. Except for a brief description of sheet apshalt and sand asphalt at the end of this chapter, the remaining discussion will focus on the design and construction of asphaltic concrete pavements.

Mixes may also be designated according to use in the layered pavement system. *Surface mixes* comprise the upper layer, while *base mixes* are used for the layer directly above the subbase or subgrade. *Leveling mixes* are those used in an

---

*American Society for Testing and Materials

intermediate layer, often to eliminate irregularities in contour of an existing surface prior to construction of a new layer.

19-2 Materials for asphaltic concrete paving mixes

The coarse aggregates most frequently used in asphaltic concrete paving mixtures are crushed stone, crushed gravel, and crushed slag. Uncrushed gravel, which may also be used in some cases, is required to consist largely of rough-textured particles. The coarse aggregates are, as was indicated in Chapter 15, required to be hard, tough, clean, and durable in nature, free from an excess of flat or elongated pieces, clay material, and other deleterious substances. Aggregates known to polish under traffic are not used in surface courses.

Crushed stone and gravel are generally permitted to have percentages of wear in the Los Angeles abrasion test of not more than 40, although this figure is increased to 50 in some instances. They are also sometimes subjected to requirements for soundness (AASHTO M76); requirements in this respect frequently call for weight loss incurred when the material is subjected to five cycles of the sodium (or magnesium) sulfate soundness test (AASHTO Method T104) not to exceed 12 or 15 percent. Crushed slag is usually subject to the same requirements related to abrasive resistance as stone and gravel; in addition, it is generally required to weigh not less than 70 lb/ft^3. Certain additional requirements related to crushed and uncrushed gravels are found in some specifications. General requirements related to coarse aggregates to be used in asphaltic concrete base courses may be somewhat less severe than those indicated above.

Requirements for sands to be used in asphaltic concrete mixtures are usually quite general in nature. Sands are generally required to be composed of tough, clean, rough-surfaced, and angular particles, free from excess amounts of clay and other deleterious materials. Stone screenings may also be used as fine aggregate and are sometimes required to be produced from coarse aggregates that meet the requirements given above. The crushed aggregates previously listed may, of course, contain a considerable amount of material that passes a No. 10 sieve and that is classed as fine aggregate. The fine aggregate for a given mixture may consist of sand alone, stone or slag screenings alone, or a blend of sand and stone or slag screenings.

A variety of materials are used for mineral filler, when required, and specifications in use by a given agency generally list the specific materials that may be used for this purpose. Commonly used materials include limestone dust, dolomite dust, portland cement, slag dust, and other similar inert mineral substances. The mineral filler is generally required to be dry and free from lumps. It is desirable that this material be definitely hydrophobic in nature.

The bituminous material used in the preparation of hot-mix, hot-laid asphaltic concrete mixtures is semisolid asphalt cement. The nature and quality of the asphalt cement are controlled by specifications like those discussed in Chapter 15. Generally speaking, the more viscous grades of asphalt cement such as AC-20 or AC-40 are recommended for highways that will carry heavy traffic in hot climates. Grade AC-10 would be suitable for pavements designed to serve medium to light traffic, except in cold climates, where AC-5 or AC-2.5 would be a better choice.

19-3 Requirements related to aggregate gradation and composition of mixtures

Separate grading requirements are usually given for the separate aggregate components of the mixture—coarse aggregate, fine aggregate, and filler—and then an overall specification is given pertaining to the composition of the paving mixture itself, including the bituminous material. The specifications of two highway agencies have been abstracted and are presented in Tables 19-1 and 19-2. These specifications are taken to be typical of those in present use.

The various specifications cited are intended to represent the requirements of typical specifications related to aggregate gradings and mix compositions. The percentages of the various sizes of aggregates and the amount of bituminous material to be used in a given mixture vary to some extent within the limits of each specification. Before the preparation of the mixture can actually begin, a single figure must be definitely established for each of the aggregate fractions contained in the specification and for the bituminous material. In other words, the composition of the mix must be definitely established. The process of doing this, which is frequently called the "design" of the mixture, results in the establishment of a definite "job-mix formula," which will then be applicable to the mixture being used on a particular project. Considerations related to the determination of the job-mix formula will be given in the following sections.

Once the job-mix formula has been established, most specifications allow small variations from it during production of the mix. The purpose of these tolerances is to obtain a uniform mixture in reasonable compliance with the specifications, since it is not possible to produce the exact job mix every time. A representative job-mix tolerance specification appears in Table 19-3.

TABLE 19-1
AGGREGATE GRADING REQUIREMENTS FOR PLANT MIX BITUMINOUS PAVEMENTS, ALABAMA HIGHWAY DEPARTMENT (2)

Sieve Designation	*Percentage Passing, by Weight*		
	Base Course	*Binder Course*	*Surface Course*
1½ in. (37.5 mm)	100	100	—
¾ in. (19.0 mm)	70−100	70−95	100
½ in. (12.5 mm)	—	—	95−100
⅜ in. (9.5 mm)	45−90	45−70	75−95
No. 4 (4.75 mm)	25−70	35−55	52−80
No. 8 (2.36 mm)	20−57	22−46	36−64
No. 50 (0.300 mm)	7−24	5−20	6−24
No. 100 (0.150 mm)	2−14	2−14	4−14
No. 200 (0.075 mm)	0−8	0−6	2−8

TABLE 19-2
AGGREGATE GRADING REQUIREMENTS FOR
PLANT MIX BITUMINOUS PAVEMENTS,
KENTUCKY DEPARTMENT OF HIGHWAYS (3)

Sieve Size	Percentage Passing, by Weight		
	Base	Binder	Surface Course
$1\frac{1}{2}$ in. (37.5 mm)	100		
1 in. (25 mm)	—		
$\frac{3}{4}$ in. (19.0 mm)	70−98	100	
$\frac{1}{2}$ in. (12.5 mm)	—	—	100
$\frac{3}{8}$ in. (9.5 mm)	44−76	57−85	80−100
No. 4 (4.75 mm)	30−58	37−68	55−80
No. 8 (2.36 mm)	21−45	25−52	35−60
No. 16 (1.18 mm)	14−35	15−38	22−46
No. 50 (0.300 mm)	5−20	5−20	5−21
No. 100 (0.150 mm)	3−10	3−10	3−14
No. 200 (0.075 mm)	—	—	2−7

TABLE 19-3
JOB-MIX FORMULA TOLERANCES (3)

Material Passing	Tolerance, Percentage Points
$\frac{1}{2}$ in. sieve and above	±6
$\frac{3}{8}$ in. and Nos. 4, 8, and 16 sieves	±5
No. 50 sieve	±4
No. 100 sieve	±2
No. 200 sieve	±1.5
Asphalt content	±0.3

19-4 Determination of the job-mix formula

The procedure that is presented here for the determination of the job-mix formula is intended to indicate the procedure that might be followed by an imaginary highway agency in determining the composition of a typical dense-graded asphaltic concrete surface mixture. Since detailed procedures in the determination of the job-mix formula vary so much, it is not deemed wise to try to present typical or "average" practice in this respect. The procedure indicated is generally similar to that followed by some agencies and has been somewhat simplified in the interests of clarity and brevity. Two general steps are involved in this process; the selection and combination of aggregates to meet the limits of the specification being used and the determination of the optimum asphalt content.

19-5 Selection and combination of aggregates

Relatively little will be said here about the selection of the aggregates to be used in a given paving mixture, although this is a very important phase of the overall design process. In the normal procedure, both coarse and fine aggregates available in the vicinity of the proposed work are sampled and carefully examined for compliance with the individual specifications for these materials. In some cases more than one suitable aggregate of each type will be found, and the selection of the particular ones to be used may then be largely a matter of economics. In other cases no suitable single aggregate may be locally available, in which case aggregates from several different sources may have to be blended to meet the requirements of the specifications, or imported aggregates may have to be used. Materials for use as mineral filler must be similarly examined and selected. Regardless of what the selection of aggregates may involve on a given project, all possible aggregates should be carefully examined and those selected that are suitable for the purpose and economically available. The aggregates finally selected for use must then be combined to provide an aggregate mixture that will comply with the limits established as to the composition of the mixture.

The procedure used in determining the proportions of coarse aggregate, fine aggregate, and mineral filler (and the optimum asphalt content) may best be illustrated by a discussion of the procedure to be followed in the design of a definitely specified asphaltic concrete mixture. This procedure will be followed in the subsequent discussion. It is assumed that the composition of the mixture that is to be designed (for which a job-mix formula is to be established) must be within the limits of the specifications in Table 19-4.

TABLE 19-4
MINERAL AGGREGATE AND MIX COMPOSITION

Passing Sieve Designation	Retained on Sieve Designation	Percent (by weight)
$\frac{3}{4}$ in. (19.0 mm)	$\frac{1}{2}$ in.	0−6
$\frac{1}{2}$ in. (12.5 mm)	$\frac{3}{8}$ in.	9−40
$\frac{3}{8}$ in. (9.5 mm)	No. 4	9−45
No. 4 (4.75 mm)	No. 10	8−27
Total coarse aggregate	No. 10	50−65
No. 10 (2.00 mm)	No. 40	6−22
No. 40 (0.475 mm)	No. 80	8−27
No. 80 (0.177 mm)	No. 200	5−17
No. 200 (0.75 mm)	—	5−8
Total fine aggregate and filler	(Passing No. 10)	35−50
Total mineral aggregate	—	100
Total mix		
Total mineral aggregate		92−95
Asphalt cement		5−8
Total mix		100

It will be assumed that the aggregates that are available for use in this mixture have been thoroughly tested and examined for compliance with the specifications related to these materials. Sieve analyses of the aggregates that can be most economically used in this case, as determined by AASHTO methods of tests T27 and T37 (mineral filler), are shown in Table 19-5.

The problem now is to determine the proportions of the separate aggregates that must be used in order to give an aggregate combination which will meet the requirements of the specification previously given.

Not only must the proportions selected be within the specification but they must be far enough from its extremes to provide room for the job-mix tolerance, so that when it is added or subtracted the mixture will not be outside the original specification master range. This is done by trial. It is obvious that the fine aggregate and coarse aggregate only, in any combination, cannot meet the requirements of the specification for total mineral aggregate, as the specification requires from 5 to 8 percent of material passing a No. 200 sieve and the coarse aggregate contains none of this material, while the fine aggregate contains only 2 percent. Obviously, some mineral filler must be used in the mixture; for the first trial, the amount of mineral filler is arbitrarily set at 8 percent. The total coarse aggregate in the mix must be from 50 to 65 percent, and this figure is again more or less arbitrarily set at 52 percent. The remaining 40 percent must be fine aggregate.

Using the indicated proportions, calculations necessary in determining the sieve analysis of the combined aggregates are shown in Table 19-6. A comparison of the figures in the last column of this table with the requirements of the specification will show that this combination of aggregates meets the stipulated requirements. This combination will therefore be judged satisfactory, and no additional trials will be made here. Additional trials would have to be made if the combination had not been satisfactory. Various graphical methods have been devised that make the blending process somewhat simpler than

TABLE 19-5
SIEVE ANALYSIS OF AGGREGATES

Percent (by weight)		Aggregate Type		
Passing Sieve Designation	Retained on Sieve Designation	Coarse Aggregate	Fine Filler	Mineral Filler
$\frac{3}{4}$ in. (19.0 mm)	$\frac{1}{2}$ in.	5	—	—
$\frac{1}{2}$ in. (12.5 mm)	$\frac{3}{8}$ in.	32	—	—
$\frac{3}{8}$ in. (9.5 mm)	No. 4	37	—	—
No. 4 (4.75 mm)	No. 10	22	7	—
No. 10 (2.00 mm)	No. 40	4	28	—
No. 40 (0.425 mm)	No. 80	—	39	5
No. 80 (0.177 mm)	No. 200	—	24	30
No. 200 (0.075 mm)	—	—	2	65
Total		100	100	100

TABLE 19-6
CALCULATIONS FOR SIEVE ANALYSIS

Passing Sieve Size	Retained on Sieve Size	Percent (by weight)			
		Coarse Aggregate	Fine Aggregate	Mineral Filler	Total Aggregate
$\frac{3}{4}$ in.	$\frac{1}{2}$ in.	0.52× 5= 2.6	—	—	2.6
$\frac{1}{2}$ in.	$\frac{3}{8}$ in.	0.52×32=16.6	—	—	16.6
$\frac{3}{8}$ in.	No. 4	0.52×37=19.2	—	—	19.2
No. 4	No. 10	0.52×22=11.4	0.40× 7= 2.8	—	14.2
No. 10	No. 40	0.52× 4= 2.2	0.40×28=11.2	—	13.4
No. 40	No. 80	—	0.40×39=15.6	0.08× 5=0.4	16.0
No. 80	No. 200	—	0.40×24= 9.6	0.08×30=2.4	12.0
No. 200	—	—	0.40× 2= 0.8	0.08×65=5.2	6.0
Total		52.0	40.0	8.0	100.0

the arithmetic process used here. Quite frequently no possible combination of the aggregates initially selected will prove to be satisfactory, in which case others must be selected or the available ones improved by blending. This situation can be avoided with experience. Additional trials might also be made in an effort to improve the gradation of the total aggregate; these also will not be considered here. The combination selected will be used in the preparation of the trial mixes required for the determination of the optimum asphalt content.

19-6 Determination of optimum asphalt content

The steps remaining in the process of determining the job-mix formula for a given asphaltic concrete surface course are directed toward the determination of the optimum asphalt content to be used in the mix. The asphalt content established as part of the job-mix formula must, of course, be within the limits of the given specification relative to composition of the mixture.

The laboratory procedure used in determining the optimum asphalt content involves the preparation of trial mixtures, using the selected aggregates and various percentages of asphalt within the limits of the mix specification. Each trial mixture is prepared in a manner that is intended to secure a very high density. Densities that are secured in this fashion generally represent the ultimate densities that are practically attainable either in the laboratory or in the field. The density, stability, and other properties of each trial mixture are then determined and the results tabulated. Applicable criteria may then be applied to determine the optimum asphalt content.

Design criteria related to the laboratory compacted specimens frequently require that the density of satisfactory mixes be not less than a certain percentage of the "theoretical maximum density." By "theoretical maximum density" is meant the density, usually expressed in terms of the bulk specific gravity, of a voidless mixture composed of the same ingredients in the same proportions used in the actual mixture. Design criteria for density, stability, and other properties that are used by a given organization have been established by extensive laboratory and field experience and investigation and are generally applicable only to mixtures that are to be used as wearing surfaces, being regarded as unnecessary to the design of base and binder course mixtures in many cases. Although we are discussing asphaltic concrete, the same or similar design criteria may be applied to the design of the other high-type mixtures described in this chapter.

Two mix design methods are in widespread use: the Marshall method (Section 19-7) and the Hveem method (Section 19-11). The Hubbard−Field method has been widely used in the past, but is not considered here because of its present limited use.

19-7 Marshall method of design

This method of design was originally developed by Bruce Mashall, formerly of the Mississippi Highway Department, and is currently used by the Corps of Engineers in the design of paving mixtures for airfields. It is also used by several state highway departments. The method is applicable only to hot mixtures using penetration grades of asphalt cement and containing aggregates with a maximum size of 1 in. or less.

The Marshall method uses standard test specimens 2.5 in. high and 4 in. in diameter. They are prepared by use of a standard procedure for heating, mixing, and compacting the asphalt−aggregate mixtures. Density-voids determinations and calculations are made on the compacted specimens, using methods and equations presented in Section 19-8.

19-8 Preparation of test specimens (Marshall method)

In preparing test specimens for the Marshall method, it is essential that the gradation of the aggregate and the quantity of asphalt be very carefully controlled. Considerable care must be exercised in the entire conduct of laboratory testing operations so that test results will be reliable and reproducible. Tests should be planned on the basis of $\frac{1}{2}$ percent increments of asphalt content, with at least two asphalt contents above optimum and two below. In the example given, it will be assumed that, since the possible range in asphalt content is from 4 to 6 percent of the total mix, trial mixtures will be prepared with asphalt contents over a portion of this range at intervals of $\frac{1}{2}$ percent. In other words, work will be begun on the assumption that mixtures will be prepared which contain 5.0, 5.5, 6.0, 6.5, and 7.0 percent of asphalt.

As a first step in the procedure, the aggregates with the proper gradation are thoroughly dried and heated. Sufficient mixture is generally prepared at each asphalt content to form two or three specimens 4 in. in diameter and $2\frac{1}{2}$ in. in height. Each specimen will require approximately 1200 g of mixture.

The asphalt and the aggregates are heated separately and then mixed. The mixing and compaction temperatures are established by determining the temperature to which the asphalt must be heated to produce viscosities, respectively, of 170 ± 20 and 280 ± 30 centistokes kinematic.

The mixture is placed in the mold, mixed with a mechanical mixer or by hand with a trowel, and compacted. A compactive effort of 35, 50, or 75 blows is specified according to the design traffic category. The compactive effort is applied to each side of the specimen. The test specimen is then tested for stability and flow after a density determination has been made.

19-9 Density—voids analysis

The density—voids analysis for an asphalt mix begins with the determination of the bulk specific gravity, G_{mb}, of the compacted mixture. The G_{mb} is determined in accordance with ASTM Designation D2726, and it involves determining the mass of the specimen in air and the mass of the same specimen in water. The same method may be used to determine the specific gravity and density of a sample cut or cored from a completed pavement. With very open or porous mixtures, the sample may be coated with paraffin before the density determination. The bulk specific gravity of the compacted mixture (uncoated specimen) may be expressed as follows:

$$G_{mb} = \frac{W_a}{W_a - W_w}$$

(19-1)

where

G_{mb} = bulk specific gravity of compacted mixture
W_a = mass of test specimen in air (g)
W_w = mass of test specimen suspended in water (g)

For example, assume that a test specimen weighs 3041.2 g in air and 1713.2 g in water. The bulk specific gravity of the specimen would be

$$G_{mb} = \frac{3041.2}{3041.2 - 1713.2} = 2.290$$

This value of G_{mb} means that the unit weight of the compacted mixture would be 2.29(62.4) = 142.9 lb/ft^3.

The determination of the percentage of air voids in a compacted mix is complicated by the fact that a portion of the asphalt in the mix may be absorbed by the aggregate. (See Figure 19-1).

In asphalt paving technology, three different types of specific gravity are discussed. These are apparent, bulk, and effective specific gravities. Water absorption is normally used to determine the quantity of permeable voids in the aggregate, for use in specific gravity computations (ASTM Methods C127 and C128). The principal difference between bulk and apparent specific gravities is that the volume of the permeable voids is included in the volume of the aggregate for bulk specific gravity and is excluded in the volume of the aggregate for apparent specific gravity. In all cases, the bulk specific gravity is less than the apparent specific gravity. Recognizing that the absorption of asphalt by the aggregate is different from that of water, various organizations have devoted considerable effort to the development of methods of determining the effective specific gravity for use in asphalt mix calculations. The

V_{ma} = Volume of voids in mineral aggregate
V_{mb} = Bulk volume of compacted mix
V_{mm} = Voidless volume of paving mix
V_a = Volume of air voids
V_b = Volume of asphalt
V_{ba} = Volume of absorbed asphalt
V_{sb} = Volume of mineral aggregate (by bulk specific gravity)
V_{se} = Volume of mineral aggregate
(by effective specific gravity)

FIGURE 19-1 Representation of volumes in a compacted asphalt specimen. (Courtesy The Asphalt Institute.)

effective specific gravity is normally somewhere between the bulk and apparent values. At this time ASTM D-2041 has established a method for determining the maximum theoretical density of an asphaltic concrete mixture. This value attempts to give the maximum density that otherwise would have to be calculated from the specific gravities of the ingredients. This procedure makes it possible to more accurately calculate the percentage of voids in a paving mix.

The Asphalt Institute recommends the following procedure for the calculation of the percentage of voids in the mineral aggregate (VMA) and the percentage of air voids in the compacted mixture, P_a. A more complete description of this procedure is given in Ref. 1.

Step 1. Calculate the bulk specific gravity of aggregate. When the total aggregate is made up of separate fractions of coarse aggregate, fine aggregate, and mineral filler, all having different specific gravities, the bulk specific gravity of the total aggregate, G_{sb}, is calculated by the following equation:

$$G_{sb} = \frac{P_1 + P_2 + P_3}{\dfrac{P_1}{g_1} + \dfrac{P_2}{g_2} + \dfrac{P_3}{g_3}}$$

(19-2)

where

P_1, P_2, P_3 = percentages by weight of coarse aggregate, fine aggregate, and mineral filler, respectively

g_1 = bulk specific gravity of coarse aggregate, measured by method ASTM C127 or AASHTO T85

g_2 = bulk specific gravity of fine aggregate, measured by method ASTM C128 or AASHTO T84

g_3 = apparent specific gravity of mineral filler, measured by method ASTM D854 or AASHTO T100

Step 2. Calculate the effective specific gravity of aggregate. The effective specific gravity of the aggregate, G_{se}, includes all void spaces in the aggregate particles except those that absorb asphalt. It is calculated as follows:

$$G_{se} = \frac{P_{mm} - P_b}{\dfrac{P_{mm}}{G_{mm}} - \dfrac{P_b}{G_b}}$$

(19-3)

where

P_{mm} = total loose mixture (percentage by total weight of mixture = 100 percent)

P_b = asphalt (percentage by total weight of mixture)

G_{mm} = maximum specific gravity of paving mixture (no air voids), determined in accordance with ASTM D2041

G_b = specific gravity of asphalt, determined by ASTM70 or AASHTO T228

Step 3. Calculate the asphalt absorption of the aggregate. Absorption is expressed as a percentage by weight of aggregate. The absorption, P_{ba}, is calculated by the following equation:

$$P_{ba} = 100 \frac{G_{se} - G_{sb}}{G_{sb}G_{se}} G_b$$

(19-4)

where

P_{ba} = absorbed asphalt (percentage by weight of aggregate)

Step 4. Calculate the effective asphalt content of a paving mixture. The effective asphalt content, P_{be}, is the portion of the total asphalt content that remains as a coating on the outside of the aggregate particles after a portion of asphalt is lost by absorption into the aggregate particles. It is calculated by Eq. 19-5:

$$P_{be} = P_b - \frac{P_{ba}}{100} P_s$$

(19-5)

where

P_{be} = effective asphalt content (percentage by total weight of mixture)

P_s = aggregate (percentage by total weight of mixture)

Step 5. Calculate the percentage of voids in the mineral aggregate (VMA) in the compacted paving mix.   *VMA*, the voids in the mineral aggregate, is defined as the intergranular void space between the aggregate particles in a compacted paving mixture, including the air voids and the effective asphalt content. It is expressed as a percentage of the total volume. The *VMA* is based on the bulk specific gravity of the aggregate and is expressed as a percentage of the bulk volume of the compacted paving mixture. Therefore, it can be calculated by subtracting the volume of the aggregate determined by its bulk specific gravity from the bulk volume of the compacted paving mixture. This leads to the following equation:

$$VMA = 100 - \frac{G_{mb}P_s}{G_{sb}}$$

$$(19\text{-}6)$$

where
VMA = voids in mineral aggregate (percentage of bulk volume)

Step 6. Calculate the percentage of air voids in the compacted mixture.   The air voids in a compacted paving mixture consist of the small air spaces between the coated aggregate particles. The percentage of air voids in a compacted mixture, P_a, can be computed as follows:

$$P_a = 100 \frac{G_{mm} - G_{mb}}{G_{mm}}$$

$$(19\text{-}7)$$

where
P_a = air voids in compacted mixture (percentage of total volume)

A typical worksheet for the analysis of a compacted paving mixture is shown as Figure 19-2.

19-10 Stability-flow test (Marshall method)

The Marshall stability is the maximum load resistance in pounds (or newtons) which the standard test specimen will develop at 140°F when subjected to load by a standardized procedure. The Asphalt Institute has indicated that base course mixtures intended for use 4 in. or more below the pavement surface could be tested at 100° F instead of 140° F. Loading is done with the specimen lying on its side in a split loading head; loading is perpendicular to the cylindrical axis of the specimen. The flow value of the total deformation, in units of $\frac{1}{100}$ in., occurring in the specimen between no load and maximum load during the stability test. For example, if the specimen deformed 0.18 in., the flow value is 18. A typical setup for the Marshall stability test is Figure 19-3.

When testing is complete, five plots are prepared, as illustrated by Figure 19-4.

Design criteria for the Marshall method are contained in Table 19-7 (*1*).

From curves like those of Figure 19-4, asphalt contents are determined that correspond to

1. Maximum stability

WORKSHEET FOR ANALYSIS OF COMPACTED PAVING MIXTURE
(Analysis by weight of total mixture)

Sample : __FI-005-5 (4)__   Date : June 10, 1983

Identification : __REX-EAST__

Composition of Paving Mixture

Constituent Material		Specific Gravity , G		Mix Composition,% by weight of Total Mix. , P					
			Bulk		Mix or Trial Number				
					1	2	3	4	5
1. Coarse Aggregate	G_1	2.759	2.606	P_1			51.45		
2. Fine Aggregate	G_2	2.905	2.711	P_2			34.24		
3. Mineral Filler	G_3	2.697	– –	P_3			7.35		
4. Total Aggregate	G_s	– –	– –	P_s			93.04		
5. Asphalt Cement	G_b	1.010	– –	P_b			6.96		

	Equation					
6. Bulk Sp. Gr. (G_{sb}), total aggregate	19-2			2.651		
7. Max. Sp. Gr. (G_{mm}), paving mix ($^{ASTM}_{D2041}$)	–			2.438		
8. Bulk Sp. Gr. (G_{mb}), compacted mix ($^{ASTM}_{D2726}$)	–			2.344**		
9. Effective Sp. Gr. (G_{se}), total aggregate	19-3			2.726		
10. Absorbed Asphalt (P_{ba}),% by wgt. total agg.	19-4			1.05		

CALCULATIONS

11. Effective Asphalt Content (P_{be})= Line 5 P_b −(Line 10 x 1.01 x Line 4 P_s)	19-5			5.98		
12. VMA = $100 - \dfrac{\text{Line 8 x Line 4 } P_s}{\text{Line 6}}$	19-6			17.73		
13. Air Voids (P_a) = $100\left(\dfrac{\text{Line 7 - Line 8}}{\text{Line 7}}\right)$	19-7			3.86		
14. Stability (Marshall Hveem)						
15. Flow (Marshall)						
16. Swell (Hveem)						

*Equations from text.
**Average of three or more samples for Marshall Method or one sample for Hveem Method

FIGURE 19-2 Worksheet for analysis of compacted paving mixture based on weight of total mixture. (Courtesy The Asphalt Institute.)

2. Maximum unit weight.

3. Median of limits given in Table 19-7 for percentage voids, total mix.

The optimum asphalt content is the average of the asphalt contents thus determined.

FIGURE 19-3 The Marshal stability test. (Courtesy The Asphalt Institute.)

Refer again to Figure 19-4. The example data are for a dense-graded asphalt concrete mix with a $\frac{3}{4}$-in. nominal particle size aggregate to be used for heavy traffic.

For this example, the asphalt contents corresponding to the desired characteristics are

For maximum stability	4.8%
For maximum unit weight	5.1%
For median indicated value for percentage of voids, total mix	4.3%

The optimum asphalt content equals the average of the three, or 4.7 percent.
To judge the suitability of the mix at the optimum asphalt content,

TABLE 19-7
DESIGN CRITERIA FOR THE MARSHALL METHOD

	Surface and Base					
	Light Traffic		Medium Traffic		Heavy Traffic	
Marshal Method						
Mix Criteria	*Min.*	*Max.*	*Min.*	*Max.*	*Min.*	*Max.*
Compaction, number of blows each end of specimen[a]	35		50			75
Stability, N	2224		3336		6672	
(lb)	(500)	—	(750)	—	(1500)	—
Flow, 0.25 mm (0.01 in.)[b]	8	20	8	18	8	16
Percentage of air voids[c]	3	5	3	5	3	5
Percentage of voids in mineral aggregate (VMA)[d]			(see Figure 19-5)			

[a]Laboratory compactive efforts should closely approach the maximum density obtained in the pavement under traffic.
[b]The flow value refers to the point where the load begins to decrease.
[c]The portion of the asphalt cement lost by absorption into the aggregate particles must be allowed for when calculating percent air voids.
[d]Percent voids in the mineral aggregate is to be calculated on the basis of the ASTM bulk specific gravity for the aggregate.

SOURCE: *Mix Design Method for Asphalt Concrete and Other Hot-Mix Types.* Manual Series No. 2, The Asphalt Institute (1984).

examine the corresponding values of stability, flow, and percentage of voids from Figure 19-4. For an optimum asphalt content of 4.7 percent:

Stability	1800 lb
Flow	9*
Percentage of voids, total mix	2.8
Percentage of voids, mineral aggregate	14.4

By comparing these values with the limits given in Table 19-7, it is seen that the mixture is satisfactory except that the percentage of voids in the total mix is too low. Adjustments would then be made in the mixture until satisfactory results were achieved.

19-11 Hveem method of design

The Hveem method of design of paving mixtures was developed under the direction of Francis N. Hveem, formerly materials and research engineer in the California Division of Highways (*1*); it is the standard method used in California. It is generally applicable to paving mixtures using either penetration grades of asphalt cement or liquid asphalts and containing aggregates up to

*The flow value is the total movement or strain, in units of hundredths of an inch (0.25 mm).

FIGURE 19-4 Test property curves, Marshall method. (Courtesy The Asphalt Institute.)

1-in. maximum size. Therefore, it is applicable not only to the high-type paving mixtures described in this chapter but also to the intermediate mixes of Chapter 18; it has been used principally in the design of dense-graded asphaltic mixtures. The method has been standardized as ASTM D1560 and D1561.

The first step in the Hveem method is to determine the "approximate" asphalt content through determination of the centrifuge kerosense equivalent. Factors determined from the CKE test are combined with calculated surface area. The surface area of a given aggregate, or blend of aggregates, is calculated by the use of the gradation and a table of surface area factors. Surface area

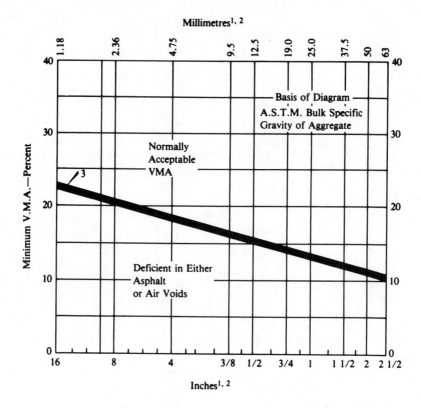

FIGURE 19-5 Minimum percentage of voids in the mineral aggregate. (Courtesy The Asphalt Institute.)

factors, expressed in units of square feet per pound, to be applied to total percentages passing the various sizes of sieves are as follows:

No. 4	2
No. 8	4
No. 16	8
No. 30	14
No. 50	30
No. 100	60
No. 200	160

An example of the calculation of the surface area of an aggregate of known gradation is given in Table 19-8. The percentage passing each sieve

TABLE 19-8
EXAMPLE OF CALCULATION OF SURFACE AREA
OF AGGREGATE OF KNOWN GRADATION,
HVEEM METHOD OF DESIGN

Sieve Size	Percentage Passing	Surface Area Factor	Surface Area
$\frac{3}{4}$ in. (19.0 mm)	100	—	—
$\frac{3}{8}$ in. (9.5 mm)	90	—	2.0
No. 4 (4.75 mm)	75	2	1.5
No. 8 (2.36 mm)	60	4	2.4
No. 16 (1.18 mm)	45	8	3.6
No. 30 (0.60 mm)	35	14	4.9
No. 50 (0.30 mm)	25	30	7.5
No. 100 (0.150 mm)	18	60	10.8
No. 200 (0.075 mm)	10	160	16.0
	Surface area = 48.7 ft²/lb		48.7

(expressed as a decimal) is simply multiplied by the applicable surface area factor and the products added. It is assumed that the surface area of the material coarser than the No. 4 sieve is 2 ft²/lb.

In determining the centrifuge kerosense equivalent in the laboratory, a special centrifuging device is used. A representative sample of 100 g of aggregate passing the No. 4 sieve is saturated with kerosene and subjected to a centrifugal force of 400 times gravity for 2 min. The amount of kerosene retained, expressed as a percentage of the dry aggregate, is the CKE.

The "surface capacity" of the coarse aggregate portion of the mixture is determined by first placing 100 g of dry aggregate that passes the $\frac{3}{8}$-in. sieve and is retained on the No. 4 into a glass funnel. The sample and funnel are immersed in SAE No. 10 lubricating oil at room temperature for 5 min. The funnel and sample are removed from the oil and drained for 15 min at a temperature of 140°F. The sample is then weighed and the amount of oil retained computed as a percentage of the dry aggregate weight.

To determine the estimated asphalt content, the following procedure is used:

1. Using the CKE value and the chart of Figure 19-6, the surface constant for fine material (K_f) is determined.

2. Using the percentage of oil retained in the surface capacity test for coaarse aggregate and Figure 19-7, the value of the surface constant for coarse material (K_c) is determined.

3. Using the values of K_f and K_c and Figure 19-8, the surface constant for the fine and coarse aggregate combined (K_m) is determined.

4. The next step is to determine the approximate asphalt content (bitumen ratio) for the mixture, based on cutback asphalts of RC-250, MC-250, and SC-250 grades. Either of two methods is used, referred to as case I or case II.

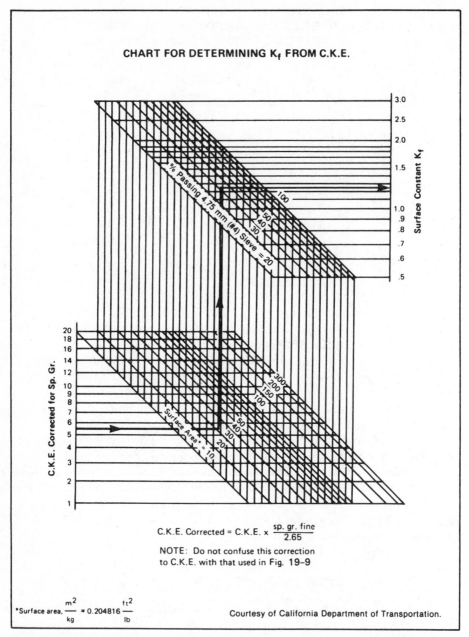

FIGURE 19-6 Chart for determining surface constant for fine material, K_f, from CKE, Hveem method of design. (Courtesy The Asphalt Institute.)

Case I applies to mixtures that have coarse and fine aggregate of similar surface and absorption characteristics; that is, K_f and K_c are approximately equal. This case is used primarily as a field method; it is used in the central laboratory of the California Division of Highways only when the sample con-

FIGURE 19-7 Chart for determining surface constant for coarse material, K_c from coarse aggregate absorption, Hveem method. (Courtesy The Asphalt Institute.)

tains 85 percent or more of material passing the No. 4 sieve. The asphalt content is determined by using the percentage of aggregate passing the No. 4 sieve, the CKE value adjusted for specific gravity (apparent specific gravities are used), and scales A, B, and C of the alignment chart of Figure 19-9.

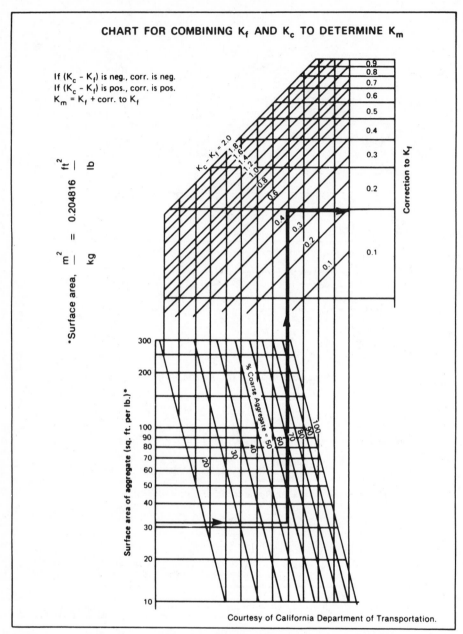

CHART FOR COMBINING K_f AND K_c TO DETERMINE K_m

If (K_c - K_f) is neg., corr. is neg.
If (K_c - K_f) is pos., corr. is pos.
K_m = K_f + corr. to K_f

Courtesy of California Department of Transportation.

FIGURE 19-8 Chart for combining K_f and K_c to determine surface constant for combined aggregate, K_m, Hveem method. (Courtesy The Asphalt Institute.)

Case II applies to mixtures of coarse aggregates and fine aggregates which have dissimilar surface and absorption characteristics; that is K_f and K_c are markedly different. In the central laboratory, this approach is used in all cases where the percentage passing the No. 4 sieve is less than 85. The asphalt

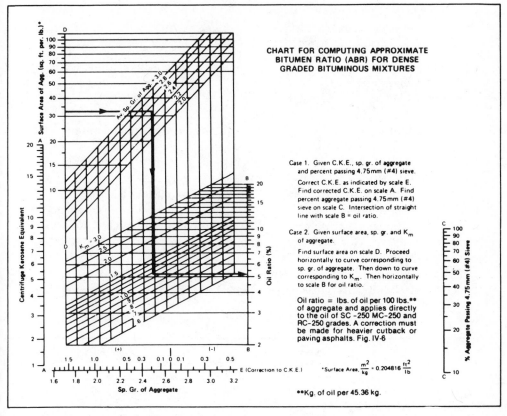

FIGURE 19-9 Chart for computing oil ratio for dense-graded bituminous mixtures. Hveem method. (Courtesy The Asphalt Institute.)

content based on a liquid asphalt is determined by using the computed surface area of the sample, the average apparent specific gravity of the aggregates, the surface constant K_m, and scales D to B on the alignment chart of Figure 19-9.

5. The asphalt content of the mixture for the particular grade of asphalt to be used is found from Figure 19-10, knowing the surface area of the sample, the grade of asphalt to be used, and the asphalt content from Figure 19-9.

In order to illustrate the procedure, consider a mixture containing aggregates that have the following properties.

Specific gravity, coarse aggregate	2.45
Specific gravity, fine aggregate	2.64
Percentage passing No. 4 sieve	45

$$\text{Average specific gravity} = \frac{100}{\dfrac{55}{2.45} + \dfrac{45}{2.64}} = 2.53$$

Surface of aggregate, ft^2/lb	32.4

Centrifuge kerosene equivalent   5.6

Percentage oil retained, coarse aggregate = 1.9
    (Corrected for specific gravity, this value is 1.7; see Figure 19-7.)

The mixture is to contain AC-10 viscosity grade asphalt cement.
From Figure 19-6, $K_f = 1.25$.
From Figure 19-7, $K_c = 0.8$.
From Figure 19-8, $K_m = 1.15$.
From Figure 19-9, and using case II, the optimum asphalt content is 5.2 percent.

From Figure 19-10, the optimum asphalt content (bitumen ratio) for AC-10 asphalt is 6.1 percent by weight of dry aggregate.

In designing a paving mixture by the Hveem method, a series of stabilometer test specimens are prepared for a range of asphalt contents, both above and below the estimated optimum asphalt content indicated by the CKE procedure. In addition, two swell test specimens are prepared at the estimated optimum asphalt content.

Samples are prepared by mixing heated aggregates and asphalt in accordance with a standardized procedure. As part of the procedure, the mixture is

FIGURE 19-10 Chart for correcting bitumen requirement due to increasing viscosity of asphalt, Hveem method. (Courtesy The Asphalt Institute.)

cured for 15 hr in an oven at 140°F. After curing, the mixture is heated to 230°F preparatory to compaction. The compaction of the speciments is accomplished by use of a mechanical compactor that consolidates the material by means of a kneading action resulting from a series of individual impressions made with a ram, the face of which is shaped as a sector of a 4-in. circle. At full pressure, each application of the ram subjects the material to a pressure of 500 lb/in.^{2} over an area of approximately 3.1 in.^{2} A detailed compaction procedure is followed, including final application of a static load to the specimen. Cylindrical specimens are formed 4 in. in diameter and approximately 2.5 in. high.

The compacted test specimens are used in the performance of four tests: swell test, stabilometer test, bulk density determination, and cohesiometer test.

In the swell test, the mold and compacted specimens are placed in a deep aluminum pan. A tripod arrangement is used to record the change in height of the specimen during the test. The upper portion of the mold is partially filled with water, which is allowed to remain on top of the specimen for 24 hr. At the end of that time, the change in height of the specimen, or *swell*, is determined. A measurement is also made of the amount of water that moves through the specimen during the 24-hr period, or the *permeability* of the sample.

Stabilometer test specimens are heated to 140°F before testing. A schematic drawing of the Hveem stabilometer is shown in Figure 19-11; the device was described briefly in Chapter 16. In testing asphaltic mixtures, the stability value resulting from the test is dependent on the ratio between the horizontal

FIGURE 19-11 Diagrammatic sketch of Hveem stabilometer. (Courtesy The Asphalt Institute.)

and vertical pressures at 400 lb/in.^{2} Detailed procedures for conducting the stabilometer test are given in Ref. 1.

Bulk density determinations are made on the stabilometer test specimens after completion of the stabilometer tests and as soon as the specimens have cooled to room temperature, using standard AASHTO procedures (4).

Design criteria for hot-mix asphaltic paving mixtures by the Hveem method are as follows:

	Light Traffic	Medium Traffic	Heavy Traffic
Stabilometer value	30+	35+	37+
Swell	Less than 0.762 mm (0.030 in.)		

An effort is also made to provide a minimum percentage of voids in the total mix of approximately 4 percent. In the application of these criteria, the optimum asphalt content of the mix should be the highest percentage of asphalt the mix will accommodate without reducing the stability or void content below minimum values. Generally, the voids in the total mix should approach the 4 percent minimum; however, in cases where the mixture shows a marked decrease in stability for slight increases in asphalt content, it is advisable to adjust the aggregate grading or reduce the asphalt content to provide a less critical mixture with more air voids.

The Hveem method of mix design is a lengthy and complex procedure. Here we have provided an abbreviated description of the most significant steps that are involved in the procedure. Readers interested in more detail are referred to The Asphalt Institute's Manual Series No. 2 (1).

19-12 Construction methods for hot-mix, hot-laid asphaltic concrete

All the mixtures described in this chapter are, of course, proportioned and mixed in central mixing plants. Construction procedures discussed here are intended to be generally applicable to all the hot-mix, hot-laid pavements described in this chapter and are also largely applicable to the cold mixtures that are discussed. The necessary construction steps are described in some detail in this section; deviations from the basic procedure indicated will be mentioned in subsequent sections of this chapter. The specific procedures described here are primarily intended to be applicable to the construction of wearing surfaces, although the same procedures are also generally applicable to the construction of base and leveling courses. The fundamental steps in the construction of a high-type bituminous pavement may be listed as follows:

1. Preparation of the mixture.
2. Preparation of base or leveling course.
3. Transportation and placing of the surface-course mixture.
4. Joint construction.
5. Compaction and final finishing.

19-13 Preparation of mixture

The essential elements of a central mixing plant have been described in Chapter 18. Plants used in the preparation of high-type hot mixtures are essentially the same as those previously described, except that certain additional units may be added to the plant and the entire setup geared to the necessity for volume production of uniform hot mixtures with very close control over the proportioning and mixing steps.

Central mixing plants for the preparation of high-type bituminous paving mixtures may be of continuous or batch type. Figure 19-12 shows the flowchart for a continous-flow plant. The gradation unit shown is a four-bin element, providing for four sizes of aggregate. This is an older type of continuous plant which utilizes a pugmill type of mixer as a final step in the mixing process. These units are increasingly being replaced by drum dryer plants, which were described in Section 18-14.

Refer again to Figure 18-9, which shows a cutaway view of the heart of a batch-type hot-mix plant. Operation of this plant, as with certain plants produced by other manufacturers, is almost completely automatic. Batch plants are widely used for asphalt production; however, like the older models of continuous plants, batch plants are gradually being replaced by drum dryer plants. A photograph of a drum dryer mixing plant is shown as Figure 19-13.

Two of the units that are frequently included in a hot-mix plant and that have not previously been mentioned are the dust-collecting system and a device for feeding mineral filler or "dust" into the mixture. The dust collector is generally operated adjacent to and in conjunction with the aggregate drier and is necessary for efficient plant operation in many cases. The collector serves to eliminate or abate the dust nuisance that exists around the plant and is especially important in plants that are located in or adjacent to municipalities. Modern dust-collection systems are highly efficient; frequently the fine material collected, if uniform in gradation, is fed back into the plant at the entrance to the hot elevator. To meet air pollution control requirements, many urban plants have "wet wash" systems in whch exhaust from the dust collector enters a tower and passes through a series of water sprays that remove the remaining dust.

Mineral filler that is added to the mixture is not normally passed through the aggregate drier, this material being fed by a separate device directly into the mixing unit in some cases or into the aggregate batchng unit. Separate feeder units are also used in the preparation of cold mixtures to supply liquefier, other fluxes, or hydrated lime to the mix. Central mixing plants for asphaltic mixtures are quite complex in nature and no further detailed description is deemed desirable in this text. The description of the basic units given in the preceding chapter and the information given in this chapter should suffice to give the reader a good understanding of the basic nature of these plants.

Control of the uniformity of hot mixes of this type is important, as any appreciable change or variation in gradation or asphalt content will be reflected in a change in some other characteristic of the mix. The control of the plant to produce the job mix chosen is a cooperative effort of the plant operator and inspector and requires continual inspection and careful control.

FIGURE 19-12 Flow diagram for single-aggregate, continuous-flow hot-mix plant. (Courtesy Barber–Greene Company.)

FIGURE 19-13 Photograph of a drum-dryer mixing plant. (Courtesy Cedarapids, Inc.)

Sampling and testing are among the most important functions in plant control. Data from these tests are the tools with which to control the quality of the product. Samples may be obtained at various points in the plant to establish if the processing is in order up to those points. A final check of the mixture is needed to determine its uniformity, its gradation, and its asphalt content. An extraction test measures the asphalt content and provides aggregate from which the gradation may be determined.

The results of extractions and gradations should fall within the job-mix tolerance specified; if they do not, corrective measures must be taken to bring the mix within the uniformity tolerance. Adjustments are frequently necessary, particularly during the early stages of plant operation or if changes occur in the aggregates. For a complete discussion of this phase of construction, see Ref. 5.

The temperature—viscosity relationship for the asphalt should be used to arrive at a proper mixing temperature, since asphalts vary as to source and type and grade. Generally speaking, higher viscosities are usually more suitable for coarse aggregate mixes and the lower viscosities for fine aggregate mixes. Many agencies now specifiy a mixing temperature, based on viscosity, as part of the job-mix formula and allow a variation of ±20°F from that figure.

It is likely that a majority of mixtures of this type are now prepared at temperatures between 250° and 275°F and the overall range is perhaps 225° to 325°F. Modern plants incorporate recording temperature gages that provide a continuous record of asphalt and aggregate temperatures.

19-14 Temperatures and mixing times
As has been indicated, the aggregate and bituminous material are handled separately in the plant, prior to their combination in the mixing unit. The

bituminous material is uniformly heated to a specified temperature in a tank or "kettle," while the aggregates are heated and dried in the drier unit. The temperature of the asphalt is carefully controlled in order to prevent overheating or "burning" of the material, with consequent destruction of certain desirable qualities. The temperature of the asphalt prior to its entrance into the mixer is generally from 225° to 350°F, and many modern specifications place the upper limit of this temperature at 300° or 325°F.

In batch-type plants, the dry aggregates are proportioned into the mixer unit and thoroughly manipulated for a brief time prior to the introduction of the bituminous material. Time for this dry-mixing operation generally would be from 15 to 20 sec. The bituminous material is then introduced into the mixer and the material mixed for a period which generally is a minimum of 45 sec. Total time in the mixer in a batching plant would then generally be from 55 to 70 sec and may be longer if necessary to produce a uniform and homogeneous mix. Similar times are required in continuous mixing units. Time in the mixer should be held to a minimum consistent with proper mixing.

19-15 Automatic control systems

Several states now require automatic controls on hit-mix plants used to produce mixtures on state highway contracts, and contractors in other states have installed automatic controls even though they are not required. Batch plants may be manual, semiautomatic, or automatic depending on the degree of automation. In manual plants, air cylinders or hydraulic cylinders, actuated by the operator by electric switches, control bin gates, fines feeders, asphalt supply and spray valves, the weigh box discharge gate, and the pugmill discharge gates (5).

In semiautomatic batch plants, the various operations comprising each mixing cycle are under automatic control. Thus, the quantities of asphalt and aggregate introduced into the mixes, the mixing times, the sequencing of the mixing functions, and the operation of the pugmill discharge gate are controlled automatically.

The fully automatic batch plant repeats the weighing and mixing cycle until the operator stops it or until it stops itself because of material shortage or some other extraordinary event. Automatic plants also normally provide a record of the amount of material incorporated into each batch.

In drum-mix plants, aggregate must be properly graded prior to its entry into the mixing drum. This is normally accomplished automatically with precision belt feeders that control the amounts of each aggregate being fed into the mixer. The aggregate is weighed before drying, and the weights of aggregate are electronically corrected to dry aggregate readings.

The asphalt being delivered to the drum mixer is controlled by a metering system that is interlocked with the aggregate weigh system to ensure the exact asphalt content in the mix.

A sensing device located at the discharge end of the drum mixer continuously monitors the temperature of the mix. The flow of asphalt mix is delivered to a surge silo, where a weigh system monitors the amount of material delivered into each truck (5).

19-16 Transportation of mixture

The mixture is discharged from the plant into dump trucks or trailers for transportation to the job site. The vehicles that are used are required to have tight and smooth metal beds that have previously been cleaned of all foreign material. The vehicle bed may be sprayed with a light coat of limewater, soap solution, or some similar substance to prevent adherence of the mixture. Fuel oils should not be used as they have a detrimental effect on the mixture. Sometimes the vehicle is required to be insulated against excessive heat loss in the mixture during hauling and frequently is covered with a canvas to protect the mix from the weather.

19-17 Preparation of base

Surface courses of asphaltic concrete are frequently placed on new or existing bases that require very little preparaton prior to the placing of the new surface other than thorough sweeping and cleaning to remove loose dirt and other foreign materials. In other cases, the existing base or surface on which the mixture is to be placed may require extensive corrective measures. Quite frequently, when the existing surface is disintegrated, broken, or irregular in nature, specific defects may be corrected by the application of "patches" of asphaltic concrete. Excessive joint sealing compound and fatty areas should also be removed. In certain cases it may be desirable to place a leveling course of asphaltic concrete to correct irregularities in the existing surface. At points where the asphaltic concrete mixture is to be placed in contact with surfaces of manholes, curbs, gutters, and such, these surfaces are generally painted with a light coat of hot asphalt cement or liquid asphaltic material.

19-18 Placing of mixture

Placing of the asphaltic concrete mixture is permitted only when the base is dry and under favorable weather conditions. Placing of hot mixtures is usually suspended when the air temperature becomes less than 40°F, except under unusual circumstances. Reference 6 gives specific laydown temperatures required for various conditions. It specifies minimum temperatures of the surface of the base for mixture placement, which vary with the mixture temperature measured just prior to dumping in the spreader and the lift thickness. The values range from 35° to 120°F, depending on these variables.

Base, leveling, and surface courses are placed and compacted in separate operations. In certain cases, unusually thick layers composed of the same mixture may be placed in two or more courses.

The steps of spreading and finishing are, in the large majority of cases, accomplished by the use of mechanical spreading and finishing machines (pavers) such as were described in the preceding chapter. These machines generally process one lane width at a time and must be capable of producing a smooth, even surface that has the desired elevation, grade, and cross section. This is best accomplished if the speed of laying is adjusted to the output of the plant, thus avoiding frequent long stops. Figure 19-14 shows a Blaw−Knox paver with a 19-ft screed engaged in the spreading and finishing of an asphaltic concrete wearing surface. The surface is checked during rolling and any

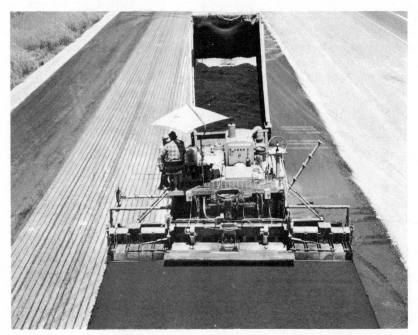

FIGURE 19-14 Bituminous paver with 19-ft screed. (Courtesy Blaw–Knox Company.)

irregularities, "fat" spots, or similar defects are corrected. In areas where the mechanical paver cannot operate satisfactorily, the mixture is placed on dump boards, spread, and raked by hand.

In placing the asphaltic concrete mixture, special attention must be given to the construction of joints between old and new surfaces or between successive days' work. It is essential that a proper bond be secured at longitudinal and transverse joints between the newly placed mixture and an existing surface, regardless of its nature, and special procedures, generally performed by hand, are utilized to ensure the formation of adequate joints.

The best longitudinal joint is formed when the material in the edge being laid against is still warm enough for effective compaction. This means that two finishers working in echelon can produce the most satisfactory joint. When one finisher is used, a procedure of moving from one lane to the other may be used so that the edge of the previously placed lane is still hot when the new lane is placed. Various other procedures are used when laying against a cold edge, such as cutting the cold edge back and painting with bituminous material. Some agencies allow the use of an infrared joint heater attached to the front of the finisher, which heats the cold material along the edge prior to placing a new lane.

19-19 Compaction of the mixture

When the spreading and finishing operations described above have been completed, and while the mixture is still hot, rolling is begun.

Rolling may be carried out by steel-wheel or pneumatic-tired rollers or by a combination of the two. Steel-wheel rollers may be of three types: three-wheel rollers of 10 to 12 tons in weight, two-axle tandem rollers of 8 to 12 tons, and three-axle tandem rollers of 12 to 18 tons.

Pneumatic-tired rollers are becoming more popular as they provide a closely knit surface by kneading aggregate particles together. They will achieve, during the rolling operation, a more uniform density than usually develops after vehicular traffic has used the asphalt surface for some time. The contact pressure exerted by a pneumatic-tire compactor is dependent on tire size, ply rating, wheel load, and tire inflation pressure. Tire contact pressures generally range from 40 to 90 lb/in.[2] Use of a self-propelled pneumatic-tire roller for this purpose is shown in Figure 19-15.

Use is now being made of vibratory rollers for compacting bituminous pavements. These vibratory rollers provide a centrifugal force of up to 47,000 lb and vibrations up to 3000 per minute. These vibratory rollers may replace other conventional compaction equipment to some extent in the future.

Rolling of the bulk of the pavement is done in a longitudinal direction, beginning at the edges and gradually progressing toward the center, except on superelevated curves, where rolling begins on the low side and progresses toward the high side; successive trips of the roller overlap to ensure complete and uniform coverage. Rolling progresses at a uniform, slow rate of speed, with the drive roll or wheel nearest the paver.

Rolling procedures vary with the properties of the mixture, thickness of layer, and other factors. In modern practice, rolling is divided into three phases, which follow closely behind one another: initial or "break-down" rolling, intermediate rolling, and finish rolling. The breakdown and intermediate phases provide needed density and the final rolling gives the required final smoothness. A procedure used by most organizations calls for the use of three-wheel or tandem rollers for breakdown, pneumatic-tired rollers for intermediate rolling, and tandem rollers for finish rolling. Use of a steel-wheel

FIGURE 19-15 Pneumatic-tired roller on a Kansas paving project. (Courtesy *Engineering News-Record.*)

roller in compaction of a hot-mix asphaltic pavement is illustrated in Figure 19-16.

Experiments by the Washington State Highway Commission (7) showed the effectiveness of compaction done when the asphalt mixture was still hot (above 200°F), with thick lifts (up to 6 in. or more loose thickness), and using high-pressure (90 lb/in.^{2}) pneumatic-tired rollers operating directly behind the paver.

Standards applied to the completed wearing surface are as high as is deemed practically possible. Requirements related to the finished surface frequently stipulate that the surface shall be smooth, even, and true to the desired grade and cross section, with no deviation of more than $\frac{1}{8}$ in. from a line established by a template or a 10-ft straightedge being permitted. This seems to be a very high standard, but such surfaces are readily attainable in most cases with modern construction equipment. Deviations from the specified thickness of surface course are generally permitted to be no more than $\frac{1}{4}$ in. The density to be obtained in the completed wearing surface may be stipulated in terms of a percentage of the theoretical maximum density or a percentage of the density of the laboratory-compacted mixture. A typical requirement is that the density of the completed surface course should not be less than 95 percent of the laboratory compacted density for the same mixture. Requirements related to the density of the completed pavement are very important and must be such as to guarantee, on the one hand, the attainment of a satisfactory density, and on the other, that a density is not specified that cannot be economically secured or that is impossible of attainment without crushing the aggregates.

FIGURE 19-16 Three-axle tandem roller compacting an asphaltic mixture. (Courtesy Galion Iron Works and Manufacturing Company.)

The density of the compacted mixture is determined on samples cored or sawed from the completed mat by the procedure discussed in Section 19-9. Many agencies, in addition to determining density on compacted mixtures, determine for control purposes the percentage of voids and voids filled and obtain large enough samples to determine the gradation and asphalt content of the compacted mixture. Experimental use is being made of nuclear density gages and airflow meters (air permeability measuring devices) to measure density and control the field rolling process.

Rolling completes the construction of most asphaltic concrete pavements, and traffic may be permitted on the surface as soon as the compacted mixture has adequately cooled. In some cases, a seal coat is also applied to the newly completed pavement, as described in a previous chapter.

SHEET ASPHALT

Sheet asphalt is a type of paving mixture that is composed of carefully graded fine aggregate (sand), mineral filler, and asphalt cement. It is used almost exclusively as a surface mixture, being laid on a binder course which is generally 1 to 1.5 in. thick or directly on a prepared base course, such as an old concrete or brick pavement. The sheet-asphalt pavement itself may be from $\frac{1}{2}$ to 3 in. thick, although the majority of these surfaces are from 1 to 1.5 in thick. A binder course used in conjunction with a sheet-asphalt surface is generally composed of a mixture that closely resembles an open-graded asphaltic concrete mixture, as no mineral filler is generally used in the binder course.

At the present time, sheet asphalt is principally used in the surfacing of city streets, and many of the principal streets in the larger cities of the United States are paved with this material. Properly constructed, it is capable of withstanding the effects of very heavy wheel loads and very heavy volumes of traffic without damage and with very low maintenance costs. Surfaces of this type are frequently preferred in cities because of their added virtue of being extremely smooth, noiseless riding surfaces that are waterproof and easily cleaned. Sheet-asphalt surfaces are somewhat more difficult to design, and their construction is slightly more difficult to control than that of asphaltic concrete mixtures of comparable stability and durability. One of the principal faults of sheet asphalt is its susceptibility to contraction cracking in cold weather.

Sheet-asphalt mixtures are distinguished from the asphaltic concrete mixtures that have been discussed by the complete lack of coarse aggregate and very close control of the gradation of the sands used. More mineral filler is used in these mixes than in somewhat comparable asphaltic concretes, and the percentage of mineral filler frequently approaches 16 in a sheet-asphalt mix. The asphalt cement that is used may be slightly harder than that used in asphaltic concrete, asphalt cements with penetration ranges of 50 to 60 and 60 to 70 having been quite commonly used. Asphalt contents are also generally somewhat higher than for dense-graded asphaltic concrete mixtures. Construction procedures related to the preparation and placing of sheet-asphalt mixtures are practically the same as those described for asphaltic concrete pavements. Somewhat higher temperatures are sometimes used in heating the sand and asphalt cement that are to be used in a sheet-asphalt mix. As a final step, the surface of a sheet-asphalt pavement may be lightly brushed with portland cement or limestone dust immediately after completion of the rolling.

SAND–ASPHALT

In some sections of the country, a type of hot plant mixture that is of importance is sand–asphalt. "Sand–asphalt" may be defined as a surface or base course formed from a mixture of sand and asphalt cement, with or without mineral filler. This type of mixture has been extensively used in areas where suitable sands are economically available, on highways that carry moderate amounts of traffic. The construction of sand–asphalt pavements is accomplished by methods that closely resemble those used in building sheet-asphalt surfaces, the chief difference between the two being that sand–asphalt mixtures are not as carefully specified or controlled and consequently are usually somewhat less dense and less stable than comparable sheet asphalts.

The usual thickness of a base course of this type is from 3 to 4 in., while a surface course would generally be about 2 in. in compacted thickness. Surface courses of sand–asphalt usually contain from 5 to 10 percent of asphalt, and base courses may contain slightly less. Requirements related to the composition of sand–asphalt mixtures vary somewhat with different organizations, as might be expected, since it is usually desired to use local sand in a mixture of this kind. Some specifications are quite open in nature, while others quite closely control the grading of the sand used. A requirement related to density and stability is also frequently included as part of a specificaiton for sand–asphalt, with requirements in this respect being generally somewhat less severe than for sheet-asphalt or asphaltic concrete mixtures.

TABLE 19-9
GRADATION REQUIREMENTS FOR SAND–ASPHALT

	Percentage Passing (by weight)	
Sieve Size	Type I	Type II
$\frac{1}{4}$ in. (6.25 mm)	100	100
No. 8 (2.36 mm)	75–100	50–90
No. 16 (1.18 mm)	60–90	25–65
No. 30 (0.600 mm)	45–75	15–45
No. 50 (0.300 mm)	15–45	5–30
No. 100 (0.150 mm)	5–15	3–20
No. 200 (0.075 mm)	2–6	2–6

SOURCE: *Kentucky Standard Specifications for Road and Bridge Construction*. Transportation Cabinet, Department of Highways, Frankfort (1985).

The variation in specifications of this sort makes it undesirable to attempt to include a "typical" specification for this type of mix. One specification has been selected, however, which is fairly representative of those in current use. Table 19-9 gives the requirements related to gradation of sand asphalt as contained in the *Kentucky Standard Specifications for Roads and Bridges*.

PROBLEMS

19-1 Given below are the requirements of a specification related to the grading of the mineral aggregates in an asphaltic concrete mixture, and the sieve analyses of two aggregates (A and B) that are economically available for this use. Determine the range of blends of aggregates A and B that will produce a combined aggregate which will meet the limits of the specification, and given the gradings of the aggregate combinations selected.

Sieve Designation	Percentage Passing (by weight)		
	Mix	Aggregate A	Aggregate B
1 in. (25 mm)	100	100	—
$\frac{3}{4}$ in. (19 mm)	95−100	95	100
$\frac{3}{8}$ in. (9.5 mm)	60−75	72	76
No. 4 (4.75 mm)	40−55	54	52
No. 10 (2.00 mm)	30−40	43	38
No. 40 (0.425 mm)	12−22	30	18
No. 200 (0.075 mm)	3−6	8	5

19-2 A design is being prepared for an asphalt concrete paving mixture. The following ingredients are to be used in the preparation of a trial mixture:

	Percentage of Total Mix by Weight	Specific Gravity
Coarse aggregate	56.0	2.611 (bulk)
Fine aggregate	30.0	2.690 (bulk)
Mineral filler	7.0	3.100
Asphalt cement	7.0	1.030

By ASTM D2041, the maximum specific gravity of the paving mixture $G_{mm} = 2.478$, and by ASTM D2726, the bulk specific gravity of the compacted paving mixture sample $G_{mb} = 2.384$. Compute the percentage of voids in the compacted mineral aggregate and the percentage of voids in the compacted mixture. What percentage of the air voids are filled with asphalt?

19-3 A sample cut from the completed pavement constructed by using the mix of Problem 19-2 weighs 3540 g in air and 1962 g in water. Compare the density obtained in the field with that of a laboratory-compacted specimen of the same mixture. Has the rolling process achieved a satisfactory degree of density in this case?

19-4 An asphaltic concrete surface course mixture is being designed by the Marshall method. The aggregate combination in a trial mixture contains 40 percent coarse aggregate, 52 percent fine aggregate, and 8 percent mineral filler; specific gravities of these materials are 2.68, 2.72, and 2.83, respectively. Asphalt cement to be used has a specific gravity of 1.02.

Results of tests on specimens prepared from this aggregate combination and various percentages of asphalt (by weight of total mixture) are as follows:

Percentage	Specimen Weight (g)		Stability	Flow
of Asphalt	In Air	In Water	(lb)	($\frac{1}{100}$ in.)
4.5	1308.7	765.0	1722	8
5.0	1314.2	772.3	1765	9
5.5	1318.9	779.7	1788	11
6.0	1323.3	782.1	1632	14
6.5	1328.6	782.9	1406	18

The mixture is to be designed for very heavy traffic. Determine the optimum asphalt content for this mixture. Is the mixture satisfactory?

19-5 A paving mixture is to be designed by the Hveem method. Gradation of the aggregate is as follows:

Sieve	Percentage Passing (by weight)
$\frac{3}{4}$ in. (19.0 mm)	100
$\frac{1}{2}$ in. (12.5 mm)	94
$\frac{3}{8}$ in. (9.5 mm)	78
No. 4 (4.75 mm)	62
No. 8 (2.36 mm)	45
No. 16 (1.18 mm)	34
No. 30 (0.60 mm)	25
No. 50 (0.30 mm)	18
No. 100 (0.150 mm)	14
No. 200 (0.075 mm)	8

The centrifuge kerosene equivalent for the aggregate is 3.8. Specific gravity of the coarse aggregate is 2.62; fine aggregate, 2.72. The percentage oil retained by the coarse aggregate is 2.5.

Estimate the quantity of asphalt required for this mixture, assuming that the binder is to be (a) a liquid asphalt, grade MC-800; (b) an asphalt cement, penetration grade 85-100.

Suppose the stabilometer value of the mixture of (b), using the computed quantity of asphalt, is 42; is the mixture satisfactory for heavy traffic? What other requirements pertain to the mixture?

19-6 Visit a plant in your area which is producing hot-mix asphaltic concrete or a similar high-type mixture. Prepare a report describing the controls that are applied to ensure production of a mixture of uniformly high quality. Include such factors as aggregate gradings and proportions, quantity of bituminous material, temperatures, mixing times, and so on.

19-7 Describe in detail the laydown procedure used for a surface of asphaltic concrete, or a similar type of mixture, in your area. Give facts related to temperature of the mix, spreader, rolling equipment and procedure, variations in the surface of the compacted layer, density achieved by rolling, and so on.

REFERENCES

1. *Mix Design Methods for Asphalt Concrete and Other Hot-Mix Types.* Manual Series No. 2, The Asphalt Institute, College Park, Md., (May 1984).

2. *Standard Specifications for Highways and Bridges.* Alabama Highway Department, Montgomery (1985).

3. *Kentucky Standard Specifications for Road and Bridge Construction.* Transportation Cabinet, Department of Highways, Frankfort (1985).

4. *AASHTO Materials, Part II, Tests,* 13th ed. American Association of State Highway and Transportation Officials, Washington, D.C. (1982).

5. *Asphalt Plant Manual.* Manual Series No. 3, The Asphalt Institute, College Park, Md. (1983).

6. *Standard Specifications for Construction of Roads and Bridges on Federal Highway Projects.* Federal Highway Administration, Washington, D.C. (1985).

7. Minor, Carl E. Placing and Compacting Thick Lifts of Asphalt Concrete. Paper presented at the 11th Annual Convention of the National Asphalt Paving Association (1966).

chapter twenty

CONCRETE PAVEMENTS

The many and varied uses of portland cement concrete in the United States have made this material such an integral part of our everyday lives that it hardly seems necessary to define it. However, "portland cement concrete" may be defined as a plastic and workable mixture composed of mineral aggregate such as sand, gravel, crushed stone, or slag, interspersed in a binding medium of cement and water. When first combined, the materials listed form a plastic, workable mass that may be easily handled and shaped into any desired form. A short time after mixing, the concrete begins to stiffen or "set" because of chemical action between the cement and water in the mixture, and in a relatively short time it forms a dense, hard mass that possesses considerable compressive and flexural strength.

When properly designed and constructed, concrete roads and streets are capable of carrying almost unlimited amounts of any type of traffic with ease, comfort, and safety. Surfaces of this type are smooth, dust-free, and skid-resistant, having a high degree of visibility for both day and night driving and generally having low maintenance costs. They are economical in many locations because of their low cost of maintenance and their relative permanence. They are, of course, classed as high-type pavements. The principal use of surfaces of this type has been in the construction of heavily traveled roads and city streets, including those in residential, business, and industrial areas. It is the standard material for urban expressways, even in states where asphalt surfaces are widely used. A wearing surface of portland cement concrete usually consists of a single layer of uniform cross section that has a thickness of 6 to 11 in. and that may not require a separate base course, often being constructed directly on a prepared subgrade or subbase. A new concrete base course may be constructed to serve as a support for one of the several types of bituminous wearing surfaces. Old concrete pavements have been extensively used as bases for new bituminous wearing surfaces in many areas.

A further distinction may be noted between the surfaces and bases that are discussed in this chapter and the so-called flexible pavements that have previously been described in this book. Concrete surfaces and bases are frequently classed as "rigid" pavements, the term "rigid" implying that pavements constructed of this material possess a certain degree of "beam strength" that permits them to span or "bridge over" some minor irregularities in the subgrade or subbase on which they rest. Thus minor defects or irregularities in the supporting foundation layer may not be reflected in the surface course, although, of course, defects of this type are certainly not desirable, as they may lead to failure of the pavement through cracking, breaking, or similar distress.

The design and construction of concrete pavements is a fairly complex subject and it seems desirable to list the major topics that will be covered in this chapter. Major subjects that will be discussed in subsequent sections include materials, proportioning of concrete mixtures, structural pavement design, and the construction of portland cement concrete pavements.

In the interests of simplification, in the remainder of this text the term "concrete" will be taken to be synonymous with "portland cement concrete."

MATERIALS

The materials included in concrete, as generally used in highway construction, are coarse aggregate, fine aggregate, water, cement, and one or more admixtures. These materials will be discussed in turn in the following sections, along with a discussion of typical requirements of specifications related to each.

20-1 Coarse aggregate

Coarse aggregates most frequently used in portland cement concrete include crushed stone, gravel, and blast-furnace slag. Other similar inert materials may also be used, and the listed materials may be used singly or in combination with one another. Specific requirements related to coarse aggregates to be used for this purpose may be divided into five groups: deleterious substances, percentage of wear, soundness, weight per cubic foot (slag), and grading.

As mentioned in Chapter 15, highway agencies specify the maximum percentages of various deleterious substances which are permitted to be present in aggregates. Specific requirements related to the cleanliness of the aggregate vary somewhat with different agencies. The requirements in Table 20-1 related to deleterious substances are indicative of those in current use (*1*).

In certain areas, considerable difficulty has been experienced with aggregates that contain deleterious substances which react harmfully with the alkalies present in the cement. Such reactions generally result in abnormal expansion of the concrete. Methods have been devised (ASTM Methods C227 and C289) for detecting aggregates with these harmful characteristics, and suitable stipulations are included in typical specifications (for example, ASTM C33).

The ability of the coarse aggregate to resist abrasion is generally controlled by inclusion of the specifications of a maximum permissible percentage of wear in the Los Angeles abrasion test. Maximum permissible percentages of wear in this test (AASHTO Designation T96) range from as low as 30 to as high as 65 in the specifications of the various state highway departments. On the average, however, the maximum permissible loss varies from 40 to 50 percent for all three of the principal types of coarse aggregate, although many specifications do not contain a requirement related to the percentage of wear of slag.

A number of states, particularly those in the northern portion of the country, include requirements related to soundness, as measured by the use of

TABLE 20-1
LIMITS FOR DELETERIOUS SUBSTANCES AND PHYSICAL PROPERTIES OF COURSE AGGREGATE FOR CONCRETE (_1_)

Typical Uses	Weathering Exposure	Clay Lumps and Friable Particles	Chert (less 2.4 sp. gr. SSD)[a]	Sum of Clay Lumps, Friable Particles, and Chert[a]
		Maximum Allowable (percent)		
Bridge decks and other uses where	Severe	2.0	3.0	3.0
surface disfigurement due to popouts	Moderate	3.0	3.0	5.0
is objectionable	Negligible	5.0	5.0	7.0
Pavements, base courses, and side-	Severe	3.0	3.0	5.0
walks where a moderate number of	Moderate	5.0	5.0	7.0
popouts can be tolerated	Negligible	5.0	8.0	10.0
Concealed concrete not exposed to weather	All types	10.0	—	—

General requirements, all types of construction and weathering regions:
  Maximum allowable material finer than No. 200 sieve (0.075 mm) = 1.0%
  Maximum allowable coal and lignite = 0.5%
  Maximum wear, Los Angeles abrasion test = 50%
  Maximum loss, magnesium sulfate soundness test (five cycles) = 18%

[a]These limitations apply only to aggregates in which chert appears as an impurity. They are not applicable to gravels that are predominantly chert. Limitations on soundness of such aggregates must be based on service records in the environment in which they are used. (SSD = saturated, surface dry.)

SOURCE: *AASHTO Materials*, Part I, *Specifications*, American Association of State Highway and Transportation Officials (1982).

sodium (or magnesium) sulfate, in their specifications for coarse aggregate. The soundness test used is an accelerated weathering test. The weighted loss in the sodium sulfate soundness test (AASHTO Designation T104) may generally be said to have a maximum permissible value ranging from 10 to 15 percent, as measured at the end of five cycles of alternate wetting and drying. A small number of states use a freezing and thawing test for the same purpose. In states in which slag is used as a coarse aggregate, specifications generally require that the unit weight of this material be not less than 70 (or 75) lb/ft.^{3}

As would be expected, requirements related to the grading of the coarse aggregates used in portland cement concrete vary considerably with different organizations. The material is generally required to be well graded from coarse to fine, with the maximum size of coarse aggregate permitted by most highway organizations being from 3 to 2 in. In this work the division between coarse and fine aggregate is made on the No. 4 sieve. That is, material which is retained on a No. 4 sieve is classed as coarse aggregate. Grading requirements are generally quite open, permitting the use of a large number of locally available aggregates which have proved to be satisfactory for use in portland cement concrete.

Typical gradings for coarse aggregate are given in AASHTO Designation M43 (*1*).

20-2 Fine aggregate

The fine aggregate that is most generally used in concrete is <u>sand</u> (either natural or manufactured) composed of tough, hard, and durable grains.

The cleanliness of the sand used may be controlled by the inclusion of requirements related to the maximum amounts of various kinds of deleterious substances that may be present, in a fashion similar to that indicated above for coarse aggregate. Particular emphasis is frequently placed upon the maximum percent of "silt" (material passing the No. 200 sieve) that may be contained in the fine aggregate. This amount must generally not exceed 2 to 5 percent of the total. An additional cleanliness requirement is generally imposed on sands, in that excess amounts of organic impurities must not be present. The amount of organic material is generally controlled by subjecting the material to the colorimetric test for organic impurities (AASHTO Designation T21).

Further requirements of specifications related to sands may include stipulations as to soundness. The soundness of these materials is controlled by specifying the maximum loss that will be permitted in five cycles of alternate wetting and drying in the sodium sulfate soundness test. A typical requirement is that the loss in the sodium sulfate soundness test may not exceed 10 percent. Modern specifications also frequently include stipulations intended to control the potential alkali reactivity of fine aggregates.

The grading requirements of the various highway agencies related to fine aggregate are somewhat more uniform than similar requirements related to coarse aggregate. Many of the state highway departments utilize grading requirements that are the same as, or very similar to, those given in the "Standard Specifications for Fine Aggregate for Portland Cement Concrete," AASHTO Designation M6 (see Table 20-2).

20-3 Water

Almost any normal source of water may be used in the manufacture of portland cement concrete. Specifications for this material frequently require simply that water to be used in the mixture is to be suitable for drinking, although other water may be used in some instances if it has been demonstrated by laboratory test or field experience that it is suitable for this purpose. Water that is used must be free from an excess of alkalies, acids, oil, or organic matter. Detergents in water will cause high entrained air content. Seawater has been successfully

TABLE 20-2
GRADING OF FINE AGGREGATE
(AASHTO DESIGNATION M6-81)

Sieve Designation	Percentage Passing (by weight)
$\frac{3}{8}$ in. (9.5 mm)	100
No. 4 (4.75 mm)	95–100
No. 16 (1.18 mm)	45–80
No. 50 (0.30 mm)	10–30
No. 100 (0.150 mm)	2–10

used in some concrete mixtures, although it is not generally used in paving operations.

20-4 Portland cement

Portland cement is produced in five basic categories, designated type I through type V. Three of these classes of cement are commonly used in highway construction: types I, II, and III. These types are defined in Chapter 15. Where an air-entraining agent is introduced into these compositions, they are designated types IA, IIA, and IIIA, respectively.

The American Association of State Highway and Transportation Officials has published specifications (*1*) for portland cement giving the requirements for such properties as fineness, soundness, strength, time of setting, and air content of the mortar. AASHTO has also published standard methods for sampling and testing of portland cement for proper classification and quality control (*2*).

Cements conforming to the specifications for all of these types may not be readily available in some areas. The engineer should determine whether the proposed type is available before specifying any type of portland cement other than type I.

20-5 Admixtures

By an "admixture" is meant, in this case, any substance other than aggregate, water, or portland cement that may be added to a concrete mixture. A large number of admixtures may be used in conjunction with the standard ingredients of portland cement concrete for various purposes and in various ways. Only a few of these admixtures are of importance in highway construction, and certain of these will be discussed in the following paragraphs.

Perhaps the most important admixtures that may be added to concretes used in highway construction are those used to produce air-entrained concrete. Numerous materials can be used as air-entraining agents, including natural wood resins, fats, various sulfonated hydrocarbons, and oils. Some of these materials are insoluble in water and must be saponified before they can be used as admixtures. ASTM C233 gives a program of testing for the evaluation of materials proposed for use as air-entraining agents, while ASTM C260 is a tentative specification for these materials.

Another admixture that is of some consequence is calcium chloride, which is generally used as an accelerating agent. That is, calcium chloride materially decreases the time of hardening of the cement and thus leads to increased early strength of the concrete mixture. The amount of calcium chloride is generally limited to less than 2 percent by weight of standard cement. Other advantages claimed for the use of small amounts of calcium chloride include improved curing of the concrete by reducing the loss of moisture during the early period of hydration, slight increases in the ultimate strength of the concrete, improved workability of the plastic mixture, increased resistance to abrasion, and decreased shrinkage. Caution should be exercised to avoid excessive use of calcium chloride, which could reduce the durability of concrete. The same properties obtained by the use of calcium chloride usually can be obtained by using type III cement or by increasing the cement factor.

Powdered materials such as diatomaceous earth, pumice, fly ash, and hydrated lime are occasionally used as admixtures in concrete, primarily as workability agents. Researchers have reported that the use of fly ash in concrete has a number of advantages including better protection against alkali–aggregate reaction and increased workability and strength from the use of this material. Widespread use is being made of various admixtures in casting concrete bridge decks to control the setting time, thus controlling dead load deflections and helping to ensure composite action between the deck and the underlying steel girders.

Admixtures may also be used to produce colored concrete. For example, red iron oxide has been used in concrete in some states for the purpose of marking traffic lanes. Similarly, the center strip in a concrete pavement is sometimes formed from concrete into which black iron oxide has been mixed.

PROPORTIONING OF CONCRETE MIXTURES

As has been indicated, a concrete mixture contains four basic ingredients, coarse aggregate, fine aggregate, cement, and water, with possible addition of admixtures to obtain specific properties. Before concreting operations may begin on any given project, the proportions of these materials must be established. The procedure of determining the amount of each of these materials that must be used to produce a concrete mixture of the desired characteristics is known as the "design of the mix." In the brief discussion that follows, no admixtures will be considered, and the design of air-entrained mixtures will be given mention in a later portion of this chapter. The objective of any procedure that may be utilized for the design of a given mixture is to determine the most practical and economical combination of materials that will produce a workable concrete which will have the desired properties of durability and strength after it has hardened.

Certain basic factors govern the design of satisfactory concrete mixtures, regardless of the detailed method that may be used. Detailed procedures that may be used may be quite complex and lengthy, involving considerable laboratory and field testing, or they may be very simple in nature, depending largely on the size of the project, the desired qualities of the hardened mixture, and the degree of experience which has been attained by the individual designer or organization in the use of materials suitable for concrete in a given area. The procedure outlined here is largely based on that contained in a standard of the American Concrete Institute (3). It involves the selection of a trial mixture to be used as a starting point for field concreting operations or for a detailed laboratory study. The procedure presented here is not necessarily one that is typical of the present practice of a majority of highway agencies, but it is given here because of its exposition of the fundamental factors involved and because of its universal applicability. A procedure of this type may be regarded as being especially desirable when concrete of a given flexural strength is desired.

Concrete should be placed with the minimum quantity of mixing water consistent with proper handling. Proportions should be selected to produce concrete

1. Of the stiffest consistency (lowest slump) that can be placed efficiently to provide a homogeneous mass.
2. With the maximum size of aggregate economically available and consistent with adequate placement.
3. Of adequate durability to withstand satisfactorily the weather and other destructive agencies to which it may be exposed.
4. Of the strength required to withstand the loads to be imposed without danger of failure.

A basic step to be taken in the design of a trial mixture is to select the water−cement ratio that is consistent with the desired durability and compressive strength desired in the hardened mixture. The water−cement ratio is a quantity of fundamental importance in the design of concrete mixture and is defined as the ratio of the amount of water, exclusive of that absorbed by the aggregate, to the amount of cement in a concrete mixture. It is preferably expressed as a decimal by weight of the water and cement. Formerly, it was commonly expressed in terms of gallons per sack of cement. A water−cement ratio of 6 gallons per sack of cement is equivalent to a water−cement ratio of 0.53 by weight, calculated as follows: $w/c = 6$ gal/sack $= 6(8.34)/94 = 0.53$. Generally speaking, both the durability and compressive strength of a given concrete mixture increase with a *decrease* in the water−cement ratio; that is, the lower the water−cement ratio, the higher will be the compressive strength and the greater the durability of a given concrete mixture, all other things being the same and assuming that a workable, homogeneous mixture is secured. Maximum values of the water−cement ratio to be used in designing concrete using average materials for a desired compressive strength and to resist known conditions of exposure are well established. State highway department specifications related to concrete to be used in paving establish maximum permissible water−cement ratios varying from 0.47 to 0.55, depending on the experience of the organization concerned. The average value used is probably between 5.5 and 6.0 gallons per sack. They also require a minimum of 470 to 610 pounds of cement per cubic yard, feeling that the use of less cement than this will result in surface abrasion. The relationship between water–cement ratio and flexural strength is not so clearly established and should be determined by laboratory or field testing, although the flexural strength of a given mixture may generally be expected to be from 15 to 20 percent of the compressive strength for comparable conditions of curing and age. Compressive strengths of 3000 to 4000 lb/in^2. (28-day curing) are commonly specified in highway work, while a typical required flexural strength is 650 lb/in^2. (28-day curing, third point loading). Lower minimum flexural strengths may be specified for opening to traffic.

Another step required in the design of the mix, as outlined in the publication cited above, is to select the desired consistency of the concrete to permit proper handling and placing of the mixture under the anticipated job conditions. The consistency of the fresh concrete mixture may be measured by the slump test (AASHTO Designation T119). The slump is measured by placing the fresh concrete mixture into a galvanized metal mold formed into the shape of the frustum of a cone, which is 8 in. in diameter at the base, 4 in. in diameter at the top, and 12 in. high. The concrete is placed in the mold in three layers, each of which is rodded by 25 strokes of a $\frac{5}{8}$-in. diameter rod 24 in. long. The top layer is struck off even with the top of the mold and the mold raised vertically away from the concrete. The concrete mass subsides, and its height after subsidence is measured immediately. The difference between this height and the original height of the mass (12 in.) is the "slump." The slump test is illustrated in Figure 20-1. The desired slump for concrete to be used in pavements is from 1 to 3 in. Consistency may also be measured in the field by observing the penetration of a 6-in., 30-lb metal ball into the fresh concrete (ASTM C360).

FIGURE 20-1 Equipment for slump test. (Courtesy Soiltest, Inc.)

Another step that must be taken in the general design procedure that is being explained is the selection of the maximum size of aggregate that may be used. The maximum size of aggregate used in paving generally varies from 2.5 to 1.5 in. in slabs that are of normal thickness and that contain relatively small amounts of reinforcing. The maximum size of aggregate may be considerably less when the slab is to be heavily reinforced.

Additional variables that must be established in the design of a trial concrete mixture are the unit water content of the mixture and the proportion of coarse aggregate to be used in the aggregate combination. Various methods have been presented for determining these quantities, and detailed design procedures now in use by the various organizations show considerable deviation at this point in the design process. The general method being discussed defines the unit water content as the amount of water required per cubic yard of concrete.

The quantity of water per unit volume of concrete necessary to produce a mixture of the desired consistency is influenced by the maximum size, particle shape, and grading of the aggregate and by the amount of entrained air. It is relatively unaffected by the quantity of cement.

The approximate quantities of mixing water for different slumps and maximum sizes of aggregates are as follows:

Slump (in.)	Water (lb/cu yd) for Max. Size of Aggregate			
	1	*1½*	*2*	*3*
Non-Air-Entrained Concrete				
1–2	300	275	260	220
3–4	325	300	285	245
Air-Entrained Concrete				
1–2	270	250	240	205
3–4	295	275	265	225

These quantities may be used with sufficient accuracy for computing cement factors for trial batches. They are maximum quantities that should be expected for fairly well-shaped, angular coarse aggregates graded within the usual limits. If aggregates which are otherwise suitable lead to higher water requirements than given in the tabulation, the cement content should be increased to maintain the desired water–cement ratio.

The minimum amount of mixing water and maximum strength will result for given aggregates when the largest quantity of coarse aggregate is used consistent with desired workability. Precise quantity of coarse aggregate for a given mixture is best determined by laboratory investigation, with later adjustment in the field. However, in the absence of laboratory data, an estimate of the proper proportions can be made for aggregates of conventional grading from empirical relationships shown in Table 20-3 (*3*).

Volumes are based on aggregates in dry-rodded condition as described in Method of Test for Unit Weight of Aggregate (ASTM Designation C29).

TABLE 20-3
VOLUME OF COARSE AGGREGATE PER UNIT
VOLUME OF CONCRETE

Maximum Size of Aggregate (in.)[a]	Volume of Dry-Rodded Coarse Aggregate per Unit Volume of Concrete for Different Fineness Moduli of Sand			
	2.40	2.60	2.80	3.00
1	0.71	0.69	0.67	0.65
$1\frac{1}{2}$	0.75	0.73	0.71	0.69
2	0.78	0.76	0.74	0.72
3	0.82	0.80	0.78	0.76

[a] 1 in = 2.54 cm.

These volumes are selected from empirical relationships to produce concrete with a degree of workability suitable for usual reinforced construction. For less workable concrete, such as required for concrete pavement construction, they may be increased about 10 percent.

The information presented has been taken from the publication of the American Concrete Institute which was previously cited. Additional information of this type is contained in that publication, and expanded tables of this sort are also to be found in the Design and Control of Concrete Mixtures, published by the Portland Cement Association.

Because of lack of space, no attempt will be made here to illustrate the specific use of the fundamental data that have been presented. Suffice it to say that various trial mixtures may be designed and examined in the laboratory, following the determination of certain basic "design properties" of the cement and aggregates involved. The mix may be designed for a given flexural strength, compressive strength, or both, and for the desired consistency. Some adjustment must generally be made in the selected mix on the basis of field conditions in order to secure concrete of satisfactory quality and economical proportions.

20-6 Air-entrained concrete mixtures

Air-entrained concrete, as has been indicated, contains a small amount of entrapped air which is present in the form of small, disconnected air bubbles uniformly distributed throughout the mass. The desired amount of air is generally from 4 to 8 percent of the total mix. The chief advantage accruing from the inclusion of this amount of air in the mixture is increased durability, including resistance to calcium chloride and other salts that are widely used in northern states for ice control and that have frequently resulted in surface scaling and spalling of concrete pavements. It also increases resistance to sulfate action and to freezing and thawing. Air-entrained concrete mixes have a high degree of workability compared to similar regular mixtures, being generally somewhat sticky and plastic in nature and showing little tendency to segregation. They may be placed at slightly lower slumps than regular mixtures.

Air-entrained concrete mixtures should also be designed by the trial-mix method. It is necessary to determine and control the amount of air that is actually incorporated in the mixture. Several methods have been devised to measure the air content of air-entrained concrete mixtures, including the gravimetric method (AASHTO T121 and ASTM C138), the volumetric method (ASTM C173), and the pressure method (AASHTO T152 and ASTM C231). The last of these is most widely used and is based on the principle of Boyle's law ($P_1V_1 = P_2V_2$); a known pressure is applied to a known volume of concrete and the change in volume determined. A pressure device for determining the air content is shown in Figure 20-2.

STRUCTURAL DESIGN OF CONCRETE PAVEMENTS

The following several sections of this chapter will be devoted to what has been called the "structural design" of concrete pavements. Included in these sections is the design of the various elements of the pavement structure itself, including considerations related to the various types of joints and their spacing, the use of reinforcing steel, and the thickness required in the slab cross section.

JOINTS AND JOINT SPACING

Joints are installed in concrete pavements to control the stresses induced by volume changes in the concrete. These stresses may be produced in a concrete slab because of (1) its contraction due to a uniform temperature drop or a

FIGURE 20-2 Determination of air content by the pressure method. (Courtesy Soiltest, Inc.)

decrease in moisture, (2) its expansion due to a uniform temperature increase, and (3) the effects of the "warping" of pavements due to a vertical temperature or moisture differential in the slab.

20-7 Contraction stresses

As do most construction materials, concrete changes in volume with a change in temperature. Thus a concrete slab, if it is free to move, contracts with a drop in temperature. However, if the contraction or "movement" of the slab is wholly or partially prevented, tensile stresses are developed. In a pavement, resistance to movement of the slab is caused by friction between the bottom of the slab and the subgrade.

It is also known that concrete generally contracts or "shrinks" with a decrease in moisture content and expands to some extent with an increase in moisture. Since a concrete slab tends to decrease in moisture content with the passing of time, the slab might be expected to contract slightly with a resultant opening of transverse joints. This circumstance would somewhat tend to offset later expansion of the slab due to temperature change and thus further tend to substantiate the general omission of expansion joints.

20-8 Expansion stresses

If a concrete pavement slab is subjected to a uniform increase in temperature, the slab will increase in length. Under extreme combinations of circumstances it may be imagined that a long length of concrete slab may buckle or "blow up" if this expansion is prevented. In order to prevent blow-ups, relief may be provided by the installation of transverse expansion joints. These joints, which are described in more detail later, generally provide a space an inch or so in width in which expansion may take place without damage to the slab.

However, friction between the slab and the subgrade prevents much of this expansion, and the compressive stresses created by this restraint are generally quite small as compared with the compressive strength of the concrete. This fact, coupled with the fact that concrete shrinks somewhat during hardening, has led many designers to the conclusion that expansion joints may be spaced at very great intervals, provided only that adequate contraction joints are provided and that the concrete does not have unusual expansion qualities and was not placed at exceedingly low temperatures.

20-9 Warping stresses

Very frequently a differential in temperature exists between the top and bottom of a concrete pavement. For example, during the day the temperature of the top portion of the slab may be considerably greater than that of the bottom. Under this circumstance the top portion of the slab tends to expand more than the bottom. This expansion and its effect in producing a slightly convex slab surface are resisted by the weight of the slab, with the consequence that fibers in the top portion of the slab are placed in compression and those in the bottom in tension. Stresses in the pavement from this case are termed "temperature warping stresses." Conversely, at night the top of the slab is frequently cooler than the bottom, with the result that the top of the slab is in tension and the bottom in compression.

Temperature warping is of importance in the design of concrete pavements in two ways. First, the flexural stresses due to this cause may, under certain conditions, be of considerable magnitude in themselves, without regard to their combination with stresses due to traffic loads. Second, the warping that does take place due to a temperature differential may partially destroy the subgrade support beneath portions of the slab, with the result that stresses due to traffic loads may be considerably increased over those that would exist if the pavement received uniform subgrade support.

If a difference in moisture content exists between the top and the bottom of the slab, the effect is somewhat similar to that produced by a temperature differential. In perhaps the majority of cases the moisture content in the bottom of the slab is greater than in the top, with the result that the bottom tends to expand, resulting in compression of the bottom fibers and somewhat offsetting stresses resulting from temperature warping during the daytime, when the temperature differential might be expected to be greatest.

Joint design

To control the stresses resulting from the combined effects of temperature and moisture changes and wheel loadings, four types of joints are commonly provided for concrete pavements:

1. Transverse contraction joints.
2. Transverse construction joints.
3. Transverse expansion joints.
4. Longitudinal joints.

20-10 Transverse contraction joints

Transverse contraction joints are used for two purposes: to control cracking of the slab resulting from contraction and to relieve warping stresses. The type of contraction joint that is most widely used at the present time is the weakened-plane joint, one design of which is shown in Figure 20-3. As contraction occurs,

Sawed 1/8" to 1/4" width or premolded insert

Sealant reservoir, see text for dimensions

1/4 h

h

Sawed or Premolded Insert

For reinforced pavements, smooth dowel bars installed at depth h/2.
See text for use of dowel bars.

FIGURE 20-3 Weakened-plane contraction joint. (Courtesy Portland Cement Association.)

the lower portion of the slab cracks at the weakened plane and the structural integrity of the joint is maintained by the dowel bars and aggregate interlock. This type of joint is also called "dummy" contraction joint.

Details of the designs of contraction joints vary to some extent, with bar diameters varying from $\frac{5}{8}$ to $1\frac{1}{2}$ in. being used (commonly one-eighth of the slab thickness) and with lengths varying from 12 to 20 in. Bar spacings generally are 12 to 15 in. from center to center of the bars. The groove in the pavement is frequently either of constant $\frac{1}{8}$- or $\frac{1}{4}$-in. width, as shown in Figure 20-3. This groove is most often formed by sawing but may be formed by use of a premolded strip inserted in the fresh concrete. A double saw cut may be used to construct a wider joint sealant reservoir at the top of the pavement. Figure 20-4 illustrates the wider reservoir which may be needed to provide a better shape factor for the sealant material. The groove is later filled with one of a number of materials, including field molded or poured sealants or preformed compression seals. When field molded sealants are used, a tape or metal foil should be placed at the bottom of the sealant space to prevent bonding between the sealant and the concrete.

Dowel bars used in this type of joint (and in expansion joints) are not bonded to the concrete on one side of the weakened plane, and freedom of movement is ensured by painting or lubricating one end of the dowel, by enclosing one end in a sleeve, or by other similar methods. It is essential that freedom of movement be ensured in the design and placing of the joint, since the purpose of the joint will be largely destroyed if movement is prevented. The dowels must be carefully placed so that they are horizontal and parallel to the centerline of the slab. They must also be designed to ensure their proper functioning under the stresses imposed on them in their capacity as load-transfer devices.

For plain concrete pavements, the spacing of contraction joints must be close enough to control cracking. Spacing is dependent on local experience and often is 15 or 20 ft. If the spacing is above 30 ft, load transfer devices should be used at the joints.

Engineers have discovered two ways to improve joint performance and extend the life of plain concrete pavements. One way is to construct transverse joints that are skewed slightly with respect to a line at right angles to the

FIGURE 20-4 Detail of a weakened-lane contraction joint. (Courtesy Portland Cement Association.)

centerline of the pavement (4 or 5 ft in 24 ft). A skewed joint causes the wheel loads of each axle to cross the joint one at a time, thereby reducing the deflection and stress at the joints. Another technique for improving the performance of plain concrete pavements is to construct the joints at random spacings (13, 19, 18, and 12 ft) to avoid rhythmic effects of automobiles associated with spacings that are exact multiples of 7.5 ft.

Distance between transverse contraction joints may be somewhat greater when the pavement is reinforced, since the distributed steel aids in controlling cracking. Joint spacing is dependent on several factors including the quantity of steel and the thickness of slab. The Portland Cement Association (4) recommends joint spacings of not more than 40 ft for reinforced pavements.

Continuously reinforced pavements, which have relatively heavy continuous steel reinforcement in the longitudinal direction, tend to develop small transverse cracks at close intervals. A high degree of load transfer is provided at these crack faces, which are restrained by the steel and by aggregate interlock. Such pavements therefore are normally built without transverse contraction joints.

20-11 Transverse construction joints

An additional type of joint that may be used in a concrete pavement may be called a "construction joint." Transverse construction joints may be placed at the end of a day's "run" or when work ceases due to some other interruption. If construction is stopped at the location of a transverse joint, the joint assembly may be installed as usual, concrete placed on one side of the joint, and the other side of the joint protected in an appropriate fashion until concreting operations can be resumed. If a transverse construction joint falls within the middle third of the regular joint interval, it should be a tied joint. This prevents the joint from opening and possibly causing "sympathetic" cracking of the pavement in the adjacent lane. Longitudinal construction joints may be made in a similar fashion when, for example, the pavement is constructed with a width of one lane. Joints of this type may also be formed by the use of steel forms that incorporate a keyway or by fastening a tongue-and-groove premolded or metal strip on the inside of the forms; tie bars may be bent back out of the way or tie bolts may be used.

20-12 Transverse expansion joints

Expansion joints that are used at the present time are usually from $\frac{3}{4}$ to 1 in. wide and extend the full depth of the slab. The joint space is filled with some compressible, elastic, nonextruding material. A variety of materials have been used as expansion joint fillers, including bituminous material, cork, rubber, cork–rubber compounds, bitumined fabrics, wood, and many others. Dowel bars are normally used in expansion joints, and their diameter, spacing, and length would generally be similar to that indicated for contraction joints. The dowels must be adequate to perform their load-transfer function and they must be designed and placed so as not to interfere with proper functioning of the joint. Figure 20-5 shows a somewhat typical design of a transverse expansion

FIGURE 20-5 Typical doweled expansion joint.

joint. Whenever salt corrosion is anticipated to be a problem, coated dowel bars may be specified. Specifications for corrosion-resistant coated dowel bars are covered in AASHTO M254 and testing of coated dowel bars is covered in AASHTO T253.

After many years of laboratory and field studies, practice relative to the spacing of transverse joints has at last reached some degree of uniformity over the country. Details vary, but the following discussion reflects current practice for both plain and reinforced concrete pavements.

The Portland-Cement Association (4) has reported that, except at intersections and structures, no expansion joints need be provided in a concrete pavement when

1. The pavement is constructed of materials that have normal expansion characteristics.
2. The pavement is constructed during those periods of the year when ambient temperatures are well above freezing.
3. The pavement is divided into relatively short panels by contraction joints so spaced as to prevent the formation of intermediate cracks.
4. The contraction joints are properly maintained to prevent the infiltration of relatively incompressible materials.

If concrete pavements are built in cold weather, or of materials that have high coefficients of expansion, expansion joints may be necessary at intervals of several hundred feet. Otherwise, they are required only adjacent to structures and intersections.

20-13 Longitudinal joints

A longitudinal joint in a concrete pavement is, of course, a joint running continuously the length of the pavement. The joint divides, for example, a two-lane pavement into two sections, the width of each being the width of a traffic lane. The purpose of longitudinal joints is simply to control the magnitude of temperature warping stresses in such a fashion that longitudinal cracking of the pavement will not occur. Longitudinal cracking has been almost completely eliminated in concrete pavements by the provision of adequate longitudinal joints. In two-lane pavements, the two slabs are generally tied together by means of steel tie bars extending tranversely across the joint and spaced at intervals along the length of the joint. The purpose of the tie bars in a longitudinal joint is to prevent movement of one slab with respect to the other, since, in addition to serving the function of relieving temperature warping stresses, the joint functions in transmitting a portion of the load carried by one slab to the other, thus reducing the stresses in the loaded slab. Adequate "load transfer" is dependent on maintaining close contact between the two slabs (or portions of the joint) across the longitudinal joint. In multiple-lane pavements, no more than four lanes should be tied together.

Longitudinal joints used in concrete pavements are of several different types, including both patented and nonpatented varieties. Two of the most commonly used types are those designated as "deformed" or "keyed" and "weakened-plane" joints. A typical longitudinal joint of the former type is shown in Figure 20-6. This type of joint often is used when the pavement is built one lane at a time; a weakened-plane joint generally is used if two or more lanes are paved at one time. Joints of this type are generally staked in position, prior to placing of the concrete, by steel pins driven at intervals through the joint assembly along its length. In this type of joint, concrete in adjacent slabs is separated by the materials used in the joint, which may, for example, be sheet metals or asphalt mastic board. The weakened-plane type of joint is formed by grooving, sawing, or insertion of a premolded strip and is described in detail in the portion of the discussion related to contraction joints.

Steel tie bars are used to tie the two slabs together and are firmly "bonded" to the concrete; the behavior of these bars should not be confused with that of the dowel bars mentioned in later sections. Tie bars that are used in longitudinal joints are generally designed to withstand the entire stress created by contraction of the slab in a transverse direction.

The cross-sectional area, A, in square inches, of steel required per foot length of longitudinal joint can be computed by the following equation:

$$A = \frac{bC_f wh}{12f_s}$$

$$(20\text{-}1)$$

KEYWAY DIMENSIONS

Groove 1/8" to 1/4" wide and 1" deep.

Seal

First slab

1:4 Slope

0.2 h

h

0.1 h

Deformed tiebar

FIGURE 20-6   Keyed longitudinal construction joint with tie bars. (Courtesy Portland Cement Association.)

where

b = distance between joint and the nearest untied joint or free edge (ft)

C_f = coefficient of subgrade (or subbase) resistance to slab movement, usually taken as 1.5

w = weight of concrete (lb/ft^3; 150 lb/ft^3 for normal-weight concrete)

h = slab thickness (in.)

f_s = allowable working stress in steel (lb/in.^{2}) (usually taken as approximately two-thirds of the yield strength)

This equation provides sufficient steel in the tie bars to overcome the resistance of the subgrade or subbase to horizontal movement when the slab is contracting. It assumes that this resistance is developed over the distance between the tied joints and the nearest free edge.

The tie-bars should be long enough so that the anchorage on each side of the joint will develop the allowable working strength of the tie bar (4). Allowing 3 in. for inaccurate placement of the tie bar, the length of the tie bar, L_t, in inches, becomes

$$L_t = \tfrac{1}{2}\,\frac{f_s \times d_b}{350} + 3$$

(20-2)

where

f_s = allowable working stress in steel (lb/in.^{2}) (same as in Eq. 20-1)

d_b = diameter of tie bar (in.)

This equation assumes that the maximum working stress for the bond in deformed tie bars is equal to 0.1 times the compressive strength of the concrete, up to a maximum of 350 lb/in^2.

Equations 20-1 and 20-2 can also be used to determine the size, spacing, and length of tie bars used for transverse emergency construction joints.

Additional information on the design of joints for concrete pavements is given in Ref. 4.

20-14 Pumping of joints

Special attention is given in this section to a problem that may be encountered in the design (and maintainance) of joints in concrete pavements. This problem is "pumping." The phenomenon of pumping may be defined as follows. With a certain combination of factors present, the movement of slab ends under traffic loads causes the extrusion or "pumping" of a portion of the subgrade material at joints, in cracks, and along the edges of the pavement. Pumping through joints will only occur under the following circumstances: (1) frequent occurrence of heavy wheel loads, (2) the existence of a surplus of water in the subgrade soil, and (3) the presence of a subgrade soil that is susceptible to pumping. All three of these elements must be present in order for pumping to occur. The amount of soil removed by pumping may be sufficient to cause a sizable reduction in subgrade support for the slab and may result in eventual failure of the pavement.

Design procedures intended to prevent pumping will be briefly discussed here, while maintenance operations intended to correct pumping that occurs in existing pavements will be discussed in the next chapter. Since much of the difficulty in this respect has been encountered at expansion joints, an obvious solution is to minimize the number of these joints or eliminate them altogether; this consideration has doubtless been a contributing factor in the present trend to minimize the use of expansion joints. Also, since pumping is associated with fine-grained soils, another obvious solution is to replace or improve these so-called pumping soils. This may be done, for example, by the use of a nominal thickness of granular material or a selected sandy soil, or by stabilization of the existing subgrade soil through the use of bituminous material or portland cement.

In modern practice, a granular subbase from 3 to 6 in. thick is employed by many organizations in locations where the subgrade soil is such as to be susceptible to pumping. Much thicker subbases may be used in areas subject to frost action or on very weak soils. Requirements for subbase materials vary; a suitable specification is AASHTO M147 (Table 18-1). A desire to prevent pumping is one factor in the increasing use of cement-treated base courses beneath concrete pavements. Some states now use asphaltic plant mixtures under concrete pavements for the same purpose.

USE OF DISTRIBUTED STEEL REINFORCING

Distributed steel reinforcing is used primarily to control cracking of a concrete pavement and to maintain the structural integrity of the slab between transverse joints. In accordance with principles previously stated, some organizations believe that reinforcing steel is unnecessary where contraction joints are spaced at intervals of 15 to 20 ft. If the distance between contraction joints is appreciably greater than this, it may generally be regarded as desirable, with the amount of steel required being a direct function of the spacing between joints. Cracking of the slab will occur even though the steel is present, but the steel serves to hold the edges of the cracks close together, thus preventing the progressive opening of these cracks and accompanying detrimental effects. Distributed steel, as commonly used in concrete pavements, is not counted on to contribute to the flexural strength of the slab. By holding cracks tightly closed, however, it maintains the shearing resistance of the slab and, consequently, its load-carrying capacity. These properties are greatly reduced if cracks are permitted to open.

Distributed reinforcing is used in the form of welded wire "fabric" or bar mats, both of which contain both longitudinal and transverse elements. Different sizes and spacing of the elements of the reinforcing "mat" are used by the various highway agencies in order to conform with requirements related to the area of steel required across the width and length of the slab. The steel is usually placed in the slab at or slightly above the middle of the slab. The amount of steel required in a given case is based on the assumption that it must be strong enough to drag both ends of the slab over the subgrade toward its center. This assumption leads to the following equation for the determination of the amount of steel:

$$A = \frac{LC_fW}{2f_s} \tag{20-3}$$

where

A = amount of steel cross section (in.^{2}/ft of slab width)

L = length of slab between joints (ft)

C_f = coefficient of friction between the slab and the subgrade, typically 1.5

W = weight of slab per square foot of pavement surface (lb)

f_s = working stress in the steel (lb/in.^{2}, typically 30,000 to 45,000 lb/in.^{2})

Generally, the cross-sectional area of longitudinal steel ranges from about 0.1 to about 0.2 percent of the cross-sectional area of the slab. Weights required typically range from about 55 to 80 lb per 100 ft.^{2} of pavement surface.

20-15 Continuously reinforced pavement

Following World War II, interest was focused on the possiblity of building continuously reinforced pavements without joints, with the construction of pavements of this type in Illinois and New Jersey. Longitudinal steel was used continuously over the length of 1 mile in the New Jersey experiment and 3500 ft in Illinois. The amount of steel was varied in different sections of these test roads, ranging from 0.3 to 1.0 percent. Subsequently, test sections of continuously reinforced concrete pavement were built in Texas, California, Pennsylvania, and several other states.

Since 1959, when there were less than 100 miles of continuously reinforced concrete highway pavement in the United States, use of this type of pavement has grown rapidly. By 1979 there were approximately 13,000 miles of such pavements in service in the United States.

The required amount of longitudinal reinforcing steel in this kind of pavement is not in direct proportion to the length of the slab. Observations of long (1 mile or more) continuously reinforced slabs indicate that only 400 or 500 ft of the pavement at each end is subject to longitudinal movement; for all practical purposes, the long central section is fully restrained. Seasonal movements at the slab ends total less than 2 in. regardless of the slab length.

It is believed that the optimum quantity of reinforcing steel is that which results in a crack spacing of 3 to 10 ft. Crack intervals in this range typically occur with steel ranging from 0.5 to 0.7 percent of the cross-sectional area of the slab.

Recommended slab thickness varies from 6 to 9 in., depending on traffic conditions. Many engineers believe that the thickness of continuously reinforced pavements should be the same as that of conventional concrete pavements.

Both welded wire fabric and bar mats are used for continuous reinforcement. For wire fabric, spacing of the longitudinal wires should be not less than

4 in. nor more than 12 in.; spacing of transverse wires should not exceed 24 in.; and the clearance between the reinforcement and the slab edges should not be less than 2 in. nor more than 6 in.

DIMENSIONS OF THE CROSS SECTION (THICKNESS DESIGN)

It is the purpose of this section to discuss the determination of the design thickness of a concrete pavement. Several different factors enter into the thickness determination, as will be outlined in following sections. Basically, however, the pavement must have a thickness that is adequate to support the loads which will be applied to it during its service life, and the design must be economical.

Many organizations and individuals have given their attention to the problem of thickness design of concrete pavements during the last 50 years or so. Space will not permit the chronicling of the many contributions that have been made by these investigators and others. Rather, it has been decided to present the essentials of only one design method, in the belief that this method is typical of modern approaches to this subject. The method presented is that contained in the publication, "Thickness Design for Concrete Pavements," published by the Portland Cement Association (PCA) in 1984 (5).

According to PCA engineers, design considerations that are vital to satisfactory performance and long life of a concrete pavement are reasonably uniform support for the pavement, elimination of pumping by the use of a thin treated or untreated base course, adequate joint design, and a thickness that will keep load stresses within safe limits.

General objective of the design procedure is to determine the minimum thickness that will give the least annual cost.

Since direct compressive stresses in concrete pavements due to wheel loads are very small in relation to the compressive strength of the material, they can be ignored. Flexural strength is a key factor in thickness design, since the flexural stress produced by a heavy wheel load often is more than half the flexural strength.

Flexural strength is measured by modulus of rupture tests on beams subjected to third-point loading. PCA recommends the use of 28-day to 90-day strengths for roads and streets, since very few stress repetitions occur during the first 90 days of pavement life, compared with the millions of repetitions that occur after that time. Also, the strength of concrete increases with age.

Another major factor in thickness design is subgrade or subbase support, which is measured by the Westergaard modulus of subgrade reaction, k. To determine k, a plate bearing test generally similar to that described in Section 16-8 is performed, generally using a 30-in. plate. Typical values of k for the groups of the AASHTO soil classification system are given in Table 20-4.

The Portland Cement Association recommends that k values used in design be those determined in normal summer or fall weather (not during the "spring breakup" or when the subgrade is frozen). The k value can be increased by using a treated or untreated subbase, although use of a subbase for this

TABLE 20-4
TYPICAL VALUES OF THE MODULUS OF SUBGRADE REACTION, k

AASHTO Soil Group	$k \, (lb/in.^3)^a$
A-1-a	400 plus
A-1-b	250 plus
A-2-4, A-2-5	300 plus
A-2-6, A-2-7	175−325
A-3	200−325
A-4	100−300
A-5	50−175
A-6	50−225
A-7-5, A-7-6	50−225

^{a}1 lb = 0.4536 kg; 1 in = 2.54 cm

purpose alone seldom would be economical. For a subgrade with a k value of 100, a 4 in. cement-treated subbase should increase the k value to at least 280, according to PCA designers.

The PCA recommends two design criteria:

1. Fatigue, to keep pavement stresses from repeated loads within acceptable limits to prevent fatigue cracking.
2. Erosion, to limit the effects of pavement deflections at joints, corners, and the edges of slabs in order to control the erosion of foundation materials.

Concrete is subject to fatigue, as are other construction materials. In the PCA design concept, a fatigue failure occurs when a material ruptures under continued repetitions of loads that cause stress ratios of less than 1. The stress ratio is the ratio of flexural stress to the modulus of rupture.

Flexural fatigue research on concrete has shown that the number of stress repetitions to failure increases as the stress ratio decreases; studies show that, if the stress ratio is less than 0.55, concrete will withstand virtually unlimited stress repetitions without loss in load-bearing capacity. To be conservative, designers reduced this ratio to 0.50.

In the past, thickness design methods for concrete pavements included an allowance for impact of moving loads. The PCA method discards this concept, substituting "load safety factors" in the design. The following load safety factors are recommended: Interstate and other multiple-lane projects where there will be an uninterrupted flow of heavy truck traffic, 1.2; highways and streets with moderate volumes of truck traffic, 1.1; and streets and highways that carry small volumes of truck traffic, 1.0.

Comprehensive analyses of concrete stresses and deflections at pavement joints, corners, and edges have established that the most critical axle load positions are (5).

1. For stresses, when the truck wheels are placed at or near the pavement edge and midway between the joints.
2. For pavement deflections, at the slab corner when an axle load is placed at the joint with the wheels at the corner.

It is known that only a small percentage of trucks travel with their outside wheels placed at the edge. Most truck drivers drive with their outside wheels placed about 2 ft from the pavement edge. The Portland Cement Association design procedure is based on the assumption that 6 percent of the trucks travel with the outside tire at or beyond the pavement edge. The maximum stresses and deflections result from those loadings. At increasing distances inward from the pavement edge, the magnitudes of stresses and deflections decrease, but the frequency of load applications increases.

PCA engineers analyzed the distributions of truck placements and accounted for the variability of positioning of loadings in their design figures and tables. They computed the fatigue incrementally for various load placements inward from the edge of the slab. They expressed the results in terms of an equivalent load factor which, when multiplied by the edge-load stress, gives the same degree of fatigue consumption that would result from a given load distribution. A similar approach was used to account for the effects of load placements on deflections.

Procedures recommended by the PCA are based on a design period of 20 years.

The PCA design procedure takes into account the effects of the erosion of foundation and shoulder materials due to pavement deflections at slab edges, joints, and corners. This analysis recognizes modes of pavement distress such as pumping, faulting, and shoulder problems that are unrelated to fatigue.

Another major factor in designing a concrete pavement is, of course, the traffic to which it will be subjected during its service life. The number and magnitude of heavy vehicle loads are critical.

The PCA manual (5) details two methods for determining traffic volumes and the number and magnitude of heavy wheel load repetitions; neither method is explained here. One approach is to use figures based on traffic volume and classification counts plus special loadometer surveys. Such information is generally available from the state highway departments or other traffic planning agencies for the design of any major route. Another way is to base the design on traffic capacity considerations for the route in question; capacity is discussed in Section 5-6.

PCA engineers worked out a number of design examples, one of which is summarized below.

EXAMPLE 20-1. *Thickness Design, Rigid Pavements.* A four-lane rural Interstate project is to be built in rolling terrain. The design period for the concrete pavement is 20 years. The current traffic volume (*ADT*) is 12,900 and the projection factor for 20 years is 1.5. Designers estimate that truck traffic volume (*ADTT*) is 19 percent of the total.

The design *ADT* is 12,900 x 1.5 = 19,350 vehicles per day, and the *ADTT* is 19,350(0.19) = 3680. Truck traffic each way is 3680/2 = 1840. Data not shown here reveal that 81 percent of the trucks will be in the right lane. Thus, for this lane and a design period of 20 years, there will be 1840 x 0.81 x 365 x 20 = 10.88 million trucks.

PCA engineers evaluated five designs (1A through 1E). For design 1A, the k of the subgrade was taken to be 100 lb in.^{3} with a 4-in. untreated granular subbase. Combined k for the subgrade and subbase was taken to be 130 lb/in.^{3} The load safety factor is 1.2, with a modulus of rupture of the concrete of 650 lb/in.^{2}

The engineers assumed a trial depth of 9.5 in. Key calculations are shown on the design worksheet, Figure 20-7.

For the fatigue analysis, equivalent stresses are determined from Table 20-5 for single and tandem axles and entered on the design worksheet as items 8 and 11, respectively. The equivalent stresses are then divided by the concrete modulus of rupture and entered as items 9 and 12 (stress ratio factors). Next, the allowable repetitions are determined for each loading category from Figure 20-8. The percentage of fatigue is then calculated by dividing the expected repetitions (column 3) by the allowable repetitions (column 4). The sum of the percentages of fatigue used is totaled for all loading categories, and for this example is 62.8 percent.

In a similar way, the erosion analysis involves determining the allowable repetitions to ensure that harmful erosion of foundation and shoulder materials does not occur from pavement deflections at slab edges, joints, and corners.

To perform this analysis, erosion factors are determined from Table 20-6 and recorded on the design worksheet as items 10 and 13 for single- and tandem-axle loadings, respectively. Allowable repetitions are then determined from Figure 20-9 and recorded in column 6 for each loading category.

The damage due to erosion is then compiled for each loading condition by dividing the expected repetitions (column 3) by the allowable repetitions (column 6). The total erosion damange (percent) is determined by summing the incremental damages (column 7). For this example, the erosion damage is 38.9 percent.

The totals of fatigue use and erosion damage use of 62.8 and 38.9 percent, respectively, show that the 9.5-in. thickness is satisfactory for the given conditions.

In the preceding example, we have illustrated the Portland Cement Association design procedure for one set of conditions: doweled joints, untreated subbase, and no concrete shoulder. Reference 5 provides similar illustrations and examples for designs of other combinations of subbase support, shoulder treatment, and joint design.

20-16 Present status of design

It should be noted at this stage that, although design principles and methods such as those illustrated above are in existence at the present time, these methods are not commonly applied as a part of the routine operations of a majority of highway agencies, including many state highway agencies. Thicknesses of concrete pavements, as used in this country, commonly vary from 7 to

Calculation of Pavement Thickness

Project _Design IA, four-lane Interstate, rural_

Trial thickness ___9.5___ in.   Doweled joints: yes ✓ no ____

Subbase-subgrade k ___130___ pci   Concrete shoulder: yes ____ no ✓

Modulus of rupture, MR ___650___ psi   Design period __20__ years

Load safety factor, LSF ___1.2___

4-in. untreated subbase

Axle load, kips	Multiplied by LSF 1·2	Expected repetitions _throughout life of pavement_	Fatigue analysis		Erosion analysis	
			Allowable repetitions	Fatigue, percent	Allowable repetitions	Damage, percent
1	2	3	4	5	6	7

8. Equivalent stress ___206___   **10.** Erosion factor ___2.59___
9. Stress ratio factor ___0.317___

Single Axles

30	36.0	6,310	27,000	23.3	1,500,000	0.4
28	33.6	14,690	77,000	19.1	2,200,000	0.7
26	31.2	30,140	230,000	13.1	3,500,000	0.9
24	28.8	64,410	1,200,000	5.4	5,900,000	1.1
22	26.4	106,900	Unlimited	0	11,000,000	1.0
20	24.0	235,800	''	0	23,000,000	1.0
18	21.6	307,200	''	0	64,000,000	0.5
16	19.2	422,500			Unlimited	0
14	16.8	586,900			''	0
12	14.4	1,837,000			''	0

11. Equivalent stress ___192___   **13.** Erosion factor ___2.79___
12. Stress ratio factor ___0.295___

Tandem Axles

52	62.4	21,320	1,100,000	1.9	920,000	2.3
48	57.6	42,870	Unlimited	0	1,500,000	2.9
44	52.8	124,900	''	0	2,500,000	5.0
40	48.0	372,900	''	0	4,600,000	8.1
36	43.2	885,800			9,500,000	9.3
32	38.4	930,200			24,000,000	3.9
28	33.6	1,656,000			92,000,000	1.8
24	28.8	984,900			Unlimited	0
20	24.0	1,227,000			''	0
16	19.2	1,356,000				
				Total **62.8**	Total	**38.9**

FIGURE 20-7 Design worksheet for example problem using PCA procedure. (Courtesy Portland Cement Association.)

TABLE 20-5
EQUIVALENT STRESS—NO CONCRETE SHOULDER (SINGLE AXLE/TANDEM AXLE)

Slab Thickness (in.)	\k of Subgrade−Subbase (lb/in.^{3})						
	50	100	150	200	300	500	700
4	825/679	726/585	671/542	634/516	584/486	523/457	484/443
4.5	699/586	616/500	571/460	540/435	498/406	448/378	417/363
5	602/516	531/436	493/399	467/376	432/349	390/321	363/307
5.5	526/461	464/387	431/353	409/331	379/305	343/278	320/264
6	465/416	411/348	382/316	362/296	336/271	304/246	285/232
6.5	417/380	367/317	341/286	324/267	300/244	273/220	256/207
7	375/349	331/290	307/262	292/244	271/222	246/199	231/186
7.5	340/323	300/268	279/241	265/224	246/203	224/181	210/169
8	311/300	274/249	255/223	242/208	225/188	205/167	192/155
8.5	285/281	252/232	234/208	222/193	206/174	188/154	177/143
9	264/264	232/218	216/195	205/181	190/163	174/144	163/133
9.5	245/248	215/205	200/183	190/170	176/153	161/134	151/124
10	228/235	200/193	186/173	177/160	164/144	150/126	141/117
10.5	213/222	187/183	174/164	165/151	153/136	140/119	132/110
11	200/211	175/174	163/155	154/143	144/129	131/113	123/104
11.5	188/201	165/165	153/148	145/136	135/122	123/107	116/98
12	177/192	155/158	144/141	137/130	127/116	116/102	109/93
12.5	168/183	147/151	136/135	129/124	120/111	109/97	103/89
13	159/176	139/144	129/129	122/119	113/106	103/93	97/85
13.5	152/168	132/138	122/123	116/114	107/102	98/89	92/81
14	144/162	125/133	116/118	110/109	102/98	93/85	88/78

SOURCE: *Thickness Design for Concrete Highway and Street Pavements.* Portland Cement Association (1984).

10 in., and each organization normally has developed "standard designs" that are satisfactory for its purposes. Usually, a concession is made to the principles outlined in the preceding section by the use of greater thicknesses for heavily traveled routes and areas of poor subgrade support.

20-17 Composite pavements

In broad terms, a composite pavement is one which combines dissimilar pavement types, that is, "flexible" and "rigid" pavements. Usually, a composite pavement comprises a concrete- or cement-treated base course with a wearing surface of asphaltic concrete.

Comparatively few composite pavements have been designed and built as new pavements, although New York city has for many years built city streets with a portland cement concrete base, an asphaltic concrete binder course, and a sheet-asphalt wearing surface. Other cities build similar pavement structures.

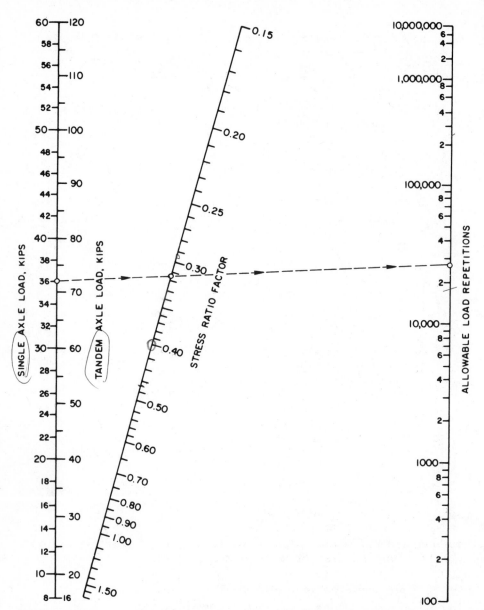

FIGURE 20-8 Nomograph for fatigue analysis—allowable load repetitions based on stress ratio factor (with and without concrete shoulder). (Courtesy Portland Cement Association.)

Many composite pavements have, of course, been produced by stage construction or by resurfacing operations. Overlays of existing pavements are discussed in Section 21-5.

Relatively little design information is available on composite pavements, although some experimental installations have been built and the Corps of

TABLE 20-6
EROSION FACTORS—DOWELED JOINTS, NO CONCRETE SHOULDER
(SINGLE AXLE/TANDEM AXLE)

Slab Thickness (in.)	k of Subgrade−Subbase (lb/in.³)					
	50	100	200	300	500	700
4	3.74/3.83	3.73/3.79	3.72/3.75	3.71/3.73	3.70/3.70	3.68/3.67
4.5	3.59/3.70	3.57/3.65	3.56/3.61	3.55/3.58	3.54/3.55	3.52/3.53
5	3.45/3.58	3.43/3.52	3.42/3.48	3.41/3.45	3.40/3.42	3.38/3.40
5.5	3.33/3.47	3.31/3.41	3.29/3.36	3.28/3.33	3.27/3.30	3.26/3.28
6	3.22/3.38	3.19/3.31	3.18/3.26	3.17/3.23	3.15/3.20	3.14/3.17
6.5	3.11/3.29	3.09/3.22	3.07/3.16	3.06/3.13	3.05/3.10	3.03/3.07
7	3.02/3.21	2.99/3.14	2.97/3.08	2.96/3.05	2.95/3.01	2.94/2.98
7.5	2.93/3.14	2.91/3.06	2.88/3.00	2.87/2.97	2.86/2.93	2.84/2.90
8	2.85/3.07	2.82/2.99	2.80/2.93	2.79/2.89	2.77/2.85	2.76/2.82
8.5	2.77/3.01	2.74/2.93	2.72/2.86	2.71/2.82	2.69/2.78	2.68/2.75
9	2.70/2.96	2.67/2.87	2.65/2.80	2.63/2.76	2.62/2.71	2.61/2.68
9.5	2.63/2.90	2.60/2.81	2.58/2.74	2.56/2.70	2.55/2.65	2.54/2.62
10	2.56/2.85	2.54/2.76	2.51/2.68	2.50/2.64	2.48/2.59	2.47/2.56
10.5	2.50/2.81	2.47/2.71	2.45/2.63	2.44/2.59	2.42/2.54	2.41/2.51
11	2.44/2.76	2.42/2.67	2.39/2.58	2.38/2.54	2.36/2.49	2.35/2.45
11.5	2.38/2.72	2.36/2.62	2.33/2.54	2.32/2.49	2.30/2.44	2.29/2.40
12	2.33/2.68	2.30/2.58	2.28/2.49	2.26/2.44	2.25/2.39	2.23/2.36
12.5	2.28/2.64	2.25/2.54	2.23/2.45	2.21/2.40	2.19/2.35	2.18/2.31
13	2.23/2.61	2.20/2.50	2.18/2.41	2.16/2.36	2.14/2.30	2.13/2.27
13.5	2.18/2.57	2.15/2.47	2.13/2.37	2.11/2.32	2.09/2.26	2.08/2.23
14	2.13/2.54	2.11/2.43	2.08/2.34	2.07/2.29	2.05/2.23	2.03/2.19

SOURCE: *Thickness Design for Concrete Highway and Street Pavements.* Portland Cement Association (1984).

Engineers has extended methods used to design composite pavements for airfields to apply to roads and streets. The California method described in Chapter 16 is applicable to pavement structures that include cement-treated base courses. Various aspects of composite pavement design are discussed in Ref. 6.

CONSTRUCTION OF CONCRETE PAVEMENTS

The construction of portland cement concrete pavements is marked by the use of a large number of specially designed machines, each of which performs a specific function in the construction process. Once paving operations are begun, the various steps in the construction procedure are arranged in the form of a continuing series of separate operations that are planned and coordinated so that the construction proceeds with a minimum loss of time and effort.

FIGURE 20-9 Nomograph for erosion analysis—allowable load repetitions based on erosion factor (without concrete shoulder). (Courtesy Portland Cement Association.)

Each of the separate steps must be done carefully and precisely so that the completed pavement will meet the exacting standards for structural strength and smoothness that are applied to it. The exact methods and machines used in the construction process vary somewhat from job to job, and no attempt will be made here to discuss all the possible variations in job methods and procedures. However, the following is the sequence of separate steps on typical projects:

1. Preparation and preliminary finishing of the subgrade.
2. Placing of forms (where used).
3. Final finishing of the subgrade.
4. Installation of joints.
5. Batching of aggregates and cement.
6. Mixing and placing concrete.
7. Placing and finishing concrete.
8. Slipform paving.
9. Curing.

20-18 Preparation and preliminary finishing of the subgrade

The preparation of the subgrade on which a concrete pavement is to rest is, of course, a very important step in the overall construction process. It is essential that uniform subgrade support be provided for the completed pavement during its service life and that the pavement be free from other detrimental effects associated with unsatisfactory subgrade soils. Included among the most commonly encountered problems that are related to the character and condition of the subgrade soil are pumping and frost action, while other difficulties may become evident when soils of inadequate shearing strength, high-volume-change soils, organic soils, alkali soils, soils that are difficult to drain, and others are encountered. Provision must be made for elimination of these defects in the subgrade soil before the pavement is placed.

In some cases the existing subgrade soil may be entirely suitable as a foundation and thus require only compaction and adequate drainage prior to the placing of a concrete slab. In other instances a layer of imported soil may be used between the existing soil and the slab, or the existing soil may be stabilized, as with portland cement, bituminous material, or granular materials. It is common practice to provide a blanket or "insulation" course of suitable granular material between the existing soil and the pavement itself. Various materials are used to form this layer or "subbase," including gravel, sand, sand and gravel, crushed rock, and so on. The depth of treatment varies from a minimum of 3 to 6 in. or so if the primary purpose is to prevent pumping; very much thicker layers of granular material may be used to prevent frost action in northern states.

In recent years many states, particularly those in the Far West, Southeast, and Southwest, have gone to the use of treated materials for subbases beneath concrete pavements. Most frequently used is a cement-treated base course, described more fully in Section 20-29, but some states now use asphaltic mixtures for this purpose.

Whatever the solution used in a specific case, the layer immediately beneath the slab must be brought to a high degree of density and stability and must be adequately drained. Preliminary finishing of this layer will generally be accomplished by the use of motor graders or similar equipment that shape the layer to the desired cross section, elevation, and line in what have previously been termed "finegrading operations." In most cases the subgrade layer will be brought very close to final shape before construction associated with the slab itself actually begins.

20-19 Placing of forms

Most concrete highway pavements are constructed with the slipform paver (see Section 20-24); however, steel forms are used for many city streets and some small highway projects. Where steel forms are used, they must be very carefully placed and secured in position so that the desired position, width, elevation, and grade may be secured in the final slab. Forms commonly used in highway work are straight 10-ft sections that are aligned both vertically and horizontally by slip joints and are held in position by three or more steel stakes driven at intervals at the back of the form. Forms of this type vary somewhat in size and weight, being available in heights of 6 to 12 in., with corresponding base widths over a similar range. One frequently used size, for example, has a height of 9 in. and a base width of 8 in. Special forms are available for curbs, curb and gutter sections, and for the construction of sharp curves, including both rigid and flexible-type curved forms.

Several machines are frequently used in connection with the placing of the forms. One of these is the so-called "form grader" that cuts a trench of exact size in the correct position to receive the forms, as shown in Figure 20-10. With this machine the trench is cut with a hydraulic cutter, which is adjustable hydraulically by the operator. Forms are then generally aligned and placed by hand, with the steel pins being driven either by hand or with the aid of an air hammer. A second machine may then be used to tamp the forms securely in place. The placing of the forms on most jobs is characterized by extreme care, as many of the machines used in later operations ride on the forms and it is imperative that the forms be securely and carefully aligned and placed. Their elevation and alignment are, of course, carefully checked prior to continuance of construction operations. Concrete pavements generally are constructed in two-lane widths; forms are used on both sides. Sufficient forms are generally required to be available on the job to permit the forms to remain in place during preliminary curing of the slab and to provide several hundred feet of forms in place ahead of the paver.

20-20 Final finishing of the subgrade

The next step in the indicated sequence of operations at the job site is the final shaping of the subgrade to the exact dimensions required by the plans and specifications. This operation is generally accomplished by a machine called a "subgrader" or "fine-grader," which rides on the forms and cuts off the subgrade to the exact shape desired. The subgrade is generally left a little high prior to this step, and the earth (or subbase material) excavated by this machine is thrown outside the forms. Extensive use also is now being made of electroni-

FIGURE 20-10 Cutting of trench to exact line and grade, prior to placing of forms. (Courtesy Portland Cement Association.)

cally controlled fine-graders without forms, particularly when slipform pavers are used. This machine operates from a taut wire. Trimming is done by cutting blades on a rotating drum. In some instances, trimming of the subgrade is accompanied by final rolling with steel rollers, particularly where a granular subbase is being built. A scratch template is generally employed at this stage to check the final finish of the subgrade.

20-21 Installation of joints

The installation of the various types of joints that may be used in a concrete pavement is also a very important step in the construction process. A portion of the process of construction of the required joints, which we may call the "installation of joint assemblies," normally takes place between the final finishing of the subgrade and the beginning of actual concreting operations. Such assemblies may not be required, as for a plain concrete pavement in which transverse joints are sawed and dowel bars are not used. Sawing of joints is covered in Section 20-26. Extreme care must be used in all operations accompanying the construction of joints if they are to function properly. Dowel bars, as commonly used in transverse joints, must be carefully placed and aligned parallel to the centerline and subgrade so that they will not inhibit free movement of the slab ends in a longitudinal direction and so that they will properly perform their load-transfer functions.

   Joint assemblies which are used to facilitate the construction of joints are quite varied in nature, depending on the exact type of joint that is to be used.

Typical joint assemblies are shown in Figure 20-11, which illustrates the use of the dowel unit manufactured by the Bethlehem Steel Company. The removable cap shown in Figure 20-12a is removed after the concrete has been placed and subjected to preliminary finishing. The slot shown in Figure 20-11b may be made just after the concrete has been finished by pressing a thin steel plate of appropriate dimensions into the concrete at the desired location or by sawing the hardened concrete. The spaces above the joint filler or the surface slot in a weakened plane joint, Figure 20-11b, are filled with one of several sealers previously discussed after the concrete has hardened. The sealer is forced into the joint space by special machines. Edges of the surface groove except for sawed joints are usually finished to a curve that has a $\frac{1}{4}$-in. radius. It should be noted that dowel bars used in transverse joints are commonly treated with paint, grease, or other materials on one side of the joint to prevent bonding of the bar to the concrete; the other end of the bar will, of course, be bonded. In transverse joints in which dowel bars are not used, which are commonly contraction joints of the weakened-plane type, the only steps involved in the construction are the formation of the surface groove, subsequent edging of the joint, and installation of joint sealer. Longitudinal joints are installed in a similar fashion, except that the tie bars are bonded to the concrete on both sides of the joint.

FIGURE 20-11 Typical dowel assemblies. (Courtesy Bethlehem Steel Co.)

20-22 Mixing of concrete

At this point in our discussion of the construction of concrete pavements, all preparations have been made at the job site for the placing of concrete between the forms. Let us now examine the steps of preparing and mixing the concrete.

Formerly, aggregates and cement were batched "dry" at a convenient location some distance away from the job site, brought to the site in trucks, and mixed with water on the job, usually in a boom-and-bucket paver. This method is rarely used today and is not described here.

Today, on most highway projects, all ingredients of the concrete, including the water, are batched and mixed at a central location. The fresh concrete is then transported to the paving site in transit mixers or in trucks with agitating or nonagitating bodies.

Central mixing plants vary widely. In urban areas, the plant often is a highly sophisticated, stationary "ready-mix" plant capable of producing automatically many mixtures of varying composition for many different purposes. Other plants are portable or semiportable; they are automatically controlled, can be moved from one location to another quite easily, and have production rates up to about 500 cu yd/hr with a single mixer.

Nearly all modern concrete plants are fully automatic and equipped with recording systems. A single operator at a control panel performs simple steps that control all the functions of charging, mixing, and discharging. Ingredient proportions can be controlled by preset dials, tapes, or batch plugs. The control system incorporates an automatic moisture-sensing device in the fine aggregate; adjustments are made automatically in batch quantities to correct for changes in moisture content. A plant also may be equipped with a "slump meter," which measures the current required to turn the mixer; the meter is calibrated with slump tests and gives an immediate check on the consistency of the concrete.

Permanent plants have more elaborate control systems that sense and govern the flow of all materials from the aggregate stockpiles through the plant to discharge of the concrete; ready-mix trucks often are dispatched from a central location by two-way radio.

Figure 20-12 shows a plant that has two tilting-drum mixers; the mixers discharge into trucks with side-dump bodies.

Figure 20-13 shows a close-up view of a tilting-drum mixer discharging concrete into a truck with an elevating discharge, agitating truck body.

20-23 Placing and finishing

Placing of the fresh concrete at the job site and finishing the slab to required smoothness are important operations that are done in a variety of ways. The number and kind of machines used vary from state to state and among contractors.

If a boom-and-bucket paver is used, the bucket dumps the concrete on the subbase ahead of a spreader. The most commonly used spreader is one that incorporates a long screw or auger across the front of the machine; the screw

FIGURE 20-12 Concrete mixing plant with two tilting-drum mixers.

distributes the concrete uniformly across the subbase between the forms. The spreader may carry vibrators and often has a transverse screed at the rear to give the slab a preliminary strike-off.

In a conventional paving train, the spreader is followed by a finishing machine that incorporates two transverse, oscillating screeds. Vibrators may be mounted on the front of this machine. This unit often pulls along a long-wheelbase, float-type finisher. The float, which is suspended from a frame, operates independently of the side forms to iron out any irregularities left by the transverse screeds.

Also available are "one-pass" finishing machines that ride on the side forms, combine in one unit a spreader, vibrators, transverse screeds, and longitudinal float. These units have been described as "slipform pavers riding on side forms." (See Section 20-24.)

Contractors often use special machines to place central mix concrete between forms. These machines use hoppers to receive the concrete. In one machine of this type, the hopper travels across the grade to deposit one fresh concrete between the forms. Another machine of this type is shown in Figure 20-14, on this project, two agitator-body trucks dumped concrete simultaneously into hoppers on either side of the grade. A transverse screw moved the concrete across the grade.

FIGURE 20-13 Tilting-drum mixer discharging concrete into agitating body truck. (Courtesy Wire Reinforcement Institute.)

FIGURE 20-14 Agitating-body trucks discharging concrete into hopper-type spreader. (Courtesy *Engineering News-Record*.)

Simple box-type spreaders often are used with slipform pavers.

The placing of distributed steel reinforcing in the form of either bar mats or wire mesh, complicates the spreading operation but generally leaves the finishing operations unchanged. One way of placing the steel is to lay one course of concrete with a spreader; workmen then place the sheets or mats of steel on top of the fresh concrete. The spreader then makes a second pass over the steel (or a second spreader may be used).

A second method, and one that is growing in popularity, is to spread the concrete full depth, lay the steel on top, and then use another machine (a mesh installer) to force the steel into the concrete to the required depth. This operation is shown in Figure 20-15.

Special methods are used to place continuous reinforcement.

The pavement is generally checked during final finishing operations by means of long wooden, steel, or aluminum straightedges and any irregularities removed.

The final finish is applied to the surface by belting, brooming, or use of a burlap drag. Belting is done by the use of a narrow canvas or rubber belt that is moved longitudinally along the surface with a slight transverse motion; two people handle the belt, one on either side of the slab. In securing a broomed finish, which is considered to have slightly greater skid resistance than a belted surface, long-handled fiber brooms from 18 to 24 in. wide are used. The finish is obtained by placing the broom at one edge of the pavement and drawing it transversely across the surface in such a way that corrugations are produced

FIGURE 20-15 Blades of mesh placer push and vibrate reinforcing steel into fresh concrete. (Courtesy Portland Cement Association.)

that are not more than $\frac{1}{8}$ in. in depth. Mechanical brooms are now common. A burlap drag is pulled longitudinally along the surface of the pavement, thus producing shallow surface corrugations; this is the method most commonly used in modern practice. The joints and edges of the pavement may be given a final finish, generally by the use of hand tools, with workmen operating from bridges as required.

The surface of the pavement is generally checked again after the concrete has hardened, with surface irregularities up to $\frac{1}{8}$ in. (as measured by a 10-ft straightedge) being permitted. High spots detected at this stage may be removed by grinding. Use is also being made of "profilographs" or "road roughness indicators," which measure surface variations over a length of pavement by an accumulative measurement of deflections of a test tire. Roughness is usually expressed in terms of inches per mile.

20-24 Slipform paving

Introduction of the first production models of slipform pavers in 1954 revolutionized concrete paving in the United States (7). By 1985, slipform paving was being used almost exclusively for highway concrete paving projects.

Principal advantage of the slipform paver is the fact that one machine, under the control of a single operator, replaces the several machines in a conventional paving train. Hand finishing is held to a minimum. Since there are no side forms used, the labor of setting and handling the forms is eliminated. Pavements of outstanding smoothness have been built by the slipform method.

Several manufacturers produce slipform pavers, which vary in details of design and operation. All operate on the same general principle, however, to combine several operations in the one machine. Each of the machines has an electronic guidance system that operates from a taut wire to maintain line and grade; the taut wire often is the same one used to guide the subgrader (Section 20-20) during final finishing of the subbase.

Concrete is dumped on the subbase ahead of the paver and spread to a reasonably uniform depth, then struck off by an oscillating screed. The concrete is then vibrated heavily and forced through the space formed by the traveling side forms that are part of the paver, the subbase, and a heavy transverse pan or beam (or beams) at the rear of the machine. The beam forms the surface of the slab and can be adjusted to give the desired cross section. Some machines of this type have trailing side forms of variable length behind the machine to restrain the fresh concrete while the machine moves slowly forward; others carry no trailing forms. Careful control of the consistency (usual slump is 1 to $1\frac{1}{2}$ in.) and entrained air content is essential for successful results.

Figure 20-16 shows a slipform paver in use on a highway project; conventional dump trucks supply concrete to the paver, which has no trailing forms. The edge of the slab behind the end of the trailing forms (on another make of paver) is shown in Figure 20-17; a mason is touching up the slab edge.

Finishing operations vary, but handwork is held to a minimum. Figure

FIGURE 20-16 Slipform paver at work on an Interstate project. (Courtesy Portland Cement Association.)

FIGURE 20-17 Pavement edge left by trailing form of slipform paver. (Courtesy Wire Reinforcement Institute.)

FIGURE 20-18 Alumium tubular float finishing a concrete pavement. (Courtesy Wire Reinforcement Institute.)

20-18 shows a long aluminum tube to float the surface of the slab on some projects; the keyed joint is to provide vertical load transfer for an additional lane to be placed in a subsequent operation.

20-25 Curing

Curing of a concrete paving slab is necessary in order that the concrete harden properly. It should be noted that water is absolutely necessary for the proper hydration of the cement and that the hardening of the concrete is not a drying-out process. Steps then must be taken to prevent loss of moisture from the concrete during the curing period. A large number of different curing methods are available, and the specifications of highway agencies related to this phase of concrete construction may permit several different alternative procedures to be used.

Earth or straw may be spread over the surface of the pavement and kept constantly wet during the curing period. Another "wet-curing" process involves the spreading of burlap, felt, or cotton mats over the surface; the mats are sprinkled and kept constantly wet. Waterproof paper may be placed over the slab to retain the moisture and is widely used in some areas. Burlap, cotton mats, and paper are commonly furnished in rolls that cover the entire width of the pavement.

By far the most popular method in current use involves the spray application of light-colored fluid to the entire area of the wet concrete. This is the commonly used "membrane" method. The fluid forms a film over the pavement that prevents moisture loss. The color in the fluid disappears after the passage of time. Membrane fluid is commonly applied by special spray machines that ride on the forms, or on rubber tires if a slipform paver is used, and ensure a uniform application of material over the entire area. The pavement edges as well as the riding surface must be cured.

Preliminary wet curing is required by some highway departments and the period of preliminary curing generally varies from 1 to 3 days. Total curing time, as evidenced by the period which must elapse before the pavement may be opened to traffic, commonly varies from 7 to 14 days. Opening of the pavement to traffic is very frequently based on the attainment of a certain minimum flexural strength of the hardened concrete rather than on an arbitrary time period.

20-26 Joint sawing

Most states in which a substantial mileage of concrete pavements is being built either permit or require the sawing of transverse joints; longitudinal joints are also being sawed in some cases. Sawing is done by the use of self-propelled, manually guided single-blade concrete saws, as illustrated in Figure 20-19, or multiblade saws which ride directly on the pavement or the side forms. To make joint sealing more effective, some states now require a step-down joint with a wider groove at the surface; such joints are cut by two or more saws operating in tandem. The geometry of the joint varies with the type of sealant.

Joints are sawed a short time after the concrete has been given its final finish. The time at which sawing is done is critical. On the one hand, the sawing must be done before random cracking occurs; however, premature sawing causes excessive spalling, water erosion, and excessive blade wear. The best way to determine the best time for sawing probably is to make a short trial cut a few

FIGURE 20-19 Concrete saw with a diamond blade cutting a pavement joint. (Courtesy Portland Cement Association.)

hours after final finishing operations; appearance of the cut is then evaluated. Time of sawing varies from as little as 4 to as much as 24 hours after placing.

20-27 Other considerations

Forms may generally be removed from the hardened concrete after 12 to 24 hr. Special devices are usually used to pull the forms so that the pavement will not be damaged in this process. Edges of the slab may be given their final finish at this time. At the end of the curing period and before opening the road to traffic, the surface grooves in longitudinal and transverse joints are cleaned and filled with joint sealing compound or some other material, like the pre-formed compression seal described in Section 20-10.

Some mention may be made of concreting in cold weather. Most highway organizations suspend concreting operations when the air temperature is 40° F (or less) and falling and resume operations when the temperature is 35° F and rising. In many sections of the country concrete construction is completely suspended during the winter months. When concrete must be placed in cold weather, the heating of aggregates and water to moderate temperature is quite common practice. In such circumstances the temperature of the concrete prior to placing may generally be required to be from 50° to 80° F. Calcium chloride frequently is used as an accelerator to hasten the initial set of concrete placed in cold weather.

20-28 Concrete base courses

Concrete base courses are used occasionally to provide a support for a high-type bituminous pavement (as in a vehicular tunnel). Their principal use at the present time is in urban areas. They are generally built of uniform thickness, with thickness of 6 to 9 in. being common.

Concrete bases have been built of both plain and reinforced concrete, with and without joints to control cracking of the base. Authorities differ as to the exact structural behavior of concrete bases that support a bituminous wearing surface. The combined thickness of the base and wearing surface must, of course, be adequate to support the loads placed on the pavement structure. Many designers believe that a bituminous wearing surface of nominal thickness contributes but little to the structural strength of the pavement. It is customary in some organizations, however, to use concrete in the base course that has slightly less strength than that required of concrete for paving. A requirement of a flexural strength of 450 lb/in.^{2} is perhaps typical of such practice. This decrease in strength is generally justified by the fact that temperature warping stresses should be somewhat less in concrete bases than in pavements.

Another consideration related to structural behavior of the base is the fact that extensive cracking that occurs in the base will generally be reflected in the wearing surface over a period of time. With this fact in mind, many designers feel that enough joints and reinforcing steel should be used in the base to effectively control cracking. Again because of lessened temperature warping stresses, longitudinal joints may not be necessary in a concrete base which has a width of 20 ft or so. Contraction joints of the weakened-plane type are generally provided in concrete bases to control cracking, while transverse expansion joints are less commonly used. It is recommended that transverse joints in concrete base courses be placed at short intervals (15 to 20 ft) in order to prevent excessive opening of the joints and resultant cracking of the surface course.

Methods used in the construction of concrete base courses are very similar to those used in the construction of concrete pavements. Requirements relative to the surface finish of the completed base may be somewhat less severe than similar requirements relative to pavements. The surface of the concrete is usually given a roughened, broomed finish where a bituminous wearing surface is to be placed. The base course must be cured as indicated previously for concrete pavements.

20-29 Cement-treated bases

Cement-treated bases are used widely beneath concrete pavements and to a lesser extent beneath asphaltic wearing surfaces. Principal purposes of a CTB are to provide a stable base for the construction of the concrete slab, particularly when a slipform paver is used, and to prevent pumping.

A cement-treated base course mixture is very similar to the soil–cement mixtures described in Chapter 18. In many cases, however, there are two major differences; the material with which the cement is mixed is a crushed or uncrushed granular material of moderate to low plasticity, and cement contents are relatively low (from $2\frac{1}{2}$ to 6 percent by dry weight of aggregate in California, where bases of this type have been built for many years).

Specifications for materials vary, but typically a clean graded aggregate with a maximum size of 1 in. and from 3 to 15 percent passing a No. 200 sieve would be specified for a class A cement-treated base. Cement-treated bases vary from about 4 to 7 in. in thickness. State highway agencies usually specify a minimum compressive strength at 7 days. The Portland Cement Association (8) recommends that enough cement be mixed with the granular materials to develop an unconfined compressive strength of not less than 300 lb/in.^{2} at 7 days.

Construction of a cement-treated base course is by either road mixing or plant mixing, using the general methods described in Chapter 18. The mixture is compacted at or near optimum moisture to maximum density. It is cured as is soil-cement, often by applying a bituminous material to the top surface.

PROBLEMS

20-1   Determine the proper spacing of $\frac{1}{2}$-in round tie bars along the longitudinal joint of a 10-in. thick concrete pavement. The slab width is 12 ft and the allowable working stress of the steel is 36,000 lb/in.^{2}

20-2*   Develop an equation similar to Eq. 20-1 using SI (metric) units.

20-3   Determine the required length of the tie bars described in Problem 20-1.

20-4*   Express Eq. 20-2 in SI (metric) units.

20-5   Determine the total area of reinforcing steel (wire mesh) required to control the cracking at contraction joints of a 9-in. thick concrete pavement. The joints are spaced 30-ft apart and the working stress of the steel is 30,000 $lb/in.^2$

20-6   A concrete pavement is to be built on a subgrade with a k value of 70 $lb/in.^3$ The combined k of the subgrade and a 4-in. thick untreated granular base is 90 $lb/in.^3$ The design period is 20 years. Over this period, the expected loadings are tabulated below:

Single Axle Load, kips	Expected Repetitions
30	20,000
26	75,000
22	300,000
18	700,000
14	900,000

Tandem Axle Load, kips	Expected Repetitions
48	50,000
40	480,000
32	1,700,000
24	2,000,000

(a)   Determine the minimum thickness of slab that can be used safely for this rural Interstate route if the modulus of rupture of the concrete is 650 $lb/in.^2$ (28 days). Assume doweled joints and no concrete shoulder.

(b)   Can a thinner slab be used for this pavement if the modulus of rupture of the concrete is increased to 700 $lb/in.^2$ at 28 days?

20-7   Solve Problem 20-6 if a cement-treated base with a design k of 300 $lb/in.^3$ is substituted for the untreated granular base.

REFERENCES

1. *Standard Specifications for Transportation Materials and Methods of Testing, Part I, Specifications.* American Association of State Highway and Transportation Officials, Washington, D.C. (1982).

2. *Standard Specifications for Transportation Materials and Methods of Testing Part II, Tests.* American Association of State Highway and Transportation Officials, Washington, D.C. (1982).

3. *ACI Manual of Concrete Practice 1984, Part I, Materials and General Properties of Concrete.* American Concrete Institute, Detroit, Mich. (1984).

4. *Joint Design for Concrete Highway and Street Pavements.* Portland Cement Association, Skokie, Ill. (1980).

5. *Thickness Design for Concrete Highway and Street Pavements.* Portland Cement Association, Skokie, Ill. (1984).

6. *Composite Pavement Design.* Highway Research Board Record No. 37, Transportation Research Board, Washington, D.C. (1963).

7. *Symposium on Slip Form Paving.* Highway Research Record No. 98, Transportation Research Board, Washington, D.C. (1965).

8. *Subgrades and Subbases for Concrete Pavements.* Portland Cement Association, Skokie, Ill. (1971).

chapter twenty-one

HIGHWAY MAINTENANCE AND REHABILITATION

When a highway construction project has been completed, accepted from the contractor, and final payment has been made to the contractor, a new facility is available for use by the traveling public. By the same token, a new responsibility is created to preserve the new investment and to serve and protect the interests of the traveling public. This is true not only for newly constructed highways but also for all highways of our road system. Sudden failures, damage by storms, gradual deterioration, and unexpected obstructions can cause personal injury, death, or delay.

"Highway maintenance" is defined as the function of preserving, repairing, and restoring a highway and keeping it in condition for safe, convenient, and economical use. "Maintenance" includes both physical maintenance activities, such as patching, filling joints, mowing, and so forth and traffic service activities including painting pavement markings, erecting snow fences, and removing snow, ice, and litter. It does not include major rehabilitation or reconstruction activities, such as widening the roadbed, or extensive resurfacing projects.

Highway maintenance programs are designed to offset the effects of weather, vandalism, organic growth, and traffic wear and damage, as well as deterioration due to the effects of aging, material failures, and design and construction faults (1). As Figure 21-1 illustrates, in spite of vigilant and determined efforts to maintain the serviceability of a highway, there comes a time when a major rehabilitation of the facility is required. The rehabilitation work, termed "betterment," is not considered as maintenance, but the periodic routine activities before and after rehabilitation are. The *AASHTO Maintenance Manual* (1) includes a table that clearly distinguishes between the four general classes of roadway maintenance and construction activities:

1. Maintenance
   a. Traffic services
   b. Physical maintenance
2. Construction
   a. Betterment
   b. Construction and reconstruction

For example, resurfacing a pavement with bituminous material $\frac{3}{4}$ in. thick or more for a length of 500 continuous feet or more is considered to be "betterment." Resurfacing a pavement with material less than $\frac{3}{4}$ in. thick or replace-

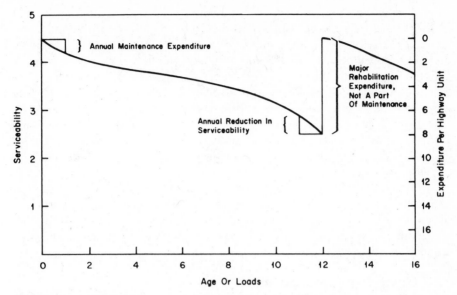

FIGURE 21-1 Typical highway performance curve. (Courtesy Transportation Research Board.)

ment of pavement for a length less than 500 continuous feet is defined as "physical maintenance."

21-1 Importance and challenge of the highway maintenance function

The importance of highway maintenance lies mainly in the need to protect the substantial investment that has been made in the highway plant. Furthermore, the significance of the maintenance function is seen in the vital need to continue the economic and safety benefits of the 3.8 million mile public road system. The traveling public is acutely aware of the condition of public streets and highways and is inclined to be quite vocal in its displeasure about potholes, ragged shoulders, illegible signs, and highway drainage structures that do not function properly.

In recent years, the highway maintenance function has increased in both complexity and scope. These changes have been brought about by the addition of significant mileage to the system, notably the interstate system, by increases in vehicle mileage of travel, and , in many states, increases in the maximum size of allowable wheel loads. In addition, the traveling public now expects a higher level of maintenance than it tolerated previously. Systematic mowing, large-scale landscaping, and litter control are examples of the escalation of maintenance standards that has occurred, largely during the past decade. Furthermore, the maintenance must frequently be performed under conditions of heavy traffic, entailing much greater difficulties as well as potential hazards to both workers and the traveling public.

Maintenance personnel face an enormous challenge in adequately maintaining the vast system of public roadways. The magnitude of the problem is seen in the size of the maintenance budgets for the various state highway and transportation agencies. In 1983, more than $5.5 billion was spent by states for maintenance and traffic services (2), representing about 17.3 percent of the total highway dollars disbursed. It is estimated that an additional $7.7 billion was spent for maintenance by counties, townships, and municipalities.

During the past decade, annual maintenance costs have increased at an average rate of about 7.9 percent, aggravating the maintenance engineer's difficult task of coping with rising traffic volumes and public demands for increased services.

Figure 21-2 shows the approximate services and cost distribution for highway maintenance in a typical year, on a national basis. There is, of course, considerable variation among the various states in how the maintenance dollar is spent, reflecting differences in climate, topography, and maintenance practices.

21-2 Maintenance management

To increase the productivity of labor and equipment used in maintenance operations, highway agencies are increasingly applying modern management methods to the maintenance functions. By this technique, maintenance budgets are based on the estimated costs of accomplishing specific work programs to stated levels of service and performance standards. Maintenance management includes control procedures that are applied to ensure that the work accomplished by field personnel is in line with the objectives of the program and the budgeted funds available.

The basic purpose of maintenance management systems is to capture information about maintenance activities performed and resources expended.

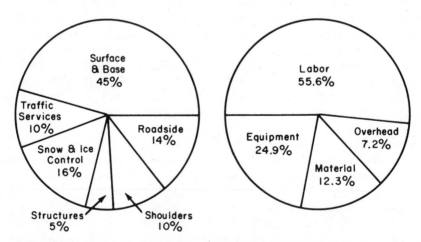

FIGURE 21-2 The maintenance dollar, showing approximate services and cost distribution for a typical year. (Courtesy Highway Users Federation for Safety and Mobility.)

Maintenance management systems do not manage programs, reduce costs, or improve performance; rather, they provide maintenance engineers and managers with the information and analytical tools needed to allow them to do so (3).

Although the details of maintenance management systems vary a great deal among the various highway agencies, there is considerable agreement as to the basic concepts and objectives of such programs. In the following paragraphs, some of the essential elements of a maintenance management program are described. The subject is covered in more detail in Refs. 3, 4, and 5.

Typical features of a maintenance management system include

1. Development of an annual work program.
2. Budgeting and allocating resources.
3. Work authorization and control.
4. Scheduling.
5. Performance evaluation.
6. Fiscal control.

Work program development

This phase of the program normally involves the following types of activities or steps:

1. Defining work activities.
2. Establishing work quantity planning values.
3. Establishing maintenance performance standards.
4. Conducting a road inventory and inspection.
5. Estimating the size of the work program.

It is first necessary to define the various work activities in terms that are specific and measurable. Studies have shown that in a typical maintenance operation only 20 to 30 activities will account for 75 to 80 percent of the workload. Thus, the number of activities defined is relatively small and therefore manageable. Typical work activity definitions are shown in Figure 21-3.

For each work activity, work quantity planning values must be established. These are the number of units of work required per year to provide a desired level of service. Such quantities are usually based on appearance, safety, and protection of the highway investment. Examples of work quantity planning values include 0.25 tons of asphaltic mix per lane mile per year, three mowings per acre per year, and so forth.

It is then necessary to establish, for each activity, performance standards indicating productivity of a typical crew. Since productivity depends on the size and makeup of the crew and the work method and equipment employed, such details constitute an essential part of the performance standard. The performance standard will normally show the productivity in terms of work units such as those indicated in Figure 21-3, for example, tons of premix placed per day, miles of shoulders bladed per day, and so forth. (See Figure 21-4.)

Code	DESCRIPTION	Work Unit
	ROADWAY SURFACES	
101	PATCH WITH PREMIX. Patching of road and shoulder surfaces using commercial or shop prepared mixes. Associated deficiencies include: • bituminous surfaces: potholes, upheavals; slippage cracks; raveling. • PCC surfaces: blowups; spalling.	Ton of Premix
102	LEVEL WITH PREMIX. The application of a leveling course to distorted bituminous road and shoulder surfaces. Associated deficiencies include: • channels or ruts • grade depressions • utility cuts	Ton of Premix
	SHOULDERS AND SIDE APPROACHES	
201	BLADE OR RESHAPE. Blade or reshape unpaved shoulders without adding material. Associated deficiencies include: • earth and sod shoulders: growth buildup; loss of shoulder slope; shoulder drop-off; ruts or soft spots. • gravel shoulders: vegetation growth; incorrect cross slope; shoulder drop-off, edge ruts; potholes or soft areas; windrows; excess material on pavement edge.	Shoulder mile bladed
	DRAINAGE	
301	CLEAN AND RESHAPE DITCHES. Emergency and preventive maintenance of slope and side ditches. Associated deficiencies include: • emergency maintenance: cracks; • breaks; erosion and deterioration from storms. • preventive maintenance; erosion; non-uniformity in grade line or cross section; rubbish; slough or debris accumulation; cracks, breaks and lining deterioration.	100 lin. ft. lgths. of ditch cleaned or reshaped

Code	DESCRIPTION	Work Unit
	TRAFFIC SERVICES	
701	SIGN MAINTENANCE. Emergency and preventive sign maintenance of any kind. Associated deficiencies include: • emergency maintenance: any damage that negates the function of regulatory or warning signs. • preventive maintenance: minor damages to signs or posts; paint deterioration; burned out or broken lamps on illuminated signs; etc.	Each sign
702	SIGNAL MAINTENANCE. Emergency and preventive signal maintenance of any kind. • emergency maintenance: signal failure from any cause. • preventive maintenance: electrical maintenance; shop repairs and overhaul; replace and clean lamps and lenses; repairs to poles; repairs to pull box covers; etc.	Each signal
	SERVICES AND OVERHEADS	
901	EQUIPMENT SERVICING/REPAIR. Work performed by field maintenance crews on equipment: • haul/tow equipment • pick up and deliver equipment or parts • install/change attachments • lubricate, wash or servicing of any type.	None
951	FIELD SUPERVISION. The supervision time of Superintendents and Foremen not physically engaged in carrying out work activity.	None
952	LEAVES. Leave time for all personnel • annual vacations • holidays • sick leave • military leave • compensatory leave • etc.	None

FIGURE 21-3 Examples of typical maintenance work activity definitions and work measurement units. (Courtesy National Association of County Engineers.)

A road inventory and inspection may be required to identify and describe field conditions in need of maintenance. Many agencies maintain roadway inventory records, and it may only be necessary to supplement existing data. The inventory data provide up-to-date information on the types, amounts, and locations of roadway maintenance work.

Budgeting and allocating resources

The next step involves estimating the amount of labor, equipment, and materials required to carry out the program and computing the annual costs of the work. In most highway agencies, average costs of labor, equipment, and materials can normally be provided by the accounting division. It is then a simple matter to convert resources requirements of each work activity to the overall estimated activity costs. For example, the procedure for computing the cost of premix patching might be

PERFORMANCE STANDARD	ACTIVITY: __PATCH WITH PREMIX__ ACTIVITY CODE: __101__ EFFECTIVE DATE: _____

DESCRIPTION AND PURPOSE:

Patching of road and shoulder surfaces using commercial or shop prepared mixes. Associated deficiencies include:

- bituminous surfaces: potholes, upheavals, slippage cracks, raveling.
- PCC surfaces: blowups, spalling.

PERFORMANCE CRITERIA:

Work program category is: Routine–Unlimited
Authority for performing work activity is: Foreman
Quantity control category is: Unlimited
Major attention should be placed during: Spring and Fall

CREW, EQUIPMENT, MATERIAL	WORK METHOD
Crew Size (incl. flagmen): 1 driver/workman 1 workman 1* flagman * Adjust to traffic conditions.	1. Place safety devices and signs. 2. Square up and clean potholes, depressions. 3. Tack sides and bottom of hole. 4. Shovel mix into hole and level. 5. Compact with hand tamper or truck wheels. 6. Place top layer. 7. Screed and compact again.

Equipment:

Description	No.	Class
3T truck	1	678

Material:

01 — hot or cold premix
misc.   tack

PRODUCTIVITY	WORK UNIT
Daily Production: 3.0 to 4.5 tons	ton of premix

FIGURE 21-4 Sample performance standard. (Courtesy National Association of County Engineers.)

Type	Resource Requirements		Average Unit Cost	Annual Cost
Labor	960 person-hr	×	$3.00/hr	$2880.
Equipment	320 dump-truck-hr	×	$1.50/hr	480.
Materials	150 tons	×	$6.00/ton	900.
Total (activity code 101)				$4260.

Similar calculations are performed for each work activity and the results are summarized for each class of work activities, for example, roadway surface, shoulders and approaches, and so forth.

Typically, cost data of this type are used by the highway agency's accounting staff to prepare a formal budget in accordance with established policy. After appropriate review and modification, the budget is approved. The approved budget then becomes the basis for legislative appropriation and authorization.

Work authorization and control

"Historically, there has been reasonable control of budgeted expenditures, but little control of operations. There is abundant and consistent evidence of great and unwarranted variations in the quantities of work performed for comparable situations. Also, the ways in which work is performed and the productivity of work forces have varied tremendously" (4). Operational control of the work is required if performance budgeting is to be meaningful.

In a state agency, the authorization to proceed with the maintenance work must be passed along the various levels of administration; from top management, to the district engineer, to the subdistricts, and areas where the work is to be performed.

The recommended system of authorizing work employs a work-order approach in which crew schedule cards are distributed through upper management personnel to first line supervisors (foremen) of work crews. An example crew schedule card is shown in Figure 21-5. Each card should represent a single day's work for a standard size crew performing the specified work. Crew schedule cards also provide a convenient means for supervisors to report on the work performed. Typically, the area foreman records the actual times required, the equipment used, the travel time, materials used, and so forth. Information from the crew schedule cards are periodically summarized and evaluated by the maintenance engineer and other maintenance managers. By such means, it is possible to determine the progress of work in each area and how well foremen are adhering to recommended crew sizes and levels of performance.

Scheduling

In order to provide a reasonable balance of workload throughout the year, a seasonal schedule of maintenance operations should be developed to facilitate the schedule of appropriate work activities during specific periods of the year. (See Figure 21-6.) Normally, each maintenance foreman schedules a designated number of crew-days for each work activity under his supervision. He will also usually be required to make appropriate arrangements to ensure availability of needed materials and equipment.

Typically, each road maintenance division has equipment scheduling boards that indicate reservations for and locations of available equipment at all times. Some agencies assign a priority rating to each project to ensure proper scheduling of the maintenance equipment.

Performance evaluation

The work performance of maintenance crews, as reported on the crew schedule cards, should be periodically summarized and reviewed by manage-

CREW SCHEDULE CARD	CREW SIZE 3	ACTIVITY 101

RD. CLASS *RRI* AREA *01*   CARD NO. *26-1*

ACTIVITY *Patch with Premix*
LOCATION *Route # 68 & 57*
FOREMAN *Smith*   DATE *1-16-*

LABOR		EQUIPMENT	
HRS	EMPLOYEE	NUMBER	HRS.
8	*Al Ray*	*678-0609*	*7*
8	*Joe Dobb*		
8	*Pete Jones*		
24	TOTAL		

MATERIAL		
DESCRIPTION	UNIT	AMOUNT
Bituminous Material	*Tons*	*3*

ACCOMPLISHMENT *3*   UNIT *Tons of Bit*

(front)

PATCH WITH PREMIX

DESCRIPTION AND PURPOSE:

Patching on road and shoulder surfaces using commercial or shop prepared mixes.
Associated deficiencies include:

- bituminous surfaces: potholes, upheavals, slippage cracks, edge raveling.

- PCC surfaces: blowups, spalling.

WORK METHOD:

1. Place safety devices and signs.

2. Square up and clean potholes, depressions.

3. Tack sides and bottom of hole.

4. Shovel mix into hole and level.

5. Compact with hand tamper or truck wheels.

6. Place top layer.

7. Screed and compact again.

(Back)

FIGURE 21-5 Sample crew day card. (Courtesy National Association of County Engineers.)

ment personnel. Such summaries typically include, for each maintenance activity: (1) the number of crew schedule cards used to date, (2) the number of person-days or person-hours reported, and (3) the work units accomplished. This permits the maintenance engineer and other managers to know the progress of work, the productivity of the various work crews, and how well the foremen are adhering to established practices.

Fiscal control

Most highway agencies also have fiscal control procedures to ensure that budgetary limitations are observed. Control is exercised by the issuance of timely periodic status reports to persons with the authority for expending money or obligating expenditures. Maintenance managers should carefully monitor the status of expenditures and costs and their relationship to the performance budget. Careful fiscal review of maintenance operations will make it possible to avoid overexpenditures of budgetary allotments and to recommend adjustments in such allotments when unexpected contingencies arise.

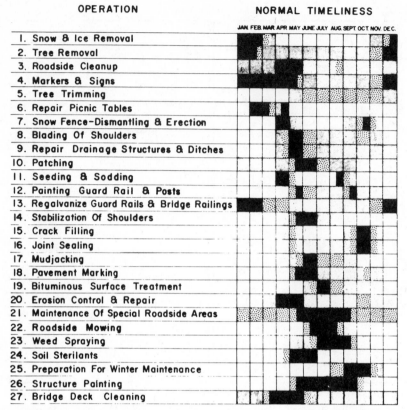

FIGURE 21-6 Seasonal schedule of maintenance operations; depth of tone indicates timeliness of operation. (Courtesy Michigan Department of State Highways and Transportation.)

MAINTENANCE OPERATIONS

In the following paragraphs, we will briefly describe typical maintenance operations in highway agencies in the United States. It is convenient to categorize our discussion of these practices in terms of the following types or classes of highway maintenance, the maintenance of (1) roadway surfaces, (2) shoulders and approaches, (3) roadsides, (4) bridges, tunnels, and drainage structures, (5) traffic control and safety devices, and (6) the control of snow and ice. For further guidance on these matters, the reader should consult Ref. 1.

MAINTENANCE OF ROAD SURFACES

21-3 Road surfaces

This classification of roads is taken to include those constructed from natural earth, sand–clay mixes, and various coarse graded aggregate materials such as gravel, crushed stone, crushed slag, chert, and caliche. Water-bound and traffic-bound macadam surfaces are included in this classification.

To maintain this type of road properly, it is necessary that the surface be kept smooth, firm, and free from excess loose material, with the proper crown for adequate drainage. A crown of $\frac{1}{2}$ in. per foot sloping outward from the centerline is recommended (*1*). The use of a dust palliative is frequently necessary in order to minimize the loss of material and to eliminate as much as possible the hazards due to dust. Maintenance of this type of road can include (1) patching, (2) blading, (3) scarifying and resurfacing, (4) stabilization and dust control, and sometimes (5) application of a bituminous surface treatment.

Patching

Failures of soil–aggregate roads are principally due to improper drainage, poorly mixed materials, or an inadequate foundation. A careful investigation should be made to determine which of these may be the cause of failure. Surface drainage can be ensured through the use of the proper amount of crown. Excessive moisture in the subgrade may be eliminated by lateral or side drains to intercept free water or to lower the water table. Perforated pipe underdrains may also be installed.

When road failure is due to a poor mixture or gradation of materials, it is usually necessary to remove the unsuitable material and replace it with suitable materials of the proper gradation. The edges of the area removed from the traveled roadway should be squared and the patching material placed and thoroughly tamped, adding water when necessary to obtain the proper compaction. When the area to be repaired is large, or where the surface is worn away, it may be better to scarify, reshape the surface, and add the necessary additional material.

Blading

Blading is a general operation carried out on all soil and stabilized surfaces. Its purpose is to fill in the ruts and smooth out any irregularities in the surface.

Blading is done by a motor grader. The versatile motor grader also can do many other maintenance operations such as shaping of shoulders, cleaning and shaping of ditches, and similar work. Blading also may be done by trucks with underbody blades. This equipment is usually attached to four-wheel-drive trucks and hydraulically operated.

Blading should be done as soon as practicable after a rain, when the surface materials are moist. The work is accomplished by dragging or blading the surface materials from the edges toward the center and then drawing the loosened material back again, depositing it in depressions in the surface. Some blading may be required during dry weather in order to remove surplus loose materials from the traveled roadway. This material is bladed to the side of the road and bladed back during wet weather or when the material is moist.

Scarifying and resurfacing

When the areas to be patched are numerous, or when they extend over a considerable area, it is often more efficient to recondition the entire surface. The usual procedure is to scarify the surface of the roadway to the full depth of the surface material, proper care being exercised so that any undesirable material from the shoulder or the subgrade is not mixed with the surface-course materials. When additional material must be added, it should be selected wth care. A visual inspection and a knowledge of local materials may aid in the selection of the necessary materials. However, if the section being repaired is of considerable extent, laboratory and field testing of the separate materials and their combinations may be necessary.

When the scarifying of the old road surface has been completed and any necessary material added and mixed, a blade grader is used to shape the road surface to proper crown and thickness. The necessary compaction may be accomplished by permitting traffic to use the surface while continuing blading or dragging operations, or by rolling and watering if necessary until the surface is thoroughly compacted.

A familiar failure on gravel roads is the formation of corrugations running transverse to the traveled way. This "washboard" effect can be temporarily corrected by blading and dragging. Usually the construction of a more suitable wearing surface is the only permanent solution to the problem. The addition of a fine binder where the surface mixture is too coarse will tend to lessen the formation of corrugations. Likewise, where an excess of fine material exists, the addition of angular coarse material and a reduction of clay binder tend to alleviate this condition.

Stabilization and dust treatment with palliatives

Stabilization of roads with the use of binder soil such as clay, the chlorides, or other additives is often done by maintenance forces. The methods of stabilization are similar to those discussed in Chapter 18. The use of dust palliatives serves two purposes: (1) elimination of dust and (2) preservation of the soil−aggregate surface. Tests have indicated that the annual loss of material on gravel roads is equivalent to $\frac{1}{2}$ to 1 in. of surface. In the case of an 18-ft surface, this would be 150 to 300 cu yds of material per mile of road. The use of palliatives reduces this loss and at the same time aids in the stabilization of the surface.

The materials most commonly used as dust layers are calcium chloride, sodium chloride, and bituminous substances. Calcium chloride is most generally used because of its "hygroscopic" nature, in that it attracts moisture, and because it is "deliquescent," that is, it dissolves itself in the moisture which it attracts. When calcium chloride is scattered over the road, it absorbs and holds two to five times its weight in water and then serves to keep the surface damp and dustless for prolonged periods.

In dry weather, calcium chloride will absorb moisture from the air during periods of high humidity, such as at night, in sufficient amounts to maintain a dustless surface during the day. Calcium chloride is susceptible to dilution by rain and gradually leaches into the soil, where, if present in sufficient concentration, it tends to reduce the severity of frost heaves.

Calcium chloride is available in flake and granular form, and best results are obtained when mechanical spreaders are used. Either a lime-drill type or a centrifugal spreader may be used. Overlapping of the spreads in the middle of the road will safeguard against leaving an untreated portion in the middle of the roadway.

The best time to apply the material is directly after a rain. The first application is usually made in the spring when dust begins to appear, followed by periodic applications during the summer months when needed. The initial application is made at a rate of $\frac{3}{4}$ to $1\frac{1}{2}$ lb per square yard. Later application requirements will vary. Some roads require as little as $1\frac{1}{2}$ lb per square yard through the season, and others require a total of 2 to $2\frac{1}{2}$ lb per square yard, depending on the texture of the surface, previous amount of stabilization, dryness of the season, and amount of traffic. Shaded areas require smaller amounts of palliatives than do areas exposed to the sun. The same is true for low-lying sections of road. Frequent applications may be desirable on curves, grades, intersections and other places where excessive wear occurs.

Use is also made of various bituminous materials for laying dust on soil–aggregate roads. The surface to be treated must be bladed to proper crown before the material is added, and any loose material must be removed. Materials used may be tars RT-1 and RT-2; the slow-curing road oils SC-70 and SC-250; or the medium-curing cut-back MC-70. Asphalt emulsions are also used.

The roadway should be fairly dry and free from dust when the bituminous materials are applied, except when an emulsion is used, when the presence of moisture is desirable. Emulsions require a certain period for breaking and curing. If traffic must follow soon after the use of the emulsion, the surface is covered with a light coating of clean coarse sand.

The amount of bituminous material for dust-laying ranges from 0.10 to 0.25 gallon per square yard. Emulsions often are diluted, one part emulsion to three parts of water. Application should be made with a pressure distributor.

Bituminous surface treatment

Bituminous surface treatment operations, which are sometimes done by maintenance forces, have been described in Chapter 18. As traffic increases, the road surface may be improved in various ways. Care should be exercised to segregate maintenance costs from those of betterments as much as possible.

21-4 Maintenance of bituminous surfaces

The various types of bituminous surfaces and their construction have been described in Chapters 18 and 19. Weaknesses develop in these surfaces due to weathering, the action of alternate freezing and thawing, or failure of the base or subgrade. The maintenance of these surfaces may be divided into five distinct operations; (1) patching, (2) paint patching, (3) scarifying, (4) resealing, and (5) nonskid surface treatment. These operations are described in Ref. 6 from which much of the following discussion has been abstracted. This also applies to the maintenance of concrete surfaces, which is described later. Reference 7 is another good source of information on the maintenance of asphalt pavements.

Before maintenance repairs are made to the surface, it is important to make the usual investigation for the cause of failure. If failure indicates a poor subgrade or faulty drainage, these items should be corrected before any repairs are made on the surface. Some states are now using road roughness measurements and the pavement serviceability concept as an aid in setting up maintenance programs.

Patching may consist simply of an application of a thin coat of bituminous material to an area where cracking or raveling appears and spreading a coarse mineral aggregate over the affected area. Care must be exercised in the application of the proper amount of bituminous material in order to prevent bleeding, with resulting slipperiness of the repaired surface.

In areas where the surface shows raveling and disintegration, a bituminous mixture may be placed over the weakened area and then compacted. Both hot and cold bituminous mixtures are used, with asphaltic concrete and sheet asphalt being the most common. Where the surface has disintegrated to a depth of 2 or 3 in., the failed area should be cut out, all loose material removed, the base course made true and firm, and the bituminous patching material placed in the hole. The edge of the hole should be cut vertically and should receive a coat of bituminous material to seal out any moisture that may enter the subgrade. The final patch should conform to the grade and contour of the existing surface. The steps involved in making a permanent repair of an asphalt pavement are illustrated by Figure 21-7.

Patches may also be made by a penetration method that consists of filling the hole with aggregate, compacting the aggregate, and then applying the bituminous material. Crushed stone is then applied and the patch is rolled. Many different patching methods are used with good results. The method used will depend on the type of bituminous surface and current practice prevailing in the particular locality.

Paint patching is essentially a preventive maintenance operation. Its purpose is to seal the surface to prevent penetration of water and to strengthen the wearing surface. The work consists of applying a thin coating of bitumen over the surface and placing fine stone chips, pea stone, or coarse sand over the patched area. The patch is then broomed and rolled with a light roller.

On low-type bituminous surfaces, such as bituminous treated gravel, where large areas are in poor condition, the entire surface should be scarified and reworked. This can be accomplished by the use of a motor grader with a scarifier attachment or with a scarifier and a mechanical pulverizer or mixer. A combination of a ripper (not shown) and the cutter−crusher−compactor shown in Figure 21-8 also may be used to salvage old bituminous roads and streets. The material is reworked, using additional bituminous material as required. It is then bladed to proper crown and grade and compacted. The compacted surface is then sealed with a bituminous material.

The seal coats and surface treatments described in Chapter 18 are widely used to rejuvenate old bituminous surfaces or to correct a slippery condition. Slipperiness is usually caused by an excess of bitumen on the road surface. This may be due to the application of too much bituminous material or an improper gradation of aggregate, or it may be caused by the kneading action of heavy traffic pushing the asphalt binder to the surface. Skidproofing may be achieved by application of a surface treatment using sharp, angular aggregate such as blast-furnace slag or by disking the surface. Slurry seal coats (Chapter 18) are widely used for this purpose. In severe cases the complete surface may have to be reworked in order to correct slipperiness.

Based on several experimental projects by the Arizona Department of Transportation, a novel approach to the maintenance of bituminous surfaces has evolved that uses an asphalt rubber paving composition. The composition contains a high percentage of ground rubber from discarded automobile tires. Research has shown that application of the asphalt rubber over badly cracked pavement surfaces can delay the need for pavement resurfacing.

Resurfacing of old bituminous pavements

When maintenance operations become excessively expensive, it is difficult to maintain an adequately smooth surface, the pavement has to be widened, or for other reasons, the old pavement may be resurfaced. Before undertaking a resurfacing project, a careful study of its economic justification should be made. A general rule of thumb is that resurfacing will probably be economical when from 2 to 3 percent of the total pavement area requires patching each year. Both the availability of money and the priority of the project will influence the decision. A comprehensive study of the condition of the existing surface and related factors is desirable before undertaking a resurfacing project. Included among the factors that must be given detailed consideration are the following:

1. History of pavement.
2. Existing physical condition.
3. Character of existing and anticipated traffic.
4. Relation of geometric design to safe and efficient operation of anticipated traffic.

Corrective measures that may be required in the preparation of an old surface for resurfacing may be very considerable in extent or somewhat minor, depending on the type and condition of the existing surface and its adequacy to serve its new function as a base course. For example, major reconstruction measures may be required when substantial changes in grade, profile, or crown are called for in the new surface. Recommended reductions in grade and crown may be accomplished by scarifying and recompacting old surfaces such as water-bound macadam or stabilized soils; higher types of surfaces may require removal and reconstruction to the desired grade and section. Changes in width are generrly accomplished by duplicating the original construction as closely as possible, with particular attention being paid to the development of adequate supporting power in the widened section so that this section will not subside under service to a lower elevation than exists in the original pavement.

The old surface, regardless of its type, may be rough, with waves and

(a) Removing surface and base

(b) Applying tack coat to vertical surfaces

(c) Backfilling hole with plant—mix

FIGURE 21-7 Patching an asphalt pavement. (Courtesy The Asphalt Institute.)

(d) Spreading the mix

(e) Compacting the mix

(f) Straightedging the patch

FIGURE 21-7(*continued*)

FIGURE 21-8 Compaction–cutter–crusher attachment processes an old bituminous surface. (Courtesy American Tractor Equipment Company.)

bumps that are too considerable in magnitude to be compensated for by a realtively thin wearing surface. These bumps and waves may be removed by scarifying and recompacting certain surface types. When old bituminous wearing surfaces have these defects and are to be resurfaced, the bumps and waves may be removed by cutting or burning away the old surface with a heater planer. Burners in a big, insulated furnace heat and soften the surface, which is then scarified and remixed if desired, with additional bituminous material added as needed. Sometimes it may be more desirable to correct existing surface irregularities by the construction of a special leveling course, which may be the same bituminous mixture used in the original construction or a new mix. The construction of a leveling course is a major operation that usually is quite similar to the placing of a bituminous wearing surface. The course will have a variable thickness and will be constructed to the desired section and grade. The leveling course will have a lower bitumen content than the wearing course.

Holes and depressions that occur in the existing surface must also be corrected before resurfacing (or the placing of a leveling course) may begin. The holes are filled with the same mixture used in the original construction or with a special patching mixture.

In many cases the preparation of the surface may consist simply of the removal of loose material on the surface by the use of rotary brooms or power blowers. If excessively dusty or unconsolidated areas remain after this treatment, they may have to be primed before construction may proceed.

When preparations are complete, the new surface is constructed by methods that are very similar to those described previously for new surfaces. Multiple-lift surface treatments and hot asphaltic paving mixtures, particularly asphaltic concrete and sheet asphalt, are used most frequently. Road-mix surfaces of various types are used to some extent.

21-5 Maintenance of portland cement concrete surfaces

The maintenance of concrete pavements consists for the most part of (1) filling and sealing joints and cracks in the pavement surface; (2) repairing spalled, scaled, and map-cracked areas; (3) patching areas where failure has occurred; (4) repairing areas damaged by settlement or pumping; and (5) treating buckled pavements.

The purpose of filling and sealing joints and cracks is to prevent the seepage of moisture to the subgrade and to preserve the original joint space. At the time the joint is sealed, the concrete must be dry and the joint space thoroughly cleaned of all scale, dirt, dust, and other foreign matter, including old joint sealer. Power cutters are used to cut and groove the joints prior to resealing; a mechanically driven wire brush is used to thoroughly clean the joint. Immediately before sealing, the joint is blown out by the use of a jet of compressed air. Hot-applied rubber–asphalt sealing compounds are often used; they are generally heated and melted in a portable (rubber-tired) melting kettle of the double-boiler, indirect heating type which utilizes oil or other suitable material as a heat transfer medium. The recommended type of equipment for applying the sealer is a mechanical pressure-type applicator. Mechanical applicators of the gravity type are also used, as are hand pouring pots. Cold-applied, two-component elastomeric polymers also are used as joint sealers. This maintenance operation is carried on throughout the year. The most favorable time, however, is during the warm summer months, when working temperatures are higher and better results can be obtained.

The repair of spalled, scaled, and map-cracked surfaces is a common maintenance problem for this type of pavement. Map-cracking is distinguished by irregular cracking over the pavement surface. Spalling is a chipping or splintering of sound pavement and usually occurs along the joints or cracks in the pavement. Scaling is caused by the deterioration or disintegration of the concrete and may occur anywhere on the pavement surface.

Minor cracks in a portland cement concrete surface are usually repaired by filling with a joint sealing compound. AASHTO (1) recommends that cracks more than $\frac{1}{8}$ in. wide be filled with a material that will allow for expansion and contraction of the concrete. A concrete or epoxy repair should be made when cracked, chipped, or splintered areas are too large to control by routine joint sealing methods.

When only fine cracks are present linseed oil has been used with apparent success as a preventive maintenance measure for scaling of portland cement concrete. When a considerable area . . . has been affected by scaling or excessive wear, a surface treatment of hot plant mix or slurry seal is recommended.(*1*)

Failed areas in concrete pavement should be repaired with cement concrete. The failed areas are marked out by a person qualified for this type of work. A crew then follows with jackhammers or other mechanical equipment, breaks and removes the broken concrete, and prepares the area for the new surface. At the edges of the patch the old slab of concrete should be undercut and the patch placed as shown in Figure 21-9. High early strength cement should be used in the new concrete. The thickness of the patch should never be less than that of the existing slab, and the use of reinforcing steel is very often desirable. When patches are made adjacent to expansion joints, the expansion joint should be replaced. Contraction or construction joints may be omitted, however, if the patch extends across the entire width of pavement.

Surfacing irregularities can be removed from concrete pavements by grinding or cutting the surface with a special machine known as a "bump-cutter." Such machines have a cutting head with diamond blades that produce a series of shallow, parallel grooves in the surface, giving it a nonskid texture.

Pumping of concrete pavements

One of the major problems in the maintenance of concrete surfaces is correcting settlement of the slab caused by pumping. Pumping is usually indicated by (1) spalling of the pavement near the centerline and a transverse joint or crack, (2) ejection of water from joints and cracks, (3) discoloration of

FIGURE 21-9 Concrete patching. (Courtesy Federal Highway Administration.)

pavement surface by subgrade soil, (4) the presence of mud boils at the pavement edge, and (5) breakage of the pavement.

Pumping of concrete pavements may be prevented by maintaining adequate drainage, correcting faulty drainage, and sealing joints and cracks. Where pumping has progressed to an appreciable degree, this condition is corrected by mud-jacking or subsealing. This procedure consists of drilling holes into the slab and forcing in a suitable slurry to fill the voids between the subgrade and the slab.

The usual method involves the drilling of one hole in the lowest portion of the settled pavement. This hole is about 1.5 in. in diameter and usually placed about 12 in. from the point. If needed, an additional hole is drilled about 30 in. from the edge. Water and sand are blown out of the area through the drilled hole by use of compressed air. This is necessary to avoid dilution of the slurry to be forced under the pavement. As the slurry is forced into the drilled hole, pumping is continued until all voids are filled and the settled slab is raised to its proper position. The holes are then plugged.

The slurry mixtures used vary a great deal, but for the most part they consist of a combination of loam, cement, and water or of asphaltic materials. Two widely used mixtures are as follows (*1*):

1. Soil, 60 to 84 percent; cement, 16 to 40 percent.
2. Soil, 77 percent; cement, 16 percent; and cut-back asphalt (SC-250), (MC-70), (RC-250), 7 percent.

The mix used should be fluid enough to spread rapidly and flow into small cavities.

The buckling or blow-up of old concrete pavements is a continuing problem; it is usually caused by longitudinal expansion and the failure of transverse expansion joints to function properly. When this occurs, it is often necessary to remove the damaged portion of the pavement and replace it with a patch of concrete or bituminous material. Buckling or blow-ups may be prevented, when evidence of extreme compression is noted in the pavement, by cutting a wider expansion joint if one already exists or providing a new joint across the pavement.

Resurfacing of old concrete pavements

When justified, old concrete pavements may be completely resurfaced. Three materials are in common use: asphaltic concrete, asphaltic concrete reinforced with welded wire fabric, and portland cement concrete.

In preparation for resurfacing with asphaltic concrete, joints and cracks may be cleaned and filled with asphaltic concrete. Bituminous patches that show an excess of bituminous material must be removed. Major defects must be corrected, and the cause of the failure must be corrected before the new surface is placed. A common problem with this type of resurfacing is that, over a period of time, joints and cracks in the old concrete pavement show up in the new wearing surface; this phenomenon is called "reflection cracking." In attempts to prevent reflection cracking, several different approaches have been used;

these include covering cracks and joints with metal plates, paper, or wire mesh reinforcing; construction of a layer of gravel or crushed stone several inches thick on the old pavement; use of an open-graded bituminous leveling course of considerable thickness; use of rubber in the asphaltic mixture to provide additional flexibility in the pavement structure; and seating of the old slab with heavy rollers and impact breakers. None of these methods has been entirely satisfactory. When preparations have been completed, the asphaltic concrete is generally placed in two layers—a binder course and a surface course—by conventional methods.

Reflection cracking in asphaltic concrete mixtures laid on old concrete base courses has been prevented in a number of cases by the incorporation of a layer of welded wire fabric (like that used in reinforced concrete pavements) in the pavement structure (8). The style of fabric most commonly used is 3 x 6—10/10, which has a mesh opening of 3 by 6 in. with No. 10 gage wire. The sheets are placed so that the 6-in. spaced wires are parallel to the centerline. The entire pavement area normally is covered with mesh, but some success has been reported with the mesh laid in strips over the joints and cracks. The fabric is placed either directly on the existing pavement or on a leveling course. Sheets of fabric may be lapped for a short distance or butted up against one another. Sheets are placed in position and the asphaltic concrete placed with a conventional paver equipped with a special hold-down device. Fine, dense-graded asphaltic concrete mixtures, with a maximum size of aggregate of 1 in., have given the best results.

Old concrete pavements may be resurfaced with portland cement concrete. Considerable success has been achieved in recent years with thin overlays, from 1 to 3 in. thick. The old pavement must be thoroughly cleaned and all unsound material removed; the bumpcutter described earlier may be used to scarify the surface. The surface then may be washed down with detergent and acid baths (muriatic acid is often used). Best results are achieved if a thin grout layer is used to provide bond between the old pavement and the new surface. A 1:1 sand–cement grout is used; it is brushed onto the pavement just before the new concrete is applied. In some cases the grout has been placed pneumatically. Concrete must be of high quality and resistant to freezing and thawing, particularly where it will be subjected to salt applications for ice control. In the northern states, air-entrained concrete is normally used. To obtain adequate bond, thorough compaction of the fresh concrete is essential. This is usually done by the use of special vibratory screeds. Whenever the new surface extends over a joint, or butts against an expansion joint in the old pavement, a joint will be necessary in the new concrete. Joints may be made by tooling the plastic concrete or by sawing. Adequate curing of the new surface is essential. In some cases, particularly where the old pavement is excessively cracked, wire mesh reinforcing may be used in the new surface.

MAINTENANCE OF SHOULDERS AND APPROACHES

21-6 Shoulder maintenance

The importance of properly maintaining road shoulders becomes apparent when they are considered as a continuation of the traveled surface that must be used in cases of emergency. Adequate shoulders are also necessary if the full capacity of the roadway surface is to be utilized.

Shoulders may comprise many types of bases and surfaces including soil aggregate, soil surfaces that are capable of supporting vegetation and that may be seeded, and various bituminous surfaces. It is now almost standard practice to pave shoulders on heavily traveled roads; often the shoulders differ in color and texture from the travel lanes. Each of these approaches presents a slightly different type of maintenance problem.

A well-graded gravel shoulder will provide a satisfactory all-weather shoulder for light traffic; the maintenance of shoulders of this type will follow a pattern similar to that of soil–aggregate surfaces. This will include the dragging and blading of the shoulder surface for the proper slope and the filling of any ruts. It may be necessary to replace worn-out and lost materials at periodic intervals. This work can be done with a light motor grader.

Turf shoulders require the filling of holes and ruts, rolling or blading down high spots, and seeding, sodding, and fertilizing to maintain a satisfactory turf. It is also necessary to mow and clean the shoulders and provide some means of weed control or weed eradication. Mowing is usually accomplished with light equipment manufactured for this purpose. Treated bituminous and paved shoulders require maintenance that is similar to that for surfaces made of these same materials, which were previously described, including patching, surface treatment, and so on.

21-7 Maintenance of approaches

Closely related to shoulder maintenance is that of road approaches. Approaches include public side roads, private driveways, ramps, speed-change lanes, and turnouts for mailboxes and bus stops. The methods of maintaining approach areas do not differ much from those used for regular shoulder maintenance. However, because of the traffic in these areas, special maintenance efforts may be required to keep approaches free of potholes, ruts, and other types of deterioration.

MAINTENANCE OF ROADSIDES

Roadside maintenance in general may include the care of the area between the traveled surface and the limits of the right-of-way. This will include any median strip or landscaped areas on divided or dual-lane highways; roadside parks and picnic and recreational areas; and various appurtenances such as right-of-way fences, picnic tables, and the like.

In the following paragraphs, we group roadside maintenance activities into three classes: (1) vegetation management and control, (2) maintenance of rest areas, and (3) litter control.

21-8 Vegetation management and control

This class of maintenance includes mowing; weed eradication and control; seeding, sodding, and the planting of vegetation; and care of trees and shrubs.

The mowing of roadsides is done to provide better sight distances, improve drainage, reduce fire hazards, and improve the appearance of the roadway. Mowing should be started as soon as the grass or weeds are high enough to be cut, and it should be continued at periodic intervals, when necessary, throughout the growing season. Mowing can be accomplished with machines or by hand. Some heavy-duty mowing machines are available that are capable of mowing roadside brush up to 2.5 in. in diameter. Cleaning of roadsides after mowing can be accomplished by the use of hand or power rakes. Grass cuttings can be left as a mulch, but brush cuttings should be removed. The extent of mowing recommended by AASHTO (*1*) is given in Table 21-1.

Chemicals may be used on roadside areas to retard vegetation growth and lessen the frequency of mowing, especially in areas that are difficult to mow. Chemicals are also used to control weeds (herbicides), insects (insecticides), and fungi (fungicides). The spraying of chemicals along roadsides should be done by qualified personnel following the manufacturer's recommendations and guidelines of the Environmental Protection Agency. Precautions are required to avoid harm to native plants and wildlife and to ensure the safety of the traveling public, nearby residents, and maintenance personnel.

Seeding, sodding, and the planting of vegetation are important maintenance operations for the prevention of erosion. Seeding may be done on relatively flat areas, while sodding is necessary on steeper slopes. On steep slopes where seeding or sodding is not practical, the ground surface may be protected by the planting of vines or similar ground cover. The planting of dune grasses on sandy slopes to prevent wind erosion has been quite successful.

The care of trees and shrubs along the roadsides may include planting, trimming, fertilizing, spraying, and the construction of tree wells. Major tree surgery and the removal of broken limbs caused by storms are often necessary. A large number of highway departments have forestry departments that take care of all trees and shrubs along major highways.

21-9 Maintenance of rest areas

Special roadside areas that have been developed for picnic and recreational areas, such as roadside parks and turnouts, should be maintained in a manner necessary to meet minimum health and aesthetic standards. Tables, benches, and refuse containers should be placed at convenient locations. The grounds should be kept cleaned, trees and shrubs trimmed, and stoves kept in good condition. Toilet facilities should be kept in a sanitary condition, and any drinking water should be frequently tested for purity. The parking areas and access roads must also be properly maintained.

21-10 Litter control

Removal of rubbish and debris (litter) is one of the vexing and expensive problems of roadside maintenance. Debris such as fallen branches, rocks, landslides, and articles that have fallen from trucks should be removed immediately in order to protect the traveling public. The remains of animals killed by motor vehicles should also be removed promptly and buried at a convenient location. Garbage and trash dumped within the highway right-of-way must be removed because of its unsightliness and for reasons of health. Generally

TABLE 21-1
MOWING RECOMMENDATIONS OF AASHTO

	On slopes $2\frac{1}{2}$:1 or less	Measured Along Surface of Slopes More than 1:1 slopes within 10 ft[a] outside of pavement edge
Backslopes and downslopes	Mow 15 ft outside of pavement edge	Mow brush and as required for sight dist.
Inslopes	Mow 15 ft or entire width up to 25 ft outside of pavement edge or to bottom of ditch line	Now brush up to 15 ft if entire width and as required for sight distance
Median areas	Mow 15 ft outside of pavement edge or entire width if less than 50 ft	Mow brush and as required for sight dist.
Interchange areas	Mow 15 ft from pavement edge and any areas that obstruct sight distances or areas used for snow storage	Mow brush and as required for sight dist.

Source: *AASHTO Maintenance Manual* (1976).

Note 1: On slopes more than $2\frac{1}{2}$:1 up to 1:1 within 20 ft outside of pavement edge, one swath width past slope that does not exceed $2\frac{1}{2}$:1 should be mowed using side-mounted units.

Note 2: Generally, when guard rail is present, an area 3 ft in width should be mowed behind the guard rail, and brush should be removed as required for sight distance.

[a] 1 ft = 0.3048 m.

speaking, roadsides should be thoroughly cleared of litter in the spring and periodically as needed thereafter (*1*).

MAINTENANCE OF BRIDGES, TUNNELS, AND DRAINAGE STRUCTURES

21-11 Bridge maintenance

Highway agencies continually inspect bridges and elevated structures, then take any necessary steps to minimize deterioration or repair damage caused by accidents, floods, or other unforeseen events.

Bridge superstructures require periodic maintenance to preserve and protect the investment made in them. For instance, steel bridges (unless they are made of special alloys or have been galvanized) are cleaned and painted regularly to prevent corrosion.

A serious problem for the maintenance engineer is the deterioration of concrete bridge decks, particularly on heavily traveled routes in northern states, where chlorides are used extensively for snow and ice control. Concrete normally protects the reinforcment bars from corrosion. However, the penetration of water and deicing chemicals results in the corrosion of the reinforcement bars and causes the concrete to crack.

Bridge decks that have experienced only minor deterioration may be patched with specialized concretes. Advantages and disadvantages of some of the products used for bridge deck patching are given in Table 21-2. Spalled areas may be temporarily repaired with asphaltic patching materials.

When the deterioration is severe, it may be necessary to cover the bridge deck with an overlay or to remove and reconstruct the entire deck. Two types of overlays have been successfully used (*9*): (1) a bituminous concrete wearing surface placed over a waterproofing membrane and (2) a portland cement mortar or concrete (possibly including a liquid latex admixture). In such work, it is recommended that all bad concrete be removed and that a grout be carefully applied to the surface before the new surface is constructed (*10*).

Bridge inspections should be complete enough to discover any undermining of the bridge footings or any damage to the substructure. Timber bulkheads should be repaired as soon as evidence of failure is shown. Piling that shows deterioration caused by erosion, corrosion, or attack by organisms should be replaced when necessary.

The stream bed should be kept clean and free from debris in order that the free flow of water may be maintained. Provision for the installation of some sort of barrier to prevent floating debris from damaging the bridge floor during periods of high water must sometimes be made.

The maintenance of pumping facilities at highway underpasses is also a very important operation.

The maintenance of bridges is usually carried on by a special bridge crew trained and skilled in this type of work. Some state highway departments have heavy equipment that is necessary for all phases of bridge maintenance, including mobile shops.

TABLE 21-2
COMPARISON OF PRODUCTS USED
FOR BRIDGE DECK PATCHING

Patch Type	Advantages	Disadvantages
Regular portland cement	Ease of handling; low cost	Slow cure; edge cracks
High early strength cement	Reduced cure; low cost	Shrinkage
Metallic aggregates	Reduce shrinkage	High cost; may contain salt
Latex additives	Increase bond	High cost
Epoxy resin	Fast cure; good bond	High cost; edge cracks

SOURCE: *Concrete Bridge Deck Durability.* National Cooperative Highway Research Program Synthesis of Highway Practice 4 (1970).

The maintenance of bridges is an extremely important and complex phase of highway maintenance. Additional information on this important topic is given in Ref. 9.

21-12 Maintenance of tunnels

It is important that tunnels be inspected by qualified maintenance personnel on a regular basis. Areas requiring special concern by inspectors are (1) supports, (2) tunnel walls and ceilings, (3) portals, (4) lighting, and (5) ventilation. AASHTO (*1*) recommends that timber supports be inspected annually for evidence of decay, insect infestation, and other deterioration. Tunnel walls and ceilings should be examined from time to time for evidence of water leaks. When used, ceramic tiles should be washed annually and kept in proper repair. Tunnel portals should be kept free of loose rocks, snow deposits, or other materials that could pose a hazard to users of the facility.

Special attention should be given to tunnel lighting to ensure that an adequate level of visibility is maintained. Generally, burned-out lamps should be replaced immediately, and group replacement of lamps should be scheduled before 90 percent of the rated life elapses (*1*).

Ventilation systems for tunnels should be inspected at least monthly, and airflow meters should be employed to ensure that the ventilation system is functioning as designed.

21-13 Drainage and drainage structures

All drainage structures and appurtenances have to be kept in good working condition so as to provide free and unobstructed flow. Generally, maintenance operations include those related to (1) surface drainage, (2) ditches, and (3) culverts. Modern roads are generally well developed for drainage, but many of the older ones lack adequate drainage. Observation of conditions at times of heavy rains and high water will give the best indications of the ability of the drainage system to perform its function properly without erosion or damage. The entire system must be so maintained as to handle heavy rainfalls and to function properly under all conditions, even under winter conditions of ice and snow.

The first objective of surface drainage is to remove the water quickly so that it will not interfere with the use of the road. To accomplish this, the roads are crowned and elevated. On the high-type road surfaces, these crowns are more permanent than on the lower type of road surfaces. Thus, on gravel roads and similar types, it is necessary to maintain a proper crown by blading.

Wide shallow ditches are preferable for maintenance and are less dangerous when vehicles have to leave the road. Certain types of roadside ditches may be kept clean and the proper slope maintained by the use of a power grader. Care should be exercised to preserve seeded and sodded areas and shrubs during maintenance operations. The original line or grade of the ditch should also be maintained. When ditches become eroded owing to excessive grade, handling of large volumes of water, or a combination of both, it may be necessary to retard the flow or to line them, as described in Chapter 10.

All culverts should be checked regularly to see that they are free from obstructions. Inlets and outlets should be kept open and free from refuse, and culverts should be cleaned and repaired when necessary.

MAINTENANCE OF TRAFFIC CONTROL AND SAFETY DEVICES

21-14 Maintenance of traffic control devices

Signs, signals, and markings for the direction, warning, and regulation of traffic are essential to traffic safety and must be properly maintained. This work includes the installation, repair, and painting of signs and the care of pavement markings. Because of their location, many pavement markings must necessarily be renewed practically every year. Many departments have mechanized marking equipment for the application of the painted stripe. Large machines of this type can apply simultaneously as many as three stripes, either solid or broken, and use two colors of paint, if desired. Reflectorized glass beads may be added to the painted surface during the striping process. The beads have proved beneficial for night driving. Thermoplastic compounds are used for pavement markings, also, as are raised plastic disks cemented to the road surface. Many highway departments maintain shops for the fabrication and repair of signs and other traffic control devices. Automatic traffic signals have to be checked, repaired, and often readjusted to traffic conditions. Highway departments are continually standardizing and modernizing traffic signs and signals, and much of this work is done by maintenance forces. During the winter season in the northern states, special attention is given to warning signs to keep them free from snow and visible as much as possible.

21-15 Maintenance of safety devices

A variety of appurtenances within highway rights-of-way are provided for the safety of the users. Such features include guardrails, barriers, impact attenuators, pedestrian overpasses and underpasses, and fences to restrict access of pedestrians and animals. Highway safety devices should be frequently and systematically inspected and repaired on a high-priority basis. Because of variations in the design and the materials used in the construction of such devices, each type of appurtenance has its special maintenance procedures. Guidance on the maintenance of various types of highway safety devices is given in Ref. 1.

THE CONTROL OF SNOW AND ICE

Considering the public's complete dependence on the uninterrupted use of highways throughout the year, winter maintenance has become one of the most essential activities of many state highway departments. This is particularly true in the northern states, where a program for snow removal and ice control has to be considered when the annual maintenance budget is prepared. The amount of money expended for these services will vary a great deal in the many states, and it is dependent in large measure on the length and intensity of storms and the amount of snowfall in a particular area.

Measures preparing for snow removal and ice control are taken before the start of the winter season. The most important of these are (1) the preparation of the necessary equipment; (2) stockpiling of ice-control chemicals and abrasives, if used; (3) placing of snow fences, snow guides, and containers for ice-control abrasives; and (4) organization arrangements.

The necessary equipment needed for snow-removal operations consists of large trucks, various types of snowplows, power graders, bank-slicers, and sanders. Some trucks are equipped with underbody blades, others with "Vee" plows or with bank-slicers. All equipment should be put into working order before the winter season starts. An ample supply of snowplow shoes and cutting edges should be on hand for each type of plow. Tire chains, flags, blinker lights, tow chains, and other safety devices should also be checked and placed, ready for use.

Ice-control chemicals and abrasives should be stockpiled previous to the first storm in the fall—the quantities to be stockpiled may be estimated on the basis of past experience in a particular district or division. Stockpiles should be located with care at convenient places along the road or at central locations, such as main or branch garage yards, gravel or sand pits, or other places where overhead bins or mechanical loaders are available. Generally, a coarse heavy sand is the best abrasive. Cinders or other abrasives, however, may be desirable, depending on availability and cost. Calcium chloride is frequently incorporated and thoroughly mixed with the abrasive at the time the stockpiles are made. Fifty to 100 lb of calcium chloride per cubic yard are used, and a light application of calcium chloride is usually spread over the surface of the pile to prevent freezing.

Snow fences are generally placed parallel to the road at a distance of 75 to 125 ft from the centerline of the road in areas where observations show that drifts constantly form. A snow fence usually consists of fabricated wooden slats mounted on steel posts, which can be removed, rolled in bundles, and stored when not in use. The purpose of a snow fence is to change the velocity of the snow-laden wind so that the snow is deposited before it reaches the roadway. As the snow accumulates near the fence, the fence is raised in order that it may effectively gather snow. Extension posts are provided when this is necessary. In certain areas where right-of-way is available, natural windbreaks or snow fences may be formed by planting evergreens parallel to the road.

Snow guides may be made from peeled saplings approximately 12 ft long with a maximum top diameter of 2 in. They are sometimes painted with alternate bands of black and white or some other suitable color, and they are often fitted with reflectorized buttons. They are used for marking the roadway or road hazards in areas of heavy snowfall. Guides marking the locations of guardrails, culverts, headwalls, and curbs are essential during plowing operations when these structures are covered with snow.

The placing of containers for abrasives at strategic locations has proved very beneficial to the traveling public, who can avail themselves of these materials if needed before maintenance crews arrive. The containers are usually barrels equipped with a utensil, such as a gallon can, for removing the abrasive.

To properly carry on the work of snow removal and ice control, good organization is necessary. Many states maintain a weather reporting service, with all districts of a division reporting adverse weather or road conditions to a central office. These give such information as temperature, snowfall, and other adverse weather conditions, such as rain, fog, mist, and smog. Road conditions are then rebroadcast to serve as a source of information for the traveling public. Maintenance crews are generally subject to call at all hours and provisions are usually made for dual crews to operate for alternate 12 hr periods during emergencies. The use of radio communications has facilitated work during snow-removal and ice-control operations. Mobile and fixed station units assist in the rapid dispatching of equipment, the reporting of adverse conditions, and the general control of operations. Use is made of both one-way and two-way radio communication systems for other maintenance operations as well.

The procedure followed to keep trunk highways clear of snow varies. One method is to start plowing with trucks equipped with underbody blades after $\frac{1}{2}$ to 1 in. of snow has fallen, using a conventional curved cutting edge to remove the snow before it becomes packed by traffic. These units are generally kept working during a continuous snowfall up to the time when the snow becomes too deep for their effective operation. When this occurs, light trucks equipped with side delivery plows start operating and keep in operation until the storm has subsided or until sufficient snow has accumulated to require heavier trucks equipped with side delivery plows or light trucks with "Vee" plows. When this equipment proves inadequate, heavy "Vee" plows or "snogos" are brought into operation. When the storm has abated, snow removal operations are continued until the snow is removed to a point beyond the outer edge of the shoulder. On trunk lines in urban areas, use is often made of mechanical loaders to speed snow removal. Use is sometimes made of heavy rotary plows for opening up roads after storms of such intensity that snow-removal operations have to be temporarily suspended. Care must be exercised at railroad crossings to remove snow between the rails so that ice does not form.

Ice control means the application of quick-action emergency measures to counteract slippery conditions as early as possible and the removal of all ice from the pavement after skid control has been established. These operations may consist of removing sleet or thin ice, thick ice or frozen slush, and hard-packed snow from the surface of the roadway. Salt, calcium chloride, a mixture of the two, and abrasives mixed with chlorides are widely used for ice control in urban and rural areas. Many organizations use a mixture of 1 part calcium chloride and 2 parts sodium chloride, by volume. Application rates vary from 200 to 500 lb of mixture per mile of two-lane roadways. They depend largely on straight chemicals in urban areas, because abrasives may clog storm sewers. Guidelines for chemical application rates are shown as Table 21-3.

To remove sleet or thin ice from the roadway, treated abrasive may be placed on the surface at the rate of 2 to 3 cu yd/mile, as the ice forms. Chlorides may also be used alone. First consideration in ice-control treatment is given to approaches to railroad crossings, road intersections, and dangerous horizontal and vertical curves. This kind of work is usually done by light trucks in advance of other ice-control measures.

Sleet or thin ice on gravel roads should first be serrated. This process forms grooves, and in many cases will bring up enough gravel to control skidding. If this does not occur, chloride may be added. Thick ice or frozen slush is also scored wth a serrated blade in order to concentrate the chemical on the road. When the ice or slush has been loosened, it is bladed off the road.

Abrasives treated with calcium or sodium chloride, or chlorides alone, should be used as sparingly as possible on portland cement concrete pavements. This is because repeated freezing and thawings of concrete in contact with these salts are conductive to scaling.

On urban expressways, the overwhelming need for getting a bare pavement quickly leads to excessive use of salts, with attention being focused on building pavements that can withstand this treatment.

PAVEMENT REHABILITATION

By timely and proper maintenance, highway engineers can extend a pavement's usefulness. Eventually, however, even the best-maintained pavement will begin to disintegrate and will need to be rehabilitated.

The traditional approach to pavement rehabilitation has been to either

TABLE 21-3
GUIDELINES FOR CHEMICAL APPLICATION RATES

| Weather conditions | | | Application rate (pounds[a] of material per mile[b] of two-lane road or two lanes of divided) | | | |
Temperature	Pavement Conditions	Precipitation	Low- and High-Speed Multilane Divided	Two- and Three-Lane Primary	Two-Lane Secondary	Instructions
30°F and above	Wet	Snow	300 salt	300 salt	300 salt	Wait at least 0.5 hr before plowing
		Sleet or freezing rain	200 salt	200 salt	200 salt	Reapply as necessary
25°–30°F	Wet	Snow or sleet	Initial at 400 salt Repeat at 200 salt	Initial at 400 salt Repeat at 200 salt	Initial at 400 salt Repeat at 200 salt	Wait at least 0.5 hr before plowing; Repeat as necessary
		Freezing rain	Initial at 300 salt Repeat at 200 salt	Initial at 300 salt Repeat at 200 salt	Initial at 300 salt Repeat at 200 salt	
20°–25°F	Wet	Snow or sleet	Initial at 500 salt Repeat at 250 salt	Initial at 500 salt Repeat at 250 salt	1200 of 5:1 sand salt; repeat same	Wait about 0.75 hr before plowing; repeat
		Freezing rain	Initial at 400 salt Repeat at 300 salt	Initial at 400 salt Repeat at 300 salt		Repeat as necessary
15°–20°F	Dry	Dry snow	Plow	Plow	Plow	Treat hazardous areas with 1200 of 20:1 sand/salt
	Wet	Wet snow or sleet	500 of 3:1 salt/ calcium chloride	500 of 3:1 salt/ calcium chloride	1200 of 5:1 sand	Wait about 1 hr before plowing; continue plowing until storm ends; then repeat application
Below 15°F	Dry	Dry snow	Plow	Plow	Plow	Treat hazardous area with 1200 of 20:1 sand/salt

[a] 1 lb = 0.4536 kg.
[b] 1 mile = 1.6093 km.

SOURCE: *AASHTO Maintenance Manual.* American Association of State Highway and Transportation Officials (1976).

reconstruct it with all new materials or to patch and overlay it with a new wearing surface. Both asphaltic and portland cement concrete mixes have been used for pavement resurfacing (*11*).

Increasing construction costs have caused highway engineers to search for new methods of rehabilitating highway pavements. Two such methods, milling and pavement recycling, are briefly discussed in the following paragraphs.

21-16 Milling of pavements

In recent years, equipment manufacturers have developed pavement milling machines that are capable of correcting pavement and surface distresses such as rutting, raveling, bleeding, and aggregate polishing. (See Figure 21-10.) Such machines are equipped with rotating drums which are fitted with special cutting teeth for milling asphalt and concrete pavements to predetermined depths (*12*). Figure 21-11 illustrates a milling drum from one of these machines. Pavement milling machines can be used in a number of ways to rehabilitate worn and distressed pavements. Typical uses of these machines are listed in Table 21-4.

21-17 Pavement recycling

Hot-mix recycling is a proven approach for the rehabilitation of worn-out asphaltic pavements. It involves (1) removing the existing pavement to full or partial depth, (2) reducing the reclaimed materials to a suitable size for reprocessing, (3) blending the reclaimed material with virgin aggregates and

FIGURE 21-10 Pavement milling machine. (Courtesy Cedarapids. Inc.)

FIGURE 21-11 Rotating drum with cutting teeth in a pavement milling machine. (Courtesy Cedarapids, Inc.)

liquid asphalt, and (4) relaying the materials as a base, binder, or surface course.

Advantages of hot-mix recycling include

1. Economy.
2. Conservation of natural resources.
3. Improvement of the structural strength of the pavement with little or no change in thickness.
4. Correction of existing deficiencies in the pavement mix.

Before hot-mix recycling can proceed, all of the materials must be tested and evaluated to determine the optimum blend. Samples of the reclaimed material are randomly selected for testing, preferably from the cold feedstock piles at the plant site. Laboratory tests of the pavement samples are made to determine the aggregate gradation, asphalt content, and asphalt viscosity at 140° F (60° C). The old asphalt is extracted from the sample by a standard procedure such as ASTM Designation D2172, providing a measure of the quantity of asphalt in the old mix. The extracted asphalt is reclaimed and tested by ASTM Method D2171 to determine its viscosity at 140° F (60° C). The reclaimed aggregate is subjected to a sieve analysis to determine the aggregate gradation and to reveal any deficiencies in aggregate gradation. It will normally be necessary to add virgin aggregate to the reclaimed asphalt pavement to produce a mix with the desired gradation.

TABLE 21-4
TYPICAL USES AND ADVANTAGES OF COLD MILLING MACHINES

Problem	Solution	Advantages
Bleeding or flushing of asphalt pavement and polishing of aggregates in flexible and rigid pavements cause skid hazards.	Restore surface texture to slick pavements by skimming off the surface to a depth of $\frac{1}{2}$ in. in flexible pavements and $\frac{1}{4}$ in. in rigid pavements.	Skid-resistant surface can produced at a low cost.
Portland cement concrete pavement faulting with differing elevations of slabs at expansion joints creates poor rideability and danger to road users.	At the expansion joint, mill the high slabs flush to the adjacent slabs, tapering off to zero, leaving a textured drivable surface.	This provides a more economical solution than the traditional approach of replacement and patching.
Light rutting and corrugating in the upper region of a wearing course causing poor drainage and poor rideability.	Mill wearing course down to the deepest wheel rut and corrugation, maintaining grade and slope and leaving a new textured riding surface.	A better drained and smoother surface can be produced without the need for new materials.
Successive overlays have filled the curbs of an urban street, causing poor drainage and rideability.	Mill the pavement to predetemined grade and slope, reexposing curb and restoring drainage. Resurface with new or recycled material.	Solution provides better drainage and rideability.

SOURCE: *Adapted from Cold Planing.* Bulletin CP-1, Iowa Manufacturing Company (1985).

New low-viscosity asphalt must normally be added to the mixture of reclaimed pavement and new aggregate to increase the total asphalt content to the desired quantity and to provide an asphalt with a desired viscosity. In addition, organic materials called recyling agents are usually added to restore the old asphalt to desired specifications. The detailed procedures for determining the mix design for a recycled pavement are not given here but may be found in Ref. 13.

Old pavements may be removed (1) by cold milling or planing, as described in Section 21-16, or (2) by ripping and crushing. Where the ripping and crushing procedure is used, scarifiers, grid rollers, or rippers are used to break up the asphalt pavement. It is then loaded and hauled to a crushing and screening plant, where it is pulverized. After pulverization by one of these methods, the material is stockpiled at a mixing plant for subsequent processing.

Recycled hot mixes may be prepared either by batch plants or by drum-mix plants. The plants must be modified to allow the reclaimed asphalt pavement material to be heated and dried without exposing it directly to the high-temperature flame and combustion gases in the dryer. Without these modifications, the asphalt in the reclaimed material will harden excessively, and it will not be possible to comply with regulations governing harmful exhaust stack emissions.

In batch plants, the aggregate is proportioned from cold-feed bins, heated in a conventional aggregate dryer, and then conveyed to the hot storage bins in the usual manner. The reclaimed asphalt pavement material is not heated or dried but is carried directly from the stockpile to a special cold-feed bin. From there, the material is conveyed to the weigh hopper, where it is weighed as a fifth material in a normal four-bin batch plant. The cold reclaimed material is mixed in the pugmill with the superheated untreated aggregate. After mixing, the blended material goes to a storage silo and is subsequently loaded into trucks.

Contractors and plant manufacturers have developed several schemes to modify drum-mix plants to ameliorate problems with air pollution. One approach has been to use a split feed in which the untreated aggregate enters the drum at the burner end and the reclaimed material is brought in some distance downstream from the burner in order to avoid exposure to the flame and extremely hot gases. This arrangment can be seen in Figure 18-11.

Another type of hot-mix recycling plant is a "drum-in-a-drum" design. In this design, the burner fires into a smaller drum that is partially inserted into the upper end of the main drum. The new aggregate enters the inner drum and is exposed to the flame, but the reclaimed material is fed into the main drum, where it is protected from the flame.

Still another type of hot-mix recylcing plant uses a combustion chamber with a conical heat shield inserted between the burner and the drum to protect the reclaimed material from the flame (*13*).

Hot-mix recycled mixtures are spread and compacted by using conventional procedures such as those described in Chapter 19.

Looking to the future, it is expected that greater emphasis will be placed on pavement recycling as engineers continue to search for cost-effective ways to provide safe and efficient transportation and protect the vast investment in the nation's highway system.

REFERENCES

1. *AASHTO Maintenance Manual.* American Association of State Highway and Transportation Officials, Washington, D.C. (1976).

2. *Highway Statistics.* Federal Highway Administration, Washington D.C. (1983).

3. *Street and Highway Maintenance Manual.* American Public Works Association, Chicago, Ill. (1985).

4. *Performance Budgeting System for Highway Maintenance Management.* National Cooperative Highway Research Program Report 131, Roy Jorgensen Associates, Transportation Research Board, Washington, D.C. (1972).

5. *Maintenance Management Systems.* National Cooperative Highway Research Program Synthesis of Highway Practice 110, Transportation Research Board, Washington, D.C. (1984).

6. *Policy on Maintenance of Roadway Surfaces.* American Association of State Highway and Transportation Officials, Washington, D.C. (1965).

7. *Asphalt in Pavement Maintenance.* Manual Series No. 16, The Asphalt Institute, College Park, Md. (1967).

8. *Reinforced Bituminous Concrete Overlays.* Wire Reinforcement Institute, Washington, D.C. (1962).

9. *AASHTO Manual for Bridge Maintenance.* American Association of State Highway and Transportation Officials, Washington, D.C. (1976).

10. *Concrete Bridge Deck Durability.* National Cooperative Highway Research Program Synthesis of Highway Practice 4, Transportation Research Board, Washington, D.C. (1970).

11. *Guide to Concrete Resurfacing Designs and Selection Criteria.* Portland Cement Association, Skokie, Ill. (1981).

12. *Cold Planing.* Bulletin CP-1, Iowa Manufacturing Company, Cedar Rapids (1985).

13. *Asphalt Hot Mix Recycling.* Manual Series No. 20, The Asphalt Institute, College Park, Md. (1981).

APPENDIX

Equivalents among Metric (SI) and Traditional Units of Measurement

Traditional to Metric (SI)	*Metric (SI) to Traditional*

Lengths

1 in = 2.54 cm

1 ft = 0.3048 m

1 mi = 1.6093 km

1 cm = 0.3937 in.

1 m = 3.2808 ft

1 km = 0.6214 mi

Areas

1 in. = 6.4516 cm^2

1 ft^2 = 0.0929 m^2

1 mi^2 = 2.590 km^2

1 acre = 4.047 m^2

1 acre = 0.404 hectare

1 cm^2 = 0.1550 in.^{2}

1 m^2 = 10.7639 ft^2

1 m^2 = 0.0002471 acre

1 hectare = 2.471 acres

1 km^2 = 0.3861 mi^2

Volumes

1 gal = 3.78541

1 ft^3 = 28.31691

1 ft^3 = 0.02832 m^2

1 l = 0.2642 gal

1 l = 0.035313 ft^3

1 m^3 = 35.3133 ft^3

Velocities

1 mi/hr = 1.6093 km/hr

1 km/hr = 0.6214 mi/hr

Mass

1 lbm = 0.4536 kg

1 ton (2000 lbm) = 907.18 kg

1 kg = 2.2028 lbm

1 kg = 0.0011 ton (2000 lbm)

Force

1 lbf = 4.4482 N

1 N = 0.2248 lbf

Pressure or Stress

1 lbf/ft^2 = 47.880 N/m^2

1 N/m^2 = 0.02088 lbf/ft^2

Densities

1 lb ft^3 = 0.01602 g/cm^3

1 lb ft^3 = 16.0183 kg/m^3

1 g/cm^3 = 62.430 lb/ft^3

1 kg/m^3 = 0.062430 lb/ft^3

Flow Rates

1 cfs = 0.02832 m^3/sec

1 m^3/sec = 35.3133 cfs

INDEX